Textile Science and Clothing Technology

Series Editor

Subramanian Senthilkannan Muthu, SgT Group & API, Hong Kong, Hong Kong

This series aims to broadly cover all the aspects related to textiles science and technology and clothing science and technology. Below are the areas fall under the aims and scope of this series, but not limited to: Production and properties of various natural and synthetic fibres; Production and properties of different yarns, fabrics and apparels; Manufacturing aspects of textiles and clothing; Modelling and Simulation aspects related to textiles and clothing; Production and properties of Nonwovens; Evaluation/testing of various properties of textiles and clothing products; Supply chain management of textiles and clothing; Aspects related to Clothing Science such as comfort; Functional aspects and evaluation of textiles; Textile biomaterials and bioengineering; Nano, micro, smart, sport and intelligent textiles; Various aspects of industrial and technical applications of textiles and clothing; Apparel manufacturing and engineering; New developments and applications pertaining to textiles and clothing materials and their manufacturing methods; Textile design aspects; Sustainable fashion and textiles; Green Textiles and Eco-Fashion; Sustainability aspects of textiles and clothing; Environmental assessments of textiles and clothing supply chain; Green Composites; Sustainable Luxury and Sustainable Consumption; Waste Management in Textiles; Sustainability Standards and Green labels; Social and Economic Sustainability of Textiles and Clothing.

Rouf Ahmad Bhat · Vandana Singh ·
Gowhar Hamid Dar · Syed Maqbool Geelani
Editors

Biocentric Approaches in Textile Waste Management

Paradigm for Sustainable and Greener Development

Springer

Editors
Rouf Ahmad Bhat
Department of School Education
Government of Jammu and Kashmir
Srinagar, Jammu and Kashmir, India

Vandana Singh
Department of Microbiology, SSAHS
Sharda University
Greater Noida, Uttar Pradesh, India

Gowhar Hamid Dar
Department of Environmental Science
Government Degree College Kulgam
Kulgam, Jammu and Kashmir, India

Syed Maqbool Geelani
Department of Environmental Science
Cluster University Srinagar
Srinagar, Jammu and Kashmir, India

ISSN 2197-9863 ISSN 2197-9871 (electronic)
Textile Science and Clothing Technology
ISBN 978-3-032-04973-5 ISBN 978-3-032-04974-2 (eBook)
https://doi.org/10.1007/978-3-032-04974-2

This Springer imprint is published by the registered company Springer Nature Switzerland AG
The registered company address is: Gewerbestrasse 11, 6330 Cham, Switzerland

Dedicated to My Dad…

"Dad, Some heroes wear uniforms not just in service, but in life. Though they soar beyond our sight, their legacy gives us wings."

To my beloved father, Mr. B. Singh, an Indian Air Force officer, my hero, and the wind beneath my wings. You didn't just serve the skies, you taught us to dream beyond them. You showed us that discipline and humility can coexist, that courage can be quiet, and that true strength lies in lifting others, even when you carry your own burdens in silence. You inspired me to reach higher, to rise after every fall, and to never be afraid of flying toward the unknown. Though you are no longer with us, your voice still echoes in my heart, reminding me, "Look up—your sky is waiting." This book is a tribute to the man who gave me roots in values and wings of purpose. It carries every lesson you taught, every sacrifice you made, and every dream you believed for me. You may have taken your final flight, but you remain the pilot of my soul.

With endless love and pride,
Your Daughter
Vandana Singh

Preface

The textile industry, while serving as a cornerstone of global industrial and economic activity, has been identified as a major source of environmental pollution. Effluents from textile manufacturing processes are often rich in synthetic dyes, heavy metals, surfactants, and recalcitrant organic pollutants. These contaminants are chemically stable, biologically persistent, and ecotoxic, posing severe threats to aquatic ecosystems, terrestrial habitats, and public health. Conventional physicochemical treatment methods, although widely used, suffer from critical limitations including incomplete pollutant removal, high operational costs, and the generation of secondary waste streams such as toxic sludge. In contrast, biocentric approaches rooted in microbial metabolism, enzymatic catalysis, phytotechnology, and bioengineered nanomaterials present ecologically viable, mechanistically diverse, and sustainable alternatives. These biological systems exhibit high specificity, adaptability, and the ability to degrade or immobilize diverse pollutants under ambient conditions, aligning with the principles of green chemistry and ecological engineering.

This book, *Biocentric Approaches in Textile Waste Management: Paradigm for Sustainable and Greener Development*, presents a comprehensive view of biological textile waste management strategies, underpinned by current research, mechanistic insights, and applied technologies. The volume opens with Chapter "Hazardous Chemicals in Textile Industries, Risk and Assessment", which discusses the types and toxicological impacts of hazardous chemicals in textile processes and the need for structured risk assessment. Chapter "Biopolymers in Textile Waste Management: Sources, Properties, and Prospects" introduces biopolymers from renewable sources as eco-compatible substitutes for synthetic polymers in textile applications. Chapter "Nano-bioremediation for Dye Effluents" details nano-bioremediation techniques for dye-laden effluents, while Chapter "Circular Economy in Textile Waste Management: Principles and Practices" explores circular economy frameworks and how biocentric design thinking can be integrated into textile lifecycle management. Enzymatic and microbial transformations of textile dyes are elaborated in Chapters "Biotransformation of Textile Dyes: Mechanism and Application", "Biological Decolorization of Textile Dye Wastewater Using Microbial Consortia", "Enzyme-Based Bioremediation of Textile Dyes and Effluents: A Bird's-Eye View on Current Status and Limitations", and "Biodecontamination of Textile Waste Through Microbial Enzymes", providing both theoretical background and practical advancements in microbial consortia engineering and enzymatic degradation mechanisms. Heavy metal remediation through bioremediation and biosorption is discussed in Chapter "Bioremediation of Heavy Metals in Textile Industry Effluents for Sustainable Future", and the role of biochar as a multifunctional remediation agent is highlighted in Chapters "Application of Biochar for Remediation of Textile Industry Waste" and "Biochar for Remediation of Textile Industrial Waste". Chapters "Biocentric Approaches to Textile Waste Management: Sustainable Solutions and Environmental Perspectives" and "Phytoremediation of Textile Industrial Effluents for Sustainable Future" discuss integrated biocentric strategies, emphasizing environmental perspectives, phytoremediation applications, and circular bioeconomy integration. Chapters "Phycoremediation of Toxic Azo Dyes from Contaminated Environment" and "Biological Treatment of Textile Bleaching and Printing Effluents" extend this discussion to phycoremediation and the treatment of complex effluents from textile bleaching and printing processes. Chapter "Phytogenic Nanoparticles for Waste Water Treatment" introduces phytogenic nanoparticles as a cutting-edge, green nanotechnology solution for wastewater treatment. Finally, Chapter "Metabolic Pathways of Microbial Detoxification of Toxic Dyes" addresses microbial metabolic pathways in dye detoxification, elucidating enzymatic routes and the influence of environmental variables on treatment efficacy.

Concomitantly, these chapters present a multi-dimensional, systems-level perspective that is urgently needed to address the multifaceted environmental challenges posed by the textile sector. This volume is designed to serve as a vital resource for researchers, practitioners, policy makers, and students involved in environmental biotechnology, industrial ecology, and textile engineering. Furthermore, by compiling the latest scientific advances in biocentric textile waste management, this book aims to catalyze interdisciplinary collaboration and foster innovation toward a cleaner, safer, and more sustainable future for the textile industry.

Srinagar, India	Dr. Rouf Ahmad Bhat
Greater Noida, India	Dr. Vandana Singh
Kulgam, India	Dr. Gowhar Hamid Dar
Srinagar, India	Dr. Syed Maqbool Geelani

About This Book

The book *Biocentric Approaches in Textile Waste Management—Paradigm for Sustainable and Greener Development* offers a scientifically rich exploration of sustainable strategies for addressing the environmental impact of textile waste. The textile industry, being one of the major contributors to global pollution, generates wastewater containing persistent dyes, heavy metals, and toxic organic compounds. Conventional physicochemical methods for waste treatment are often costly, inefficient, and environmentally damaging, thereby necessitating innovative biological approaches. This book highlights various biocentric solutions, including bioremediation, enzyme-based degradation, biochar application, phytoremediation, and phycoremediation as environmentally friendly alternatives. The opening chapters emphasize the risks associated with hazardous chemicals used in textiles and advocate for integrated risk assessment and biological treatment strategies. Special focus is given to biopolymers derived from renewable resources, which not only replace synthetic materials but also serve in waste remediation due to their biodegradable nature. Nanotechnology, especially nano-bioremediation and phytogenic nanoparticles, emerges as a promising tool for removing complex pollutants from textile effluents. These methods improve the adsorption and catalytic degradation of dyes and heavy metals, while minimizing secondary pollution. Enzyme-mediated processes involving laccases, peroxidases, and azo-reductases are also explored for their role in decolorizing and detoxifying textile dyes under optimized environmental conditions. Furthermore, the book also underscores the role of microbial consortia and engineered microorganisms in enhancing biodegradation efficiency. Metabolic pathways responsible for dye detoxification are examined in detail, showing how microbial systems can transform hazardous compounds into less toxic forms. Similarly, biochar, with its high surface area and functional groups, is presented as an effective adsorbent for effluent remediation and carbon sequestration. Additionally, the text integrates circular economy principles tailored to the textile sector, promoting waste valorization, resource recovery, and the design of bio-based polymers to supplant environmentally deleterious materials. This multidisciplinary treatise not only elucidates the molecular and process-level dynamics of biocentric textile waste management but also underscores translational aspects including industrial scalability, economic feasibility, and regulatory frameworks essential for advancing sustainable textile manufacturing and effluent treatment globally.

Highlights

- Provides in-depth analysis of hazardous chemicals and pollutants in textile waste and their environmental impact.
- Explores biological remediation methods including microbial degradation and enzymatic biotransformation of textile dyes.
- Evaluates the role of nanotechnology and biochar in enhancing textile wastewater treatment efficiency.
- Integrates circular economy principles focusing on biopolymer development, recycling, and sustainable waste management.
- Discusses translational and policy perspectives essential for industrial implementation of biocentric textile waste solutions.

Contents

About the Editors

Dr. Rouf Ahmad Bhat (Ph.D.) has pursued his doctorate at Sher-e-Kashmir University of Agricultural Sciences and Technology Kashmir (Division of Environmental Sciences) and is presently working in the Department of School Education, Government of Jammu and Kashmir. Dr. Bhat has been teaching for the past 09 years. He is an author of more than 55 research articles (h-index 34; i-index 59; total citation 4800) and 50 book chapters and has published more than 50 books with international publishers (*Springer, Elsevier, CRC Press Taylor and Francis, Apple Academic Press, John Wiley, IGI Global*). He has his specialization in toxicology, phytochemistry, and phytoremediation. Dr. Bhat has presented and participated in numerous state, national, and international conferences, seminars, workshops, and symposium. Besides, he has worked as an associate environmental expert in World Bank-funded Flood Recovery Project and also the environmental support staff in the Asian Development Bank (ADB)-funded development projects. He has received many awards, appreciations, and recognition for his services in the field of environmental sciences. He has served as an editorial board member and a reviewer of reputed international journals. Dr. Bhat is still writing and experimenting with diverse capacities of plants for use in aquatic pollution remediation.

Dr. Vandana Singh (Ph.D.) is an Assistant Professor in the Department of Microbiology at the School of Allied Health Sciences, Sharda University, India. She earned her M.Sc. in Microbiology from the University of Pune and completed her Ph.D. in Microbiology at Noida International University. With more than 14 years of teaching and research experience, she has developed expertise across diverse domains of Microbiology and Biotechnology, including Industrial Microbiology, Medical and Environmental Microbiology, Food Microbiology, Microbial Nanotechnology, Nanomaterials, and Microbial Biotechnology. Her research is directed towards the application of microbiological and biotechnological approaches for sustainable environmental and health solutions, with a strong emphasis on microbial fuel cells, wastewater and biological treatment technologies, bioremediation of industrial and agricultural pollutants, biodegradation pathways, microbial food technology, and the molecular basis of biofilm and biosurfactant production. Her work integrates fundamental microbiology with applied nanotechnology and biotechnology, enabling the design of microbial and nano-based systems for environmental and biomedical applications. She has made significant scholarly contributions, reflected in her portfolio by many patents, research/review papers, edited book, and book chapters with leading reputed publishers such as Elsevier, Springer, Wiley, and CRC Press. She has also contributed extensively to curriculum design, academic training, and evidence-based microbiological education, while guiding numerous research projects at undergraduate and postgraduate levels. Her long-term research vision focuses on advancing sustainable microbial and nanobiotechnological interventions for food technology, energy generation, pollutant remediation, and microbial resource management, thereby addressing critical challenges in health, food security, and environmental sustainability.

Dr. Gowhar Hamid Dar (Ph.D.) is an Assistant Professor in the Higher Education Department, Government of Jammu and Kashmir and is currently posted in Government Degree College Kulgam where he serves as the Head, Department of Environmental Science, Coordinator, Committee for Promotion and Research & Nodal Officer for All India Survey on Higher Education (AISHE). Prior to this Dr. Dar worked at Government Sri Pratap College, Srinagar, the oldest higher education institution of the Jammu and Kashmir, India. He has a Ph.D. in environmental science with specialization in environmental microbiology with special focus on fish microbiology, fish pathology, industrial microbiology, taxonomy, and limnology. He has been teaching post-graduate and graduate students for over a decade and has more than 90 research articles (**h**-index **19**; **i**-index **27**; total citation > **1300)** in international and national journals of repute. Further, Dr. Dar has more than 25 books with international publishers (*Springer, Elsevier, CRC Press Taylor and Francis, Apple Academic Press, John Wiley, IGI Global*) to his credit. Moreover, he supervises a number of students for the completion of degrees (Ph.D./PG). Dr. Dar has been working on the Isolation, Identification and Characterization of microbes, their pathogenic behaviour and also studies the impact of pollution on the development of diseases in fish species for the last several years in the Kashmir Himalaya. He has received many awards and appreciations for his services towards the science and development. Besides, he also acts as a member of various research and academic committees and boards. Further, Dr. Dar is a principal investigator and co-principal investigator for different Research and Development projects sanctioned by Government of India and Government of Jammu and Kashmir. Dr. Dar also acts as reviewer for various internationally acclaimed scientific journals.

Dr. Syed Maqbool Geelani has specialized in Natural Textile Dye Extraction and Application, Water Quality Assessment, Water Management, Wastewater Treatment, Nature-based Solutions (NbS), Sustainability, Environmental Pollution, Water Harvesting, Teaching and Environmental Engineering and Consulting. He has been teaching graduate and post-graduate students for over 11 years and has established himself as a dedicated academic and environmental professional. He served as National Coordinator for MOOC Courses under the Ministry of Education, Government of India. He is the author of 09 books, 14 book chapters and 34 research articles published with leading International Publishers such as Springer, Elsevier and Taylor and Francis. His scholarly work has received over 260 citations with an h-index of 10 and i10 index of 10. He has organized, presented and participated in numerous state, national and international conferences, seminars, workshops and symposiums and has earned several awards. Alongside his academic contributions, he serves as an Environmental Engineer and Consultant with various industries and organizations and contributes to the scientific community as an editorial board member and reviewer for reputed international journals.

Hazardous Chemicals in Textile Industries, Risk and Assessment

Zahid Bashir, Suhail Bashir, Syed Safoora Andrabi, Rashmi Ranjan Mandal, Deep Raj, Rangabhashiyam Selvasembian, Aamir Nazir Lone, Mehraj Ud din Bhat, Mudasir Ahmad Dar, Shah Khalid Ahmad, and Nurul Huda Abd Kadir

Abstract The textile industry is a significant source of environmental pollution due to its heavy reliance on hazardous chemicals throughout production processes. Wastewater generated from these operations often contains toxic dyes, heavy metals, and other pollutants, posing severe risks to aquatic ecosystems and human health.

Z. Bashir · R. R. Mandal · D. Raj · R. Selvasembian · A. N. Lone
Department of Environmental Science and Engineering, School of Engineering and Sciences, SRM University-AP, Amaravati, Andhra Pradesh, India
e-mail: zahidbashir5175@gmail.com

R. R. Mandal
e-mail: rrmandal007@gmail.com

D. Raj
e-mail: deepraj2587@gmail.com

R. Selvasembian
e-mail: rambhashiyam@gmail.com

A. N. Lone
e-mail: aamirnazir_lone@srmap.edu.in

S. Bashir
ENS Environmental Consultancy, Sharjah, UAE
e-mail: suhailbashir1272@gmail.com

S. S. Andrabi
Department of Environmental Science, University of Kashmir, Srinagar, India
e-mail: andrabisafoora13@gmail.com

M. Ud din Bhat
Department of Environmental Science Amity School of Life Science, Amity University, Gwalior, Madhya Pradesh, India
e-mail: mehrajbhat490@gmail.com

M. A. Dar
Centre of Research for Development (CORD), University of Kashmir, Srinagar, India
e-mail: dmudasir551@gmail.com

S. K. Ahmad
Cluster University Srinagar Gogji-Bagh, Srinagar, Jammu & Kashmir, India
e-mail: shahkhalidahmed@gmail.com

R. A. Bhat et al. (eds.), *Biocentric Approaches in Textile Waste Management*, Textile Science and Clothing Technology, https://doi.org/10.1007/978-3-032-04974-2_1

This study examines the presence and impact of hazardous chemicals in textile industry waste, emphasizing the importance of risk assessment and effective treatment strategies. The persistence of untreated effluents in water bodies highlights the urgent need for integrated waste management solutions. Biological treatment methods such as bioremediation and anaerobic digestion have emerged as sustainable, cost-effective approaches for degrading organic contaminants. Additionally, engineering controls can be implemented to reduce exposure and mitigate risks at the source. A systematic, multifaceted approach combining biological, physical, and chemical techniques is essential for comprehensive risk management, ensuring environmental protection and public safety.

Keywords Textile pollution · Hazardous chemicals · Risk assessment · Environmental protection · Anaerobic digestion

1 Introduction

The textile industry is one of the largest and most important sectors worldwide, contributing significantly to economic growth and job creation [4, 29]. It is also one of the oldest and most established industries globally, with a profound impact on human civilization and modern economies [24]. Industrialization and economic progress are key drivers for the development of any nation, and the textile sector plays a central role in this process [14]. Textiles are an essential part of daily life and a vital component of the global economy, making it hard to imagine a world without them. The textile Industry is expanding rapidly across the globe. China leads in production, followed by India and the United States. India is also the second-largest exporter of textiles and garments, contributing 5% to global trade. The industry significantly influences the global economy and is one of the key pillars of economic development in many countries. In 2018, the global textile market was valued at approximately \$920 billion and is expected to grow to \$1230 billion by 2024, with an annual growth rate of nearly 5% [24]. Despite its economic importance, the current linear production model of the textile industry poses severe environmental challenges. The sector consumes 20% of the world's water resources (79 billion cubic meters) and contributes 10% of global carbon emissions (1.7 billion tons of CO_2). Additionally, over half of clothing items are discarded within a year, creating 92 million tons of waste annually. Recycling rates are alarmingly low, with less than 1% of used textiles being recycled into new garments. This inefficiency results in material losses worth \$500 billion annually. If these trends continue, the environmental and societal impacts could become catastrophic by 2050 [112].

N. H. A. Kadir (✉)
Faculty of Science and Marine Environment, Universiti Malaysia Terengganu, Terengganu, Malaysia
e-mail: nurulhuda@umt.edu.my

Textile processing involves several stages, each contributing to environmental pollution through the release of specific pollutants. During desizing, effluents with high Biochemical Oxygen Demand (BOD), Chemical Oxygen Demand (COD), and lubricants are discharged (Fig. 1). Scouring produces pollutants like sodium hydroxide, detergents, fats, oils, waxes, and residual lubricants. Bleaching releases hydrogen peroxide and sodium silicate, while mercerizing generates highly alkaline effluents containing sodium hydroxide [17]. Dyeing contributes to pollution through heavy metals, salts, surfactants, and effluents with high BOD and COD [40]. Printing adds to the contamination by releasing suspended solids, urea, metals, and solvents. These pollutants, including organic matter, chemicals, and solid residues, pose significant environmental challenges, highlighting the importance of proper wastewater treatment to mitigate their impact. Textiles processing and associated pollutants are presented in Fig. 1.

The textile industry is a significant contributor to environmental pollution, primarily due to the hazardous chemicals used throughout its production processes [69]. These chemicals pose serious risks to both human health and ecosystems, creating a pressing need for effective management and reduction strategies [57, 163]. Common hazardous chemicals in textiles include formaldehyde, azo dyes, heavy metals, phthalates, chlorobenzenes, polyfluorinated substances (PFAS, and brominated flame retardants [3, 52, 76, 155].Textile dyes used in various manufacturing processes are harmful to human health, the environment, and animals. The release of azo dyes into water systems has been identified as particularly dangerous, prompting studies on the potential toxicity of synthetic dyes. Research has shown that these dyes are harmful to many organisms, including aquatic life [59]. Formaldehyde, a well-known carcinogen, can cause DNA damage and mutations in mammalian cells. Recognized by the International Agency for Research on Cancer (IARC) as

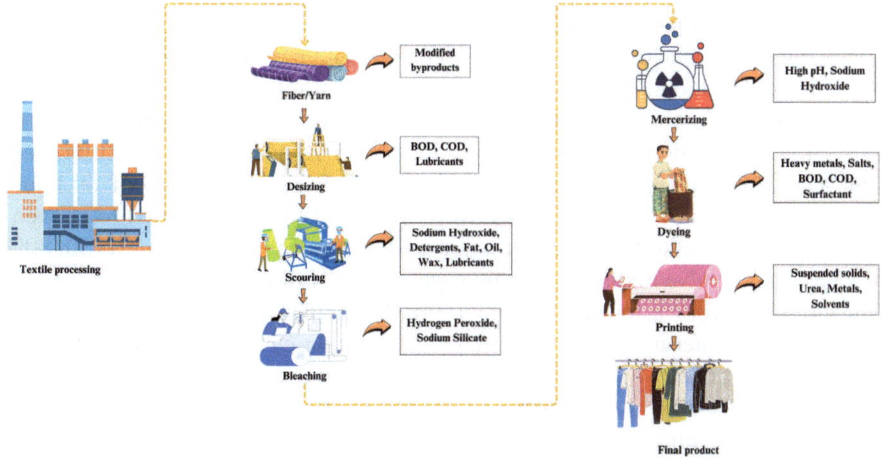

Fig. 1 Circular economy model. *Source* European Parliament Research service

a Group 1 carcinogen, it poses serious health risks, including an increased likelihood of nasopharyngeal cancer (NPC) and leukemia in humans [129]. Wastewater from the textile industry contains harmful trace elements like zinc (Zn), copper (Cu), arsenic (As), and chromium (Cr), which negatively impact the environment. The highly colored wastewater is also associated with several health problems, such as skin ulcers, inflammation, nausea, rashes, and hemorrhaging [69]. Additionally, polyfluorinated substances have been linked to various health issues in both animal studies and humans, including cancer, immune system damage, reproductive and developmental toxicity, liver toxicity, and thyroid problems [87].

The treatment of textile pollutants requires a multifaceted approach that incorporates various techniques to effectively remove or neutralize these contaminants from wastewater. Biological treatment methods, including bioremediation and anaerobic digestion, utilize microorganisms to degrade organic pollutants, making them eco-friendly and cost-effective options [77]. Singh et al.[125] demonstrated a sustainable bioremediation approach using acclimatized *Spirulina* sp. for the removal of heavy metals (cadmium and chromium) and dyes (methyl orange and crystal violet) from wastewater. The study achieved complete removal of cadmium and chromium within 24 h. Additionally, the system demonstrated partial degradation of methyl orange (~ 60%) and higher degradation efficiency for crystal violet (~ 85%), highlighting the effectiveness of microalgal bioremediation for industrial wastewater treatment. Furthermore, [15] explored the use of PersiLac3, a thermo-halotolerant laccase, for removing toxic azo dyes from textile wastewater under high-salt conditions. The enzyme showed impressive activity (112.25%) even in very salty environments, with up to 3 M NaCl, demonstrating its suitability for textile effluents. PersiLac3 efficiently broke down Congo red dye, achieving 82.9% removal within just 15 min, even in high concentrations (5000 mg/L). The presence of typical wastewater ions improved the dye removal to 92.3%.

Chemical treatment techniques, such as Fenton's reagent and coagulation, involve the use of reagents to oxidize or precipitate contaminants, while physical methods like membrane filtration and reverse osmosis separate pollutants based on size and ionic properties [48]. Huong et al. [56] demonstrated the efficacy of Cu-doped graphitic carbon nitride (Cu-GCN) in degrading dye pollutants. Degradation efficiencies of 86.9% for methylene blue (D-MB) and 75.1% for Rhodamine B (D-RhB) were achieved, with real textile wastewater samples showing efficiencies of 69.1 and 55.3% for D-MB and D-RhB, respectively, within 60 min. Compared to conventional chemical methods like coagulation and oxidation, which achieve 50–80% efficiency but involve higher chemical usage, Cu-GCN offers a sustainable and efficient alternative for wastewater treatment. Javed et al. [61] investigated the treatment of textile wastewater using a hybrid catalytic ozonation-electrocoagulation (HCOP-EC) process with ZIF-67 as a catalyst. Under optimal conditions, the HCOP-EC process achieved 79.6% decolorization, 73.3% COD removal, and 69.04% BOD removal after 30 min. Advanced oxidation processes (AOPs) combine chemical oxidation with other techniques to generate powerful oxidizing agents that break down complex organic molecules into less harmful substances [116]. Recent innovations in treatment technologies, such as the use of biosurfactants combined with membrane technology, show promise in enhancing pollutant removal while reducing operational costs [139]. By employing a combination of these diverse techniques, the textile industry can significantly mitigate its environmental impact and improve water quality in affected ecosystems, paving the way for more sustainable practices in textile production and wastewater management.

Regulatory challenges further complicate efforts to manage hazardous chemicals effectively. Risk assessment serves as a critical tool in managing the environmental and health impacts of hazardous chemicals in textile waste. By systematically identifying, quantifying, and evaluating risks, it provides a scientific basis for decision-making. This process encompasses hazard identification, exposure assessment, dose–response analysis, and risk characterization. Effective risk assessment enables stakeholders, including industry players, policymakers, and environmental agencies, to implement targeted mitigation strategies. Furthermore, it fosters the development of safer chemical alternatives and cleaner production technologies, aligning the textile industry with global sustainability goals.

2 Types of Hazardous Chemicals in Textile Waste

The textile industry is a significant source of pollution due to the various chemicals used in the production and dyeing processes [59, 69, 143]. Among these, dyes and pigments are particularly concerning as they contribute to a range of environmental and health issues [154]. Azo dyes are synthetic dyes that constitute approximately 60–80% of all textile colorants, prized for their vibrant hues and cost-effectiveness, however, they can decompose into aromatic amines, some of which are carcinogenic, leading to health risks such as skin allergies and dermatitis [28, 143]. Heavy metals like lead, cadmium, mercury, and chromium (VI) are often used in dyeing processes to stabilize colors but can accumulate in the body and cause serious health issues, including neurological damage from lead and carcinogenic effects from chromium (VI),cadmium is also linked to kidney damage and bone fragility [92]. Formaldehyde, a volatile organic compound (VOC) used for wrinkle resistance in textiles, poses significant health risks such as respiratory problems and skin irritation, raising concerns for both workers and consumers [101]. Chlorinated compounds, including chlorine bleach and chlorobenzenes, are widely used for their bleaching properties but can cause respiratory issues and skin irritation upon exposure, chlorobenzenes may accumulate in the body over time, affecting liver function and the central nervous system [79, 135]. Phthalates, commonly used as plasticizers in synthetic textiles, have raised health concerns due to their endocrine-disrupting properties linked to reproductive health issues [34]. Polyfluorinated chemicals (PFAS), known for their water- and stain-resistant properties, are persistent in the environment and have been associated with various health issues [97]. Volatile organic compounds (VOCs such as toluene, benzene, xylene, and methylene chloride emitted during textile processing contribute to indoor air pollution and can cause short-term health effects like headaches and dizziness, long-term exposure is linked to severe outcomes including liver damage and cancer [67, 91]. Flame retardants added to textiles to reduce flammability have been scrutinized for their links to hormone disruption and cancers [110]. Lastly, antimicrobial agents incorporated into textiles can cause skin irritation or allergic reactions and may contribute to antibiotic resistance when released into wastewater systems during washing processes [46]. Overall, the presence of these hazardous chemicals in textile waste highlights the urgent need for better regulatory measures and sustainable practices within the textile industry to protect human health and minimize environmental impact.

3 Environmental and Health Impacts

The textile industry generates massive amounts of waste each year, significantly contributing to municipal solid waste (MSW) worldwide [114, 138]. Annually, approximately 92 million tons of textile waste are produced globally. This industry is also the second-largest source of plastic waste after packaging, generating 42 million tons of plastic waste each year [65]. In the United States, textile waste occupies around 5% of landfill space, with China and the U.S. being the top producers, generating 20 million and 17 million tons of textile waste, respectively [58].

Types of Textile Waste

Textile waste refers to fabrics or garments discarded after becoming unsuitable for their intended use [133]. It can be categorized into three main types based on its origin:

1. Pre-consumer waste: Generated during production and manufacturing processes, including fabric scraps, defective products, and unused yarns.
2. Post-consumer waste: Discarded textiles such as clothing, bedding, or curtains after use.
3. Post-industrial waste: Created by industries using textiles in products like medical supplies, automotive materials, or packaging [140].

Environmental and Economic Challenges

Textile waste poses significant environmental challenges, contributing to greenhouse gas emissions, water pollution, and soil contamination [102]. The fashion industry alone is responsible for 10% of global greenhouse gas emissions, 20% of global wastewater, and heavy consumption of water, energy, and chemicals in garment production [58, 80]. Disposal methods such as landfilling and incineration worsen the issue. Landfills consume valuable space, while incineration releases toxic substances and microplastics [43]. Economically, textile waste causes losses of approximately USD 500 billion annually due to underutilized clothing and limited recycling. Disposal also incurs additional financial costs [156].

Social Impacts

The management of textile waste intensifies social inequalities. Used clothing is often exported from wealthy nations to developing countries, where low-paid workers sort, categorize, and repackage it for second-hand markets. Items that do not sell become solid waste, adding strain to waste systems and creating environmental and health hazards [21, 156].

Barriers to Effective Management Despite the potential benefits of managing textile waste effectively, challenges persist. Developing countries often lack organized collection systems, leading to unreliable waste streams for recycling [102]. The variety of textile materials, including mixed fibers, dyes, finishes, and accessories, complicates sorting and recycling efforts [37]. Recycling technologies for natural fibers remain underdeveloped, while synthetic fiber recycling is energy-intensive and polluting [63].

Due to these limitations, disposal remains the primary method for handling textile waste, which is unsustainable as it occupies landfill space, emits greenhouse gases, and produces harmful leachates [156]. Furthermore, synthetic fibers contribute significantly to plastic pollution, as they are a major source of microfibers in the environment [137].

4 Effects on Soil and Water Quality

Soil Pollution

Textile dyeing and printing operations produce large volumes of effluent throughout various stages of textile processing, primarily due to the extensive use of chemicals and dyes. This effluent poses an ecological risk, as it degrades soil quality by altering its physicochemical properties. Consequently, the affected soils become more prone to erosion, reduced productivity, and diminished sustainability. Additionally, plant growth is adversely impacted, leading to a decrease in vegetation health and overall soil fertility.

The natural composition of soil consists of both organic and inorganic elements, which play a crucial role in supporting plant and tree growth, thereby maintaining the ecosystem's native flora and fauna. However, soil contamination occurs when textile wastewater is directly discharged into rivers and groundwater. The presence of organic and inorganic pollutants in the wastewater alters the physicochemical properties of the soil, leading to erosion, reduced productivity, and decreased sustainability, which in turn affects the quality of the food chain [35]. Additionally, industrial effluents and organic contaminants adversely affect soil quality by causing an increase in the concentration of harmful natural soil constituents, such as phosphorus, potassium, iron, and manganese, which become toxic in elevated amounts. Consequently, soil fertility is generally reduced, which later contributes to high levels of biological oxygen demand (BOD) and chemical oxygen demand (COD) can clog soil pores in agricultural fields, further diminishing soil productivity [99].

Excessive salt content in textile effluents degrades soil quality, making it unsuitable for agricultural use. Salts dissolved in the effluent increase the osmotic pressure of the soil, which in turn raises the respiration rate while reducing permeability, ventilation, and soil structure—critical factors for healthy plant growth [45]. The use of river water for irrigation has further contributed to soil degradation by causing suspended colloidal particles to block soil pores, impairing its permeability. Additionally, the application of highly salinized groundwater with a high specific absorption rate to agricultural land has led to a significant decline in agricultural production [111].

Water Pollution

The textile dyeing and finishing industry significantly contributes to water pollution because of its extensive use of chemicals and large water consumption [64]. The industry produces over 3,600 types of textile dyes and employs more than 8,000 chemicals in processes like dyeing and printing. During wet processing, massive amounts of water and chemicals are used, and untreated wastewater is often discharged into nearby drainage systems. This leads to toxic chemicals seeping into the soil, contaminating groundwater, surface water, and vegetation. The chemicals used in textile manufacturing are diverse, including inorganic compounds, organic substances, and polymers, all of which contribute to environmental pollution [9].

Additionally, textile operations often involve non-process chemicals like machine cleaners, biocides, insecticides, and boiler treatment chemicals. If these substances enter the wastewater, they further increase the pollution load [103]. The pollution level of textile effluents is assessed using the ratio of biological oxygen demand (BOD) to chemical oxygen demand (COD), which indicates the ease of biodegradation. A ratio between 1:2 and 1:3 suggests good biodegradability, and most textile effluents fall within this range. However, scouring effluents have a much higher ratio, often reaching 1:5, indicating poor biodegradability. This is due to the presence of high levels of suspended solids and grease, which hinder the degradation process [38].

Bioaccumulation and Ecotoxicology

The textile industry has long been recognized as a major contributor to environmental pollution, with its vast array of chemical-intensive processes and the subsequent release of a wide range of contaminants into the surrounding ecosystems [19]. The industry's high-water consumption and the subsequent generation of wastewater containing a cocktail of dyes, heavy metals, and other hazardous substances pose significant threats to aquatic life and human health [13, 158]. One of the primary concerns with textile waste is the issue of bioaccumulation, where toxic substances accumulate in the tissues of living organisms and become increasingly concentrated as they move up the food chain [5, 13]. Heavy metals, such as chromium, arsenic, copper, and zinc, commonly found in textile effluents, are not biodegradable and can persist in the environment for extended periods, leading to their accumulation in the bodies of aquatic organisms and potentially reaching humans through the consumption of contaminated seafood.

The ecotoxicological effects of textile waste can be wide-ranging and severe, with studies indicating that the discharge of dye-containing effluents can have detrimental impacts on a variety of aquatic organisms and ecosystems [5]. Many textile dyes and their breakdown products have been found to be toxic, carcinogenic, or mutagenic, posing significant threats to the overall health and balance of the aquatic environment. These toxic compounds can disrupt the normal physiological processes, reproductive success, and population dynamics of fish, invertebrates, and other aquatic life, leading to long-term consequences for the entire ecosystem [96]. The bioaccumulation and biomagnification of these contaminants through the food chain further exacerbate the ecotoxicological risks, potentially exposing higher trophic level organisms, including humans, to elevated levels of harmful substances of textile waste can be far-reaching, with studies indicating that the discharge of dye-containing effluents can have detrimental impacts on aquatic life. These toxic compounds can disrupt the normal physiological processes, reproductive success, and population dynamics of fish, invertebrates, and other aquatic life, leading to long-term consequences for the entire ecosystem. As the textile industry continues to grow globally, the issue of bioaccumulation and ecotoxicology of textile waste has become increasingly pressing. Effective treatment and management of textile effluents are crucial to mitigating the environmental and public health risks associated with this industry.

Human Health Risks: Carcinogenicity, Mutagenicity, and Endocrine Disruption

The textile industry consumes large quantities of fresh water and generates significant amounts of dark-colored wastewater, which is a major source of environmental pollution [75]. This wastewater, known as textile industry wastewater (TIWW), is highly toxic and complex. It contains high levels of biochemical oxygen demand (BOD) and chemical oxygen demand (COD), along with suspended and dissolved solids. Additionally, it is laden with toxic metals and harmful chemicals, including pesticides, acids, bases, salts, and auxiliary substances [167, 170]. According to the Ministry of Environment and Forests, Government of India, TIWW is the most polluting type of wastewater, containing around 72 highly toxic substances. Among these, about 30 are resistant to microbial degradation, posing serious risks to the environment and human health.

Textile dyes are highly toxic and potentially carcinogenic, posing serious health risks to both humans and animals [62, 141]. These dyes can cause a range of health issues, from skin irritation to neurological disorders, possibly by disrupting enzyme activity through the replacement of enzyme cofactors [151]. Exposure to dyes, whether by ingestion or inhalation, can lead to skin and eye irritation, especially when airborne dye particles are inhaled [30]. Workers handling reactive dyes are particularly prone to allergic conditions like contact dermatitis, conjunctivitis, rhinitis, and asthma [49]. The textile industry faces additional risks from harmful substances that may interfere with reproductive functions, such as ovulation and sperm production [136]. Azo dyes, commonly used in textiles, paper, and leather, have been linked to bladder cancer in humans [142]. In mammals, intestinal bacteria break down azo dyes into amines, which are absorbed into the bloodstream and detected in urine [11]. Specific dyes, like Reactive Green 19, Disperse Red 1, and Reactive Blue 2, have shown long-term genotoxic effects, with Reactive Green 19 displaying dose-dependent toxicity, while the others showed no such effects [82]. Dyes are widely present in the environment and can accumulate in aquatic organisms like fish and algae, entering the food chain [54]. Toxic organic compounds can concentrate in humans at levels 1000 times higher than in water [74]. During water treatment, chlorine reacts with organic materials to form trihalomethanes, byproducts intended to kill harmful microorganisms [158]. However, prolonged exposure to these byproducts can cause severe health problems, including bladder, colon, and colorectal cancers, as well as immune system suppression [141].

Certain dyes, especially azo dyes, have been found to possess mutagenic and carcinogenic properties. For example, Sudan I, a lipophilic azo dye commonly used in textiles and food, can be metabolized by intestinal microbes into harmful aromatic amines with carcinogenic potential [104]. Similarly, Disperse Red 1 dye has been shown to cause chromosomal damage in human liver (HepG2) cells and lymphocytes, increasing micronuclei frequency, which signals mutagenic activity [149]. Disperse Orange 1 dye is known to induce DNA damage through mutations such as base-pair replacements and frameshifts, altering genetic information. It also causes cell toxicity and triggers apoptosis when it interacts with HepG2 cells [44]. Moreover, Direct Blue 14 dye can be transformed into carcinogenic amines by bacteria found on human skin. Additionally, Disperse Yellow 7 dye, when degraded in natural water systems, produces amines that are recognized as carcinogens [18].

Azure B, a major metabolite of Methylene Blue, is a cationic dye that can bind to the helical structures of DNA and duplex RNA [50]. It can also integrate into the lipid membranes of animal cells, where it acts as a reversible inhibitor of monoamine oxidase A, an enzyme essential for regulating behavior in the central nervous system [83]. Research has also shown that Azure B inhibits enzymes like glutathione reductase, which plays a key role in maintaining cellular redox balance [32]. Similarly, triphenylmethane dyes, such as Basic Red 9, are widely used in industries like textiles, leather, paper, and ink but are hazardous to humans. During degradation under anaerobic conditions, these dyes produce aromatic amines, which are carcinogenic and can lead to skin irritation, allergic dermatitis, mutations, and cancer [127]. Another triphenylmethane dye, Crystal Violet, disrupts cell division by causing mitotic poisoning, leading to chromosomal damage and abnormal accumulation at the metaphase stage [89, 90]. Crystal Violet is highly carcinogenic and has been linked to reticular cell sarcoma in organs like the bladder and vagina [81]. It also causes conditions such as chemical cystitis, irritation of the skin and digestive system, and, in severe cases, respiratory and kidney failure [89, 90].

Despite the prohibition of certain dyes, more than 100 of the 4000 dyes tested for toxicity remain on the market, still posing the risk of generating carcinogenic amines [78]. Small-scale textile factories worldwide, often driven by the demand for cheap labor and export needs, continue to secretly discharge toxic dyes into waterways, contributing to environmental degradation. However, bioremediation technologies present an opportunity to address this complex issue, offering solutions that align with sustainable development. These technologies promote both eco-friendly economic growth and improved human well-being, helping mitigate the significant risks posed to ecosystems and human health.

Textile waste, especially from finishing processes, poses a significant threat to human health due to the presence of endocrine-disrupting chemicals (EDCs) [42]. The Environmental Protection Agency (EPA) defines an endocrine disruptor as an external agent that interferes with the natural hormonal functions in the body [42]. The endocrine system, which comprises glands like the thyroid, pituitary, pancreas, ovaries, and testes, along with their hormone secretions, regulates critical physiological functions such as blood sugar levels, blood pressure, metabolic rates, growth, reproduction, and aging [150]. Chemicals in textile wastewater, including micropollutants like EDCs, have been linked to disrupting these functions [78]. Advanced textile manufacturing processes introduce features such as shrink-proofing, water resistance, and flame retardancy, but these modifications come at the cost of releasing persistent and bioaccumulative EDCs into the environment [81]. Conventional wastewater treatments such as flocculation and precipitation are ineffective in removing these pollutants, necessitating advanced techniques like membrane filtration, advanced oxidation, and adsorption technologies [159].

EDCs have garnered global attention for their estrogenic effects, toxicity, and ability to bioaccumulate [142]. These compounds interfere with hormone synthesis, storage, metabolism, and receptor binding, leading to disruptions in the endocrine system, which can result in severe health issues [33]. The World Health Organization (WHO) defines EDCs as exogenous substances that alter normal hormonal functions in humans and animals, impacting both human and wildlife populations [150]. For instance, phthalates, commonly used in plastisol printing on t-shirts, are known to cause endocrine disruption, birth defects, and potentially contribute to breast cancer [55]. Similarly, benzidine-based azo dyes, banned in Europe but widely used in Asia, degrade into aromatic amines that can cause cancer, respiratory problems, and dermatitis [18]. These dyes can transfer to human skin from fabrics, further exposing consumers to these toxic substances [128]. Factory workers involved in textile production are particularly vulnerable, reporting high incidences of dermatitis, respiratory ailments, and other health complications [89, 90]. Addressing these risks requires stricter regulations and the adoption of safer, sustainable alternatives [78].

Impact on Aquatic and Terrestrial Environments

Despite covering over 71% of the Earth's surface, access to clean and safe drinking water remains a critical global issue. The rising demand for freshwater is primarily driven by agriculture (70%), industry (19%), and household use (11%) [25, 26, 47]. Among industries, the textile sector significantly contributes to water pollution, releasing harmful pollutants that contaminate both land and water ecosystems, disrupting natural habitats [131].

The textile industry discharges large amounts of wastewater, often containing untreated or treated azo dyes and other organic pollutants. Azo dyes, when degraded during or after disposal, produce toxic by-products like amino acids, while untreated wastewater has severe harmful effects on aquatic ecosystems [66, 95]. For instance, textile dyes block sunlight by absorbing and reflecting it, preventing light from reaching the photic zones where photosynthesis occurs, ultimately harming aquatic plants and ecosystems [68, 84]. This disruption can lead to decreased oxygen levels, impaired biological processes in aquatic organisms, and ecological imbalances. The impact extends to human health, as textile effluents can cause skin irritations, allergies, dermatitis, cancers, and genetic mutations. Polluted water also exhibits changes in odor, color, and toxicity [117]. Additionally, high dye concentrations in water contribute to eutrophication and long-term environmental risks due to the excessive release of minerals and the persistence of dyes that resist conventional treatment methods. Azo dyes, for instance, are notoriously difficult to degrade, with 60–70% being toxic and carcinogenic [109]. These pollutants not only endanger marine life but also compromise water quality, highlighting the urgent need for effective pollution control and sustainable practices in the textile industry.

Fish and other aquatic animals, which are important protein sources, can cause health problems like cramps, fever, and high blood pressure in humans if they absorb harmful dyes [10, 123]. Wastewater containing dyes and textile pigments not only stains the water but also causes changes in pH levels and increases the concentrations of biological oxygen demand (BOD), chemical oxygen demand (COD), total organic carbon (TOC), and suspended solids. These solids can clog fish gills, disrupting gas exchange and potentially slowing growth or causing death [20]. Long-term exposure to textile wastewater has also been linked to reduced fish appetite, leading to lower levels of protein, carbohydrates, and fats, as well as stunted growth [8]. Reactive azo dyes can damage fish DNA, causing changes in blood cells in adults and affecting gill cells in young fish, depending on the amount and duration of exposure [123, 161]. The lack of oxygen caused by these dyes, along with their impact on the immune system, makes fish more vulnerable to diseases. Consequently, fish exposed to these pollutants can pose serious health risks to humans [161].

Natural processes are unable to counterbalance the oxygen consumed by bacteria when organic materials, such as dyes and pigments, are released into water through textile waste [148]. This leads to a decrease in oxygen levels in still water bodies, where bacteria break down organic matter, consuming enough oxygen to affect one liter of water [85]. The buildup of organic materials in water causes unpleasant tastes, bacterial growth, bad odors, and discoloration [70]. Dyes become visible at concentrations as low as 5 g/L [85]. In addition to their unattractive appearance, dyes can block light in the water, preventing aquatic plants from performing photosynthesis [23]. Synthetic dyes, which are not easily broken down through regular biological methods, persist in the environment [7]. Unsaturated molecules break down more easily than saturated ones due to their higher chemical reactivity, and the stability of aromatic compounds increases with more chemical substitutes. The presence of halogen substitutes makes dyes with alkyl groups even more resistant to degradation [85]. Microalgae play a vital role in aquatic ecosystems as primary producers, benefiting both the environment and economy. However, the presence of dyes in water limits their growth, disrupting the flow of energy and nutrients in these ecosystems. The large discharge of textile dyes into water bodies like rivers, lakes, and oceans negatively impacts various factors of algal growth, such as pigment, protein, and nutrient levels. Algae are more sensitive to pollutants compared to other aquatic species, making them good indicators of pollution in toxicological studies [122]. Methylene Blue, a commonly used dye in the textile industry, releases harmful compounds like benzidine and methylene, which may have carcinogenic effects. The impact of Methylene Blue on the microalgae species Spirulina platensis and Chlorella vulgaris, known for their nutritional and ecological value, was studied [6]. Results showed that exposure to different concentrations of Methylene Blue caused a decrease in the growth rate, protein levels, and pigment content of both species. Additionally, Methylene Blue inhibited chlorophyll production, further affecting their growth and photosynthesis [6]. Aquatic plants are being used more often as natural indicators to assess the toxicity of textile dyes. To measure plant growth, factors like dry weight, number of fronds, total frond area, and chlorophyll content are considered [31, 126]. These factors are negatively affected by the toxicity of textile dyes, leading to changes in the plants' growth. Several studies have shown that dyes reduce growth rates and hinder the development of aquatic plants [2, 53, 86, 108, 126]. This is mainly due to oxidative stress or damage to the plant's photosystem, caused by a decrease in electron transport within chloroplasts [132]. In a study, Lemna minor, a key part of the food chain, was used to examine how Congo Red and Gentian Violet dyes affected the production of biogenic amines. The study found that both dyes slowed down the growth, biomass production, and chlorophyll synthesis in L. minor. Gentian Violet was more harmful than Congo Red because it disrupted the production of biogenic amines. The findings suggest that changes in decarboxylase activity and biogenic amine levels can be early signs of the harmful effects of these dyes on L. minor [2]. Textile dyes and their waste significantly affect the soil's chemistry and disrupt the natural balance of microorganisms in the soil. These pollutants also harm plant growth, particularly impacting flower blooming and seed germination. Key indicators of plant health, such as seed germination rates, seedling height, and survival rates, are negatively affected by exposure to textile dyes [16]. Healthy plants, which show active photosynthesis and high chlorophyll content, produce more dry matter. As the concentration of dye in industrial waste increases, it harms plant growth even more [165]. Polluted wastewater with high solid content lowers the oxygen available for seedlings and disturbs the osmotic balance in seeds [51]. Higher dissolved solids also reduce chlorophyll levels in plants, which affects photosynthesis and slows down plant growth [121]. High levels of abscisic acid can damage chlorophyll and inhibit its production, further stunting plant growth [164]. Additionally, textile dye waste causes plants to accumulate more proline, a sign of stress [41]. Genotoxic effects, such as chromosomal changes and micronuclei formation, have been seen in seeds of Alliumcepaexposed to textile effluents with azo dyes. Even low amounts of Black Dye Commercial Product cause cell damage, chromosomal changes, and cell death in the meristem of A. cepa [147].

5 Risk Management Strategies

The term Risk is defined as the occurrence of any potential event that could affect the receptor. However, it is observed that the risk level is considered based on the event's likelihood and its potential impact. All the activities related to industries, or any other sector has specific risk associated with their aspects. However, to address these risks it is essential to develop risk management strategies to manage the identified risks associated to the specific industries or any other sector.

The risk management involves detailed assessment of likelihood of injury to personnel, damage to assets, or exposure to health and safety, pollution or contamination risk and the severity of the effects of the associated injury, damage or exposure. It can be also defined as a tool that is used to provide the information required for necessary action and decisions regarding the requirements for management of the environment and the health and safety [120]. The primary objective for performing risk assessment is to ensure that receptors like human health, assets or environment etc. are protected from these risks and the necessary resources are provided to make sure that any risks associated with in a system or activity are identified reduced to acceptable levels. It imperative to primarily identify the risks associated with the activity or industry etc. and maintain the record within a Risk Register. Additionally, Evaluate the level of significance of the risk (e.g. Extreme Risk, High Risk, Moderate Risk and Low Risk) [153]. It has been observed that the within the textile industry, risk assessment is an important step due the sector's internal complexities of the industry, which includes the hazardous chemicals usage, increased energy consumption and significant use of water for production activities. The industry has potential occupational hazards, that includes exposure to dyes, chemicals and particulate matter that has potential to pose significant risks to worker health and safety [130]. However, considering the major risks due to the industrial activities, it has been observed that increased water footprint is among the primary risk. Environmental risks, including water contamination from untreated effluents etc., further emphasize the implementation of risk management strategies to reduce the impact and for the necessary action.

The untreated effluents have substantial portion of chemicals that are potential for water contamination that will adversely affect ecosystems and human health. Therefore, identifying and mitigating these risks is essential to make sure compliance with environmental regulations and safeguard both the workforce and the surrounding community [152]. However, when the risks are categorized, the industry shall identify corrective actions to manage hazards and to reduce the industry specific associated risks to an acceptable level, often referred to as "As Low As Reasonably Practicable" (ALARP). For low-risk hazards, actions may not be necessary. The other risks that include moderate, high and extreme risks, it is important to implement appropriate control measures to reduce the risk. The controlled measures shall be structured to managed risks in an effective manner. Wherever possible, elimination of risk shall be prioritized, followed by alternatives, such as using less hazardous substances or environmentally friendly materials. Apart from these, Engineering controls can be also implemented for reduction or mitigation of risk related to the industries etc. This systematic approach ensures comprehensive risk management and minimizes exposure to hazards effectively.

6 Regulatory Framework and Standards

International Regulations

The textile industry is one of the largest contributors to water pollution, greenhouse gas emissions, and waste production. Due to the significant environmental impact, global policies and standards have been established to promote sustainability in the sector. The Paris Agreement, an international treaty, aims to limit the rise in global temperatures to no more than 2 °C. Governments encourage textile manufacturers to adopt more eco-friendly production methods, including using renewable energy sources in their operations [102]. Another initiative, the Global Fashion Agenda, brings together top leaders to help transform the fashion industry into a more sustainable business. Additionally, the "Fashion Industry Charter for Climate Action" outlines commitments to reducing greenhouse gas emissions. Industry organizations, like the OECD Guidelines for Multinational Enterprises, provide recommendations for responsible business practices across industries, including textiles. These guidelines emphasize environmental protection and ethical supply chain management [117].

REACH, which stands for Registration, Evaluation, Authorisation, and Restriction of Chemicals, is a set of rules created by the European Union to protect people's health and the environment from the dangers of chemicals. This regulation requires companies that produce or import chemicals into the EU to provide detailed information about the substances, including their properties, uses, and potential risks [27]. The goal is to make sure that chemicals are safe to use and that there are steps in place to reduce any harm. The European Chemicals Agency (ECHA) can review the chemicals and ban or restrict those that are especially dangerous. REACH plays an important role in controlling hazardous chemicals by requiring companies to prove that the chemicals are safe at every stage, from production to use.

ZDHC (Zero Discharge of Hazardous Chemicals) is a global initiative aimed at eliminating hazardous chemicals from the textile and footwear industry's supply chain. Launched in 2011, it provides a comprehensive framework for brands and suppliers to collaborate on safe and sustainable chemical practices [12]. The textile industry uses over 8000 chemicals, many of which pose serious health and environmental risks [98]. A key aspect of ZDHC is the Manufacturing Restricted Substances List (MRSL), which guides manufacturers in selecting safer chemicals for production [105]. The MRSL lists harmful substances and sets limits on restricted chemicals in the finishing process, helping reduce environmental damage caused by textile production [12]. This list is essential in preventing harmful chemicals from entering wastewater and solid waste, which can lead to significant pollution [100, 143].

The Basel Convention is a vital international agreement that focuses on regulating the movement and disposal of hazardous waste across borders. Its main aim is to protect human health and the environment from the harmful effects of hazardous waste [72, 73]. This treaty is particularly relevant to the textile industry, which generates significant amounts of hazardous waste during certain production processes [124, 143]. The textile sector uses a lot of chemicals, many of which are considered hazardous. The Basel Convention sets rules to ensure that these dangerous substances are managed safely, reducing their potential harm to health and the environment [106]. For instance, the Convention mandates that hazardous waste be disposed of in a way that is safe for the environment, which is especially important for textile manufacturers who often struggle to manage their waste properly [72, 73].

National Policies on Waste Management and Control of Chemicals

Waste management and chemical control are essential for safeguarding both human health and the environment. To address growing concerns about waste disposal and safe chemical handling, countries worldwide have developed various policies. India, like other nations, has implemented several policies aimed at managing hazardous substances, industrial waste, and chemical safety. These regulations promote sustainable practices, recycling, and reducing environmental damage. In India, the Environment Protection Act (EPA) of 1986 is the primary law governing environmental protection. In the chemical sector, strict regulations control the handling of hazardous materials, such as the Chemical Accidents (Emergency Planning, Preparedness, and Response) Rules of 1996, The Manufacture, Storage, and Import of Hazardous Chemicals Rules of 1989, and The Factories Act of 1948, which set safety standards in industries to ensure safe chemical handling and waste disposal, protecting workers from chemical risks. Globally, international policies also aim to manage waste and control chemicals to reduce environmental risks and promote sustainability. The Basel Convention, adopted in 1989, regulates the international movement and disposal of hazardous waste, requiring consent from the receiving country before transport [129]. It encourages environmentally responsible disposal methods. The Stockholm Convention, 2001, seeks the complete elimination of persistent organic pollutants (POPs) by banning or limiting their use [144]. Similarly, the Minamata Convention, 2013, focuses on reducing mercury pollution through safe storage, disposal practices, and eliminating mercury from industrial processes [154]. The Rotterdam Convention of 1998 ensures that countries are informed before importing hazardous chemicals through a Prior Informed Consent (PIC) procedure [155].

Standards for Textile Industry Effluents

Many countries have set limits on pollutants in textile wastewater to protect the environment. These regulations focus on key factors like COD, BOD, and heavy metals. For example, the European Union has strict rules for wastewater treatment to ensure pollutant levels do not exceed set limits [39, 72]. Similarly, the American Dye Manufacturers Institute has guidelines for controlling the color of effluent, which helps maintain the quality of wastewater released from dyeing plants (Ponnusamy et al.) [162]. Additionally, countries that are part of the Basel Convention are required to develop national plans that support the convention's objectives, making the management of hazardous waste more effective [22, 105] (Table 1).

Table 1 Regulatory frameworks and standards that different regions use to manage textile wastewater, ensuring environmental sustainability and minimizing pollution

Regulation/ organization	Pollutants monitored	Standards/guidelines	Region/countries affected
European Union Directives	COD, BOD, Heavy metals, EQS for pollutants	Stringent EQS limits for wastewater treatment	EU Member States
American Dye Manufacturers Institute (ADMI)	Effluent color limits	Guidelines for acceptable levels of effluent color from dyeing plants	USA
Basel Convention	Hazardous waste, including textile waste	Requires countries to develop strategies for handling hazardous waste	Countries that are signatories of the Basel Convention
National Regulations (Various Countries)	COD, BOD, Heavy metals, dyes	Country-specific allowable limits for textile wastewater pollutants	Varies by country

Future Directions and Sustainable Practices

Green Chemistry in Textile Manufacturing

Green chemistry has become a key approach in the textile industry, as it helps reduce environmental harm during production. One innovation in this area is the use of supercritical carbon dioxide ($scCO_2$) for dyeing textiles. This method utilizes $scCO_2$ as a solvent, which is non-toxic, easy to recover, and reusable, reducing water consumption and chemical waste [88, 146]. The $scCO_2$ dyeing process not only saves water but also decreases the use of harmful chemicals that are commonly used in traditional dyeing methods, offering a more sustainable option [88]. Another major advancement is the use of bio-based dyes, which are derived from natural sources. These dyes reduce the dependence on synthetic dyes that often contain toxic substances, aligning with the growing demand for eco-friendly products and lowering the environmental impact [115].

Circular Economy Approaches for Textile Waste

The circular economy focuses on reusing and recycling textile materials to reduce waste and minimize environmental harm. It promotes designing products to last longer, supporting repair and reuse, and implementing efficient recycling systems throughout the textile supply chain. To make circular economy models work in the textile industry, effective recycling systems are essential. This includes setting up facilities for collecting and sorting textile waste and investing in advanced recycling technologies that can handle mixed materials [118]. Challenges in recycling often stem from consumer awareness and proper disposal practices, which need more focus to ensure a smooth shift to a circular economy [113].

Innovative Hazardous Waste Minimization

New technologies in the textile industry are helping reduce hazardous waste while supporting sustainability. One of the most effective methods is the use of zero liquid discharge (ZLD) systems. ZLD systems treat and recycle wastewater efficiently, ensuring no liquid waste is released into the environment. These systems not only conserve water but also significantly reduce the environmental impact of wastewater [162] (Fig. 2).

Along with ZLD, advanced wastewater treatment methods, such as biological treatment, membrane filtration, and advanced oxidation, help improve water quality by removing contaminants and pollutants from textile wastewater [162]. These technologies are especially important in areas with limited water resources, as treated wastewater can be reused in manufacturing processes.

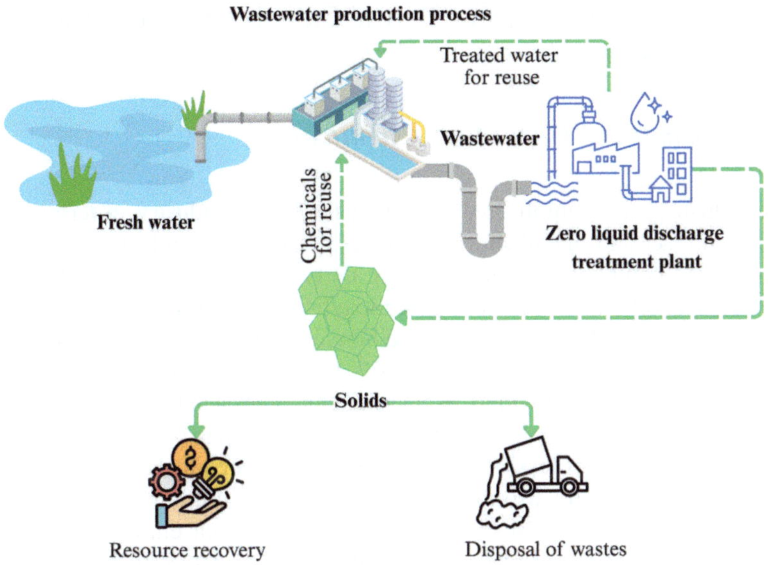

Fig. 2 Zero liquid discharge (ZLD) treatment plant

7 Policies and Stakeholder Role Sustainable Management

The textile industry can achieve sustainability only through collaboration among key stakeholders, including manufacturers, policymakers, and consumers. Manufacturers play a crucial role by adopting eco-friendly practices, such as using natural dyes, biodegradable materials, and advanced technologies like zero liquid discharge systems for wastewater management [36]. Policymakers can support this transition by creating an enabling environment through regulations and incentives. For example, tax benefits for eco-friendly production and grants for sustainable technology research can motivate manufacturers to embrace greener methods [36]. Effective policies, such as tax incentives for investing in cleaner technologies or using sustainable materials, can drive companies toward sustainable practices (Akter et al., 2022). Waste management and reducing environmental impacts remain significant challenges for the textile industry. However, by adopting sustainability-focused approaches, complying with regulations, and fostering innovation, the industry can move toward a more environmentally responsible future. Achieving sustainability in the textile sector requires international regulations, supportive national policies, and collaborative efforts among all stakeholders to meet sustainable development goals.

8 Hazardous Chemicals in Textile Industry Waste, Risk and Assessment

Risk Assessment of Hazardous Chemicals

All the industries are complicated including the textile industry in terms of environmental impact and manufacturing practices. The textile industry is one among the largest and complicated manufacturing units presently. However, these industries generate enormous amount of waste including hazardous chemicals that have a significant impact on the environment and human health. It was observed that, there is systematic approaches available to identify the hazard within an industry [71].

Risk assessment is used widely as a systematic approach to identify and control the hazard within a system. Where the risk assessment has been seen as a process that assist associated professionals to identify what chemicals are released from any particular activity and also the elements that may cause impact to occupational health [1]. However, risk assessment is not a process that eliminates all hazards within any industry or workplace, neither it is used as a tool to identify the risk and to prevent the danger associated with the risk. The various hazards associated with any activity or industry are identified and critically assessed through standard methods and procedures of Qualitative Risk Assessment and QRA (Quantitative Risk Assessment).

The are many industries that are operating from last many decades, where textile industry is also among the oldest industries of human civilization. The cloth is among the basic necessity after food for survival. The industries are involved in many processes that involves significant amount of water and chemicals during the wet processes [94]. While after the processes the effluents from these industries enter the different ecosystems and is potential to cause pollution (soil and water contamination) and disruption to an ecosystem along with its impact on human health. It is evident that, the textile effluents contain dyes, heavy metals, inorganic salts, surfactants, organic contaminants, oil, and grease [69]. These contaminants fall under the hazardous category.

Hazard Identification

The concept of Hazard identification is defined as most significant aspects of a health and safety management system. While the hazard identification acts as the base for defining safety in processing works, implementing prevention programs and considering other necessary measures to eliminate or curtail dangers. However, danger is defined as any source of possible damage, harm or adverse health effects on receptor. A hazard's risk is the probability that it will cause harm. Identifying existing and potential workplace dangers, assessing risks, selecting and implementing controls, and reviewing hazards are all part of the hazard identification process [60].

It is important to mention that Hazard identification is among the initial steps towards environmental regulation and compliance along with development of recommendation and mitigation strategies. However, the effective hazard identification step requires active participation of all associated stakeholders to ensure an effective implementation. However, the techniques such as risk assessments, safety audits and job safety analysis are commonly used to identify significant hazards within an industry or workplace.

Development of Risk Management Strategies

The term Risk is defined as the occurrence of any potential event that could affect the receptor. However, it is observed that the risk level is considered based on the event's likelihood and its potential impact. All the activities related to industries or any other sector has specific risk associated with their aspects. However, to address these risks it is essential to develop risk management strategies to manage the identified risks associated to the specific industries or any other sector. The risk management involves detailed assessment of likelihood of injury to personnel, damage to assets, or exposure to health and safety, pollution or contamination risk and the severity of the effects of the associated injury, damage or exposure. It can be also defined as a tool that is used to provide the information required for necessary action and decisions regarding the requirements for management of the environment and the health and safety [120]. The primary objective for performing risk assessment is to ensure that receptors like human health, assets or environment etc. are protected from these risks and the necessary resources are provided to make sure that any risks associated with in a system or activity are identified reduced to acceptable levels. It imperative to primarily identify the risks associated with the activity or industry etc. and maintain the record within a Risk Register. Additionally, Evaluate the level of significance of the risk (e.g.- Extreme Risk, High Risk, Moderate Risk and Low Risk) [153]. It has been observed that the within the textile industry, risk assessment is an important step due the sector's internal complexities of the industry, which includes the hazardous chemicals usage, increased energy consumption and significant use of water for production activities. The industry has potential occupational hazards, that includes exposure to dyes, chemicals and particulate matter that has potential to pose significant risks to worker health and safety [130]. However, considering the major risks due to the industrial activities, it has been observed that increased water footprint is among the primary risk. Environmental risks, including water contamination from untreated effluents etc., further emphasize the implementation of risk management strategies to reduce the impact and for the necessary action [152].

References

1. Ab Rahim, M. S., Reniers, G., Yang, M., & Bajpai, S. (2024). Risk assessment methods for process safety, process security and resilience in the chemical process industry: A thorough literature review. *Journal of Loss Prevention in the Process Industries*, 105274.
2. Adomas, B., Sikorski, Ł., Bęś, A., & Warmiński, K. (2020). Exposure of *Lemna minor* L. to gentian violet or Congo red is associated with changes in the biosynthesis pathway of biogenic amines. *Chemosphere, 254*, 126752.
3. Agnihotri, S., Sheikh, J. N., Singh, S. P., & Behera, B. K. (2024). Flame-retardant textile structural composites for construction application: A review. *Journal of Materials Science, 59*(5), 1788–1818.
4. Alam, M. F. B., Hosen, M. I., Mridha, J. H., Chowdhury, S. E., & Rahman, M. A. (2023). Assessing the barriers of integrating technological innovations in textiles sector: Implications towards sustainable production. *Green Technologies and Sustainability, 1*(3), 100039.
5. Aldalbahi, A., El-Naggar, M. E., El-Newehy, M. H., Rahaman, M., Hatshan, M. R., & Khattab, T. A. (2021). Effects of technical textiles and synthetic nanofibers on environmental pollution. *Polymers, 13*(1), 155.
6. Ali, S. K., & Saleh, A. M. (2012). Spirulina—An overview. *International Journal of Pharmacy and Pharmaceutical Sciences, 4*(3), 9–15.
7. Allaoui, M., Berradi, M., & Elharfi, A. (2016). Synthesis of a new asymmetric semi-permeable membrane and based alloy of two polymers PVC and PSU: Application in the treatment of a colored solution. *Moroccan Journal of Chemistry, 4*(1), 1–4.
8. Almroth, B. C., Cartine, J., Jönander, C., Karlsson, M., Langlois, J., Lindström, M., Lundin, J., Melander, N., Pesqueda, A., Rahmqvist, I., & Renaux, J. (2021). Assessing the effects of textile leachates in fish using multiple testing methods: From gene expression to behavior. *Ecotoxicology and Environmental Safety, 207*, 111523.
9. Al-Tohamy, R., Ali, S. S., Li, F., Okasha, K., Mahmoud, Y.A.-G., Elsamahy, T., Jiao, H., Fu, Y., & Sun, J. (2022). A critical review on the treatment of dye-containing wastewater: Ecotoxicological and health concerns of textile dyes and possible remediation approaches for environmental safety. *Ecotoxicology and Environmental Safety, 231*, 113160.
10. Amer, O. A., Ali, S. S., Azab, M., El-Shouny, W. A., Sun, J., & Mahmoud, Y. A. G. (2022). Exploring new marine bacterial species, *Alcaligenes faecalis* Alca F2018 valued for bioconversion of shrimp chitin to chitosan for concomitant biotechnological applications. *International Journal of Biological Macromolecules, 196*, 35–45.
11. Amin, T., Afrin, M., Haque, Z., & Islam, M. R. (2016). Effects of textile dye wastewater on reproductive system of mice and their progeny. *International Journal of Scientific Engineering and Research, 7*, 565–570.
12. Amutha, K. (2017). Sustainable chemical management and zero discharges. *Sustainable Fibres and Textiles*, 347–366.https://doi.org/10.1016/b978-0-08-102041-8.00012-3
13. Ananthashankar, A. G. R. (2013). Production, characterization and treatment of textile effluents: A critical review. *Journal of Chemical Engineering & Process Technology, 5*(1).
14. Appiah, M., Gyamfi, B. A., Adebayo, T. S., & Bekun, F. V. (2023). Do financial development, foreign direct investment, and economic growth enhance industrial development? Fresh evidence from Sub-Sahara African countries. *Portuguese Economic Journal, 22*(2), 203–227.
15. Ariaeenejad, S., Kavousi, K., Zeinalabedini, M., Jahanshahi, D. A., & Beh-Afarin, S. R. (2024). Enhanced bioremediation of saline azo dye effluents using PersiLac3: A thermo-halotolerant laccase from a tannery metagenome. *Results in Engineering, 24*, 103172.
16. Arshad, H., Imran, M., & Ashraf, M. (2020). Toxic effects of Red-S3B dye on soil microbial activities, wheat yield, and their alleviation by pressmud application. *Ecotoxicology and Environmental Safety, 204*, 111030.
17. Azanaw, A., Birlie, B., Teshome, B., & Jemberie, M. (2022). Textile effluent treatment methods and eco-friendly resolution of textile wastewater. *Case Studies in Chemical and Environmental Engineering, 6*, 100230.

18. Balakrishnan, V. K., Shirin, S., Aman, A. M., de Solla, S. R., Mathieu-Denoncourt, J., & Langlois, V. S. (2016). Genotoxic and carcinogenic products arising from reductive transformations of the azo dye, disperse yellow 7. *Chemosphere, 146*, 206–215.
19. Behera, M., Nayak, J., Banerjee, S., Chakrabortty, S., & Tripathy, S. K. (2021). A review on the treatment of textile industry waste effluents towards the development of efficient mitigation strategy: An integrated system design approach. *Journal of Environmental Chemical Engineering, 9*(4), 105277. Elsevier.
20. Berradi, M., Hsissou, R., Khudhair, M., Assouag, M., Cherkaoui, O., El Bachiri, A., & El Harfi, A. (2019). Textile finishing dyes and their impact on aquatic environs. *Heliyon, 5*(11), e02711.
21. Bick, R., Halsey, E., & Ekenga, C. C. (2018). The global environmental injustice of fast fashion. *Environmental Health, 17*, p.92.
22. Camilleri, M. A. (2020). European environment policy for the circular economy: Implications for business and industry stakeholders. *Sustainable Development, 28*(6), 1804–1812.
23. Cardoso, N. F., Lima, E. C., Royer, B., Bach, M. V., Dotto, G. L., Pinto, L. A., & Calvete, T. (2012). Comparison of *Spirulina platensis* microalgae and commercial activated carbon as adsorbents for the removal of reactive red 120 dye from aqueous effluents. *Journal of Hazardous Materials, 241*, 146–153.
24. Chand, S., Chand, S., & Raula, B. (2023). Textile and apparel industries waste and its sustainable management approaches. *Journal of Material Cycles and Waste Management, 25*(6), 3132–3143.
25. Chen, G., An, X., Li, H., Lai, F., Yuan, E., Xia, X., & Zhang, Q. (2021a). Detoxification of azo dye direct black G by thermophilic *Anoxybacillus* sp. PDR2 and its application potential in bioremediation. *Ecotoxicology and Environmental Safety, 214*, 112084.
26. Chen, Y. G., Huang, J. H., Luo, R., Ge, H. Z., Wołowicz, A., Wawrzkiewicz, M., Gładysz-Płaska, A., Li, B., Yu, Q. X., Kołodyńska, D., & Lv, G. Y. (2021). Impacts of heavy metals and medicinal crops on ecological systems, environmental pollution, cultivation, and production processes in China. *Ecotoxicology and Environmental Safety, 219*, 112336.
27. Chen, T.-L., Kim, H., Pan, S.-Y., Tseng, P.-C., Lin, Y.-P., & Chiang, P.-C. (2020). Implementation of green chemistry principles in circular economy system towards sustainable development goals: Challenges and perspectives. *Science of The Total Environment, 716*, 136998.
28. Chien, J. R. C., & Ganesan, J. J. (2024). Advancing sustainable approaches for the removal and recycling of toxic dyes from the aquatic environment.
29. Christian, D., Gaekwad, A., Dani, H., Shabiimam, M. A., & Kandya, A. (2023). Recent techniques of textile industrial wastewater treatment: A review. *Materials Today: Proceedings, 77*, 277–285.
30. Clark, M. (2011). *Handbook of textile and industrial dyeing: Principles*. Elsevier.
31. Colin, N., Maceda-Veiga, A., Flor-Arnau, N., Mora, J., Fortuño, P., Vieira, C., Prat, N., Cambra, J., & De Sostoa, A. (2016). Ecological impact and recovery of a Mediterranean river after receiving the effluent from a textile dyeing industry. *Ecotoxicology and Environmental Safety, 132*, 295–303.
32. Couto, C. M., Almeida, C. A., & Silva, J. M. (2016). Enzyme inhibition by Azure B: A potential environmental and health concern. *Toxicology Letters, 254*, 12–19.
33. Couto, N., Wood, J., & Barber, J. (2016b). The role of glutathione reductase and related enzymes on cellular redox homoeostasis network. *Free Radical Biology and Medicine, 95*, 27–42.
34. Czarnywojtek, A., Jaz, K., Ochmaåƒska, A., Zgorzalewicz-Stachowiak, M., Czarnocka, B., Sawicka-Gutaj, N., Ziółkowska, P., Krela-Kaźmierczak, I., Gut, P., Florek E., & Ruchała, M. (2021). The effect of endocrine disruptors on the reproductive system-current knowledge. *European Review for Medical & Pharmacological Sciences, 25*(15).
35. Dave, D., & Vyas, N. (2022). Impact of textile effluents on soil in and around Pali, Western Rajasthan, India. *The Scientific Temper, 13*(1), 150–153.

36. De Oliveira, C. R. S., da Silva Júnior, A. H., Mulinari, J., & Immich, A. P. S. (2021). Textile re-engineering: Eco-responsible solutions for a more sustainable industry. *Sustainable Production and Consumption, 28*, 1232–1248.
37. Dissanayake, D. G. K., & Weerasinghe, D. U. (2021). Fabric waste recycling: A systematic review of methods, applications, and challenges. *Materials Circular Economy, 3*, 24.
38. Diwyanjalee, G. R., Bellanthudawa, B. K. A., De Silva, D. K. N. S., & Gunawardena, A. R. (2024). Biodegradability index (BDI) as an indicator for effluents quality measurement: A case study based on different industry sectors in Matara District, Sri Lanka. *Water Practice & Technology, 19*(8), 3092–3108. Available at: https://doi.org/10.2166/wpt.2024.183.
39. Drakvik, E., Altenburger, R., Aoki, Y., Backhaus, T., Bahadori, T., Barouki, R., Brack, W., Cronin, M. T. D., Demeneix, B., Hougaard Bennekou, S., van Klaveren, J., Kneuer, C., Kolossa-Gehring, M., Lebret, E., Posthuma, L., Reiber, L., Rider, C., Rüegg, J., Testa, G., van der Burg, B., & Bergman, Å. (2020). Statement on advancing the assessment of chemical mixtures and their risks for human health and the environment. *Environment International, 134*, 105267.
40. Dutta, P., Rabbi, M., Abu Sufian, M., & Mahjebin, S. (2022). Effects of textile dyeing effluent on the environment and its treatment: A review. *Engineering and Applied Science Letters, 5*(1), 1–1.
41. Ebency, C. I. L., Rajan, S., Murugesan, A. G., Rajesh, R., & Elayarajah, B. (2013). Biodegradation of textile azo dyes and its bioremediation potential using seed germination efficiency. *International Journal of Current Microbiology and Applied Sciences, 2*(10), 496–505.
42. Environmental Protection Agency (EPA). (2021). *Endocrine disruptors*. Available at: https://www.epa.gov/endocrine-disruption
43. European Parliament. (n.d.). *The impact of textile production and waste on the environment*, https://www.europarl.europa.eu/pdfs/news/expert/2020/12/story/20201208STO93327/20201208STO93327_en.pdfat
44. Ferraz, E. R., Grando, M. D., & Oliveira, D. P. (2011). The azo dye Disperse Orange 1 induces DNA damage and cytotoxic effects but does not cause ecotoxic effects in *Daphnia similis* and *Vibrio fischeri*. *Journal of Hazardous Materials, 192*(2), 628–633.
45. Goyal, P., & Chauhan, S. S. (2014). Impact of textile wastewater on agricultural fields: A case study of Sanganer town, Jaipur. *International Journal of Geology, Earth and Environmental Sciences, 4*(1), 37–44.
46. Gulati, R., Sharma, S., & Sharma, R. K. (2022). Antimicrobial textile: Recent developments and functional perspective. *Polymer Bulletin, 79*(8), 5747–5771.
47. Gupta, V., Garg, S., Capalash, N., Gupta, N., & Sharma, P. (2015). Production of thermo-alkali-stable laccase and xylanase by co-culturing of *Bacillus* sp. and *B. halodurans* for biobleaching of kraft pulp and deinking of waste paper. *Bioprocess and Biosystems Engineering, 38*(5), 947–956.
48. Hanafi, M. F., & Sapawe, N. (2020). A review on the current techniques and technologies of organic pollutants removal from water/wastewater. *Materials Today: Proceedings, 31*, A158–A165.
49. Hanger, K. (2003). *Industrial dyes: Chemistry, properties and applications, health and safety aspects*.
50. Haq, I., & Raj, A. (2018). Biodegradation of Azure-B dye by *Serratia liquefaciens* and its validation by phytotoxicity, genotoxicity and cytotoxicity studies. *Chemosphere, 196*, 58–68.
51. Hassan, M. M., & Carr, C. M. (2018). A critical review on recent advancements of the removal of reactive dyes from dyehouse effluent by ion-exchange adsorbents. *Chemosphere, 209*, 201–219.
52. Hemashenpagam, N., & Selvajeyanthi, S. (2023). Textile dyes and their effect on human beings. In *Nanohybrid materials for treatment of textiles dyes* (pp. 41–60). Singapore: Springer Nature Singapore.
53. Hocini, I., Benabbas, K., Khellaf, N., Djelal, H., & Amrane, A. (2019). Can duckweed be used for the biomonitoring of textile effluents? *European Journal of Environmental Integration, 4*(1), 1–8.

54. Hossain, M., Rakib, M., & Amin, N. (2018). Toxicity and bioaccumulation of endocrine disruptors in textiles. *Journal of Environmental Biology, 39*(3), 293–300.
55. Hossain, M. S., Das, S. C., Islam, J. M., Al Mamun, M. A., & Khan, M. A. (2018). Reuse of textile mill ETP sludge in environmentally friendly bricks–effect of gamma radiation. *Radiation Physics and Chemistry, 151*, 77–83.
56. Huong, N. T. M., Hoai, P. T. T., TaeYoung, K., & Huong, P. T. (2024). Exploring Cu-doped graphitic carbon nitride for treatment of dye pollutants in textile wastewater: Benefits and limitations. *Diamond and Related Materials, 146*, 111160.
57. Hussain, S., & Reza, M. (2023). Environmental damage and global health: Understanding the impacts and proposing mitigation strategies. *Journal of Big-Data Analytics and Cloud Computing, 8*(2), 1–21.
58. Huun, K. (2021). Textile waste. *Environmental Center*. Available at: https://www.colorado.edu/ecenter/2021/10/05/textile-waste
59. Islam, T., Repon, M. R., Islam, T., Sarwar, Z., & Rahman, M. M. (2023). Impact of textile dyes on health and ecosystem: A review of structure, causes, and potential solutions. *Environmental Science and Pollution Research, 30*(4), 9207–9242.
60. Janizadeh, R., Kakaei, P., Nickdel, T., Tahmasebi, M., Tamoradi, N., & Nezhad, A. (2019). Textile industry hazard identification and risk assessment by using HAZAN method. *Archives of Occupational Health, 3*. https://doi.org/10.18502/aoh.v3i4.1554.
61. Javed, F., Tariq, M. F., Ikhlaq, A., Munir, H. M. S., & Altaee, A. (2025). Remediation of textile wastewater by hybrid technique using ZIF-67 catalyzed ozonation coupled with electrocoagulation. *Journal of Water Process Engineering, 69*, 106604.
62. Jin, X., Wu, C., Tian, X., Wang, P., Zhou, Y., & Zuo, J. (2021). A magnetic-void-porous $MnFe_2O_4$/carbon microspheres nano-catalyst for catalytic ozonation: Preparation, performance and mechanism. *Environmental Science & Ecotechnology, 7*, 100110.
63. Kamble, Z., & Behera, B. K. (2021). Upcycling textile wastes: Challenges and innovations. *Textile Progress, 53*, 65–122.
64. Kant, R. (2012). Textile dyeing industry: An environmental hazard. *Natural Science, 4*(1), 22–26. Available at: https://doi.org/10.4236/ns.2012.41004
65. Kasavan, S., Yusoff, S., Guan, N. C., Zaman, N. S. K., & Fakri, M. F. R. (2021). Global trends of textile waste research from 2005 to 2020 using bibliometric analysis. *Environmental Science and Pollution Research, 28*, 44780–44794.
66. Khalaj, M., Kamali, M., Khodaparast, Z., & Jahanshahi, A. (2018). Copper-based nanomaterials for environmental decontamination—An overview on technical and toxicological aspects. *Ecotoxicology and Environmental Safety, 148*, 813–824.
67. Khan, A., Kanwal, H., Bibi, S., Mushtaq, S., Khan, A., Khan, Y. H., & Mallhi, T. H. (2021). Volatile organic compounds and neurological disorders: From exposure to preventive interventions. *Environmental contaminants and neurological disorders* (pp. 201–230). Springer International Publishing.
68. Khan, M. D., Abdulateif, H., Ismail, I. M., Sabir, S., & Khan, M. Z. (2015). Bioelectricity generation and bioremediation of an azo-dye in a microbial fuel cell coupled activated sludge process. *PLoS ONE, 10*(10), e0138448.
69. Khan, W. U., Ahmed, S., Dhoble, Y., & Madhav, S. (2023). A critical review of hazardous waste generation from textile industries and associated ecological impacts. *Journal of the Indian Chemical Society, 100*(1), 100829.
70. Khattab, T. A., Rehan, M., & Hamouda, T. (2018). Smart textile framework: Photochromic and fluorescent cellulosic fabric printed by strontium aluminate pigment. *Carbohydrate Polymers, 195*, 143–152.
71. Khoshsepehr, Z., Alinejad, S., & Alimohammadlou, M. (2023). Exploring industrial waste management challenges and smart solutions: An integrated hesitant fuzzy multi-criteria decision-making approach. *Journal of Cleaner Production, 420*, 138327.
72. Kienzler, A., Bopp, S. K., van der Linden, S., Berggren, E., & Worth, A. (2016). Regulatory assessment of chemical mixtures: Requirements, current approaches and future perspectives. *Regulatory Toxicology and Pharmacology, 80*, 321–334.

73. Kirchherr, J., Piscicelli, L., Bour, R., Kostense-Smit, E., Muller, J., Huibrechtse-Truijens, A., & Hekkert, M. (2018). Barriers to the circular economy: Evidence from the European Union (EU). *Ecological Economics, 150*, 264–272.
74. Korpi, A., Järnberg, J., & Pasanen, A. L. (2009). Microbial volatile organic compounds. *Critical Reviews in Toxicology, 39*(2), 39–193.
75. Kumar, A., Sharma, G., Naushad, M., Ala'a, H., García-Penas, A., Mola, G. T., Si, C., & Stadler, F. J., (2020). Bio-inspired and biomaterials-based hybrid photocatalysts for environmental detoxification: A review. *Chemical Engineering Journal, 382*, p.122937.
76. Kumari, U. (2024). Textile dyes and their impact on the natural environment. In *Dye pollution from textile industry: Challenges and opportunities for sustainable development* (pp. 17–30). Singapore: Springer Nature Singapore.
77. Kuppan, N., Padman, M., Mahadeva, M., Srinivasan, S., & Devarajan, R. (2024). A comprehensive review of sustainable bioremediation techniques: Eco friendly solutions for waste and pollution management. *Waste Management Bulletin.*
78. Lacasse, K., & Baumann, W. (2012). *Textile Chemicals: Environmental Data and Facts.* Springer Science & Business Media.
79. Lee, H., Kim, K., Park, J., & Lee, J. G. (2024). Contamination of trichlorobenzene isomers in food: Toxicity, analytical methods, occurrence in food, and risk assessments. *Applied Biological Chemistry, 67*(1), 89.
80. Lee, K. K., & Tang, K. H. D. (2020). Agaricales (gilled mushrooms) as biosorbents of synthetic dye. *Malaysian Journal of Medical and Health Sciences, 16*(Suppl. S11), pp.10–17.
81. Lellis, B., Fávaro-Polonio, C. Z., Pamphile, J. A., & Polonio, J. C. (2019). Effects of textile dyes on health and the environment and bioremediation potential of living organisms. *Biotechnology Research and Innovation, 3*(2), 275–290.
82. Leme, D. M., Primo, F. L., Gobo, G. G., da Costa, C. R. V., Tedesco, A. C., & de Oliveira, D. P. (2015). Genotoxicity assessment of reactive and disperse textile dyes using human dermal equivalent (3D cell culture system). *Journal of Toxicology and Environmental Health A, 78*(7), 466–480.
83. Li, H., Zhang, R., Tang, L., Zhang, J., & Mao, Z. (2014). Evaluation of *Bacillus* sp. MZS10 for decolorizing Azure B dye and its decolorization mechanism. *Journal of Environmental Sciences, 26*(5), 1125–1134.
84. Liang, J., Ning, X. A., Kong, M., Liu, D., Wang, G., Cai, H., Sun, J., Zhang, Y., Lu, X., & Yuan, Y., (2017). Elimination and ecotoxicity evaluation of phthalic acid esters from textile-dyeing wastewater. *Environmental Pollution, 231*, 115–122.
85. Lim, S. L., Chu, W. L., & Phang, S. M. (2010). Use of *Chlorella vulgaris* for bioremediation of textile wastewater. *Bioresource Technology, 101*(19), 7314–7322.
86. Lobiuc, A., Olaru, S., Hancu, E. I., Costica, N., Fortuna, M. E., Zamfirache, M. M., & Constantinescu, G. (2018). Toxicity and removal of direct red 28 diazo dye in living polymeric systems. *Revue de Chimie, 69*(7), 1628–1635.
87. Lohmann, R., Cousins, I. T., DeWitt, J. C., Gluge, J., Goldenman, G., Herzke, D., Lindstrom, A. B., Miller, M. F., Ng, C. A., Patton, S., Scheringer, M., & Wang, Z. (2020). Are fluoropolymers really of low concern for human and environmental health and separate from other PFAS?. *Environmental Science & Technology, 54*(20), 12820–12828.
88. Mahmud, I., & Kaiser, S. (2020). Recent progress in waterless textile dyeing. *Journal of Textile Science & Engineering, 10*(6), 1–3.
89. Mani, S., & Bharagava, R. N. (2016a). Exposure to crystal violet, its toxic, genotoxic and carcinogenic effects on environment and its degradation and detoxification for environmental safety. *Reviews of Environmental Contamination and Toxicology, 237*, 71–104.
90. Mani, S., & Bharagava, R. N. (2016b). Textile dyeing industry effluents and their impact on the environment. *Environmental Science and Pollution Research, 23*(15), 14865–14879.
91. Manickam, P., & Vijay, D. (2021).Chemical hazards in textiles. In *Chemical management in textiles and fashion* (pp. 19–52). Woodhead Publishing.
92. Mariappan, S., & Issac, R. (2022). Removal of heavy metals from textile industries with natural adsorbents. *Journal of Current Science and Technology, 12*(2), 372–390.

93. Ministry of Environment, Forest and Climate Change (MoEFCC), Government of India. (2016). *The hazardous and other wastes (management and transboundary movement) rules*. 621-78.
94. Mohan, D., & Shukla, S. P. (2022). Hazardous consequences of textile mill effluents on soil and their remediation approaches. *Cleaner Engineering and Technology, 7*, 100434.
95. Mudhoo, A., Ramasamy, D. L., Bhatnagar, A., Usman, M., & Sillanpää, M. (2020). An analysis of the versatility and effectiveness of composts for sequestering heavy metal ions, dyes and xenobiotics from soils and aqueous milieus. *Ecotoxicology and Environmental Safety, 197*, 110587.
96. Nayak, S., Sahoo, G., Das, I. I., Mohanty, A. K., Kumar, R., Sahoo, L., & Sundaray, J. K. (2023). Poly- and Perfluoroalkyl substances (PFAS): Do they matter to aquatic ecosystems? *Toxics, 11*(6), 543. [online] Available at: https://doi.org/10.3390/toxics11060543
97. Newland, A., Khyum, M. M., Halamek, J., & Ramkumar, S. (2023). Perfluoroalkyl and polyfluoroalkyl substances (PFAS)—Fibrous substrates. *Tappi Journal, 22*(9), 559–572.
98. Nimkar, U. (2018). Sustainable chemistry: A solution to the textile industry in a developing world. *Current Opinion in Green and Sustainable Chemistry, 9*, 13–17.
99. Nolz, R., & Kammerer, G. (2017). Evaluating a sensor setup with respect to near-surface soil water monitoring and determination of in-situ water retention functions. *Journal of Hydrology, 549*, 301–312.
100. Oelze, N. (2017). Sustainable supply chain management implementation-enablers and barriers in the textile industry. *Sustainability, 9*(8), 1435.
101. Patra, A. K., & Pariti, S. R. K. (2022). Restricted substances for textiles. *Textile Progress, 54*(1), 1–101.
102. Pensupa, N., Leu, S.-Y., Hu, Y., Du, C., Liu, H., Jing, H., Wang, H., & Lin, C. S. K. (2017). Recent trends in sustainable textile waste recycling methods: Current situation and future prospects. *Topics in Current Chemistry, 375*, 76.
103. Periyasamy, A. P. (2024). Recent advances in the remediation of textile-dye-containing wastewater: Prioritizing human health and sustainable wastewater treatment. *Sustainability, 16*, 495. [online] Available at: https://doi.org/10.3390/su16020495. Accessed January 3, 2025.
104. Piątkowska, M., Jedziniak, P., Olejnik, M., Zmudzki, J., & Posyniak, A. (2018). Absence of evidence or evidence of absence? A transfer and depletion study of Sudan I in eggs. *Food Chemistry, 239*, 598–602.
105. Plakantonaki, S., Kiskira, K., Zacharopoulos, N., Chronis, I., Coelho, F., Togiani, A., Kalkanis, K., & Priniotakis, G. (2023). A review of sustainability standards and ecolabeling in the textile industry. *Sustainability, 15*(15), 11589.
106. Popescu, A. (2024). Enhancing environmental standards in textiles: The role of OEKO-TEX® standard 100. *Bulletin of the Transilvania University of Brasov. Series V: Economic Sciences*, 59–68.
107. Puri, M., Gandhi, K., & Kumar, M. S. (2023). Emerging environmental contaminants: A global perspective on policies and regulations. *Journal of Environmental Management, 332*, 117344.
108. Rápó, E., Posta, K., Csavdári, A., Vincze, B. É., Mara, G., Kovács, G., Haddidi, I., & Tonk, S. (2020). Performance comparison of *Eichhornia crassipes* and *Salvinia natans* on Azo-Dye (Eriochrome Black T) phytoremediation. *Crystals, 10*(7), 565.
109. Rawat, D., Sharma, R. S., Karmakar, S., Arora, L. S., & Mishra, V. (2018). Ecotoxic potential of a presumably non-toxic azo dye. *Ecotoxicology and Environmental Safety, 148*, 528–537.
110. Reynolds, K. J., Zagho, M. M., Robertson, M., Qiang, Z., & Nazarenko, S. (2022). Environmental, health, and legislation considerations for rational design of nonreactive flame-retardant additives for polymeric materials: Future perspectives. *Macromolecular Rapid Communications, 43*(24), 2200472.
111. Rott, U., & Minke, R. (1999). Overview of wastewater treatment and recycling in the textile processing industry. *Water Science and Technology, 40*(1), 137–144.
112. Saha, K., Dey, P. K., & Kumar, V. (2024). A comprehensive review of circular economy research in the textile and clothing industry. *Journal of Cleaner Production*, 141252.

113. Saha, K., Dey, P. K., & Papagiannaki, E. (2022). Implementing circular economy in the textile and clothing industry. In *Supply chain sustainability in small and medium sized enterprises* (pp. 239–276). Routledge.
114. Salman, M., Demir, M., Tang, K. H. D., Cao, L. T. T., Bunrith, S., Chen, T.-W., Darwish, N. M., AlMunqedhi, B. M., & Hadibarata, T. (2022). Removal of cresol red by adsorption using wastepaper. *Industrial and Domestic Waste Management, 2*, 1–8.
115. Samanta, A. K. (2018). Application of natural dyes to cotton and jute textiles: science and technology and environmental issues. In *Handbook of renewable materials for coloration and finishing* (pp. 213–277). Wiley.
116. Saravanan, A., Deivayanai, V. C., Kumar, P. S., Rangasamy, G., Hemavathy, R. V., Harshana, T., Gayathri, N., & Alagumalai, K. (2022). A detailed review on advanced oxidation process in treatment of wastewater: Mechanism, challenges and future outlook. *Chemosphere, 308*, 136524
117. Sarvajith, M., Reddy, G. K. K., & Nancharaiah, Y. V. (2018). Textile dye biodecolourization and ammonium removal over nitrite in aerobic granular sludge sequencing batch reactors. *Journal of Hazardous Materials, 342*, 536–543.
118. Schumacher, K. A., & Forster, A. L. (2022). Textiles in a circular economy: An assessment of the current landscape, challenges, and opportunities in the United States. *Frontiers in Sustainability, 3.*
119. Secretariat of the Basel Convention. (2023). *Basel convention on the control of transboundary movements of hazardous wastes and their disposal.*
120. Settembre-Blundo, D., González-Sánchez, R., Medina-Salgado, S., & García-Muiña, F. E. (2021). Flexibility and resilience in corporate decision making: A new sustainability-based risk management system in uncertain times. *Global Journal of Flexible Systems Management, 22*(Suppl 2), 107–132.
121. Sharma, A., Kumar, V., Shahzad, B., Ramakrishnan, M., Singh Sidhu, G. P., Bali, A. S., Handa, N., Kapoor, D., Yadav, P., Khanna, K., & Bakshi, P. (2020). Photosynthetic response of plants under different abiotic stresses: A review. *Journal of Plant Growth Regulation, 39*(2), 509–531.
122. Sharma, J., Sharma, S., & Soni, V. (2021). Classification and impact of synthetic textile dyes on Aquatic Flora: a review. *Regional Studies in Marine Science*, 101802.
123. Sharma, K., Sharma, P., Dhiman, S. K., Chadha, P., & Saini, H. S. (2022). Biochemical, genotoxic, histological and ultrastructural effects on liver and gills of freshwater fish *Channa punctatus* exposed to textile industry intermediate 2 ABS. *Chemosphere, 287*, 132103.
124. Shirvanimoghaddam, K., Motamed, B., Ramakrishna, S., & Naebe, M. (2020). Death by waste: Fashion and textile circular economy case. *Science of The Total Environment, 718*, 137317.
125. Singh, V. P., Godara, P., & Srivastava, A. (2024). Sustainable microalgal bioremediation of heavy metals and dyes from synthetic wastewater: Progressing towards United Nations sustainable development goals. *Waste Management Bulletin, 2*(4), 123–135.
126. Singha, K., Pandit, P., Maity, S., & Sharma, S. R. (2021). Harmful environmental effects for textile chemical dyeing practice. In *Green chemistry for sustainable textiles* (pp. 153–164). Woodhead Publishing.
127. Sivarajasekar, N., & Baskar, R. (2014). Adsorption of basic red 9 on activated waste *Gossypium hirsutum* seeds: Process modeling, analysis and optimization using statistical design. *Journal of Industrial Engineering and Chemistry, 20*(5), 2699–2709.
128. Sivarajasekar, N., & Baskar, R. (2014). Health risks associated with textile dyes. *Journal of Environmental and Analytical Toxicology, 4*(5), 215–221.
129. Soltanpour, Z., Mohammadian, Y., & Fakhri, Y. (2022). The exposure to formaldehyde in industries and health care centers: A systematic review and probabilistic health risk assessment. *Environmental Research, 204*, 112094.
130. Sonal, S., & Mishra, B. K. (2021). Role of coagulation/flocculation technology for the treatment of dye wastewater: trend and future aspects. *Water Pollution and Management Practices*, 303–331.

131. Soni, V., Keswani, K., Bhatt, U., Kumar, D., & Singh, H. (2021). In vitro propagation and analysis of mixotrophic potential to improve survival rate of *Dolichandra unguiscati* under ex vitro conditions. *Heliyon, 7*(2), e06101.
132. Sree, K. S., Keresztes, Á., Mueller-Roeber, B., Brandt, R., Eberius, M., Fischer, W., & Appenroth, K. J. (2015). Phytotoxicity of cobalt ions on the duckweed *Lemna minor*—Morphology, ion uptake, and starch accumulation. *Chemosphere, 131*, 149–156.
133. Stanescu, M. D. (2021). State of the art of post-consumer textile waste upcycling to reach the zero waste milestone. *Environmental Science and Pollution Research, 28*, 14253–14270.
134. Stockholm Convention Secretariat. (2020). *Stockholm convention on persistent organic pollutants.*1-79.
135. Stubbs, A. D. (2024). *Chlorine in water and air: The effects of sodium chloride on pollutant Photodegradation in seawater and sea ice, and indoor Hypochlorous acid levels following surface cleaning with bleach* (Doctoral dissertation, University of Saskatchewan Saskatoon).
136. Suryavathi, V., Sharma, S., Sharma, S., Saxena, P., Pandey, S., Grover, R., Kumar, S., & Sharma, K. P. (2005). Acute toxicity of textile dye wastewaters (untreated and treated) of Sanganer on male reproductive systems of albino rats and mice. *Reproductive Toxicology, 19*(4), 547–556.
137. Tang, K. H. D. (2023). Enhanced plastic economy: A perspective and a call for international action. *Environmental Science Advances, 2*, 1011–1018.
138. Tang, K. H. D., Darwish, N. M., Alkahtani, A. M., AbdelGawwad, M. R., & Karácsony, P. (2022). Biological removal of dyes from wastewater: A review of its efficiency and advances. *Tropical Aquatic Soil Pollution, 2*, 59–75.
139. Thakur, V., Baghmare, P., Verma, A., Verma, J. S., & Geed, S. R. (2024). Recent progress in microbial biosurfactants production strategies: Applications, technological bottlenecks, and future outlook. *Bioresource Technology*, 131211.
140. Tomovska, E., Jordeva, S., Trajković, D., & Zafirova, K. (2017). Attitudes towards managing post-industrial apparel cuttings waste. *Journal of the Textile Institute, 108*, 172–177.
141. Tounsadi, H., Ghazouani, A., & Boumehdi, M. (2020a). Emerging endocrine disruptors from industrial textile wastewaters. *Water Research, 173*, 115507.
142. Tounsadi, H., Metarfi, Y., Taleb, M., El Rhazi, K., & Rais, Z. (2020). Impact of chemical substances used in textile industry on the employee's health: Epidemiological study. *Ecotoxicology and Environmental Safety, 197*, 110594.
143. Uddin, F. (2021). Environmental hazard in textile dyeing wastewater from local textile industry. *Cellulose, 28*(17), 10715–10739.
144. United Nations Environment Programme (UNEP). (2023*). Minamata convention on mercury.*
145. United Nations Environment Programme. (2023). *Rotterdam convention on the prior informed consent procedure for certain hazardous chemicals and pesticides in international trade.* 1-56.
146. Varadarajan, G., & Venkatachalam, P. (2015). Sustainable textile dyeing processes. *Environmental Chemistry Letters, 14*(1), 113–122.
147. Ventura-Camargo, B. D. C., Maltempi, P. P., & Marin-Morales, M. A. (2011). The use of the cytogenetic to identify mechanisms of action of an azo dye in *Allium cepa* meristematic cells. *Journal of Environmental and Analytical Toxicology, 1*(3), 1–12.
148. Wan, Z., Jiao, Y., Ouyang, X., Chang, L., & Wang, X. (2017). Bifunctional MoS2 coated melamine-formaldehyde sponges for efficient oil–water separation and water-soluble dye removal. *Applied Materials Today, 9*, 551–559.
149. Will, Y., McDuffie, J. E., Olaharski, A. J., & Jeffy, B. D. (2016). *Drug discovery toxicology: From target assessment to translational biomarkers.* Wiley.
150. World Health Organization (WHO). (2020). *Endocrine disruptors.* Available at: WHO website.
151. Wu, L., Xu, Y., Lv, X., Chang, X., Ma, X., Tian, X., Shi, X., Li, X., & Kong, X. (2021). Impacts of an azo food dye tartrazine uptake on intestinal barrier, oxidative stress, inflammatory response and intestinal microbiome in crucian carp (*Carassius auratus*). *Ecotoxicology and Environmental Safety, 223*, 112551.

152. Wu, Z., Zhou, L., Ding, X., Wu, X., & Wang, L. (2022). Knowledge roadmap of sustainable development in the textile and apparel industry: A scientometric analysis. *Fashion and Textiles, 9*(1), 35.
153. Xiang, M., Li, Y., Yang, J., Lei, K., Li, Y., Li, F., Zheng, D., Fang, X., & Cao, Y. (2021). Heavy metal contamination risk assessment and correlation analysis of heavy metal contents in soil and crops. *Environmental Pollution, 278*, 116911.
154. Yadav, S., Tiwari, K. S., Gupta, C., Tiwari, M. K., Khan, A., & Sonkar, S. P. (2023). A brief review on natural dyes, pigments: Recent advances and future perspectives. *Results in Chemistry, 5*, 100733.
155. Yakamercan, E., & Aygün, A. (2024). Sources of persistent organic pollutants in textile industry; amounts, fate, and treatment methods. In *Pollutants and recent trends in wastewater treatment* (pp. 181–204). Cham: Springer Nature Switzerland.
156. Yalcin-Enis, I., Kucukali-Ozturk, M., & Sezgin, H. (2019). Risks and management of textile waste. In K. M. Gothandam, S. Ranjan, N. Dasgupta, & E. Lichtfouse (Eds.), *Nanoscience and biotechnology for environmental applications* (pp. 29–53). Springer International Publishing.
157. Yang, T., Zhang, Q., Wan, X., Li, X., Wang, Y., & Wang, W. (2020). Comprehensive ecological risk assessment for semi-arid basin based on conceptual model of risk response and improved TOPSIS model-a case study of Wei River Basin, China. *Science of the Total Environment, 719*, 137502.
158. Zaharia, C., & Daniel, S. (2012). Textile organic dyes—Characteristics, polluting effects and separation/elimination procedures from industrial effluents—A critical overview. In *InTech eBooks*. Available at: https://doi.org/10.5772/32373.
159. Zhao, B., Xiao, W., Shang, Y., Zhu, H., & Han, R. (2017). Adsorption of light green anionic dye using cationic surfactant-modified peanut husk in batch mode. *Arabian Journal of Chemistry, 10*, S3595–S3602.
160. Zhao, Y., Wang, C., & Du, H. (2017). Advanced treatment technologies for endocrine-disrupting chemicals in textile effluents. *Water Science and Technology, 76*(8), 2115–2123.
161. Zheng, X., Fu, D., Cheng, J., Tang, R., Chu, M., Chu, P., Wang, T., & Yin, S. (2021). Effects of hypoxic stress and recovery on oxidative stress, apoptosis, and intestinal microorganisms in Pelteobagrus vachelli. *Aquaculture, 543*, 736945.
162. Zhou, L., & Zhou, H. (2019). Evaluation of textile industry wastewater treatment and reuse in China. *Desalination and Water Treatment, 144*, 330–338.
163. Zhou, W., Li, M., & Achal, V. (2024). A comprehensive review on environmental and human health impacts of chemical pesticide usage. *Emerging Contaminants*, 100410.
164. Zhu, X., Chen, J., Qiu, K., & Kuai, B. (2017). Phytohormone and light regulation of chlorophyll degradation. *Frontiers in Plant Science, 8*, 1911.
165. Zou, H., Ning, X. A., Wang, Y., & Zhou, F. (2019). The agricultural use potential of the detoxified textile dyeing sludge by integrated Ultrasound/Fenton-like process: A comparative study. *Ecotoxicology and Environmental Safety, 172*, 26–32.

Biopolymers in Textile Waste Management: Sources, Properties, and Prospects

Sayam, Nayan Das, and Asish Sarker

1 Introduction

The increasing demand of, inadequate recycling infrastructure, and the worsening environmental impact of textile production and disposal are making textile waste management a critical global concern. The European Union's shortcomings in recycling highlight the importance for waste prevention procedures and improved management processes to foster sustainability in the textile sector [128]. Worldwide, textile industry generates waste around 1.5% of total trash, with underdeveloped countries seeing severe difficulties in implementing effective waste management system [114]. In regions such as Africa, the improper disposal of poor clothing results in notable environmental challenges, including pollution and flooding, while the overlooked potential for recycling exacerbates these difficulties [25]. Even though technological advancements in waste management, regional disparities in practices, and the pollution of textile waste, particularly from toxic chemicals and non-biodegradable fibers, present impediments to recycling initiatives [2, 55]. Consequently, tackling

Sayam
Department of Fabric Engineering, Barishal Textile Engineering College, Barishal 8200, Bangladesh

Sayam (✉)
Department of Organic Materials Science, Graduate School of Organic Materials Science, Yamagata University, Yonezawa 992-8510, Yamagata, Japan
e-mail: mahbub.sayamm@gmail.com

N. Das
Department of Wet Process Engineering, Shahid Abdur Rab Serniabat Textile Engineering College, Barishal 8200, Bangladesh
e-mail: nayandas.wpe@gmail.com

A. Sarker
Department of Yarn Engineering, Barishal Textile Engineering College, Barishal 8200, Bangladesh
e-mail: asishsarker333@gmail.com

R. A. Bhat et al. (eds.), *Biocentric Approaches in Textile Waste Management*, Textile Science and Clothing Technology, https://doi.org/10.1007/978-3-032-04974-2_2

these complex difficulties needs creative solutions and heightened consumer awareness to enhance both recovery and recycling activities. Propelled by population growth and the quick fashion trend, the textile industry's rapid augmentation has significantly intensified the collection of textile waste. This has emerged as a critical environmental concern, chiefly because of pollution arising from insufficient disposal techniques such as incineration and landfilling, which create harmful pollutants [58, 104, 106, 114]. Additionally, synthetic fibers, which are non-biodegradable and contain toxic substances that exacerbate pollution and cause potential threats to health, comprise a significant component of textile waste [58, 161]. Recycling and reuse, while considered effective waste management strategies, encounter major barriers due to the complex composition of textile materials, including fiber mixes, and the lack of structured waste management systems, especially in developing countries [23, 54, 58, 114, 145]. The execution of waste management solutions is constrained by financial limitations, particularly when it comes to smaller enterprises, and progress is also hindered by the substantial expenses related to constructing waste management infrastructures [76].

In light of these challenges, biopolymers have come to prominence as a feasible long-term solution to mitigate textile waste, offering a sustainable solution to traditional petroleum-based products. Biopolymers, such as proteins, polyhydroxyalkanoates (PHA), cellulose, chitosan and polylactic acid (PLA), have captured interest in the textile industry because of their biodegradable properties [62, 102, 110]. To inspire sustainability and minimize environmental impacts, these components can be employed in the spinning, dyeing, and finishing stages of textile manufacturing [88]. Biopolymers reduce post-consumer particle accumulation by offering biodegradable materials, unlike synthetic fibers that often contribute to landfill waste [26]. Biopolymer composites, such as those composed of recycled cotton fibers and bio-based polyethylene terephthalate, can reduce reliance on non-renewable resources and enhance mechanical properties [93]. Also, biopolymer, such as PLA, are being studied for their degradability and capacity to mitigate textile waste in environmental applications. Their utilization in pioneering fields like coatings and 3D printings offers immense possibilities for improving textile characteristics and lessening dependence on hazardous chemicals [86]. Even though biopolymers huge potential for enhancing circular economy and sustainability within the textile sector, they are not without difficulties related processing intricacies and durability. They can alleviate the ecological consequences of textile waste and help in tackling wider environmental challenges within the textile sector [76]. However, this chapter covers the origins, properties, and possible applications of biopolymers in textile waste management. In addition, this chapter emphasize the significance of biopolymers in fostering a sustainable and eco-friendly textile industry.

2 Sources of Biopolymers for Textile Waste Management

Many natural and renewable sources yield biopolymers. Figure 1 shows the sources of biopolymers. All of these provide crucial resources for advancement of groundbreaking applications in textile waste management.

Animals

Chitosan

Chitosan, a polysaccharide biopolymer synthesized from the alkaline deacetylation of chitin, is a vital element in modern technology. The structure comprises N-acetyl glucosamine and glucosamine units linked by a 1,4-glycosidic bond. Chitosan is commonly used across various industries, including textiles, because of its non-toxic, antibacterial, biodegradable, and biocompatible advantages [119]. Chitosan's appropriateness for use is determined by its molecular weight (MW); the source and synthesis process have an impact on MW. The deacetylation procedure determines

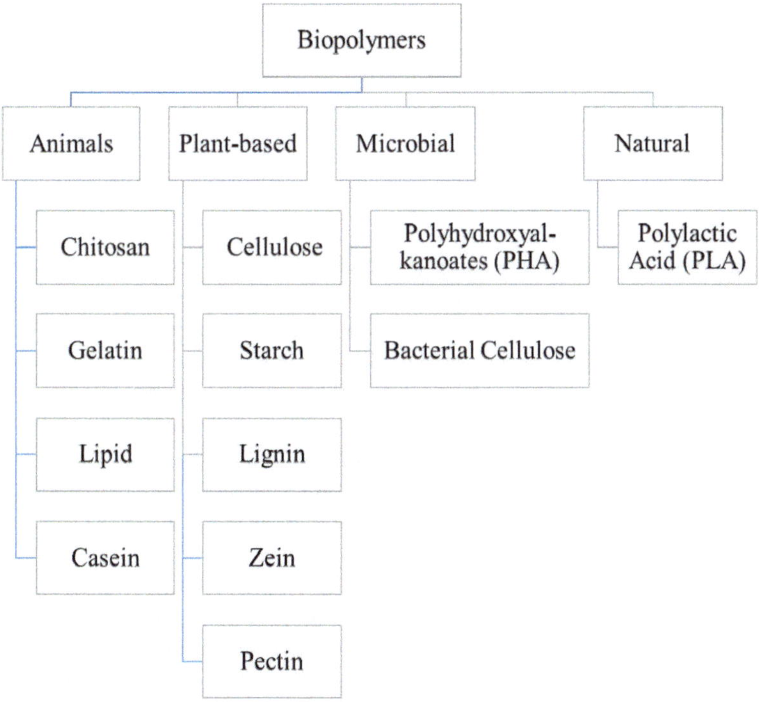

Fig. 1 Classification of biopolymer

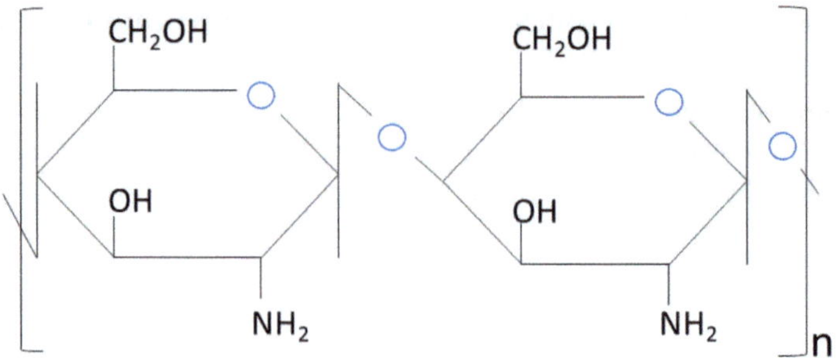

Fig. 2 Structure of chitosan

the weight of commercial chitosan, which ranges from 100 to 1200 k Dalton. Gel permeation chromatography (GPC) is the most popular way to find out the weight of chitosan molecules [132]. The majority of naturally occurring polysaccharides are either acidic or neutral, however chitosan is an exception. Chitosan is a polymer that can dissolve in various acidic liquids. Furthermore, it has been claimed that it is soluble in N-methyl morpholine N-oxide (NMMO) and water. At room temperature, aldehydes and ketones react with chitosan to make the film-forming chemicals ketimines and aldimines [90]. The chemical structure of chitosan is illustrated in Fig. 2.

Plant-Based

Starch

Starch is a notable natural bio-based polymer that has received considerable attention across multiple industrial sectors due to its advantageous chemical and physical properties. The characteristics consist of non-toxicity, biodegradability, biocompatibility, and the capacity to create cohesive films. Starch is one of the most affordable polysaccharides because it is a biopolymer that can be used repeatedly. It can be derived from a variety of plant sources, including the seeds, roots, and stalks of rice, wheat, potatoes, corn, and cassava, and is widely available in nature [99]. Starch is traditionally extracted from sources such as tapioca, sugar palm, potato, wheat, corn, and rice. Recent findings indicate that corn starch shows properties comparable to those of other biopolymers [115].

Starch is regarded as a semi-crystalline polymer that consists of repeating units of anhydroglucose, which are chemically linked by α-(1,4) glycosidic bonds. It is mainly composed of two types of monomers: amylose and amylopectin. Amylose, comprising approximately 15–30% of starch, is composed of a linear arrangement of glucose molecules and is essential to the amorphous structure of starch granules. Amylopectin, constituting 70–85% of starch, is a branching chain of glucose molecules including crystalline sections [13]. Paper, varnishes, food, textiles, bonding agents, pharmaceuticals, and dyes and paints are just a few of the industries that use starch since it is a versatile polymer [85]. The development and utilization of starch-based biodegradable materials have garnered significant attention [39, 115]. In addition to conventional textile uses, starch presents opportunities for improving sustainable textiles via the integration of magnetic nanoparticles (MNPs) [107].

Lignin

The most common biopolymer in nature is cellulose. The second most common is lignin [109]. This material is a byproduct of the paper, biofuel, and eco-chemical industries, among others that utilize wood pulp [129]. The core of trees is utilized by these industries, while the outer bark is generally regarded as waste.

Lignin is located in the outer bark. Lignin includes 15–40% of biomass, altering according to the plant species [129]. Lignin is a worthwhile biomaterial for the treatment of textile waste due to its ambiguous and convoluted chemical structure, along with its molecular weight, rich composition, and functional groups, which consists of several types of carboxylic, carbonyl, hydroxyl and methoxyl groups [3, 78, 105]. The structural attributes of lignin, such as its extensive and low molecular weight dispersion and inadequate alignment, are the main constraints in boosting the physical qualities of lignin-based fibers. Research has mainly focused on lignin homogeneity and quality [73, 74], improving its strength [66], and adjusting its structure [152, 153]. Lignin has exceptional thermal stability, high carbon content, biodegradability, and stiffness, among other attributes, all at an affordable cost [30, 87]. Lignin peroxidase is adept in the treatment of textile waste and tough compounds for the purpose of decolorization [123]. This means this treatment can be implemented at the industrial scale for the decolorization of textile dye effluents.

Cellulose

Cellulose, an organic compound represented by the formula $(C_6H_{10}O_5)_n$, stands as the most abundant renewable biological material found on the surface of our planet. These are used in many different textile applications throughout history. Cellulose is sourced from various origins, including algae, plants, microorganisms, and fungi [48, 63].

Fibers derived from cellulosic materials can be sourced from renewable natural resources such as jute, bamboo, and cotton, as well as regenerated from conventional natural cellulose sources, including viscose, lyocell, and modal [29]. The presence of non-cellulosic substances in the fiber structure, constituting around 10% by weight, including pectin, hemi-cellulose, waxes, and others, results in diminished wettability and absorbency. Consequently, additional purification techniques or fiber changes are necessary [48].

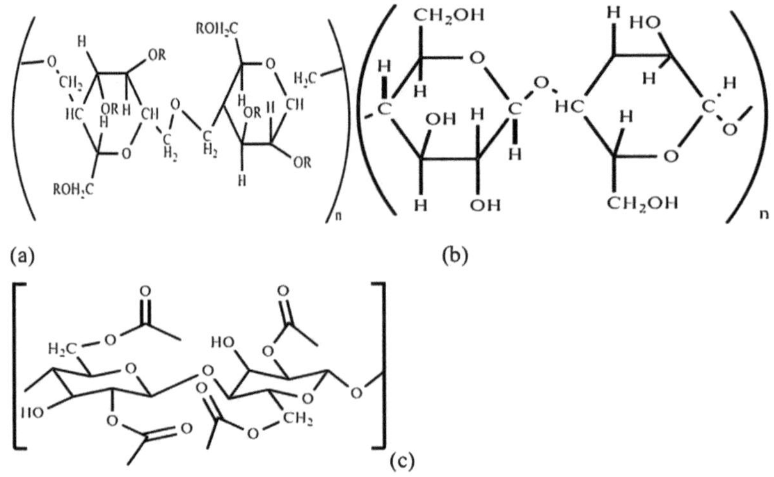

Fig. 3 Structures of **a** cellulose, **b** cellulose acetate, and **c** cellulose ether [6]

The process of cellulose acetylation yields cellulose acetates, which can also be made by spinning cellulose acetate. These substances, often referred to as acetates, can be dry-spun through the dissolution of dictates in acetone. Acetate is categorized as semi-synthetic and has superior qualities for textile production [48, 71].

With the help of sulfuric acid, cellulose is subjected to a sequence of chemical processes including acetic anhydride and acetic acid to produce cellulose triacetate. The resultant product is a crystalline triacetate with a high melting point of around 300 °C in its pure form. Additionally, triacetate undergoes hydrolysis to yield cellulose diacetate, a straightforward process facilitated by reducing the melting temperature to 230 °C [48, 71]. Cellulose ethers are water-soluble biopolymers produced by the alkali treatment of cellulose, which involves the introduction of ether functional groups. The etherification of cellulose can be achieved using various synthesis methods, including the Michael addition, Williamson ether synthesis, among others [48, 63]. Figure 3 shows how a standard cellulose molecule is put together [6].

Microbial

Polyhydroxyalkanoates (PHA)

PHA is classified as a type of polyester that is extracted intracellularly by both bacterial and fungal strains. PHA receives increased attention from both industry professionals and academic circles due to its broader applicability. Across the globe, numerous scholars have successfully isolated PHA using various strains. PHA typically consists of linear-chained repeating (R)-3-hydroxy fatty acid units linked by ester bonds. PHA is categorized as short chains (C3–C5) and medium chains (C6–C14) based on their carbon chain length. Mohapatra et al. [91] and Zheng et al. [158] obtained PHA by submerged and solid-state fermentation methods. Solid-state fermentation was responsible for the greater yield that was achieved. Additionally, a notable distinction was observed in its structural, thermal, and morphological characteristics. The selection of raw material as a carbon source impacts the quality of the microbial biopolymer. PHA acts as a storage medium and works as a terminal electron acceptor for bacteria in stressful settings [10, 162]. PHA is being utilized for the creation of bioplastic because of its ability to biodegrade [117, 144]. Othman et al. [100] investigated the effectiveness of PHA-based degradable mulch film on rice seed germination. Kelwick et al. [61] developed a protease biosensor utilizing AL-PHA beads for the identification of proteolytic activity. Amanat et al. [9] examined the long-term potential of various types of PHA-based materials for bioremediation. Tanadchangsaeng and Pattanasupong [134] investigated the biodegradability of PHA in seawater by quantifying the carbon dioxide produced during decomposition processes. The primary limitations of producing this microbial biopolymer include high feed costs and insufficient thermal and mechanical properties. Significantly better qualities can result from changes to the types of substrates, supply techniques, development conditions, and/or genetic adjustments [125].

Bacterial Cellulose

Waste streams of sugarcane jaggery from biodiesel and confectionary factories [137], corn stalk [20], oat hull [127], rice bark [36], fruit juices [69], Nylon 6-6 hydrolysate [160], cotton-based textile waste [43], fermentation wastewater [47], etc. are used to make bacterial cellulose. Bacterial cellulose is primarily extracted from various bacteria, including both Gram-positive and Gram-negative types such as *Sarcina ventriculi*, *Komagataeibacter* spp., *Acetobacter* spp., and *Pseudomonas* spp. [31, 95, 121].

The linear-chain polymer known as bacterial cellulose is composed of repeated glucose monomeric units connected by β (1–4) bonds. It is unable to dissolve in water because of its intricate structural arrangement, which includes a large number of hydroxyl groups and strong intermolecular interactions. Because it creates an intricate network of interlinked cellulose strands that are held together by hydrogen bonds, it is incredibly resistant to breaking down in water [141]. Static fermentation and agitated fermentation are both effective techniques for the production of bacterial cellulose. Zhong [159] examined that bacterial cellulose production via static fermentation is a more labor-intensive and time-consuming process than agitated fermentation. Bacterial cellulose-producing bacteria lack the ability to co-produce other compounds alongside cellulose, including pectin, lignin, and hemicellulose. Compared to cellulose made from plants, cellulose made from bacteria is much pure. Furthermore, bacterial cellulose is biocompatible due to the absence of complementary polymers and other impurities [37]. It offers a wide range of applications because it is both biocompatible and biodegradable. Alves et al. [8] developed filter membranes utilizing bacterial cellulose for industrial water treatment. Moreover, a substantial number of researchers have explored the use of bacterial cellulose-based hydrogels for the removal of dyes and heavy metals [44, 68, 70].

Xanthan Gum

The Xanthomonas campestris bacteria that are grown for sale release xanthan gum, which is an exopolysaccharide [97]. It is a high molecular weight microbial biopolymer classified as a heteropolysaccharide. The carbon source for xanthan gum production is primarily derived from the glucose and sucrose present in the substrate. Xanthan gum consists of a combination of mannose, glucuronic acid, and glucose units linked by β (1–4) bonds. This is typically achieved through an aerobic fermentation process followed by precipitation using isopropyl alcohol. The type of fermenter utilized, the culture medium, its operational mode (batch or continuous), and growth conditions, including temperature, pH level, and oxygen concentration, all influence xanthan synthesis [101]. The findings of many research described that xanthan gum is an outstanding biomaterial for the immobilization of heavy metals and the adsorption of cationic dyes due to their remarkable selectivity, recovery, and reusability upon regeneration [45, 89, 96, 133].

Algal Polymers (Alginate)

Alginate is primarily sourced from various seaweed species, such as Laminaria digitata, Sargassum cristaefolium, and Ascophyllum nodosum, as well as from certain bacterial strains, including Azotobacter spp. and Pseudomonas spp. Alginate is a linear, water-soluble hydrocolloid classified within the polysaccharide group. The arrangement and composition of monomers affect the qualities of alginate [72]. Alginates may form a cross-link (egg-box) structure via ionic interactions and the entrapment of cations. The chelating structural arrangement enables alginate to effectively remove heavy metals from wastewater. Hydrogels are formed from alginate through the addition of calcium ions, typically induced by the incorporation of acids like acetic acid. Numerous researchers have extensively studied the potential of alginate-based hydrogels for the removal of dyes and heavy metals [17, 27, 84, 120, 151, 155]. Furthermore, alginate is recognized for its ability to eliminate toxic pollutants from wastewater [42, 135]. Also, following the removal of contaminants, algal biorefinery presents a sustainable method for the recovery of biochemicals (such as acetic acid, carotenoids, eugenol, and lactic acid), bioenergy (including biomethane, biohydrogen, biogas, and bioethanol), and biomaterials (like films, biochar, coatings, and carbon fiber) derived from algal biomass [59, 150]. In the context of environmental applications, Lu et al. [82] utilized a post-cross-linking method to create sodium alginate beads for effluent treatment that were pH-sensitive and exhibited controllable swelling behavior. As current investigations explore their potential further, we can expect even more innovative advancements that leverage their distinct characteristics.

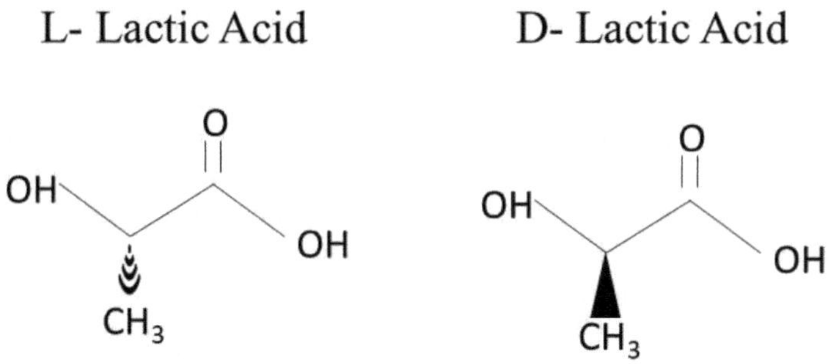

Fig. 4 Structure of lactic acid

Natural

Polylactide Acid (PLA)

Polylactide, commonly referred to as polylactic acid or PLA, is categorized within the polyester class. The composition consists of lactic acid monomers. Sugars are often transformed into lactic acid through the action of algal strains and bacterial, which are subsequently polymerized into polylactic acid [19, 118]. PLA is a biocompatible and biodegradable polymer appearing in two enantiomeric forms, L-PLA and D-PLA (Fig. 4), which can be chosen based on material qualities. PLA is a biopolymer that can be derived from lignocellulosic materials, municipal waste, and food waste streams through enzymatic and fermentation processes. The majority of waste biomass consists of the dominant natural flora of lactic acid bacteria. It is expected that sugars present in biowaste undergo natural fermentation to produce lactic acid [4, 136]. Zhang et al. [156] highlighted the potential of PLA-based films as an environmentally favorable substitute for conventional plastic mulch. Compared to petroleum-derived polymers, this microbial biopolymer is costly and exhibits lower mechanical properties. Copolymerization presents a viable solution to address this challenge and facilitate scaling at an industrial level [51].

3 Properties of Biopolymers

Biodegradability

Numerous processes, such as photodegradation, biodegradation, and thermo-oxidative degradation, cause polymer degradation. These processes all entail chemical reactions that result in crosslinking, chain scission, cross-group removal, and structural alteration [102, 139]. In biodegradation, microorganisms break down organic materials via two main mechanisms, known as aerobic and anaerobic routes, which can occur with or without oxygen [28]. Figure 5 illustrates that the final products of aerobic biodegradation include H_2O, CO_2, biomass, and byproducts resulting from the oxidation of sulfur and nitrogen. Anaerobic biodegradation produces CO_2, hydrocarbons, biomass, CH_4, and byproducts resulting from nitrogen and sulfur reduction. In contrast, anaerobic biodegradation results in CO_2, hydrocarbons, biomass, CH_4, and byproducts from sulfur and nitrogen reduction [60]. Aerobic biodegradation results in the production of water, carbon dioxide, biomass, and byproducts from sulfur and nitrogen oxidation, whereas anaerobic biodegradation yields CO_2, hydrocarbons, CH_4, biomass, and byproducts from sulfur and nitrogen reduction [28]. Upon environmental disposal, polylactide experiences hydrolysis, leading to the generation of smaller oligomers. Microorganisms subsequently transform these oligomers into water and carbon dioxide [102]. Cotton fibers mainly consist of cellulose, which can be degraded by microbes through the activity of cellulases. Enzymes known as cellulases aid in the hydrolysis and oxidation of cellulose, reducing its molecules into smaller pieces known as cellobiose. Further breakdown of these cellobiose units produces glucose and its derivatives, which are safe for the environment. In order to track changes in strength loss, weight loss, and appearance in natural topsoil and aquatic conditions, wool, the degradation of cotton, and PLA fiber was investigated during a 42-day period at 35 °C. The results indicated that degradation rates were greater in natural soil than in the aqueous medium, with cotton exhibiting the highest degradation rate, followed by wool, and then PLA fiber. Natural fibers exhibit greater susceptibility to degradation compared to PLA fibers, which are synthetic and biodegradable [102]. Egan and Salmon [28] state that the degradation rates of various textile fibers under different conditions can be summarized as follows: (i) Cotton fibers undergo biodegradation in soil within 180 days, in anaerobic environments within 30 days, and exhibit partial decomposition in fertilizer in approximately 45 days. (ii) The biodegradation of PLA (polylactic acid) fibers occurs in 45 days in compost, 30 days in anaerobic conditions, and 90 days in seawater. But within 180 days, they don't biodegrade at all in soil. (iii) PET fibers are not decomposable under any conditions.

Recyclability

Biopolymers can be recycled using natural processes in biological recycling methods, such as enzymatic breakdown and composting. Enzymatic degradation used to recycle PHAs is a noteworthy example. Researchers have engineered enzymes that can decompose PHAs into their monomer components, which can subsequently be utilized to synthesize new biopolymers [16]. This process offers environmental and sustainably beneficial alternative to traditional recycling techniques. The success of pilot projects has been underscored, showcasing the purity of recycled materials and elevated recovery rates. Composting is a useful biological recycling approach, in particular for biodegradable biopolymers, such as starch-based plastics and PLA. Commercial composting facilities can process these materials alongside organic

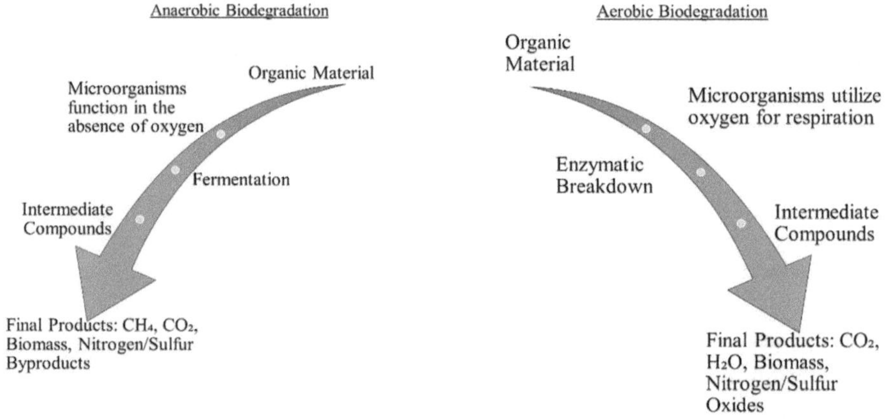

Fig. 5 Aerobic and anaerobic biodegradation pathways

waste, transforming them into compost to develop the soil quality. This technique reduces biopolymer waste designed for promotes sustainable agriculture through the production of high-quality compost [57].

Thermal Stability

Thermal stability is a pivotal criterion to assess biopolymers in textile waste management. The head degradation of polymeric composites occurs in two primary stages: (i) the depolymerization process and (ii) the release of volatile compounds [21]. Water absorption in hydrophilic cellulose can lead to hydrolysis, thereby reducing thermal stability [147]. The largest amount of weight loss during thermal decay occurs in the second stage, serving as a vital indication of the thermal stability of polymer composites [75].

Thermogravimetric analysis (TGA) is widely used to determine the thermal stability of starch-originated polymers. Accurate control of atmosphere, heating rate, and sample mass in TGA maximizes analytical results, successfully imitating both aqueous and dry process conditions. Additives are systematically utilized to improve the thermal properties of these polymers. For instance, it has been demonstrated that starch nanocomposites with clays such as kaolinite, hectorite, and montmorillonite raise the temperature at which thermal deterioration occurs. The addition of TiO_2 nanoparticles to starch leads to improved thermal stability [12].

The incorporation of Aloe Vera (AV) gel into starch-based composite films improves antimicrobial properties and thermal resistance. The highest degradation temperatures, for example, were found to be 293 °C for starch film, 298 °C for starch-chitosan film, and 302 °C for starch-chitosan film that was mixed with AV gel [11]. The enhancement is ascribed to intercalation, aggregation, and active interactions within the starch matrix [40]. To further increase thermal characteristics, starch-based polymers are often mixed with synthetic polymers, such as polyvinyl alcohol (PVA). The incorporation of ascorbic acid as a plasticizer improves thermal stability and demonstrates superior performance compared to polyols in the production of films with enhanced thermal properties [149].

Mechanical Stability

Ge et al. [33] developed a nanocomposite hydrogel composed of chitin whiskers with gelatin reinforced. They looked at how the combination enhanced gelatin's gelling capacity and mechanical qualities. Compared to tests conducted on pure gelatin, the mechanical strength, fracture stress, hardness, and strain were more stable at varying concentrations. Additionally, Nuvoli et al. [98] studied the mechanical, thermal, and structural characteristics of fish gelatin reinforced with milled tara gum and glycerol. Gelatin-based films with high glycerol content may demonstrate favorable thermal stability. However, an excessive rise in glycerol was also found to adversely affect the mechanical characteristics. This reinforcement enhanced the mechanical qualities and thermal stability of the composite [98]. The application of nonequilibrium atmospheric pressure plasma for the stabilization of gelatin nanofibers. Gelatin nanofibers underwent cross-linking through low-pressure treatment, as opposed to conventional chemical agents, to improve the solubilization issues and mechanical properties linked to gelatin. Current research is examining efficient methods and synergistic effects to enhance properties, building on earlier findings. [157] immersed cooling-induced gelatin hydrogels in a dilute $Zr(SO_4)_2$ solution to address the issue of gelatin's inadequate mechanical properties, which restricts the hydrogel's diverse applications. According to mechanical testing, the strength, elastic modulus, and compressive modulus were around 400, 476, and 1192 kPa, respectively. Compared to pure gelatin, these values are roughly 100, 5, and 11 times greater, respectively. Through atomic force microscopy-based nanoindentation tests, current research is also investigating the application of further novel techniques for graphene synthesis and mechanical properties [143]. Additionally, further research is being conducted to enhance both the physical and mechanical properties of gelatin following cross-linking with transglutaminase (MTG) [81]. The effects on texture properties will also be examined. Gelatin products have both helical and coiled conformations of gelatin chains when submerged in water; the proportion of these conformations affects the final material's mechanical qualities [111].

4 Applications of Biopolymers

Treatment of Dye-Contaminant Water

A significant contributor to pollution in textile production is the contamination of freshwater, especially due to dye treatment processes [53]. As a result, the water becomes unsafe for human consumption and loses quality. Certain dyes exhibit greater resistance and pose challenges for complete degradation through chemical decomposition, biological, photolysis and other conventional methods [50]. Advanced treatment technologies, including membrane processes, advanced oxidation processes (AOPs), and adsorption processes, demonstrate potential as effective alternatives for the removal of micropollutants; however, the generation of by-products and elevated operational expenses and concentrated residues restrict their utilization [83]. Developing new technologies for the recycling and treatment of dye-contaminated wastewater is crucial from an environmental perspective. Recently, Sirajudheen et al. [126] analyzed the applications of chitin and chitosan to the adsorption of textile dyes from water. Chitin and chitosan are among the most abundant natural biopolymers on the planet. They are endowed with distinct chemical, mechanical, optical, and physical properties as a result of their structural characteristics, such as high porosity, low density, renewability, and biodegradability. Unmodified and modified chitin and chitosan, as well as their various derivatives, are used in applications such as air pollution [92], dye adsorption [112], and heavy metal adsorption [138]. Nonetheless, the primary drawbacks are their reduced surface area, inadequate mechanochemical stability, and limited absorption capacity. To enhance these features, several experimental approaches have been utilized: mineral imprinting [124], biomaterial surface modification via crosslinkers [122], and grafting [52]. These methods resulted in increased active sites, leading to more reliable dye–adsorbent interactions.

Biopolymer on Printing

Hashemi Sanatgar et al. [41] suggested 3D printing as a cost-effective approach for textile functionalization. than traditional techniques like screen and inkjet printing, aiming to enhance the environmental sustainability of smart and functional textile production by reducing the excessive consumption of chemicals, energy and water, while minimizing waste. Using the Fused Deposition Modelling (FDM) technique, they printed nanocomposite and polymers layers directly onto textile textiles, examining the adhesion properties of these materials based on the kind of fabric and filler. Polyamide 66 (PA66) fabrics were printed with nylon, PA66, and PLA fabrics using multi-wall carbon nanotubes on PLA fabric or carbon black and polylactic acid (PLA) nanocomposites. The findings indicated that many 3D printing process variables, including, printing speed, platform temperature, and extruder temperature, significantly influence adhesion force. Increasing the extruder temperature resulted in a reduction of the tensile strength of layers produced by 3D printing. This was interpreted as an indication of enhanced brittleness resulting from elevated processing temperatures. Subsequently, an increase in printing speed resulted in a decrease in adhesive force. Improving printing speed decreased macromolecule penetration into the fabric, leading to cohesive forces that exceed adhesive forces.

Microbial Biopolymers as Adsorbents

Adsorption has become increasingly prominent in water purification in recent years. This process is a widely acknowledged as a balance separation method noted for its efficiency, effectiveness, and simplicity [116]. This process eliminates the need for additional chemicals, excessive water, or energy, resulting in reduced operating costs [94]. These man-made materials are mainly derived from petroleum and coal, which are not efficiently incorporated into recycling systems. In contrast, microbial biopolymers demonstrate superior performance compared to petroleum-based polymers regarding environmental sustainability, cost-effectiveness, and functionality [38]. Microbial biopolymers consist of various functional groups, such as phenolic hydroxyl, hydroxide, methoxyl carboxyl, and amines. The chemical chains of biopolymers made from microbes, enhance pollutant adsorption efficacy from various samples by chelating metal ions to form surface complexes [32]. Adsorption typically results from attractive forces between the compound being adsorbed, the surface of the adsorbent, and the adsorbate, the compound to which the adsorbate adheres. A range of adsorbents have been utilized, with common types classified based on their material composition. Adsorbents may originate from organic, mineral, or natural sources [116]. Microbial biopolymers and biopolymer composites, as shown in Table 1, are significant within the natural group because of their permeable structure, durability, cost-effectiveness and high specific surface area. The presence of plenty of surface functional groups on the surface enhances the efficiency of decontamination processes and facilitates the removal of greasy substances, dyes, soluble particles, and other pollutants in water. Similarly, these functional groups exhibit a tendency to interact with heavy metals such as chromium, copper, mercury, lead, nickel, zinc and cadmium. Biopolymers present a viable approach for mitigating emerging contaminants, especially surfactants, in wastewater treatment. The elimination of surfactants using biopolymers like chitosan, alginate, tannin composites, and others was covered by Biswas and Pal [15]. Da Silva Alves et al. have highlighted recent advancements in the development of chitosan-based adsorbents, highlighting their exceptional potential for the removal of a variety of pollutants, including cationic and anionic dyes and heavy metal ions [22]. Benavente et al. [14], emphasize that the plentiful hydroxyl and amino groups present on the chitosan surface facilitate the chelation of heavy metal ions, namely, Hg^{2+}, Zn^{2+}, Pb^{2+}, and Cu^{2+} demonstrating maximum adsorption capacities of 47.15 mg/g, 58.71 mg/g, 109.55 mg/g, and 79.94 mg/g, respectively.

Table 1 Biopolymers as adsorbents to treat wastewater

Biopolymer	Pollutant	References
Sodium alginate hydrogel, Carboxylated cellulose nanocrystal	Pb^{2+}	[46]
Cellulose, PVA, Graphene Composite aerogel	Cu^{2+}, Cd^{2+}, Cr^{3+}, Co^{2+}, Zn^{2+}, Pb^{2+}	[130]
n-acetylcysteine modified mica bio nanocomposite, Xanthan gum	Pb^{2+}, Cu^{2+}, Ni^{2+}	[5]
Xanthan gum, amantadine composites	Methylene blue	[77]
Chitosan	Cu^{2+}, Cr^{6+}	[103]
Carboxymethyl cellulose, chitosan, triethylenetetramine	Direct blue, Congo red	[67]
Cellulose, guar gum	Cu^{2+}, Co^{2+}	[146]
Polyacrylate, Chitosan, graphene oxide	Methylene blue, FY3	[18]
Chitosan, poly(n-vinyl-2-pyrrolidone)	Orange G	[131]
Alginate, polyethyleneimine	Cu^{2+}, Pb^{2+}	[34]
Xanthan Gum hydrogel, Poly(Acrylamide-co-Acrylic Acid)	Cd^{2+}, Ni^{2+}	[49]
Sodium alginate	Pb^{2+}	[7]
Whey protein concentrate, pullulan hydrogel	Cu^{2+}	[154]
Nanofibrillated cellulose	Cationic surfactants	[24]
Polylactic acid hydrogel	Pb^{2+}, Ni^{2+}	[142]
Cellulose, Poly(acrylic acid)	Methylene blue	[79]
Chitosan-glucose hydrogel	Co^{2+}	[80]
Cellulose, guargum	Methylene blue	[146]
Crosslinked chitosan films	SDBS	[56]

Using inexpensive sorbent hydrogels presents a promising method to enhance the capacity of biopolymer adsorbents for the removal of heavy dyes and metals. For the absorption of Pb^{2+} and Cu^{2+} ions in wastewater, a polyethylene, sodium alginate amine compound hydrogel has demonstrated an adsorption capacity of 344.8 mg/g and 322.6 mg/g, respectively [34]. In another case, a sodium alginate/ polyethyleneimine hydrogel was used for dye adsorption and demonstrated the excellent removal performance of methylene blue with a maximum absorption capacity of 400.0 mg/g. Within 30 min, approximately 99% of the dye was removed [35]. Moreover, graphene oxide (GO) was encapsulated within a composite of polyvinyl alcohol and sodium alginate. The hydrogels demonstrated remarkable efficacy as adsorbents for the removal of UO_2^{2+} and Cu^{2+}, achieving maximum absorption capacities of 403.78 mg/g and, 247.16 mg/g respectively [148]. In a separate study, graphene oxide was employed to create three-dimensional porous biopolymer gels, facilitating the formation of hydrophobic interactions and hydrogen bonds. The three-dimensional network architecture offers a plethora of active sites conducive to the efficient adsorption of heavy metal ions and cationic dyes. An intriguing technique was emphasized by [140], which involves using sodium alginate in water to make heavy metal ions gel. This study demonstrates that alginate facilitates the rapid gelation of Cd^{2+}, Cu^{2+}, and Pb^{2+} in under 10 min. The separation of gel and liquid from wastewater was accomplished seamlessly via gravitational forces. Magnetic adsorbents are gaining prominence in the field of water treatment. The increasing popularity can be linked to their ability to reduce adsorbent waste, improve adsorption efficiency, and facilitate separation by utilizing an external magnetic field [64, 65]. Abdul Rahman et al. [1] synthesized alginate, magnetic chitosan cellulose, and composite hydrogel beads. The biosorbent exhibited significant efficacy in treating heavy metal-contaminated wastewater containing Fe and Ca ions and displayed improved thermal stability relative to chitosan raw and cellulose. The mechanical strength of certain naturally sourced microbial biopolymer membranes is inadequate. This issue can be addressed by integrating it with compatible nanobiomaterials [108]. Salehi et al. [113] examined chitosan-based membranes in the context of adsorptive studies. They proposed the idea of using TiO_2 to cross-link chitosan in order to produce a dye adsorbent that regenerates itself. Furthermore, the utilization of cross-linking agents such as glutaraldehyde, glyoxal, and formaldehyde has been proposed to enhance both solubility and mechanical properties.

5 Prospects

The future of biopolymers in textile waste management is promising, with expected advancements in biopolymer production and recycling technologies likely to improve their contribution to sustainable textile systems. Biopolymers offer a viable solution to reduce textile waste, promote the circular economy, and mitigate environmental impacts in response to the increasing demand for sustainable alternatives. For sustainable textile practices in the future, biopolymer-based materials with satisfactory strength, biodegradability, and functionality is very important. Scientists should focus research on biopolymers that is both mechanically and aesthetically suitable for textile industry use. Progress in biopolymer blends, composites, and hybrid materials will boost the practical application of biopolymers, allowing greater use in fashion, technical textiles, and home textiles. Blends, composites, and hybrid materials will make biopolymers more useful in fashion, technical textiles, and home textiles. Biopolymer recycling infrastructure, especially for textile waste, is essential for bio-based material cycles. Strong recycling systems like industrial composting plants or decentralized recycling hubs will improve textile waste collection and processing. To understand the benefits of biopolymers in textile waste management, further research will explore the life cycle assessment of textiles that are made of biopolymers. When biopolymer production is energy efficient and economical, the clothing industry will have easier access to biopolymers. The incorporation of biopolymers in managing textile waste is anticipated to be strengthened by improved governmental policies and incentives that encourage sustainable practices in the textile sector. The future of biopolymers in textile waste management relies on interdisciplinary collaboration among material scientists, environmentalists, and industry stakeholders. Biopolymers will dramatically diminish the environmental effects of textiles industry by linking modern technologies with ecological goals.

6 Conclusion

Biopolymers are becoming a real option for dealing with waste in modern time, which brings both problems and prospects. Recycling, reusing, and repurposing are all important ways to handle pollution in a way that reduces waste and makes better use of resources waste and makes better use of resources as the world moves toward sustainability. The emphasis on green management practices aimed at environmental protection, resource minimization and enhanced energy efficiency aligns with the potential of biopolymers to support waste reduction efforts. The biopolymer industry is likely to gain significantly from the adoption of environmentally friendly waste management policies that emphasize efficient disposal, recycling, and reuse of these products. Understanding current management practices for textile waste is crucial for creating future systems that can effectively tackle the increasing volume of waste linked to its rising usage.

Acknowledgements The authors extend their appreciation to the editors and reviewers for their insightful feedback, recognizing their contributions in improving the quality of the manuscript.

Conflict Interests No potential conflict of interest was reported by the author(s).

References

1. Abdul Rahman, A. S., Fizal, A. N. S., Khalil, N. A., Ahmad Yahaya, A. N., Hossain, M. S., & Zulkifli, M. (2023). Fabrication and characterization of magnetic cellulose–chitosan–alginate composite hydrogel bead bio-sorbent. *Polymers, 15*(11), 1–14. https://doi.org/10.3390/polym15112494
2. Abrishami, S., Shirali, A., Sharples, N., Kartal, G. E., Macintyre, L., & Doustdar, O. (2024). Textile recycling and recovery: An eco-friendly perspective on textile and garment industries challenges. *Textile Research Journal*. https://doi.org/10.1177/00405175241247806
3. Adler, E. (1977). Lignin chemistry—Past, present and future. *Wood Science and Technology, 11*(3), 169–218. https://doi.org/10.1007/BF00365615
4. Ahmad, A., Othman, I., Rambabu, K., Bharath, G., Taher, H., Hasan, S. W., & Banat, F. (2021). Polymerization of lactic acid produced from food waste by metal oxide-assisted dark fermentation. *Environmental Technology and Innovation, 24*(November). https://doi.org/10.1016/j.eti.2021.101862

5. Ahmad, R., & Mirza, A. (2018). Application of Xanthan gum/ n-acetyl cysteine modified mica bionanocomposite as an adsorbent for the removal of toxic heavy metals. *Groundwater for Sustainable Development*, *7*, 101–108. https://doi.org/10.1016/j.gsd.2018.03.010
6. Ahmed, S., & Shabbir, M. (2024). *Biopolymers in the textile industry opportunities and limitations*. https://doi.org/10.1007/978-981-97-0684-6_3
7. Alraddadi, H. M., Fagieh, T. M., Bakhsh, E. M., Akhtar, K., Khan, S. B., Khan, S. A., Bahaidarah, E. A., & Homdi, T. A. (2023). Adsorptive removal of heavy metals and organic dyes by sodium alginate/coffee waste composite hydrogel. *International Journal of Biological Macromolecules*, *247*, 125708. https://doi.org/10.1016/j.ijbiomac.2023.125708
8. Alves, A. A., Silva, W. E., Belian, M. F., Lins, L. S. G., & Galembeck, A. (2020). Bacterial cellulose membranes for environmental water remediation and industrial wastewater treatment. *International Journal of Environmental Science and Technology*, *17*(9), 3997–4008. https://doi.org/10.1007/s13762-020-02746-5
9. Amanat, N., Matturro, B., Rossi, M. M., Valentino, F., Villano, M., & Papini, M. P. (2021). Assessment of long-term fermentability of pha-based materials from pure and mixed microbial cultures for potential environmental applications. *Water (Switzerland)*, *13*(7). https://doi.org/10.3390/w13070897
10. Aulenta, F., Fuoco, M., Canosa, A., Papini, M. P., & Majone, M. (2008). Use of poly-β-hydroxy-butyrate as a slow-release electron donor for the microbial reductive dechlorination of TCE. *Water Science and Technology*, *57*(6), 921–925. https://doi.org/10.2166/wst.2008.073
11. Bajer, D., Janczak, K., & Bajer, K. (2020). Novel starch/chitosan/aloe vera composites as promising biopackaging materials. *Journal of Polymers and the Environment*, *28*(3), 1021–1039. https://doi.org/10.1007/s10924-020-01661-7
12. Barkhad, M. S., Abu-jdayil, B., Mourad, A. H. I., & Iqbal, M. Z. (2020). Thermal insulation and mechanical properties of polylactic acid (PLA) at different processing conditions. *Polymers*, *12*, 1–16. https://doi.org/10.3390/polym12092091
13. Bastos Araruna, F., Oliveira Sousa Araruna, F., Lima Alves Pereira, L. P., Aranha Brito, M. C., Veras Quelemes, P., de Araújo-Nobre, A. R., de Oliveira, T. M., da Silva, D. A., de Souza de Almeida Leite, J. R., Fernandes Coutinho, D., da Rocha Borges, M. O., & Romão Borges, A. C. (2020). Green syntheses of silver nanoparticles using babassu mesocarp starch (Attalea speciosa Mart. ex Spreng.) and their antimicrobial applications. *Environmental Nanotechnology, Monitoring & Management*, *13*, 100281. https://doi.org/10.1016/j.enmm.2019.100281
14. Benavente, M., Moreno, L., & Martinez, J. (2011). Sorption of heavy metals from gold mining wastewater using chitosan. *Journal of the Taiwan Institute of Chemical Engineers*, *42*(6), 976–988. https://doi.org/10.1016/j.jtice.2011.05.003
15. Biswas, S., & Pal, A. (2021). Application of biopolymers as a new age sustainable material for surfactant adsorption: A brief review. *Carbohydrate Polymer Technologies and Applications*, *2*, 100145. https://doi.org/10.1016/j.carpta.2021.100145
16. Brigham, & Sinskey. (2012). Applications of polyhydroxyalkanoates in the medical industry. *International Journal of Biotechnology for Wellness Industries*, 53–60. https://doi.org/10.6000/1927-3037.2012.01.01.03
17. Cai, R., Chen, Y., Hu, J., Xiong, J., Lu, J., Liu, J., Tan, X., Liu, W., Zhou, Y., & Chen, Y. (2023). A self-supported sodium alginate composite hydrogel membrane and its performance in filtering heavy metal ions. *Carbohydrate Polymers*, *300*, 120278. https://doi.org/10.1016/j.carbpol.2022.120278
18. Chang, Z., Chen, Y., Tang, S., Yang, J., Chen, Y., Chen, S., Li, P., & Yang, Z. (2020). Construction of chitosan/polyacrylate/graphene oxide composite physical hydrogel by semi-dissolution/acidification/sol-gel transition method and its simultaneous cationic and anionic dye adsorption properties. *Carbohydrate Polymers*, *229*, 115431. https://doi.org/10.1016/j.carbpol.2019.115431
19. Cheng, Y. R., & Wang, H. Y. (2022). Highly effective removal of microplastics by microalgae Scenedesmus abundans. *Chemical Engineering Journal*, *435*, 1–23. https://doi.org/10.1016/j.cej.2022.135079

20. Cheng, Z., Yang, R., Liu, X., Liu, X., & Chen, H. (2017). Green synthesis of bacterial cellulose via acetic acid pre-hydrolysis liquor of agricultural corn stalk used as carbon source. *Bioresource Technology, 234*, 8–14. https://doi.org/10.1016/j.biortech.2017.02.131
21. Cheremisinoff, N. P., & Cheremisinoff, P. N. (1996). *Handbook of applied polymer processing technology* (Vol. 31). CRC Press. https://books.google.com.bd/books?hl=en&lr=&id=4wvCjRz7W5AC&oi=fnd&pg=PR3&dq=N.P.+Cheremisinoff,+P.N.+Cheremisinoff,+Handbook+of+Applied+Polymer+Processing+Technology,+Plastics+Engineering,+vol+38+(CRC+Press,+New+York,+1996),+p.+31&ots=apQ6c-maMa&sig=8eo8SgYjKf6q980514BBVGTBo_A&redir_esc=y#v=onepage&q&f=false
22. da Silva Alves, D. C., Healy, B., Pinto, L. A. d. A., Cadaval, T. R. S., & Breslin, C. B. (2021). Recent developments in Chitosan-based adsorbents for the removal of pollutants from aqueous environments. *Molecules, 26*(3). https://doi.org/10.3390/molecules26030594
23. Das, B., & Dwivedi, S. (2024). Post-consumer textile waste management practices and challenges in India: A systematic literature review. *Journal of Scientometric Research, 13*(2), 419–429. https://doi.org/10.5530/jscires.13.2.33
24. Das, D., & Pal, A. (2016). Adsolubilization phenomenon perceived in chitosan beads leading to a fast and enhanced malachite green removal. *Chemical Engineering Journal, 290*, 371–380. https://doi.org/10.1016/j.cej.2016.01.062
25. Davis Sumo, P. (2024). Textile waste management and recycling opportunities and challenges for Africa: A mini-review. *Holistic Approach to Environment, 14*(2), 57–70. https://doi.org/10.33765/thate.14.2.3
26. do Bem, N. A., Cardoso, F. A. R., Paccola, E. A. de S., & Rezende, L. C. S. H. (2022). 3D-printed polylactic acid biopolymer and textile fibers: comparing the degradation process. *Revista Brasileira de Ciências Ambientais, 57*(2), 302–319. https://doi.org/10.5327/z2176-94781192
27. Duc, T. H., Vu, T. K., Dang, C.-T., Nguyen, V. H., La, D. D., Kim, G. M., Chang, S. W., Bui, X.-T., Dang, T.-D., & Nguyen, D. D. (2021). Synthesis and application of hydrogel calcium alginate microparticles as a biomaterial to remove heavy metals from aqueous media. *Environmental Technology & Innovation, 22*, 101400. https://doi.org/10.1016/j.eti.2021.101400
28. Egan, J., & Salmon, S. (2022). Strategies and progress in synthetic textile fiber biodegradability. *SN Applied Sciences, 4*(1). https://doi.org/10.1007/s42452-021-04851-7
29. Eid, B. M., & Ibrahim, N. A. (2021). Recent developments in sustainable finishing of cellulosic textiles employing biotechnology. *Journal of Cleaner Production, 284*, 124701. https://doi.org/10.1016/j.jclepro.2020.124701
30. Fang, W., Yang, S., Wang, X.-L., Yuan, T.-Q., & Sun, R.-C. (2017). Manufacture and application of lignin-based carbon fibers (LCFs) and lignin-based carbon nanofibers (LCNFs). *Green Chemistry, 19*(8), 1794–1827. https://doi.org/10.1039/c6gc03206k
31. Faria, M., Cunha, C., Gomes, M., Mendonça, I., Kaufmann, M., Ferreira, A., & Cordeiro, N. (2022). Bacterial cellulose biopolymers: The sustainable solution to water-polluting microplastics. *Water Research, 222*, 118952. https://doi.org/10.1016/j.watres.2022.118952
32. Fouda-Mbanga, B. G., Prabakaran, E., & Pillay, K. (2021). Carbohydrate biopolymers, lignin based adsorbents for removal of heavy metals (Cd^{2+}, Pb^{2+}, Zn^{2+}) from wastewater, regeneration and reuse for spent adsorbents including latent fingerprint detection: A review. *Biotechnology Reports, 30*, e00609. https://doi.org/10.1016/j.btre.2021.e00609
33. Ge, S., Liu, Q., Li, M., Liu, J., Lu, H., Li, F., Zhang, S., Sun, Q., & Xiong, L. (2018). Enhanced mechanical properties and gelling ability of gelatin hydrogels reinforced with chitin whiskers. *Food Hydrocolloids, 75*, 1–12. https://doi.org/10.1016/j.foodhyd.2017.09.023
34. Godiya, C. B., Liang, M., Sayed, S. M., Li, D., & Lu, X. (2019). Novel alginate/polyethyleneimine hydrogel adsorbent for cascaded removal and utilization of Cu^{2+} and Pb^{2+} ions. *Journal of Environmental Management, 232*, 829–841. https://doi.org/10.1016/j.jenvman.2018.11.131
35. Godiya, C. B., Xiao, Y., & Lu, X. (2020). Amine functionalized sodium alginate hydrogel for efficient and rapid removal of methyl blue in water. *International Journal of Biological Macromolecules, 144*, 671–681. https://doi.org/10.1016/j.ijbiomac.2019.12.139

36. Goelzer, F. D. E., Faria-Tischer, P. C. S., Vitorino, J. C., Sierakowski, M. R., & Tischer, C. A. (2009). Production and characterization of nanospheres of bacterial cellulose from Acetobacter xylinum from processed rice bark. *Materials Science and Engineering C, 29*(2), 546–551. https://doi.org/10.1016/j.msec.2008.10.013
37. Gorgieva, S. (2023). *Sudah 96, 15*(16).
38. Gowthaman, N. S. K., Lim, H. N., Sreeraj, T. R., Amalraj, A., & Gopi, S. (2021). Chapter 15—Advantages of biopolymers over synthetic polymers: Social, economic, and environmental aspects. In S. Thomas, S. Gopi, & A. Amalraj (Eds.), *Biopolymers and their industrial applications* (pp. 351–372). Elsevier. https://doi.org/10.1016/B978-0-12-819240-5.00015-8
39. Guimarães, J. L., Wypych, F., Saul, C. K., Ramos, L. P., & Satyanarayana, K. G. (2010). Studies of the processing and characterization of corn starch and its composites with banana and sugarcane fibers from Brazil. *Carbohydrate Polymers, 80*(1), 130–138. https://doi.org/10.1016/j.carbpol.2009.11.002
40. Gutiérrez, T. J., & Álvarez, K. (2016). Physico-chemical properties and in vitro digestibility of edible films made from plantain flour with added Aloe vera gel. *Journal of Functional Foods, 26*, 750–762. https://doi.org/10.1016/j.jff.2016.08.054
41. Hashemi Sanatgar, R., Campagne, C., & Nierstrasz, V. (2017). Investigation of the adhesion properties of direct 3D printing of polymers and nanocomposites on textiles: Effect of FDM printing process parameters. *Applied Surface Science, 403*, 551–563. https://doi.org/10.1016/j.apsusc.2017.01.112
42. Hassan (El-Moushy), R. M. (2023). Novel synthesis of ultrafiltration membranes by crosslinking metal (II)-Alginate hydrogels with hexamethylene 1,6-Diisocyanate in inert solvent: Application for remediation of wastewater by removal of toxic pollutants. *Chemical Engineering Journal, 452*, 139093. https://doi.org/10.1016/j.cej.2022.139093
43. Hong, F., Guo, X., Zhang, S., Han, S., Yang, G., & Jönsson, L. J. (2012). Bacterial cellulose production from cotton-based waste textiles: Enzymatic saccharification enhanced by ionic liquid pretreatment. *Bioresource Technology, 104*, 503–508. https://doi.org/10.1016/j.biortech.2011.11.028
44. Hosseini, H., & Mousavi, S. M. (2021). Bacterial cellulose/polyaniline nanocomposite aerogels as novel bioadsorbents for removal of hexavalent chromium: Experimental and simulation study. *Journal of Cleaner Production, 278*, 123817. https://doi.org/10.1016/j.jclepro.2020.123817
45. Hu, X., Yan, L., Xu, M., & Tang, L. (2023). Photo-degradable salecan/xanthan gum ionic gel induced by iron (III) coordination for organic dye decontamination. *International Journal of Biological Macromolecules, 238*, 124132. https://doi.org/10.1016/j.ijbiomac.2023.124132
46. Hu, Z.-H., Omer, A. M., Ouyang, X., & Yu, D. (2018). Fabrication of carboxylated cellulose nanocrystal/sodium alginate hydrogel beads for adsorption of Pb(II) from aqueous solution. *International Journal of Biological Macromolecules, 108*, 149–157. https://doi.org/10.1016/j.ijbiomac.2017.11.171
47. Huang, C., Yang, X.-Y., Xiong, L., Guo, H.-J., Luo, J., Wang, B., Zhang, H.-R., Lin, X.-Q., & Chen, X.-D. (2015). Evaluating the possibility of using acetone-butanol-ethanol (ABE) fermentation wastewater for bacterial cellulose production by Gluconacetobacter xylinus. *Letters in Applied Microbiology, 60*(5), 491–496. https://doi.org/10.1111/lam.12396
48. Jacob, J., & Gopi, S. (2021). Isolation and physicochemical characterization of biopolymers. In *Biopolymers and their industrial applications* (Issue January). https://doi.org/10.1016/b978-0-12-819240-5.00003-1
49. Jafarigol, E., Afshar Ghotli, R., Hajipour, A., Pahlevani, H., & Baghban Salehi, M. (2021). Tough dual-network GAMAAX hydrogel for the efficient removal of cadmium and nickle ions in wastewater treatment applications. *Journal of Industrial and Engineering Chemistry, 94*, 352–360. https://doi.org/10.1016/j.jiec.2020.11.006
50. Javaid, R., & Qazi, U. Y. (2019). Catalytic oxidation process for the degradation of synthetic dyes: An overview. *International Journal of Environmental Research and Public Health, 16*(11), 1–27. https://doi.org/10.3390/ijerph16112066

51. Jem, K. J., & Tan, B. (2020). The development and challenges of poly (lactic acid) and poly (glycolic acid). *Advanced Industrial and Engineering Polymer Research, 3*(2), 60–70. https://doi.org/10.1016/j.aiepr.2020.01.002
52. Jiang, X., Sun, Y., Liu, L., Wang, S., & Tian, X. (2014). Adsorption of C.I. Reactive Blue 19 from aqueous solutions by porous particles of the grafted chitosan. *Chemical Engineering Journal, 235*, 151–157. https://doi.org/10.1016/j.cej.2013.09.001
53. Jo, W.-K., & Tayade, R. J. (2014). Recent developments in photocatalytic dye degradation upon irradiation with energy-efficient light emitting diodes. *Chinese Journal of Catalysis, 35*(11), 1781–1792. https://doi.org/10.1016/S1872-2067(14)60205-9
54. Juanga-Labayen, J. P., Labayen, I. V., & Yuan, Q. (2022). A review on textile recycling practices and challenges. *Textiles, 2*(1), 174–188. https://doi.org/10.3390/textiles2010010
55. Judijanto, L., & Pahrijal, R. (2024). Bibliometric analysis of the effectiveness of waste management and recycling programs in the textile industry. *West Science Interdisciplinary Studies, 02*(09), 1787–1798. https://doi.org/10.58812/wsis.v2i09.1294
56. Kahya, N., Kaygusuz, H., & Erim, F. B. (2018). Aqueous removal of sodium dodecyl benzene sulfonate (SDBS) by crosslinked chitosan films. *Journal of Polymers and the Environment, 26*(5), 2166–2172. https://doi.org/10.1007/s10924-017-1113-3
57. Kakadellis, S., Muranko, Ż., Harris, Z. M., & Aurisicchio, M. (2024). Closing the loop: Enabling circular biodegradable bioplastic packaging flow through a systems-thinking framework. *Cleaner and Responsible Consumption, 12*, 100183. https://doi.org/10.1016/j.clrc.2024.100183
58. Kamble, Z., & Behera, B. K. (2021). Upcycling textile wastes: Challenges and innovations. *Textile Progress, 53*(2), 65–122. https://doi.org/10.1080/00405167.2021.1986965
59. Kant Bhatia, S., Ahuja, V., Chandel, N., Gurav, R., Kant Bhatia, R., Govarthanan, M., Kumar Tyagi, V., Kumar, V., Pugazendhi, A., Rajesh Banu, J., & Yang, Y. H. (2022). Advances in algal biomass pretreatment and its valorisation into biochemical and bioenergy by the microbial processes. *Bioresource Technology, 358*(127437), 1–51. https://doi.org/10.1016/j.biortech.2022.127437
60. Karamanlioglu, M., Preziosi, R., & Robson, G. D. (2017). Abiotic and biotic environmental degradation of the bioplastic polymer poly(lactic acid): A review. *Polymer Degradation and Stability, 137*, 122–130. https://doi.org/10.1016/j.polymdegradstab.2017.01.009. https://e-space.mmu.ac.uk/623422/
61. Kelwick, R. J. R., Webb, A. J., Wang, Y., Heliot, A., Allan, F., Emery, A. M., Templeton, M. R., & Freemont, P. S. (2021). AL-PHA beads: Bioplastic-based protease biosensors for global health applications. *Materials Today, 47*, 25–37. https://doi.org/10.1016/j.mattod.2021.02.018
62. Kiran, S., Rehman, F. ur, Javed, S., Abrar, S., Mukhtar, N., Moied, A., Khan, N. M., Javaid, A., Rashid, Z., Hussain, M., & Nawaz, R. (2024). Ecological effects of biopolymers and their advantages for textile industry. In S. Ahmed, & M. Shabbir (Eds.), *Biopolymers in the textile industry: Opportunities and limitations* (pp. 241–272). Springer Nature Singapore. https://doi.org/10.1007/978-981-97-0684-6_10
63. Klemm, D., Heublein, B., Fink, H., & Bohn, A. (2005). Cellulose: Fascinating biopolymer and sustainable raw material. *Angewandte Chemie International Edition, 44*(22), 3358–3393. https://doi.org/10.1002/anie.200460587
64. Kolya, H., & Kang, C. W. (2023a). Bio-based polymeric flocculants and adsorbents for wastewater treatment. *Sustainability (Switzerland), 15*(12). https://doi.org/10.3390/su15129844
65. Kolya, H., & Kang, C. W. (2023b). Next-generation water treatment: Exploring the potential of biopolymer-based nanocomposites in adsorption and membrane filtration. *Polymers, 15*(16). https://doi.org/10.3390/polym15163421
66. Kubo, S., & Kadla, J. F. (2005). Lignin-based carbon fibers: Effect of synthetic polymer blending on fiber properties. *Journal of Polymers and the Environment, 13*(2), 97–105. https://doi.org/10.1007/s10924-005-2941-0

67. Kumar, A., Jeyabalan, J., Priyan V, V., Charan Patra, C., & Narayanasamy, S. (2023). Fabrication of a novel bio-polymer adsorbent with high adsorptive capacity towards organic dyes. *Industrial Crops and Products, 203*, 117166. https://doi.org/10.1016/j.indcrop.2023.117166
68. Kundu, R., Mahada, P., Chhirang, B., & Das, B. (2022). Cellulose hydrogels: Green and sustainable soft biomaterials. *Current Research in Green and Sustainable Chemistry, 5*, 100252. https://doi.org/10.1016/j.crgsc.2021.100252
69. Kurosumi, A., Sasaki, C., Yamashita, Y., & Nakamura, Y. (2009). Utilization of various fruit juices as carbon source for production of bacterial cellulose by Acetobacter xylinum NBRC 13693. *Carbohydrate Polymers, 76*(2), 333–335. https://doi.org/10.1016/j.carbpol.2008.11.009
70. Kushwaha, J., & Singh, R. (2023). Cellulose hydrogel and its derivatives: A review of application in heavy metal adsorption. *Inorganic Chemistry Communications, 152*, 110721. https://doi.org/10.1016/j.inoche.2023.110721
71. Law, R. C. (2004). 5. Applications of cellulose acetate 5.1 Cellulose acetate in textile application. *Macromolecular Symposia, 208*(1), 255–266. https://doi.org/10.1002/masy.200450410
72. Li, L., Zhu, B., Yao, Z., & Jiang, J. (2023). Directed preparation, structure–activity relationship and applications of alginate oligosaccharides with specific structures: A systematic review. *Food Research International, 170*, 112990. https://doi.org/10.1016/j.foodres.2023.112990
73. Li, Q., Serem, W. K., Dai, W., Yue, Y., Naik, M. T., Xie, S., Karki, P., Liu, L., Sue, H.-J., & Liang, H. (2017). Molecular weight and uniformity define the mechanical performance of lignin-based carbon fiber. *Journal of Materials Chemistry A, 5*(25), 12740–12746. https://doi.org/10.1039/C7TA01187C
74. Li, Q., Xie, S., Serem, W. K., Naik, M. T., Liu, L., & Yuan, J. S. (2017). Quality carbon fibers from fractionated lignin. *Green Chemistry, 19*(7), 1628–1634. https://doi.org/10.1039/C6GC03555H
75. Li, W., Wu, Q., Zhao, X., Huang, Z., Cao, J., Li, J., & Liu, S. (2014). Enhanced thermal and mechanical properties of PVA composites formed with filamentous nanocellulose fibrils. *Carbohydrate Polymers, 113*, 403–410. https://doi.org/10.1016/j.carbpol.2014.07.031
76. Li, X., Wang, L., & Ding, X. (2021). Textile supply chain waste management in China. *Journal of Cleaner Production, 289*, 125147. https://doi.org/10.1016/j.jclepro.2020.125147
77. Li, Y., Liu, Z., Wan, X., Xie, L., Chen, H., Qu, G., Zhang, H., Zhang, Y.-F., & Zhao, S. (2023). Selective adsorption and separation of methylene blue by facily preparable xanthan gum/amantadine composites. *International Journal of Biological Macromolecules, 241*, 124640. https://doi.org/10.1016/j.ijbiomac.2023.124640
78. Lin, S. Y., & Dence, C. W. (2012). *Methods in lignin chemistry*. Springer Science & Business Media. https://doi.org/10.1007/978-3-642-74065-7
79. Liu, A., He, S., Zhang, J., Liu, J., & Shao, W. (2023). Preparation and characterization of novel cellulose based adsorbent with ultra-high methylene blue adsorption performance. *Materials Chemistry and Physics, 296*, 127261. https://doi.org/10.1016/j.matchemphys.2022.127261
80. Liu, Y., Hu, L., Yao, Y., Su, Z., & Hu, S. (2019). Construction of composite chitosan-glucose hydrogel for adsorption of Co^{2+} ions. *International Journal of Biological Macromolecules, 139*, 213–220. https://doi.org/10.1016/j.ijbiomac.2019.07.202
81. Liu, Y., Weng, R., Wang, W., Wei, X., Li, J., Chen, X., Liu, Y., Lu, F., & Li, Y. (2020). Tunable physical and mechanical properties of gelatin hydrogel after transglutaminase crosslinking on two gelatin types. *International Journal of Biological Macromolecules, 162*, 405–413. https://doi.org/10.1016/j.ijbiomac.2020.06.185
82. Lu, T., Xiang, T., Huang, X. L., Li, C., Zhao, W. F., Zhang, Q., & Zhao, C. S. (2015). Post-crosslinking towards stimuli-responsive sodium alginate beads for the removal of dye and heavy metals. *Carbohydrate Polymers, 133*(April 2019), 587–595. https://doi.org/10.1016/j.carbpol.2015.07.048
83. Luo, Y., Guo, W., Ngo, H. H., Nghiem, L. D., Hai, F. I., Zhang, J., Liang, S., & Wang, X. C. (2014). A review on the occurrence of micropollutants in the aquatic environment and their fate and removal during wastewater treatment. *Science of The Total Environment, 473–474*, 619–641. https://doi.org/10.1016/j.scitotenv.2013.12.065

84. Malik, S. A., Dar, A. A., & Banday, J. A. (2023). Rheological, morphological and swelling properties of dysprosium-based composite hydrogel beads of alginate and chitosan: A promising material for the effective cationic and anionic dye removal. *Colloids and Surfaces A: Physicochemical and Engineering Aspects, 663*, 131046. https://doi.org/10.1016/j.colsurfa.2023.131046
85. Maniglia, B. C., Castanha, N., Le-Bail, P., Le-Bail, A., & Augusto, P. E. D. (2021). Starch modification through environmentally friendly alternatives: A review. *Critical Reviews in Food Science and Nutrition, 61*(15), 2482–2505. https://doi.org/10.1080/10408398.2020.1778633
86. Manvi, P. K. (2020). Biopolymers as sustainable raw material for textile industry : An opportunity or a challenge. *Journal of Advanced Chemical Engineering, 10*(2), 4568. https://api.semanticscholar.org/CorpusID:226618443
87. Mao, H., Chen, X., Huang, R., Chen, M., Yang, R., Lan, P., Zhou, M., Zhang, F., Yang, Y., & Zhou, X. (2018). Fast preparation of carbon spheres from enzymatic hydrolysis lignin: Effects of hydrothermal carbonization conditions. *Scientific Reports, 8*(1), 2–11. https://doi.org/10.1038/s41598-018-27777-4
88. Mehta, S. (2023). Biodegradable textile polymers: A review of current scenario and future opportunities. *Environmental Technology Reviews, 12*(1), 441–457. https://doi.org/10.1080/21622515.2023.2227391
89. Mohafezatkar Gohari, R., Safarnia, M., Dadvand Koohi, A., & Baghban Salehi, M. (2022). Adsorptive removal of cationic dye by synthesized sustainable xanthan gum-g p(AMPS-co-AAm) hydrogel from aqueous media: Optimization by RSM-CCD model. *Chemical Engineering Research and Design, 188*(October), 714–728. https://doi.org/10.1016/j.cherd.2022.10.028
90. Mohan, K., Ganesan, A. R., Ezhilarasi, P. N., Kondamareddy, K. K., Rajan, D. K., Sathishkumar, P., Rajarajeswaran, J., & Conterno, L. (2022). Green and eco-friendly approaches for the extraction of chitin and chitosan: A review. *Carbohydrate Polymers, 287*, 119349. https://doi.org/10.1016/j.carbpol.2022.119349
91. Mohapatra, S., Pattnaik, S., Maity, S., Mohapatra, S., Sharma, S., Akhtar, J., Pati, S., Samantaray, D. P., & Varma, A. (2020). Comparative analysis of PHAs production by Bacillus megaterium OUAT 016 under submerged and solid-state fermentation. *Saudi Journal of Biological Sciences, 27*(5), 1242–1250. https://doi.org/10.1016/j.sjbs.2020.02.001
92. Mohraz, M. H., Golbabaei, F., Yu, I. J., Mansournia, M. A., Zadeh, A. S., & Dehghan, S. F. (2019). Preparation and optimization of multifunctional electrospun polyurethane/chitosan nanofibers for air pollution control applications. *International Journal of Environmental Science and Technology, 16*(2), 681–694. https://doi.org/10.1007/s13762-018-1649-3
93. Montava-Jordà, S., Torres-Giner, S., Ferrandiz-Bou, S., Quiles-Carrillo, L., & Montanes, N. (2019). Development of sustainable and cost-competitive injection-molded pieces of partially bio-based polyethylene terephthalate through the valorization of cotton textile waste. *International Journal of Molecular Sciences, 20*(6), 1–19. https://doi.org/10.3390/ijms20061378
94. Nasir, A., Masood, F., Yasin, T., & Hameed, A. (2019). Progress in polymeric nanocomposite membranes for wastewater treatment: Preparation, properties and applications. *Journal of Industrial and Engineering Chemistry, 79*, 29–40. https://doi.org/10.1016/j.jiec.2019.06.052
95. Nguyen, Q.-D., Nguyen, T.-V.-L., Nguyen, T.-T.-D., & Nguyen, N.-N. (2022). Effects of different hydrocolloids on the production of bacterial cellulose by Acetobacter xylinum using Hestrin–Schramm medium under anaerobic condition. *Bioresource Technology Reports, 17*, 100878. https://doi.org/10.1016/j.biteb.2021.100878
96. Njuguna, D. G., & Schönherr, H. (2022). Smart and regeneratable Xanthan gum hydrogel adsorbents for selective removal of cationic dyes. *Journal of Environmental Chemical Engineering, 10*(3), 107620. https://doi.org/10.1016/j.jece.2022.107620
97. Nsengiyumva, E. M., & Alexandridis, P. (2022). Xanthan gum in aqueous solutions: Fundamentals and applications. *International Journal of Biological Macromolecules, 216*, 583–604. https://doi.org/10.1016/j.ijbiomac.2022.06.189

98. Nuvoli, L., Conte, P., Fadda, C., Reglero Ruiz, J. A., García, J. M., Baldino, S., & Mannu, A. (2021). Structural, thermal, and mechanical properties of gelatin-based films integrated with tara gum. *Polymer, 214*(July 2020). https://doi.org/10.1016/j.polymer.2020.123244
99. Ogunsona, E., Ojogbo, E., & Mekonnen, T. (2018). Advanced material applications of starch and its derivatives. *European Polymer Journal, 108*(September), 570–581. https://doi.org/10.1016/j.eurpolymj.2018.09.039
100. Othman, N. A. F., Selambakkannu, S., & Seko, N. (2022). Biodegradable dual-layer Polyhydroxyalkanoate (pha)/Polycaprolactone (pcl) mulch film for agriculture: Preparation and characterization. *Energy Nexus, 8*, 100137. https://doi.org/10.1016/j.nexus.2022.100137
101. Palaniraj, A., & Jayaraman, V. (2011). Production, recovery and applications of xanthan gum by Xanthomonas campestris. *Journal of Food Engineering, 106*(1), 1–12. https://doi.org/10.1016/j.jfoodeng.2011.03.035
102. Patti, A., & Acierno, D. (2022). Towards the sustainability of the plastic industry through biopolymers: Properties and potential applications to the textiles world. *Polymers, 14*(4). https://doi.org/10.3390/polym14040692
103. Pavithra, S., Thandapani, G., Alkhamis, H. H., Alrefaei, A. F., & Almutairi, M. H. (2021). Batch adsorption studies on surface tailored chitosan/orange peel hydrogel composite for the removal of Cr(VI) and Cu(II) ions from synthetic wastewater. *Chemosphere, 271*, 129415. https://doi.org/10.1016/j.chemosphere.2020.129415
104. Pensupa, N., Leu, S. Y., Hu, Y., Du, C., Liu, H., Jing, H., Wang, H., & Lin, C. S. K. (2017). Recent trends in sustainable textile waste recycling methods: Current situation and future prospects. *Topics in Current Chemistry, 375*(5). https://doi.org/10.1007/s41061-017-0165-0
105. Pérez, J., Muñoz-Dorado, J., De La Rubia, T., & Martínez, J. (2002). Biodegradation and biological treatments of cellulose, hemicellulose and lignin: An overview. *International Microbiology, 5*(2), 53–63. https://doi.org/10.1007/s10123-002-0062-3
106. Pervez, M. N., Mondal, M. I. H., Cai, Y., Zhao, Y., & Naddeo, V. (2021). 21—Textile waste management and environmental concerns. In M. I. H. Mondal (Ed.), *Fundamentals of natural fibres and textiles. The textile institute book series* (pp. 719–739). Woodhead Publishing. https://doi.org/10.1016/B978-0-12-821483-1.00002-4
107. Peychev, B., & Vasileva, P. (2021). Novel starch-mediated synthesis of Au/ZnO nanocrystals and their photocatalytic properties. *Heliyon, 7*(6), Article e07402. https://doi.org/10.1016/j.heliyon.2021.e07402
108. Qalyoubi, L., Al-Othman, A., & Al-Asheh, S. (2021). Recent progress and challenges on adsorptive membranes for the removal of pollutants from wastewater. Part I: Fundamentals and classification of membranes. *Case Studies in Chemical and Environmental Engineering, 3*, 100086. https://doi.org/10.1016/j.cscee.2021.100086
109. Rahman, N. H. A., Rahman, N. A., Aziz, S. A., & Hassan, M. A. (2013). Production of ligninolytic enzymes by newly isolated bacteria from palm oil plantation soils. *BioResources, 8*(4), 6136–6150. https://doi.org/10.15376/biores.8.4.6136-6150
110. Razzaq, S., Shahid, S., & Nawab, Y. (2024). Applications and environmental impact of biodegradable polymers in textile industry: A review. *International Journal of Biological Macromolecules, 282*, 136791. https://doi.org/10.1016/j.ijbiomac.2024.136791
111. Rivero, S., García, M. A., & Pinotti, A. (2010). Correlations between structural, barrier, thermal and mechanical properties of plasticized gelatin films. *Innovative Food Science and Emerging Technologies, 11*(2), 369–375. https://doi.org/10.1016/j.ifset.2009.07.005
112. Sakkayawong, N., Thiravetyan, P., & Nakbanpote, W. (2005). Adsorption mechanism of synthetic reactive dye wastewater by chitosan. *Journal of Colloid and Interface Science, 286*(1), 36–42. https://doi.org/10.1016/j.jcis.2005.01.020
113. Salehi, E., Daraei, P., & Arabi Shamsabadi, A. (2016). A review on chitosan-based adsorptive membranes. *Carbohydrate Polymers, 152*, 419–432. https://doi.org/10.1016/j.carbpol.2016.07.033
114. Sanjrani, M. A., Gang, X., & Mirza, S. N. A. (2024). A review on textile solid waste management: Disposal and recycling. *Waste Management & Research*, 0734242X241257093. https://doi.org/10.1177/0734242X241257093

115. Sapuan, S. M., & Sulaiman, S. (2020). Review of corn starch biopolymer conceptual concurrent design of bio-composite side door impact beam for passenger car. View project mechanical and thermal performances of roselle fibre reinforced thermoplastic polyurethane composites view project Walid A. November. https://www.researchgate.net/publication/345922061
116. Sarode, S., Upadhyay, P., Khosa, M. A., Mak, T., Shakir, A., Song, S., & Ullah, A. (2019). Overview of wastewater treatment methods with special focus on biopolymer chitin-chitosan. *International Journal of Biological Macromolecules, 121*, 1086–1100. https://doi.org/10.1016/j.ijbiomac.2018.10.089
117. Sashiwa, H., Fukuda, R., Okura, T., Sato, S., & Nakayama, A. (2018). Microbial degradation behavior in seawater of polyester blends containing poly(3-hydroxybutyrate-co-3-hydroxyhexanoate) (PHBHHx). *Marine Drugs, 16*(1), 1–11. https://doi.org/10.3390/md16010034
118. Sayin, S., Kohlhaas, T., Veziroglu, S., Okudan, E. Ş., Naz, M., Schröder, S., Saygili, E. I., Açil, Y., Faupel, F., Wiltfang, J., Aktas, O. C., & Gülses, A. (2020). Marine Algae-PLA composites as de novo alternative to porcine derived collagen membranes. *Materials Today Chemistry, 17*, 100276. https://doi.org/10.1016/j.mtchem.2020.100276
119. Schmitz, C., Auza, L. G., Koberidze, D., Rasche, S., Fischer, R., & Bortesi, L. (2019). Conversion of chitin to defined chitosan oligomers: Current status and future prospects. *Marine Drugs, 17*(8), 1–22. https://doi.org/10.3390/md17080452
120. Shahbazi, M., Jäger, H., Ahmadi, S. J., & Lacroix, M. (2020). Electron beam crosslinking of alginate/nanoclay ink to improve functional properties of 3D printed hydrogel for removing heavy metal ions. *Carbohydrate Polymers, 240*, 116211. https://doi.org/10.1016/j.carbpol.2020.116211
121. Sharma, P., Mittal, M., Yadav, A., & Aggarwal, N. K. (2023). Bacterial cellulose: Nanobiomaterial for biodegradable face masks—A greener approach towards environment. *Environmental Nanotechnology, Monitoring & Management, 19*, 100759. https://doi.org/10.1016/j.enmm.2022.100759
122. Shawky, H. A. (2009). Synthesis of ion-imprinting chitosan/PVA crosslinked membrane for selective removal of Ag (I). *Journal of Applied Polymer Science, 114*(5), 2608–2615. https://doi.org/10.1002/app.30816
123. Shi, Y., Chai, L., Tang, C., Yang, Z., Zheng, Y., Chen, Y., & Jing, Q. (2013). Biochemical investigation of kraft lignin degradation by pandoraea sp. B-6 isolated from bamboo slips. *Bioprocess and Biosystems Engineering, 36*(12), 1957–1965. https://doi.org/10.1007/s00449-013-0972-9
124. Shi, Y., Zhang, Q., Feng, L., Xiong, Q., & Chen, J. (2014). Preparation and adsorption characters of Cu(II)-imprinted chitosan/attapulgite polymer. *Korean Journal of Chemical Engineering, 31*(5), 821–827. https://doi.org/10.1007/s11814-014-0004-8
125. Singh, M., Kumar, P., Ray, S., & Kalia, V. C. (2015). Challenges and opportunities for customizing polyhydroxyalkanoates. *Indian Journal of Microbiology, 55*(3), 235–249. https://doi.org/10.1007/s12088-015-0528-6
126. Sirajudheen, P., Poovathumkuzhi, N. C., Vigneshwaran, S., Chelaveettil, B. M., & Meenakshi, S. (2021). Applications of chitin and chitosan based biomaterials for the adsorptive removal of textile dyes from water—A comprehensive review. *Carbohydrate Polymers, 273*, 118604. https://doi.org/10.1016/j.carbpol.2021.118604
127. Skiba, E. A., Budaeva, V. V, Ovchinnikova, E. V, Gladysheva, E. K., Kashcheyeva, E. I., Pavlov, I. N., & Sakovich, G. V. (2020). A technology for pilot production of bacterial cellulose from oat hulls. *Chemical Engineering Journal, 383*, 123128. https://doi.org/10.1016/j.cej.2019.123128
128. Solis, M., Tonini, D., Scheutz, C., Napolano, L., Biganzoli, F., & Huygens, D. (2024). Contribution of waste management to a sustainable textile sector. *Waste Management, 189*, 389–400. https://doi.org/10.1016/j.wasman.2024.08.037
129. Sudarta. (2022). Value-adding to cellulosic ethanol: Lignin polymers. *Industrial Crops and Products, 16*(1), 1–23. https://doi.org/10.1016/j.indcrop.2010.10.022

130. Sun, P., Wang, M., Wu, T., Guo, L., & Han, W. (2023). Covalent crosslinking cellulose/graphene aerogels with high elasticity and adsorbability for heavy metal ions adsorption. *Polymers, 15*(11). https://doi.org/10.3390/polym15112434
131. Sutirman, Z. A., Sanagi, M. M., Abd Karim, J., Abu Naim, A., & Wan Ibrahim, W. A. (2018). New crosslinked-chitosan graft poly(N-vinyl-2-pyrrolidone) for the removal of Cu(II) ions from aqueous solutions. *International Journal of Biological Macromolecules, 107*, 891–897. https://doi.org/10.1016/j.ijbiomac.2017.09.061
132. Svirshchevskaya, E. (2017). Correlation analysis of chitosan physicochemical parameters determined by different methods. *Organic & Medicinal Chemistry International Journal, 1*(3). https://doi.org/10.19080/omcij.2017.01.555562
133. Taktak, F. F., & Özyaranlar, E. (2022). Semi-interpenetrating network based on xanthan gum-cl-2-(N-morpholinoethyl methacrylate)/titanium oxide for the single and binary removal of cationic dyes from water. *International Journal of Biological Macromolecules, 221*, 238–255. https://doi.org/10.1016/j.ijbiomac.2022.08.139
134. Tanadchangsaeng, N., & Pattanasupong, A. (2022). Evaluation of biodegradabilities of biosynthetic polyhydroxyalkanoates in Thailand seawater and toxicity assessment of environmental safety levels. *Polymers, 14*(3). https://doi.org/10.3390/polym14030428
135. Thirumavalavan, M. (2023). Functionalized chitosan and sodium alginate for the effective removal of recalcitrant organic pollutants. *International Journal of Biological Macromolecules, 243*, 125276. https://doi.org/10.1016/j.ijbiomac.2023.125276
136. Thygesen, A., Tsapekos, P., Alvarado-Morales, M., & Angelidaki, I. (2021). Valorization of municipal organic waste into purified lactic acid. *Bioresource Technology, 342*, 125933. https://doi.org/10.1016/j.biortech.2021.125933
137. Tsouko, E., Kourmentza, C., Ladakis, D., Kopsahelis, N., Mandala, I., Papanikolaou, S., Paloukis, F., Alves, V., & Koutinas, A. (2015). Bacterial cellulose production from industrial waste and by-product streams. *International Journal of Molecular Sciences, 16*(7), 14832–14849. https://doi.org/10.3390/ijms160714832
138. Vakili, M., Deng, S., Cagnetta, G., Wang, W., Meng, P., Liu, D., & Yu, G. (2019). Regeneration of chitosan-based adsorbents used in heavy metal adsorption: A review. *Separation and Purification Technology, 224*, 373–387. https://doi.org/10.1016/j.seppur.2019.05.040
139. Vasile, C. (2000). *Handbook of polyolefins* (2nd ed.). CRC Press. https://doi.org/10.1201/9780203908716
140. Wang, F., Lu, X., & Li, X. (2016). Selective removals of heavy metals (Pb^{2+}, Cu^{2+}, and Cd^{2+}) from wastewater by gelation with alginate for effective metal recovery. *Journal of Hazardous Materials, 308*, 75–83. https://doi.org/10.1016/j.jhazmat.2016.01.021
141. Wang, T., Sun, B., Tang, K., Shen, W., Chen, C., & Sun, D. (2023). Sustainable bacterial cellulose derived composites for high-efficiency hydrogen evolution reaction. *International Journal of Biological Macromolecules, 242*, 125173. https://doi.org/10.1016/j.ijbiomac.2023.125173
142. Wang, Z.-K., Li, T.-T., Ren, H.-T., Peng, H.-K., Shiu, B.-C., Lin, J.-H., & Lou, C.-W. (2022). Construction of a novel hydrogel composite with polylactic acid nonwoven fibers: Rapid and effective removal of Pb(II) and Ni(II). *Textile Research Journal, 93*(1–2), 266–279. https://doi.org/10.1177/00405175221114637
143. Weiss, A.-V., Fischer, T., Iturri, J., Benitez, R., Toca-Herrera, J. L., & Schneider, M. (2019). Mechanical properties of gelatin nanoparticles in dependency of crosslinking time and storage. *Colloids and Surfaces B: Biointerfaces, 175*, 713–720. https://doi.org/10.1016/j.colsurfb.2018.12.005
144. Weng, Y. X., Wang, X. L., & Wang, Y. Z. (2011). Biodegradation behavior of PHAs with different chemical structures under controlled composting conditions. *Polymer Testing, 30*(4), 372–380. https://doi.org/10.1016/j.polymertesting.2011.02.001
145. Wojnowska-Baryła, I., Bernat, K., & Zaborowska, M. (2022). Strategies of recovery and organic recycling used in textile waste management. *International Journal of Environmental Research and Public Health, 19*(10). https://doi.org/10.3390/ijerph19105859

146. Yang, L., Bao, L., Dong, T., Xie, H., Wang, X., Wang, H., Wu, J., & Hao, C. (2023). Adsorption properties of cellulose/guar gum/biochar composite hydrogel for Cu^{2+}, Co^{2+} and methylene blue. *International Journal of Biological Macromolecules, 242*, 125021. https://doi.org/10.1016/j.ijbiomac.2023.125021
147. Yano, S., Hatakeyama, H., & Hatakeyama, T. (1976). Effect of hydrogen bond formation on dynamic mechanical properties of amorphous cellulose. *Journal of Applied Polymer Science, 20*(12), 3221–3231. https://doi.org/10.1002/app.1976.070201203
148. Yi, X., Sun, F., Han, Z., Han, F., He, J., Ou, M., Gu, J., & Xu, X. (2018). Graphene oxide encapsulated polyvinyl alcohol/sodium alginate hydrogel microspheres for Cu (II) and U (VI) removal. *Ecotoxicology and Environmental Safety, 158*, 309–318. https://doi.org/10.1016/j.ecoenv.2018.04.039
149. Yoon, S.-D. (2014). Cross-linked potato starch-based blend films using ascorbic acid as a plasticizer. *Journal of Agricultural and Food Chemistry, 62*(8), 1755–1764. https://doi.org/10.1021/jf4024855
150. Zhang, H., Guan, G., Lou, T., & Wang, X. (2023). High performance, cost-effective and ecofriendly flocculant synthesized by grafting carboxymethyl cellulose and alginate with itaconic acid. *International Journal of Biological Macromolecules, 231*, 123305. https://doi.org/10.1016/j.ijbiomac.2023.123305
151. Zhang, H., Han, X., Liu, J., Wang, M., Zhao, T., Kang, L., Zhong, S., & Cui, X. (2022). Fabrication of modified alginate-based biocomposite hydrogel microspheres for efficient removal of heavy metal ions from water. *Colloids and Surfaces A: Physicochemical and Engineering Aspects, 651*, 129736. https://doi.org/10.1016/j.colsurfa.2022.129736
152. Zhang, M., & Ogale, A. A. (2014). Carbon fibers from dry-spinning of acetylated softwood kraft lignin. *Carbon, 69*(April 2014), 626–629. https://doi.org/10.1016/j.carbon.2013.12.015
153. Zhang, M., & Ogale, A. A. (2016). Effect of temperature and concentration of acetylated-lignin solutions on dry-spinning of carbon fiber precursors. *Journal of Applied Polymer Science, 133*(45). https://doi.org/10.1002/app.43663
154. Zhang, M., Sun, H., Wang, Y., Piao, C., Cai, D., Wang, Y., Liu, J., & Cheng, Z. (2020). Preparation and characterization of a novel porous whey protein concentrate/pullulan gel induced by heating for Cu^{2+} absorption. *Food Chemistry, 322*, 126772. https://doi.org/10.1016/j.foodchem.2020.126772
155. Zhang, W., Ou, J., Wang, B., Wang, H., He, Q., Song, J., Zhang, H., Tang, M., Zhou, L., Gao, Y., & Sun, S. (2021). Efficient heavy metal removal from water by alginate-based porous nanocomposite hydrogels: The enhanced removal mechanism and influencing factor insight. *Journal of Hazardous Materials, 418*, 126358. https://doi.org/10.1016/j.jhazmat.2021.126358
156. Zhang, Y., Gao, W., Mo, A., Jiang, J., & He, D. (2022). Degradation of polylactic acid/polybutylene adipate films in different ratios and the response of bacterial community in soil environments. *Environmental Pollution, 313*, 120167. https://doi.org/10.1016/j.envpol.2022.120167
157. Zheng, F., Yang, X., Li, J., Tian, Z., Xiao, B., Yi, S., & Duan, L. (2022). Coordination with zirconium: A facile approach to improve the mechanical properties and thermostability of gelatin hydrogel. *International Journal of Biological Macromolecules, 205*, 595–603. https://doi.org/10.1016/j.ijbiomac.2022.02.124
158. Zheng, Y., Chen, J. C., Ma, Y. M., & Chen, G. Q. (2020). Engineering biosynthesis of polyhydroxyalkanoates (PHA) for diversity and cost reduction. *Metabolic Engineering, 58*(May), 82–93. https://doi.org/10.1016/j.ymben.2019.07.004
159. Zhong, C. (2020). Industrial-scale production and applications of bacterial cellulose. *Frontiers in Bioengineering and Biotechnology, 8*(December), 1–19. https://doi.org/10.3389/fbioe.2020.605374
160. Zhou, J., Chen, Y., Zhang, Y., Sun, S., Ullah, M. W., & Xu, W. (2021). Biotransformation of nylon-6{,}6 hydrolysate to bacterial cellulose. *Green Chemistry, 23*(19), 7805–7815. https://doi.org/10.1039/D1GC02129J
161. Zhou, Q., Le, Q. V., Meng, L., Yang, H., Gu, H., Yang, Y., Chen, X., Lam, S. S., Sonne, C., & Peng, W. (2022). Environmental perspectives of textile waste, environmental pollution

and recycling. *Environmental Technology Reviews, 11*(1), 62–71. https://doi.org/10.1080/21622515.2021.2017000

162. Zhou, W., Bergsma, S., Colpa, D. I., Euverink, G.-J. W., & Krooneman, J. (2023). Polyhydroxyalkanoates (PHAs) synthesis and degradation by microbes and applications towards a circular economy. *Journal of Environmental Management*, *341*, 118033. https://doi.org/10.1016/j.jenvman.2023.118033

Nano-bioremediation for Dye Effluents

Shivani Rawat, Avni Kumari, Afaq Majid Wani, and Syed Maqbool Geelani

1 Introduction

Environmental contamination has become a serious concern globally due to increased industrialization and the growing population. Various type of pollutants released from the industries results in degrading the quality of water bodies. Hazardous contaminants such as organic pollutants, heavy metal ions, dyes, etc., are released from chemical industries through wastewater streams [14]. Because the effluent from numerous dye industries meets water bodies, causing contamination, water pollution has become a severe threat to living beings. Water pollution occurs when harmful substances like pigments, cosmetics, food, printing, textile, leather, dyeing, etc. are released into water bodies [10]. The synthetic dyes do not degrade easily in the environment and cannot be removed by the ordinary treatment process, thereby leading to both soil and water contamination. The presence of these dyes adversely affects living organisms causing various ailments like skin irritation, nausea, vomiting, kidney and liver damage, gastrointestinal tract infection, etc. The effect can also be seen in plants where these dyestuffs inhibit seed germination, inhibit microbial activities, etc. [20].

Dye manufacturing is anticipated to be roughly 70,000 tonnes per year worldwide, with around 10,000 commercial types [5]. Synthetic textile dyes not only consists of non-biodegradable complex aromatic substances but also have nitrogen bond (–N=N–) in their structure which is carcinogenic and mutagenic [21]. The majority of these dyes are composed of complex aromatic compounds that are difficult to degrade. Dye effluents also have an impact on plant photosynthetic activities by restricting

S. Rawat · A. Kumari · A. M. Wani (✉)
Department of Forest Biology and Tree Improvement, College of Forestry, Sam Higginbottom University of Agriculture, Technology and Sciences, Prayagraj, Uttar Pradesh, India
e-mail: afaqtree@gmail.com

S. M. Geelani
Department of Environmental Engineering and Research, GuruJal Abhipsa-Foundation, Gurugram, Haryana, India

R. A. Bhat et al. (eds.), *Biocentric Approaches in Textile Waste Management*, Textile Science and Clothing Technology, https://doi.org/10.1007/978-3-032-04974-2_3

sunlight penetration. Therefore, the amount of oxygen concentration in the aquatic environment is reduced leading to damage to the water bodies. The adverse effects of dyes result in the formation of tumors, cancer, and allergies which imparts serious health issues. Dyestuffs have the structural ability, retain color, and show strong resistance to microbial degradation under sun exposure, soil, and bacteria [14]. It has been found in a study that a certain amount of dye in drinking water (1.0 mg L^{-1}) could pollute the water, and would be harmful to human health. Consuming more than 1.5 mg L^{-1} of fluoride in polluted drinking water can be harmful to humans owing to certain anthropogenic causes [1]. So the remediation of dyes is essential even though a trace quantity of dye is present in the environment.

Dyes have a complicated structure, are non-biodegradable, and last a long time in the environment. There is a need to remove these dye pollutants from the water using effective methods. Physical methods (sedimentation, filtration, adsorption, and reverse osmosis), chemical methods (neutralization, recovery, chemical oxidation, and ion exchange methods), and biological methods (using bacteria, fungi, and actinomycetes) are among the many conventional treatment methods used to remove dyes from wastewater [4]. Physical methods are too expensive, while chemical methods create some adverse effects on our environment and are harmful to all living organisms as well as human health. Physical methods have certain fundamental drawbacks such as inadequate efficiency, expensive prices, and massive consumption of energy to achieve synthesis requirements comprising high pressure and temperature [8]. Nanotechnology appears as a game-changing avenue for enhancing environmental quality, including scientific advancement that encompasses material management at the nanoscale level (a nanometer is 10^{-9} of a meter). Nanomaterials should have distinctive structures that lay somewhere between bulk materials and single atoms. Nanotechnology has recently contributed to water purification with cost-effectiveness, superior operational efficiency in removing impurities, and reuse capability due to qualities such as huge surface area, nano-size, extreme reactivity, and mechanical stability [3]. Various nanomaterials have been discovered to successfully eradicate inorganic and organic pollutants, as well as some heavy metals (such as Pb, Mo, and others) and a variety of hazardous microorganisms [22]. More advances have recently been made in nanomaterials such as nano-membranes, nano-photocatalysts, and nano-sorbents, which could be used to effectively remediate dirty water. In summary, a thorough examination of nanomaterial applications in water purification is required to assess the prospects.

2 Characteristic Properties and Types of Nanomaterials

Nanotechnology has resulted in a diverse spectrum of nanoscale materials, known as nanomaterials, with sizes ranging from 1 to 100 nm. Depending on the overall form, these elements can be zero-dimensional, one-dimensional, two-dimensional, or three-dimensional. So far, these nanoparticles are the most renowned learning resources. Because altering the size of nanoparticles allows us to quickly modify their intrinsic qualities and create a totally new material [15]. Because of their high proportion of surface area-to-volume, nanomaterials behave differently than bulk materials. All dimensions of zero-dimensional nanomaterials are measured on a nanoscale, with no dimension larger than 100 nm. Nanoparticles, such as quantum dots, are commonly used as zero-dimensional nanomaterials. One-dimensional nanomaterials, which include nanotubes, nanorods, and nanowires, have one dimension that is outside the nanoscale. Two-dimensional nanomaterials, which include plate-like structures such as graphenes, nanofilms, and nanolayers, exist outside the nanoscale. Bulk powders, nanoparticle dispersion, nanowire bundles, nanotube bundles, and other three-dimensional nanomaterials have no dimension constraints.

According to the European Commission, a nanomaterial is "an accidental, natural, or synthetic substance composed of loose particles or as an agglomeration in which 50% or more of the particles in the number size distribution have one or more exterior dimensions in the size range of 1–100 nm" [7].

On the other hand, according to US Environmental Protection Agency (EPA), several types of nanomaterials are produced and organized into four types:

1. **Carbon-based materials**: The majority of these nanomaterials are carbon-based and found in various forms of graphene, fullerene, and nanotubes. Nanotubes are cylindrical carbon nanomaterials, while fullerenes are spherical and ellipsoidal carbon nanoparticles. These particles have the potential to be used to create superior coatings materials, as well as strong and light-weighed materials and electrical appliances. These materials have piqued the attention of a various domains, including biomedical applications, because of their unusual structural dimensions and exceptional mechanical, electrical, thermal, optical, and chemical properties.
2. **Metal-based materials**: These nanomaterials include quantum dots, nanogold, nanosilver, and metal oxides such as titanium dioxide. A quantum dot is a tightly packed semiconductor crystal made up of hundreds or thousands of atoms that range in size from a few nanometers to a few hundred nanometers. The optical properties of quantum dots change as their size varies. The biological and pharmaceutical industries are interested in metal-based nanomaterials.
3. **Dendrimers**: These nanomaterials are branching nanoscale components that are used to make polymers. A dendrimer's surface has multiple chains end that can be chemically manipulated to accomplish specialized chemical tasks. This characteristic could be useful in catalysis as well. Because three-dimensional dendrimers include interior gaps into which other elements may be inserted, they could be useful for medication delivery. Polymers are most commonly employed to detect and remove contaminants. (e.g., manganese, nitrate, iron, arsenic, heavy metals, etc.), gases (e.g., CO, SO_2, NO_x), organic pollutants (e.g., aliphatic and aromatic hydrocarbons, medicines, or VOCs), as well as a broad range of biologics (e.g., bacteria, parasites, viruses, etc.).
4. **Composites**: This is the technique of combining nanoparticles or nanoparticles with other materials to create unique nanomaterials that are then used to make products. Nanoparticles, such as nanosized clays, are currently utilized

to improve mechanical, thermal, barrier, and flame-retardant qualities in goods ranging from automobile components to packaging materials.

These numerous varieties of purposefully manufactured nanomaterials are highly wanted for applications in the corporate, clinical, defense, and environmental sectors due to their distinctive electronic, enzymatic, magnetic, structural, thermal, or imaging qualities.

3 Nano-techniques for Remediation of Harmful Dyes

Nanotechnology is now being employed in wide range of applications, including targeted medicine delivery, imaging, and food hazard detection. Dye effluents from industries affect human as well as aquatic ecosystems because of their toxic effects, accumulation in living being, or persistence. These contaminants may be remediated more straightforwardly and cost-effectively by utilizing a variety of materials and processes. In recent years, to treat wastewater many nanotechnological techniques are starting getting more widespread usage [12]. Sensing and targeting, adsorption of dye onto nano sorbent, photocatalytic redox reaction employing nanocatalyst, breakdown, decontamination, and filtering of purified water are the steps followed in remediation of harmful dyes [17].

Construction of nanoparticles can be done in various ways including a reduction in the size of minute particles which in turn can be constructed from bulk materials. Some of the techniques involved in the formation of nanoparticles from this strategy include pulsed laser ablation, evaporation–condensation, spray pyrolysis, lithography, and discharge [16]. Physical approaches to form well-defined nanostructures are those in which physical forces are used to attract nanoparticles resulting in the construction of large nanostructures. The utilization of high-priced synthesis equipment large space area for setting up instruments, defective surface formation, and low production rate, high temperature, and pressure are the limitations of the physical method involved [6]. For the creation of nanoparticles, another way is to use chemical and biological processes in which atoms form clusters or new nuclei, which develop into nanoparticles. Solvothermal, pyrolysis, co-precipitation, sonochemical, and electrochemical are different techniques used for the synthesis of nanoparticles. These chemical methods are toxic and less environment friendly and result in the formation of hazardous products which creates a risk of contamination. Furthermore, the finished nanoparticles may be kept for an extensive period of time without significantly losing stability [18]. It can be concluded that the conventional approach to fabricate nano-particles and nano-structures has certain drawbacks due to their toxic and less environment-friendly nature.

Nanoparticle synthesis using green plants or microorganisms has become the most fascinating route as it has overcome all the limitations which were the major concern in other methods such as physical and chemical fabrication methods for nanoparticle synthesis. Chemicals used during the synthesis of nano-particle using chemical methods are expensive and toxic and by-products left after the process are hazardous and not eco-friendly while the physical methods consume more energy which makes them less suitable at large scale. The biological methods for the synthesis of nanoparticles came out to be a better approach due to the reduction in toxic waste production. Biological synthesis is the process of creating nanomaterials using plant extracts, bacteria, and other biological components. This technology is both environmentally benign and inexpensive since microorganisms such as yeast, viruses, fungus, and bacteria serve as platform for nanomaterial production. Microorganisms with enzymes such as reductase can convert mineral salts into nanomaterials [2].

Synthesis of nano-particles with help of microorganisms includes different microbes acting as an agent in the development of desirably sized nanoparticles due to certain factors such as the low cost of maintenance and evolution also they do not require much-specialized equipment to work with them reducing cost. Widely synthesized nano-particles using this process are Gold, silver, and cadmium sulfide as these microbial cells are considered potential bio-factories for this nanoparticle fabrication. It has been observed in several studies that several species of fungi help in fabricating mono-dispersed nanoparticles. Being an abundant source of enzymes that are helpful in the production of nanoparticles makes fungal cells a very prominent microorganism in the nano-particle fabrication process. Compared to bacteria, fungi secrete more amount of protein which is helpful in the higher production of nanoparticles. A large variety of fungal species are used in the production of silver (Ag) nano-particles such as *Verticillium* spp. [13]. For nanoparticle synthesis, a variety of green reducing and coating agents are used, all of which are cost-effective, non-hazardous, and environmentally benign. These agents and reductants help in the fabrication of nano-particles. Nano-particles produced using plant-based methods are highly stable in a very cost-effective, fast, and eco-friendly process.

4 Bioremediation of Dyes Using Nanoparticles

For the catalytic degradation of dyes, nanotechnology is an emerging technology. The employment of biological technologies for environmental remediation has been the subject of contemporary research. Bioremediation is the method used to transform hazardous compounds such as xenobiotics into less toxic substances utilizing biological processes for contaminated site remediation. Textile effluents including dyes and other organic molecules must be treated or decolored right away because of their harmful, health, and environmental effects. Zero valent iron nanoparticles are commonly utilized as an oxidizing medium for treating wastewater because iron metal is affordable, works as a powerful reducer, and has a greater efficiency and pollutant degradation ability (comprising halogenated compounds, harmful metals, and inorganic compounds). Accumulation of these nanoparticles on the bacterial interface strengthen their capacity to destroy pollutants with a –N=N– linkage. Because of their small size, large surface area, and interface-dominating features, nanoparticles are employed to decontaminate hazardous organic and inorganic chemicals from the environment. The application of iron nanoparticles has spread to numerous disciplines such as catalysis, sensors, and environmental remediation such as the adsorption and degradation of various contaminants from aqueous media due to the characteristics like small size and high density [19].

5 Microbial Nanotechnology

During the creation of nanoparticles, many microorganisms react in a number of ways. Microbes create inorganic material by diverse methods, depending on whether they are intracellular or extracellular. The microorganism's cell wall has enzymes that aid in metal ion reduction, resulting in cations that are subsequently fixed with the cell wall (negative charge) during the intracellular nanoparticle formation process. This is how nanoparticles form in the intercellular environment. Fungi also produce nitrate reductase, an enzyme that functions as a catalyst in metal ion reduction and contributes to the extracellular creation of nanoparticles. Nanoparticle production using microbes has shown to be a cost-effective and environmentally beneficial technique. Copper nanoparticles were synthesised using *Escherichia* sp. SINT7, being copper resilient. The biogenic nanoparticles have been proven to destroy azo dyes and textile wastewater [11].

Transition metals precipitated over the surface of bacteria in a unique bio-nanometal construct, generating nanoparticulate catalyst in a single process. Pd nanoparticles were made by reducing Pd(II) ions using *Clostridia pasteurianum* BC1 cultures in bacterial cell walls leading to the formation of bio-Pd nanoparticles conjugate. The dye reduction of organic dyes (Methyl Orange) by reductive hydrogenation of azo dyes at the azo linkage and with the help of molecular hydrogen generated by bacterial cells was successful. Bio-Pd is a simple and possible biocatalyst among the current nanomaterials, as opposed to other expensive and sophisticated techniques [9].

6 Conclusion

Nanotechnology has the potential to greatly contribute to the creation of better, sustainable alternatives with considerable environmental and health benefits. There is a pressing need to get rid of these contaminants as quickly as possible while causing the least amount of harm to the environment. Nanoparticles, which may be active on both tiny and big scales, can provide a more sustainable, comprehensive strategy for addressing this serious problem. Diversified nanoparticles have shown useful in eliminating certain contaminants from their original environments, such as sewage spills, dye industries, and landmines. Furthermore, they have a low hazardous effect during manufacturing rates and may thus be thoroughly explored in the future to make them more compatible with lower production costs. Environmental pollution cleanup utilizing green-manufactured nanoparticles has been a popular technique in the recent decade. There are many types of dyes, nanostructures, and nanocomposites utilized in textile dye treatment. Pigment production, food processing, cosmetics processing, textile processing, printing, leather processing, and other industries employ dyes. Advanced nanotechnology will be able to be used in the future to treat wastewater in a more environmentally friendly manner, resulting in a more sustainable environment.

References

1. Ahmad, S., Singh, R., Arfin, T., & Neeti, K. (2022). Fluoride contamination, consequences and removal techniques in water: A review. *The Royal Society of Chemistry, 1*, 620–661.
2. Alsaiari, N. S., Alzahrani, F. M., Amari, A., Osman, H., Harharah, H. N., Elboughdiri, N., & Tahoon, M. A. (2023). Plant and microbial approaches as green methods for the synthesis of nanomaterials: Synthesis, applications, and future perspectives. *Molecules, 28*, 463.
3. Baruah, S., Khan, M. N., & Dutta, J. (2016). Perspectives and applications of nanotechnology in water treatment. *Environmental Chemistry Letters, 14*, 1–14.
4. Bhatt, P., Pandey, S. C., Joshi, S., Chaudhary, P., Pathak, V. M., Huang, Y., Wu, X., Zhou, Z., & Chen, S. (2022). Nanobioremediation: A sustainable approach for the removal of toxic pollutants from the environment. *Journal of Hazardous Materials, 427*, 128033. ISSN 0304-3894.
5. Chandanshive, V., Kadam, S., Rane, N., Jeon, B.-H., Jadhav, J., & Govindwar, S. (2020). In-situ textile wastewater treatment in high rate transpiration system furrows planted with aquatic macrophytes and floating phytobeds. *Chemosphere, 252*, 126513. https://doi.org/10.1016/j.chemosphere.2020.126513.
6. Chandrasekaran, R., Gnanasekar, S., Seetharaman, P., Keppanan, R., Arockiaswamy, W., & Sivaperumal, S. (2016). Formulation of *Carica papaya* latex-functionalized silver nanoparticles for its improved antibacterial and anticancer applications. *Journal of Molecular Liquids, 219*, 232–238.
7. European Commission. (2016). Recommendation on the definition of a nanomaterial. August 29, 2020.
8. Hashmi, S. S., Shah, M., Muhammad, W., Ahmed, A., Ullah, M. A., Muhammad, N., & Abbasi, B. H. (2021). Potentials of phyto-fabricated nanoparticles as eco-friendly agents for phyto-catalytic degradation of toxic dyes and waste water treatment, risk assessment and probable mechanism. *Journal of the Chemical Society, 98*(4), 100019. ISSN 0019-4522.
9. Keerthana, V., Girigoswami, A., Harini, A., & Girigoswami, K. (2022). Waste water remediation using nanotechnology-A review. *12*(4), 4476–4495.
10. Khan, S., Naushad, M., Al-Gheethi, A., & Iqbal, J. (2021). Engineered nanoparticles for removal of pollutants from wastewater: Current status and future prospects of nanotechnology for remediation strategies. *Journal of Environmental Chemical Engineering, 9*(5), 106160. ISSN 2213–3437.
11. Mandeep, & Shukla, P. (2020). Microbial nanotechnology for bio-remediation of industrial wastewater. *Frontiers in Microbiology, 11*, 590631.
12. Marković, Z., & Marković, B. T. (2019). Treating of aquatic pollution by carbon quantum dots. In *Nanostructured materials for treating aquatic pollution* (pp. 121–145). Springer.
13. Moustafa, S. M. N., Taha, R. H., Abdelzaher, H. M. A., & Elgebaly, H. A. (2021). Novel biosynthesis of Ag-nanocomplex for controlling Verticillium wilt disease of olive tree. *Archives of Phytopathology and Plant Protection, 55*(2), 198–216.
14. Pandey, A., Shukla, P., & Srivastava, P. K. (2020). Remediation of dyes in water using green synthesized nanoparticles. *International Journal of Plant and Environment, 6*(1), 68–84.

15. Patra, S., Roy, E., Tiwari, A., Madhuri, R., & Sharma, P. K. (2017). 2-Dimensional graphene as a route for emergence of additional dimension nanomaterials. *Biosensors and Bioelectronics, 89*(Part 1), 8–27. ISSN 0956-5663.
16. Rafique, M., Sadaf, I., Rafiquea, M. S., & Tahirb, M. B. (2017). A review on green synthesis of silver nanoparticles and their applications. *Artificial Cells Nanomedicine and Biotechnology, 45*, 1272–1291.
17. Saikia, J., Gogoi, A., & Baruah, S. (2019). Nanotechnology for water remediation. In *Environmental nanotechnology* (pp. 195–211). Springer.
18. Szczyglewska, P., Feliczak-Guzik, A., & Nowak, I. (2023). Nanotechnology-general aspects: A chemical reduction approach to the synthesis of nanoparticles. *Molecules, 28*, 4932.
19. Sharma, H., & Shirkot, P. (2019). Bioremediation of azo dyes using biogenic iron nanoparticles. *7*(1).
20. Sharma, J., Sharma, S., & Soni, V. (2021). Classification and impact of synthetic textile dyes on Aquatic Flora: A review. *Regional Studies in Marine Science, 45*, 101802. ISSN 2352-4855.
21. Singh, R. L., Singh, P. K., & Singh, R. P. (2015). Enzymatic decolorization and degradation of azo dyes—A review. *International Biodeterioration & Biodegradation, 104*, 21–31.
22. Umar, K. (2018). Water contamination by organic-pollutants: TiO_2 photocatalysis. In *Modern age environmental problems and their remediation* (pp. 95–109). Springer.

Circular Economy in Textile Waste Management: Principles and Practices

Florin-Constantin Mihai

1 Introduction

Textile waste production fueled by fast fashion models based on cheap synthetic fabrics feeds the linear economy of textile industries and inhibits incentives towards a circular economy transition while natural-based fibers such as cotton are water-intensive to grow and related chemical fertilizers and pesticides pollute the soil and groundwaters [32, 35]. Furthermore, textile industries rely on cheap and labor-intensive work in developing countries raising legal, social, and ethical concerns while disrupting ecosystems through animal-based fiber sources or chemical pollution of soil and water resources [44]. The strong fashion marketing campaigns of major brands related to seasonal and annual styles encourage overconsumption of textile production while providing limited take-back programs to capture the post-consumer textile waste. Moreover, the source-separation schemes for this waste stream transform textile waste into one of the most pressing waste management issues around the globe. At the EU level, only1% of used cloth is recycled into new clothes closing the loop while textile production is responsible for 20% of clean global water pollution related to dyeing and finishing products [11] not to mention the plastic pollution associated with textile waste [27]. There is an open letter that calls for urgent measures to address textile waste management in Europe [37]. The principles of circular economy refer to both material efficiency and clean energy transition to reduce the environmental impacts of textile industries [46]. The circular economy comprises the upstream measures focusing on waste prevention (circular design) and management of secondary materials from post-industrial and pre-consumer stages and downstream measures like mechanical and chemical recycling related to pos-consumer textile waste. However, the role of consumers is important in keeping

F.-C. Mihai (✉)
Department of Exact Sciences and Natural Sciences, CERNESIM Environmental Research Center, Institute of Interdisciplinary Research, "Alexandru Ioan Cuza" University of Iasi, Iasi, Romania
e-mail: mihai.florinconstantin@gmail.com

R. A. Bhat et al. (eds.), *Biocentric Approaches in Textile Waste Management*, Textile Science and Clothing Technology, https://doi.org/10.1007/978-3-032-04974-2_4

textile products in circular economy mechanisms such as refusing to fast fashion model, reuse, donate, repair, and creative recycling. The textile waste includes both natural and synthetic fabrics posing particular challenges in driving circular economy solutions. This work aims to reveal CE principles applied to key sectors of textile industries such as clothes, shoes, and non-fashion applications (household textiles, home décor, furniture, etc.) looking to both upstream and downstream solutions to divert textile materials to end up in open environments, landfills, or incineration. The biocentric vision is related to those principles and practices that focus on natural fibers as raw materials with the lowest water, land, and chemical footprints possible as part of organic and regenerative agriculture while searching for and developing sustainable natural-based alternatives to cotton fabrics. Furthermore, ecosystem impacts must be accounted for by textile-related businesses, and some directions are developed towards sustainable animal-based fabrics (wool) or replacing them as part of a new paradigm called "cruelty-free fashion" or "vegan fashion" following the same time the CE principles [25, 31]. The chapter identifies the best practices in textile waste management and innovation of companies in the upstream sector and highlights the role of networking "business to business" and collaboration "business to consumers" in implementing CE principles in the textile sector.

2 Fast Fashion and Pollution of Open Environments

Post-consumer textile waste is poorly captured to circular economy mechanisms (reuse, repair, remanufacture, recycling) and ends up in open environments or landfills around the globe due to limitations of domestic collection and recycling capabilities or trades through legal and illegal channels to less developed regions. Fast-fashion models fed by cheap and abundant synthetic fibers invade the markets with textile products that have a short-live of usage and have poor re-capture levels through take-back programs. These garments contribute to microplastic contamination of air–water-soil compartments [6, 28]. Large quantities of textile waste are collected via municipal waste collection services in residual (mixed) bins and disposed into landfills and their contribution to CE is lost. In some geographies (e.g. Sub-Sahara Africa, Central and Eastern Asia), waste collection services are more limited [47], therefore, the uncollected textile waste is uncontrolled disposed of by open burning practices or dumping in open environments. The additional inputs of low-quality textile wastes from imports in such geographies led to an environmental crisis as shown by an investigation in Ghana [18]. For example, almost half of the textile waste from Catalonia is exported without trace while the main domestic disposal routes are landfills and incineration plants [51] similar to the whole European Union context where these disposal methods account for around 80% of (8.5 Mt) of used and textile waste [28]. Lack of effective source-separation textile infrastructure and advanced sorting centers in countries like Romania results in low recycling rates (10%) as revealed by Green [20] and illegal waste disposal practices as shown in Fig. 1.

Fig. 1 Textile and plastic waste trapped by trees on Bicaz riverbank, Romania (February 2024). *Photo* author

Textile industries argue that waste-to-energy solutions for textile waste should be regarded as intermediary solutions particularly in regions where efficient recycling infrastructure is missing because the emerging amounts of textile waste will end up in landfills [16]. Even developed countries with well-developed source separation collection schemes like Germany are facing a crisis because of high volumes of textile waste generated that exceed the storage capacity [1]. Some textile waste from Germany is sorted in the Netherlands due to insufficient operational sorting facilities [33] while textile sorting centers need to improve their performance [29]. Therefore, the current textile waste management systems in Europe seem to be overwhelmed by the pace of cheap textile product consumption [37].

3 Circular Economy in Post-Industrial and Pre-consumer Stage

The circular design is crucial in textile industries to raise the circularity of business levels and to increase the reuse, repair and re-manufacture, and recyclability of garments, shoes, and accessories [46]. Compared to post-consumer waste these post-industrial materials and pre-consumer products are clean feedstocks that could be reintegrated in the production stage in the same loop or to provide resources for other related textile industries. Pre-consumer products (unsold items, overstock products, minor defects) could be resold, redistributed to outlet markets, rental businesses, or donated to charitable organizations that take care of vulnerable communities and individuals. The secondary market development in this stage depends on networks and collaboration between diverse actors of fashion (business-to-business flows or B2B) and overall textile industries particularly for smaller and midsized enterprises (SMEs). In the latter case, the local and regional network sharing secondary materials is crucial and it has benefits to catalyze innovation and circular economy activities in the same geographies and reducing carbon footprint related to transportation. For example in Italy, 50% of textile waste is collected from textile industries, 30% from urban sorting activities and the rest of 20% from other activities while combining traditions and innovation favor circularity of wool in the Prato region [8]. Furthermore, the SMEs are regional drivers in supporting the slow fashion model as a sustainable alternative compared to the linear and consumerist model promoted by conventional and multi-national fashion businesses. Mono-material textile products are a sustainable pathway to encourage the development of closed-loop systems such as "t-shirt to t-shirt" or "jean to jean" models. Protein engineering could provide the tools to extract from the agro-industrial waste stream the raw materials to produce bio-based textile fabrics [43] and to find natural-based solutions to dyed fabrics compared to chemical-based processes. Modern recycling plants allow various textile wastes (cotton, polyester, cotton-polyester, polyester-cotton, wool, polypropylene) to breakdown to fiber level producing recycled fiber for the business sector [40].

4 Circular Economy in the Post-consumer Stage

Post-consumer textile waste is a demanding issue around the world for both businesses and municipalities. EPR policies should press the role of fast fashion industries to re-capture their post-consumer textile waste and finance reuse-repair-remanufacture and fiber-to-fiber recycling networks under the "polluter pays principle" in line with municipalities' efforts to develop source-separation schemes to feed circular activities and divert textile waste from landfills and incineration plants. The post-consumer textile wastes feed the second-hand market (clothes, shoes, accessories) while the low pricing of this sector could stimulate the rebound effects by increasing overall textile products consumption [7], but socio-demographic features could influence the disposal behavior regarding textile waste [30]. The Dutch textile industry is found to be vulnerable to rebound effects but mitigation strategies are researched through stakeholders' participation to increase the positive benefits of a circular textile industry [42]. The trades of textile waste and second-hand products feed cheap textile waste with the risk of flooding local waste management systems without sufficient textile recycling capabilities other than energy recovery solutions in incinerators. These environmental impacts are more severe in developing countries where open dumping and burning of general waste is the norm [28]. On this background, the improvement of textile waste management practices in line with CE principles is essential to decouple the current overconsumption patterns and diversion from landfills and waste to energy facilities. Strong EPR policies, tax on the fast-fashion model, and investments in advanced sorting and recycling technologies are necessary steps to tackle linear economy prevalence in textile industries.

Through cellulose carbamate technology the post-consumer textile waste with at least 88% cotton can be broken down to a molecular level after cleaning and fiber separation process resulting in Infinna fiber that mimics cotton properties and re-entered into the textile economy as raw material in clothes, home textiles, wipes, and diapers avoiding extraction of virgin materials and decreasing land and water footprints with benefits for biodiversity [14]. Furthermore, the collaboration between manufacturers, recyclers, and fashion brands is important to incorporate the highly fragmented textile waste market [44] as well as tracking textile products between regions and countries including second-hand markets through illicit channels [26] to avoid the linear economy trap as revealed in Fig. 1. Advanced sorting technologies could improve the recycling performances from "textile to textile" or "fiber to fiber" while mono-material design improves the repair, reuse, and remanufacture services assuring higher circularity levels. These fiber-to-fiber recycling technologies are required to be installed across Europe to respond to the increasing pressure of textile waste flows [22] (Fig. 2).

The biocentric vision is strongly related to source material through better agricultural practices (organic cotton, agroforestry, regenerative agriculture) in partnership

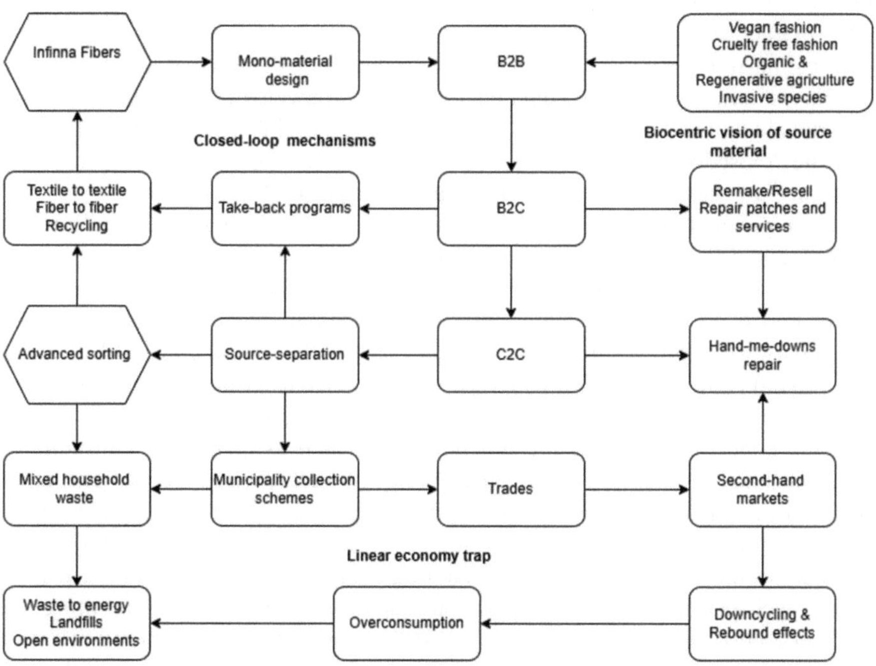

Fig. 2 Circular economy challenges in textile industries between business to business (B2B), business to consumers (B2C), and consumers to consumers (C2C) to avoid the linear economy trap

with local farmers [45] addressing invasive species issues of local ecosystems, material innovation with technology support as alternatives to plastics or animal-based materials.

CE Practices in the Clothes Sector

Take-back programs development of fast-fashion business should be mandatory as a key EPR responsibility while providing traceability for their end-of-life management. The "business to consumers" initiatives such as "repair" schemes should be facilitated to extend the product life and keep the clothes in use which could be a "win to win" solution. The closed-loop mechanisms "textile to textile " and "fabric to fabric" are desirable for clothes compared to downcycling practices (clothes to household textiles, carpets, etc.). However, fast-fashion industries rely on mixing fibers (natural based and synthetic) which is detrimental to sorting and mechanical recycling technologies [49]. Thus, the improvement of sorting technologies using infrared spectroscopy to sort with 100% accuracy the textile fibers at high speed could be a game-changer in improving recycling levels of mixed end-life products in textile industries [36]. The commercial-scale factory of Infinna fiber (30,000 t/yr) will provide sustainable alternative feedstock for fashion brands compared to virgin cotton extraction while licensing technology to reconvert pulp or viscose factory in other regions into cellulose carbamate fiber production will help to scale this circular economy solution and to shift the demand towards recycled fibers with cotton like properties [14]. Furthermore, new circular designs based on mono-material fabrics favor the "closed-loop systems" to be more efficient in the clothes sector and also for repair purposes increasing circularity ambitions as outlined in Table 1.

Repair garments is another key insight to extend the clothes' lifespans through innovative and flexible patches while cotton fibers should be procured via organic agriculture. Furthermore, some brands look to higher standards such as regenerative organic certification where cotton cultivation has lesser impacts to soil, water and biodiversity while providing better traceability along the supply chain [13]. Despite second-hand cloth benefits from a CE perspective such as prolonging the life of pre and post-consumer textile products through this global economic market [15] there are some concerns regarding the rebound effect such as increasing consumption of second-hand clothes [7] while poor quality of such products could end up in open environments in regions with waste management dysfunctionalities in terms of source-separation and logistics or exported to less developed countries. Improvement of source-separation of textile waste including clothes is mandatory in the EU from 2025. A well-performed source-separation scheme could reduce the landfilling rate to 1% in a best-case scenario in Europe [4], but some stakeholder draws attention to urgent actions [37]. In Bulgaria, 2900 tonnes of textile waste is collected through 49 municipalities based on voluntary efforts without financial support supervised by the Bulgarian Circular Textile Association [39]. Voluntary-based schemes need to be replaced with mandatory obligations under EPRs policies in many geographies

Table 1 Circular economy practices in the clothes sector

Name of organization	CE action	Key features
MUD Jeans	Jeans leasing model and "jean to jean" recycling	Jeans crafted from 40% post-consumer recycled cotton from old jeans from MUD Jeans
FabPacth	Flexible and customized repair solution	High-quality textile repair patches free of harmful substances or solvents
Infinitedfiber	Infinna™ textile fibre	High-quality fiber as a circular alternative to conventional fibers (cotton, viscose, and polyester)
Colorfix	Biomimetic solution and colour engineering	Replicating the natural coloration processes in textile dye machines with a low environmental footprint
Lindstrom	Workwear—target 100% textile recycling in 2025	HoReCa workwear from recycled materials; Winter workwear made out of plastic bottles;
Trouve	Branded resale programs, donations, sell-off, and downcycling	Rework/Remake of unselled items—ReCrafted by Patagonia and ReSewn by Eileen Fisher
Teemill	Open-access circular supply chain	Products are printed to order and made from organic cotton using renewable energy,

Source of data Sitra and organizations websites

to divert textile waste from the mixed household waste stream and prepare for reuse, repair, remanufacturing, and recycling activities that enable closed-loop systems rather than downcycling practices.

5 CE Practices in the Shoe Sector

Footwear products have a mix of elements that are difficult to process and close the loop. However, machineries are used to break the textiles into fibers that can be reused to produce recycled fabrics. The footwear business coalition to support the transition towards the CE system is important and signals the urgent need for business transformation. For example, 25 billion pairs of shoes are manufactured each year which includes 3 billion pairs of flip-flops that end up in landfills or the environment by such coalition of footwear brands [5]. Furthermore, this coalition argues that despite the high-value market the athletic footwear this sector spent close to zero on circularity initiatives [5]. For example, repair services are already put into practice by some businesses under the EU Ecolabel recommendations and Kavat succeeded in having certified 50 models in the footwear industry [9].

The consumption of shoes is 70–100 million pairs a year, or 4–6 pairs per inhabitant in the Netherlands compared to 2.5 pairs at the global level, while 10% of newly

Table 2 Bio-based materials that support circular footwear transition beside other applications

Bio-based material	CE features	Impact
PLIANT™	Alternative to synthetic rubber to support the circular footwear	Scalable with standard compression molding
TUNERA™	First-in-kind 100% biobased foam	Eliminating synthetics like ethyl vinyl acetate or polyurethane from performance foam
MIRUM®	Plastic-free alternative to leather	Diverse applications and industries
CLARUS®	Enhances virgin fibers and revitalizes recycled fibers	Create knits, wind-resistant wovens, and durable cotton canvas

Source of data https://nfw.earth/materials

produced shoes end up in incineration or recycling facilities, but key instruments under EPR policies are necessary such as standard requirements of shoes at European level and preparing for reuse, repair and recycling solutions [34]. Mechanical recycling is used to produce recycled upperparts of footwear, but this process has limitations through shredding and grinding stages to keep such materials in several recycling cycles. Some companies have started to collaborate to develop chemical footwear recycling as an alternative breaking down the upper textiles into molecules to rebuild them into high-quality lyocell and polyester prone to many recycling cycles and replacing the use of virgin materials [41]. Natural Fiber Welding supports a holistic vision in sustainable textile fabrics by replacing single-use and non-recyclable plastic with high-performance bio-based materials as shown in Table 2, while supporting regenerative agroforestry and local farmers in partnership with Terra Genesis organization [12].

Transparency and traceability are important policy measures to track the progress from the linear fashion model towards high circular ambitions while combating the threats related to counterfeits and illegal trades that could feed pollution in less developed regions [50]. A collaborative network is established to catalyze business efforts in adopting circular economy solutions in footwear industries along the supply chain and to expand the consumer's preferences towards sustainable footwear alternatives compared to the linear prevalent business-as-usual model [17].

CE Practices in Fashion Related Accessories Manufacturing

The biocentric vision requires a holistic analysis of ecosystems. In this regard, invasive species are a threat to indigenous biodiversity in diverse ecosystems. To tackle this issue innovative business model uses invasive species as a natural fiber source to produce leather-based material for textile accessories [19]. For example, invasive species in the US such as python (Everglades), non-native carp (Mississippi River Basin),and non-native lionfish (coral reefs in the Caribbean) are primary leather sources providing local jobs (hunting/fishing) and environmental benefits for rural communities while environmental footprint is significantly lower compared to conventional bovine leather [21]. An experimental study revealed that coffee grounds in combination with corn, starch, glycerine, and bio-epoxy resin are used to manufacture buttons and this application could be further developed for textile accessories such as buckles, broochs, tie pins, and clamps [2]. However, creative recycling could be spread around communities through workshops to produce new accessories based on pre-consumer textile waste [10] or post-consumer textile waste but these should be properly sorted and cleaned before processing.

CE of Textile Waste in Non-fashion Sectors (Household Textiles, Home Decor, Furniture)

Fast-fashion models feed large quantities of textile waste while close-looped systems are limited. The household textiles and home décor provide creative recycling solutions on individual, community, and business scales. Textile waste is used to provide household textiles and home décor products through a community-led business in Indonesia with training activities and raising awareness of climate change issues [3]. In Bangladesh, local entrepreneurs are using natural materials such as pineapple skins, and banana leaves to produce handcraft products such as floor covering (rugs), storage (baskets), and tableware [44]. Such initiatives provide local skills and support women's entrepreneurship while adopting e-marketing as a strategic solution to reach distant consumers. Business transition towards CE activities in textile industries could raise opportunities at local levels. For example, a family-owned business shifts the activity from the maintenance of parquet, floors, and furniture to the use of post-consumer textile waste in producing home décor, wall panels, chair seats, or backrests in Finland [24]. Textile waste and recycled fiber are used to produce yoga products such as yoga mats and bolsters by other businesses in Finland [23]. In Sweden, a variety of bags are made from 100% reused hotel textiles (e.g., bedsheets) claiming to reduce 90% of CO_2 emissions compared to virgin cotton bags [38]. Furniture design based on textile waste-starch composite is an innovative and alternative application for textile waste [48]. Furthermore, the bio-based materials (Mirum, Clarus) mentioned in Table 2 have cross-sectoral applications in providing sustainable alternative materials including for automotive industries. Testing and improving other applications for textile waste or composite materials including textiles by researchers and business actors could lead to promising alternatives and enable CE activities outside apparel industries in diverting the textile industries from linear mechanisms.

6 Conclusion

This works point out the principles of CE in textile industries while coping with deficient recycling infrastructures and emerging waste generation trends fueled by fast fashion industries. From a system perspective, the current context of textile waste management is not favorable for a transition towards a widespread functional circular economy. However, some circular economy practices are revealed for key sectors such as clothes, footwear, fashion, fashion-related accessories besides other applications that integrate pre and post-consumer textiles through business and community actions (home décor, household textiles, furniture, automative industries). The biocentric vision of the circular economy is argued by supporting innovative business models that reduce environmental footprint along the supply chain, support organic and regenerative agriculture, biobased material alternatives to conventional cotton and plastics, use invasive species as sources for leather material and adopt stricter policies regarding animal origin source of raw materials for fashion related industries such as "cruelty-free fashion" and "vegan fashion" including certifications and standards.

References

1. Armbruster, R. (2024). German waste textiles market still deep in crisis. https://www.euwid-recycling.com/news/markets/german-waste-textiles-market-still-deep-in-crisis-060824/
2. Avcioğlu, K. N. (2022). Designing textile accessories from coffee ground. *Industria Textile, 73*(3), 282–287. https://doi.org/10.35530/IT.073.03.202049
3. Alimin, N. N., Murni, E. S., Cahyani, D., Aini, N., & Mulyono, A. (2022). Fabric scrap interior accessories: A solution to textile waste in Gulon Asri waste bank community in Solo, Indonesia. *IOP Conference Series: Earth and Environmental Science, 1114*(1), 012092. https://doi.org/10.1088/1755-1315/1114/1/012092
4. Amicarelli, V., & Bux, C. (2022). Quantifying textile streams and recycling prospects in Europe by material flow analysis. *Environmental Impact Assessment Review, 97*, Article 106878.
5. Circular footwear Initiatives. Responding to a global emergency. https://circularfootwearinitiative.com/
6. Changing Markets Foundation. (2024). Fashion's plastic paralysis. How brands resist change and fuel microplastic pollution.
7. Ciechelska, A., Matel, A., Poskrobko, T., & Sidorczuk-Pietraszko, E. (2024). Circular economy rebound effect in the context of second-hand clothing consumption. *Economics and Environment*, *87*(4), 635, 1–20. https://doi.org/10.34659/eis.2023.87.4.635
8. Dress-code. (2024). Tackling textile waste in Europe—Germany, France and Italy under the loop. https://dress-ecode.com/en/2024/09/23/tackling-textile-waste-in-europe-germany-france-and-italy-under-the-loop/
9. Directore-General Environment Circular economy and sustainable footwear—Kavat's story. https://environment.ec.europa.eu/topics/circular-economy/reset-trend/get-inspired-stories-change/circular-economy-and-sustainable-footwear-kavats-story_en
10. Dhivya, R., & Subathra, B. (2023). Upcycling the pre-consumer textile waste into inventive accessories. In S. S. Muthu (Eds.), *Novel sustainable alternative approaches for the textiles and fashion industry. Sustainable textiles: Production, processing, manufacturing & chemistry.* Springer. https://doi.org/10.1007/978-3-031-37060-1_10

11. European Parliament. (2024). The impact of textile production and waste on the environment (infographics). https://www.europarl.europa.eu/topics/en/article/20201208STO93327/the-impact-of-textile-production-and-waste-on-the-environment-infographics
12. Earthshotprize. (2024). Natural Fiber Welding's remarkable natural-based circular materials replace plastic in anything from cars to clothing, while reducing carbon emissions and supporting regenerative farmers. https://earthshotprize.org/winners-finalists/natural-fiber-welding/
13. Eillen Fisher. (2024). Growing better all the time: Our regenerative organic certified® Cotton Av. https://www.eileenfisher.com/a-sustainable-life/journal/sustainability/regenerative-organic-certified-cotton.html
14. Infinitedfiber. (n.d.) https://infinitedfiber.com/our-technology/
15. Hur, E. (2020). Rebirth fashion: Second hand clothing consumption values and perceived risks. *Journal of Cleaner Production, 273*, Article 122951.
16. Fashion United. (2024). Transformative change won't come to the textile sector overnight—Here's why we need to accept intermediary solutions. https://fashionunited.com/news/business/transformative-change-wont-come-to-the-textile-sector-overnight-heres-why-we-need-to-accept-intermediary-solutions/2024102462539
17. Footwear Collective. (2023). The Footwear Collective is launched: To catalyze industry's circularity transformation. https://www.prlog.org/12990881-the-footwear-collective-is-launched-to-catalyze-industrys-circularity-transformation.html:
18. Greenpeace Africa. (2024). Fast fashion slow poison. The toxic textile waste in Ghana.
19. Gualandris, J., Branzei, O., Wilhelm, M., Lazzarini, S., Linnenluecke, M., Hamann, R., Dooley, K. J., Barnett, M. L., & Chen, C.-M. (2024). Unchaining supply chains: Transformative leaps toward regenerating social–ecological systems. *Journal of Supply Chain Management, 60*(1), 53–67. https://doi.org/10.1111/jscm.12314
20. Green Forum. (2024). Romania reuses and recycles up to 10% of textile waste. https://www.green-forum.eu/environment/20240607/romania-reuses-and-recycles-up-to-10-of-textile-waste-1165
21. Inversaleathers. Nurturing ecosystems and economies for immediate and measurable impact. https://www.inversaleathers.com/impact
22. Köhler, A., Watson, D., Trzepacz, S., Löw, C., Liu, R., Danneck, J., Konstantas, A., Donatello, S., & Faraca, G. (2021). *Circular economy perspectives in the EU textile sector*. Publications Office of the European Union.
23. LSJH. (2022). Globe Hope makes yoga products from end-of-life textile and recycled fibre. https://poistotekstiili.lsjh.fi/en/2022/03/17/globe-hope-makes-yoga-products-from-end-of-life-textile-and-recycled-fibre/
24. LSJH. (2021). Nordic Upstream utilises end-of-life textiles in furniture and wall panels. https://poistotekstiili.lsjh.fi/en/2021/11/05/nordic-upstream-utilises-end-of-life-textiles-in-furniture-and-wall-panels/
25. Lamarche-Beauchesne, R. (2023). Reviving the silenced; defining vegan fashion and classifying materials of animal origin. *Fashion Theory, 27*(7), 937–956.
26. Mihai, F. C., & Ulman, S. R. (2024). Plastic waste trade issues and environmental contamination in Romania. In *Plastic waste trade: A new colonialist means of pollution transfer* (pp. 191–216). Springer Nature Switzerland.
27. Mihai, F.-C., Meidiana, C., Elagroudy, S., Ulman, S.-R., Gutberlet, J., & Carvalho, C. (2024). Plastic waste management for zero waste to landfills: Potential, challenges, and opportunities. In *Zero waste management technologies* (pp. 97–136). Springer Nature Switzerland. https://doi.org/10.1007/978-3-031-57275-3_6
28. Mihai, F.-C., Pop, V., Ulman, S,-R., & Ozunu, A. (2025). Policy Implementation and Management of Microplastics in Freshwater and Soil. In*Handbook of Microplastic Pollution in the Environment* (pp. 367–397). CRC Press.https://doi.org/10.1201/97810003487593-14
29. Morell-Delgado, G., Peiró, L. T., & Toboso-Chavero, S. (2024). Revealing the management of municipal textile waste and citizen practices: The case of Catalonia. *Science of the total environment*907, 168093.

30. Napolano, L., Foschi, J., Caldeira, C., Huygens, D., & Sala, S. (2025). Understanding textile value chains: Dynamic probabilistic material flow analysis of textile in the European Union. *Resources, Conservation and Recycling, 212*, Article 107888.
31. Nørup, N., Pihl, K., Damgaard, A., & Scheutz, C. (2019). Evaluation of a European textile sorting centre: Material flow analysis and life cycle inventory. *Resources, Conservation and Recycling, 143*, 310–319.
32. Nencková, L., Pecáková, I., &Šauer, P. (2020). Disposal behaviour of Czech consumers towards textile products. *Waste Management, 106*, 71–76. https://doi.org/10.1016/j.wasman.2020.03.001
33. Pavani, C., Rao, P. A., Vishnu, P., Raja, H., Sriram, & Sirisha, N. (2024). Vegan leather from agricultural waste: Exploring sustainable and cruelty-free alternatives. In *From waste to wealth* (pp. 951–964). Springer Nature Singapore.
34. Pandey, R., Pandit, P., Pandey, S., & Mishra, S. (2020). Solutions for sustainable fashion and textile industry. In *Recycling from waste in fashion and textiles: A sustainable and circular economic approach* (pp. 33–72).
35. Picvisa. (n.d.) Textile waste management: The North European leadership. https://picvisa.com/textile-recycling-europe/
36. Rebelgroup (n.d.) Report: Steps towards a circular shoe chain. https://rebelgroup.com/en/projects/steps-towards-a-circular-shoe-chain/
37. Rehman, M., Petrillo, A., Ortíz-Barrios, M., Forcina, A., Baffo, I., & De Felice, F. (2024). Sustainable fashion: Mapping waste streams and life cycle management. *Journal of Cleaner Production, 444*, Article 141279.
38. Riba, J. R., Cantero, R., Canals, T., & Puig, R. (2020). Circular economy of post-consumer textile waste: Classification through infrared spectroscopy. *Journal of Cleaner Production, 272*, Article 123011.
39. Rreuse. Open letter: Textile emergency action plan needed quickly. https://rreuse.org/open-letter-textile-emergency-action-plan-needed-quickly/
40. Reusedremade. Someone else's trash is our treasure. https://reusedremade.com/pages/lifecycle
41. RPRA—Resource Productivity & Recovery Authority. (2024). Bulgarian circular textile association collects 2900 tonnes of textile waste annually. https://rpra.ca/the-hub/bulgarian-circular-textile-association-collects-2900-tonnes-of-textile-waste-annually/
42. Rester. Helping the world recover. Fiber by fiber. https://rester.fi/en/
43. Safaya, S. (2023). Vivobarefoot, Circ partner for circular footwear. https://www.just-style.com/news/vivobarefoot-circ-partner-for-circular-footwear/
44. Siderius, T., & Poldner, K. (2021). Reconsidering the circular economy rebound effect: Propositions from a case study of the Dutch circular textile valley. *Journal of Cleaner Production, 293*, 125996. https://doi.org/10.1016/j.jclepro.2021.125996
45. Schiros, T. N., Mosher, C. Z., Zhu, Y., Bina, T., Gomez, V., Lee, C. L., Lu, H. H., & Obermeyer, A. C. (2021). Bioengineering textiles across scales for a sustainable circular economy. *Chem, 7*(11), 2913–2926. https://doi.org/10.1016/j.chempr.2021.10.012
46. Switch. (2024). Assessment of manufacturers' capacity for the circular economy in Morocco and Bangladesh. https://www.switchtocircular.eu/publication/assessment-manufacturers-capacity-circular-economy-morocco-and-bangladesh?back
47. Stella McCartney. (2023). Impact report. https://www.stellamccartney.com/on/demandware.static/-/Library-Sites-StellaMcCartneySharedLibrary/default/dwa4ea4b5c/report/StellaMcCartney_ImpactReport_2023_4.pdf
48. Tanveer, H. (2023). A beginner's guide to circular textiles. The textile think tank impactful ideas & actionable insights. https://thetextilethinktank.org/a-beginners-guide-to-circular-textiles/
49. UNECE, (2024). Traceability for Sustainable Garment and Footwear. https://unece.org/trade/traceability-sustainable-garment-and-footwear
50. UNEP. (2024). Global waste management outlook. https://wedocs.unep.org/bitstream/handle/20.500.11822/44939/global_waste_management_outlook_2024.pdf?sequence=3

51. Wang, Y., Liu, C., Zhang, X., & Zeng, S. (2023). Research on sustainable furniture design based on waste textiles recycling. *Sustainability, 15*(4), 3601.
52. Wang, S., & Salmon, S. (2022). Progress toward circularity of polyester and cotton textiles. *Sustainable Chemistry, 3*(3), 376–403.

Biotransformation of Textile Dyes: Mechanism and Application

Sayam and Nayan Das

1 Introduction

The textile industry has been essential to the advancement of human civilization for millennia but uses enormous amounts of water and chemicals for dyeing and finishing. Toxic elements like Fe, Cd, Hg, Cr, Mn, Cu, Ni, Mg, and Co are often introduced by these processes. These elements build up in the biosphere, reaching levels that are hazardous and upsetting natural ecosystems. For example, synthetic dye-containing textile effluents are dumped straight into natural water bodies, seriously contaminating the soil and water. These effluents negatively impact water quality, as demonstrated by elevated biological oxygen demand (BOD), chemical oxygen demand (COD), dissolved oxygen (DO) levels, color, pH, and the presence of persistent synthetic chemicals, including azo dyes and heavy metals [3, 9]. For dyeing and printing, the textile industry uses a wide range of synthetic dyes and pigments. Each year, they use about 700,000 tonnes of synthetic dyes, such as azo, reactive, disperse, and vat dyes. Unfortunately, many of these dyes end up in bodies of water without proper treatment [80, 88]. Synthetic dyes are mainly organic compounds obtained from petrochemical sources. These dyes, which are either water-soluble or

Present Address:
Sayam (✉)
Department of Organic Materials Science, Graduate School of Organic Materials Science, Yamagata University, Yonezawa, Yamagata 992-8510, Japan
e-mail: mahbub.sayamm@gmail.com

Department of Fabric Engineering, Barishal Textile Engineering College, Barishal 8200, Bangladesh

N. Das
Department of Wet Process Engineering, Shahid Abdur Rab Serniabat Textile Engineering College, Barishal 8200, Bangladesh
e-mail: nayandas.wpe@gmail.com

R. A. Bhat et al. (eds.), *Biocentric Approaches in Textile Waste Management*, Textile Science and Clothing Technology, https://doi.org/10.1007/978-3-032-04974-2_5

insoluble, absorb quickly and give surfaces vibrant colors [36]. The chemical structure of dyes includes a chromophore, the active site responsible for absorbing visible light; an auxochrome that alters the properties of the chromophore and increases the dye's affinity for fibres; as well as conjugated aromatic structures like perylene, anthracene, and benzene [12, 72]. Moreover, these dyes contain chemically diverse groups, such as phthalocyanine structures, anthraquinone, triphenylmethane, and azo (–N=N–), which enhance their stability, brightness, and environmental resilience. Synthetic dyes containing the azo group (e.g., diazo, monoazo, polyazo, and triazo) are the predominant category of organic colourants, making up 60–70% of the total synthetic dyes utilized [24, 76, 85]. Table 1 presents the primary categories of synthetic dyes used in the textile sector, including examples, chemical groupings, application techniques, and related environmental issues.

Table 1 Common synthetic dyes, their application methods, and environmental challenge [86]

Dye type	Examples	Chemical group	Method of application	Key environmental challenge
Azo dyes	Congo Red, Methyl Orange, Direct Red 28	Azo(–N=N–)	Direct application on cellulose	Release of toxic aromatic amines during breakdown
Reactive dyes	Reactive Black 5, Procion Red H-E3B, Remazol Brilliant Blue	Halogenated aromatics	Exhaustor pad-dry-cure on cotton	High solubility and hydrolyzed dye loss in effluents
Disperse dyes	Disperse Yellow 3, Disperse Blue 56, Disperse Red 1	Nitroaromatics	High-temperature dyeing on polyester	Accumulation in sediments and low biodegradability
Vat dyes	Indigo, Vat Green 1,Vat Orange7	Polycyclic aromatics	Reduction and dyeing on cellulose	Requires strong reducing agents; insoluble residues
Acid dyes	Acid Black 1, Acid Orange 7, Acid Blue 93	Sulfonated aromatics	Exhaust on wool, silk, and nylon	Persistent coloration in acidic effluents
Basic dyes	Basic Green 4, Methylene Blue, Basic Violet 3	Cationic compounds	Exhaust on acrylic and paper	High toxicity and intense coloration
Anthraquinone dyes	Alizarin, Reactive Blue 19, Solvent Blue 59	Anthraquinone derivatives	Pad-dry-cure or exhaust	Slow degradation due to aromatic stability
Sulfur dyes	Sulfur Black 1, Sulfur Blue 15	Polysulfide linkages	Reduction and dyeing on cotton	Generation of sulfides during dyeing

Physical, biological, and chemical processes are all common ways to treat dye effluents. Each has its own pros and cons. Complex dye structures can be effectively broken down by chemical techniques like ozonation and advanced oxidation processes (AOPs), but they can also produce hazardous byproducts and require careful handling of chemical waste [39, 92]. Despite being eco-friendly, biological treatments have trouble degrading synthetic colors due to their complexity, which might inhibit microbial activity [18, 39, 61]. In response, biodegradation has become a viable and eco-friendly method for removing harmful materials, especially dyes, from textile wastewater. Microorganisms can effectively decompose synthetic dyes into harmless substances. For instance, Pseudomonas aeruginosa has shown that it can remove over 80% of its color within five days [64, 65].

Advanced oxidation processes (AOPs) integrated with anaerobic or aerobic treatments have exhibited significant removal efficiency for dyes such as Indigo Blue, Disperse Blue, and Reactive Red, attaining COD removal rates of 96.25%, 96.35%, and 96.73%, respectively [96]. Furthermore, the integration of micro-electrolysis with biological techniques has resulted in an impressive 98.90% removal of COD from refractory dye wastewater [20]. Also, the Fenton method can achieve COD elimination effectiveness of 96.6% under ideal conditions [44]. Moreover, synthesized nano ferrates have demonstrated over 96% elimination effectiveness for diverse dyes, emphasizing their fast oxidation capabilities [93]. Lastly, the treatment of fur dyeing wastewater through hydrolysis acidification combined with sequencing batch reactor (SBR) processes resulted in a COD removal efficiency of 93.75% [99].

Enzymatic biodegradation offers a very effective alternative for the treatment of textile wastewater due to its environmentally friendly characteristics and efficacy in pollutant elimination. Numerous studies have shown the efficacy of enzymes, including laccase, protease, and lipase, in decomposing synthetic dyes and diminishing COD in wastewater. For instance, bacterial isolates that produce laccase were very good at decolorizing, which led to non-toxic effluents [84]. Furthermore, eco-enzymes obtained from fruit peels have demonstrated a notable reduction in BOD and total suspended solids (TSS), exceeding traditional methods [23]. The integration of biosorption with enzymatic action, specifically through the use of immobilized laccases on diatomaceous earth, significantly improves pollutant removal, resulting in a 66.2% reduction in COD [71]. Additionally, enzyme engineering has enhanced biocatalysts for the more efficient treatment of intricate textile effluents [35]. The pursuit of economically viable and environmentally sustainable alternatives of textile dye effluent treatment are an important area of study, where biodegradation shines as a bright beacon of hope. This chapter highlights the role of enzymes, such as peroxidases and laccases, that can successfully lower pollutant levels. Therefore, supporting their use in industrial textile wastewater treatment.

2 Mechanisms of Enzymatic Biodegradation

Recently, a number of studies have used individual enzymes to look into how dyes break down [53, 69]. Analytical methods such as NMR, GC–MS, HPLC, and FTIR have been utilized to evaluate the breakdown routes from cultural extracts. The enzyme, organism, and cultural environment (anaerobic or aerobic) all impact the dye degradation process. Redox-active substances, such as reduced flavins and hydroquinone, react quickly with dyes, particularly azo dyes, in anaerobic environments [79]. Extracellular enzymes carry out the reduction since some dyes are unable to cross microbial cell membranes due to their high molecular weight and sulfonate groups. In aerobic settings, oxygen competes against dye molecules for the reduced electron carriers [41, 87]. Figure 1 depicts the suggested method for azo dye degradation by azoreductase [2]; however, no single, conclusive degradation pathway has been found.

A novel idea from Goszczynski et al. [30] was to use peroxidase, manganese peroxidase, and ligninase to help fungi break down the azo dyes 3,5-dimethyl-4-hydroxy-azobenzene-4′-sulphonic acid (I) and 3-methoxy-4-hydroxy-azobenzene-4′-sulphonamide (II). Peroxidase was obtained from Streptomyces chromofuscus, whereas manganese ligninase and peroxidase were derived from *Phanerochaete chrysosporium.* Twelve unstable intermediates were recognized, which transformed

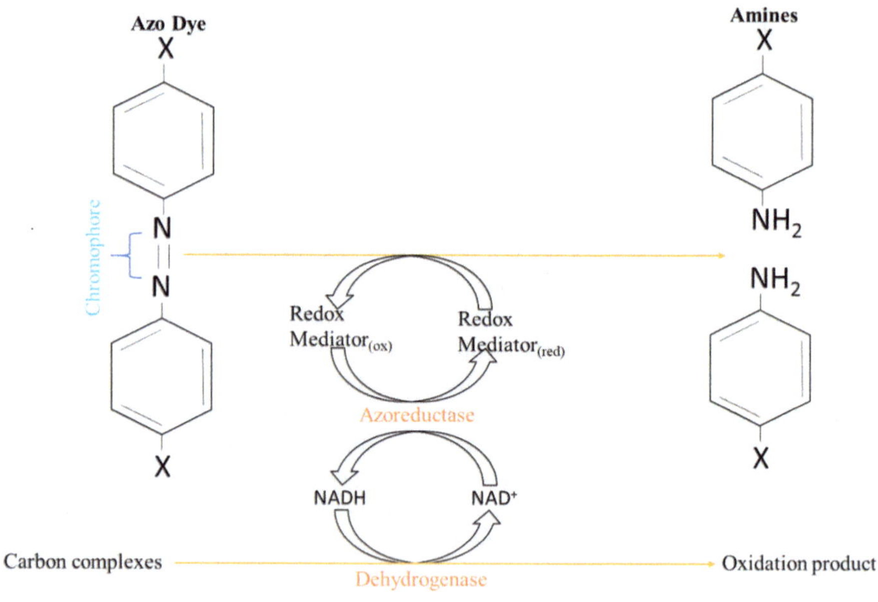

Fig. 1 Proposed mechanism for degradation of azo dyes by azoreductase [2]

into 11 organic compounds and ammonia by hydrolytic, oxidation, and redox reactions. Hydrolytic cleavage took place through two distinct mechanisms: (1) the formation of diazene and quinone derivatives in an asymmetrical manner and (2) the formation of nitroso and quinonemonoimine derivatives in a symmetric manner. López et al. [47] investigated the enzymatic degradation of Orange II using NMR and ESI-ion trap mass spectrometry to detect evolution products in the culture extract of manganese peroxidase from *Bjerkandera* sp. BOS55. They suggested three dye degradation pathways based on cleavage products: (1) symmetrical azo bond cleavage yielding 4-aminobenzenesulphonate; (2) asymmetrical naphthalene cleavage producing 1,2-naphthoquinone and 4-diazobenzenesulphonate; and (3) the formation of 2-naphthol and 4-hydroxybenzenesulphonate. Wong and Yu [91] demonstrated that oxidative enzymes, such as laccase, oxidize dye molecules into cationic radicals, which are reactive, unstable, and capable of oxidizing other non-substrate dye molecules into non-toxic compounds, accentuating the enzyme's function in dye degradation. Husain [34] indicated that specific recalcitrant dyes necessitate low molecular weight redox mediators to facilitate enzymatic degradation by improving efficiency. These mediators move electrons from biological sources to dye compounds, which speeds up the breakdown process. Heinfling et al. [33] studied the transformation of phthalocyanine and azo dyes using manganese lignin peroxidase and peroxidase acquired from *Bjerkandera adusta*. They discovered that veratryl alcohol raised the activity of lignin peroxidase by 8 to 100 times. Manganese peroxidase, on the other hand, removed colors from dyes without the need for Mn^{+} ions or redox mediators. The function of redox mediators in the degradation of dyes has been extensively investigated [14, 37, 51]. Šekuljica et al. [69] introduced a ping-pong process for the degradation of anthraquinone dye using horseradish peroxidase (HRP). HRP initially oxidizes H_2O_2 to form compound I along with water. Compound I subsequently undergoes reduction through electron transfer from the dye or co-substrate (DH), resulting in the formation of compound II and a radical. Compound II additionally diminishes, restoring enzymatic activity for an additional catalytic cycle. The reaction catalyzed by HRP is outlined below:

$$\mathrm{HRP} + \mathrm{H_2O_2} \rightarrow^{\mathrm{KI}} \text{Compound I} + \mathrm{H_2O} \tag{(i)}$$

$$\text{Compound I} + \mathrm{DH} \rightarrow^{\mathrm{K_2}} \text{Compound II} + \text{Radical} \tag{(ii)}$$

$$\text{Compound II} + \mathrm{DH} \rightarrow^{\mathrm{K_3}} \text{Radical} + \mathrm{HRP} + \mathrm{H_2O} \tag{(iii)}$$

Gomare et al. [29] suggested the enzymatic breakdown process of Golden Yellow HER via NADH–DCIP reductase, bacterial lignin peroxidase, MG reductase, and aminopyrine N-demethylase extracted from *Brevibacillus laterosporus* MTCC 2298, using GC–MS and FTIR spectral analysis. First, the researchers discovered that aminopyrine N-demethylase helped demethylate the parent dye molecules, and then reductases helped reduce the intermediate molecule. The molecule that formed, 4-(3-amino-2-hydroxy-cyclopentylamino)-benzenesulphonic-2-amino-cyclopentanol, was turned into minerals by the bacteria, yielding a safe product. A study similar to this one was done by [56], who used GC–MS, FTIR, and HNMR to figure out how the dye Reactive Red-120 breaks down. The azoreductase enzyme was demonstrated to participate in the breaking of the N=N bond, leading to the synthesis of aromatic amines via intermediate reduction and oxygenation reactions. Additionally, azoreductase enzymes facilitate dye degradation utilizing $FADH_2$, NADH, and NADPH as electron donors, which participate in the cleavage of azo bonds to yield corresponding amines [77]. Researchers Baldev et al. [8] found that the oxidative cleavage of rhodamine B by algal peroxidase enzymes led to the formation of smaller organic molecules. GC–MS and FTIR analyses of control and treated dye solutions indicated the formation of 12 distinct non-toxic molecules, which were subsequently transformed into mineralized products H_2O and CO_2. In a separate research, Khataee et al. [42] showed the degradation mechanism of Acid Blue-92 dye through enzymes, including peroxidase, superoxide dismutase, and catalase, which were extracted from Lemna minor. The authors said that higher amounts of dye caused the plant to make more reactive oxygen species. The antioxidant enzyme subsequently neutralized these substances to safeguard the plant during the initial phase of the process. Subsequently, the dye was cleaved and converted into a non-toxic compound through oxidation and reduction reactions mediated by enzymes.

3 Key Enzymes in Dye Degradation

Laccases

Laccase enzymes, which are multi-copper phenol oxidases, are preferred for the degradation of azo dyes because of their non-specific free radical mechanism that generates phenolic compounds instead of toxic aromatic amines [73]. They speed up the breakdown of azo dyes by cutting the –N=N– bond in an uneven way first, then by deamination, oxidative cleavage, dihydroxylation, and desulfonation [95, 100]. Certain bacterial laccases decompose azo dyes without cleaving the azo bond, instead generating phenolic compounds through a non-specific free radical mechanism [19, 59]. The enzyme's redox potential is associated with its inability to break the azo bond. Bacterial laccases are categorized as low-redox potential enzymes [26], whereas fungal laccases are classified as high-redox potential enzymes [48]. Fungal laccases are favored for the biodegradation of azo dyes due to their ability to prevent the formation of toxic amines [100]. For example, laccases derived from fungi such as *Fusarium soloni*, *Cladosporium cladosporioides*, *T. villosa*, and *Trametes versicolor* facilitate the transfer of four electrons from O_2 to H_2O throughout their catalytic process [2]. A diazo textile dye called Congo Red is broken down by fungal laccase. It does this by first cutting the –N=N– bond unevenly. This is followed by oxidative cleavage, desulfonation, and deamination [16, 101]. Endophytic mycoremediation mainly depends on laccases; however, other enzymes, including tyrosinase, reductase, lignin peroxidase, and manganese peroxidase, also play a substantial role in dye degradation [74]. Furthermore, the laccase enzyme of the white-rot fungus *Trametes pubescens* has been optimized to decolorize and degrade Reactive Black 5 (RB5) dye using response surface methodology [63].

Indigo Dye Degradation Mechanism

Indigo dyes are organic substances characterized by their blue color, derived from flora such as *Indigofera suffructicosa*, *Polygonum tinctorium*, *Isatis tinctoria*, and *Indigofera tinctoria*. The synthesis is currently achieved through the Baeyer-Drewson reaction. Due to hydrogen bonding, indigo dyes are not soluble in alcohols, water, or ether. For solubility, a base such as sodium hydroxide is necessary. Indigo and its derivatives are widely utilized in several different industries, although they are hazardous and can irritate skin or eyes. Laccases change Indigo dyes into oxygen by taking away four electrons from the molecule one at a time. The initial phase involves electrochemical oxidation to dehydroindigo, succeeded by a nucleophilic attack (e.g., by water) to integrate oxygen atoms into degradation products [15]. The degradation process facilitated by *Trametes hirsuta* laccase is illustrated in Fig. 2 [66]. Indigo is converted to isatin, which subsequently undergoes decarboxylation to form anthranilic acid.

Ben Younes and Sayadi [11] demonstrated that laccase from *Scytalidium thermophilum* attained 98% decolorization of Indigo Carmine within 24 h. The degradation products exhibited a lower toxicity compared to Indigo Carmine when tested against *Escherichia coli* and *Bacillus megaterium*. *T. hirsuta* and γ-proteobacterium JB effectively degraded Indigo Carmine to anthranilic acid and isatin-5-sulfonic acid [75], Fungal laccase yielded a higher percentage of anthranilic acid (98.3%) compared to bacterial laccase (56.4%), as shown in Fig. 2. Additionally, the reaction rate of *T. hirsuta* fungal laccase was greater. Although both produced the same degradation products, Campos et al. [15] discovered that *T. hirsuta* laccase degraded

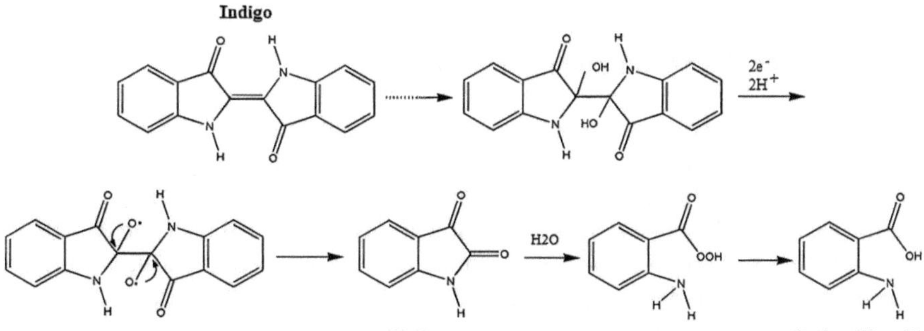

Fig. 2 Proposed mechanism for the degradation of Indigo dye by laccase [66]

the material more quickly than *Sclerotium rolfsii* laccase. This rate difference is due to the higher E_0 Cu T1 potential of fungal laccases compared to bacterial laccases [48].

Azo Dye Degradation Mechanism

Azo dyes are made up of aromatic molecules that have at least one azo bond (–N=N–). Diazonium salts are joined with phenols, amines, or naphthols to make them. When benzene or naphthalene is used instead of an azo link, functional groups like $-NO_2$, $-NH_2$, –Cl, $-CH_3$, –OH, and –CO are often added. Phenolic compounds are produced when azo dyes are broken down by enzymes through symmetric or asymmetric breakage of the azo bond through a non-specific free radical mechanism [83]. However, toxicity and mutagenicity tests are required since the procedure often produces harmful byproducts like amines [21].

Asymmetric azo bond cleavage is the first step in laccase-mediated degradation of azo dyes. This is followed by desulfonation, oxidative cleavage, demethylation, deamination, or dihydroxylation, based on the structure of the dye [1, 81, 83, 94, 100]. According to multiple studies, azo dyes degrade without bond cleavage and instead generate compounds of the phenolic type through a non-specific free radical mechanism [59]. One reason some laccases might not be able to break azo bonds could be their redox potential. For example, Pereira et al. suggested a way for *Bacillus subtilis* laccase to biotransform Sudan Orange G. This low-redox potential laccase (455 mV vs. NHE) facilitates the generation of oligomers or polymers via radical coupling while preserving the integrity of the azo bond [26]. Fungal laccases, on the other hand, are better at breaking down azo dyes because they are high-redox-potential enzymes [48]. Telke et al. [82] illustrated that laccase derived from *Aspergillus ochraceus* NCIM-1146 facilitated the degradation of Methyl Orange through the formation of carbocations, resulting in the generation of a reactive intermediate. The nucleophilic attack by –OH, $-SO_3$, or halogens led to asymmetric cleavage of the azo bond, yielding degradation products such as p-N, N′-dimethylamine phenyldiazine, and p-hydroxybenzene sulfonic acid, as illustrated in Fig. 3 [6, 82].

Anthraquinone Dye Degradation Mechanism

The anthraquinone dyes are the second most important category of dyes used in the textile industry [10]. Alizarin, the most ancient dye within this category, was initially derived from the root of Rubiatinctorum. The dyes feature a delocalized π-conjugated electron system within their core and substituents. Remazol Brilliant Blue R (RBBR) and Acid Blue 129 are two common anthraquinone dyes used in textiles. They are often used as reactive dyes in polymeric dye production, but they are known to be toxic and hard to eliminate as pollutants. Laccases demonstrate superior efficiency in the decolorization of anthraquinone dyes relative to other dye categories [97]. Zeng et al. [98] mentioned their function as redox mediators in laccase-catalyzed azo dye

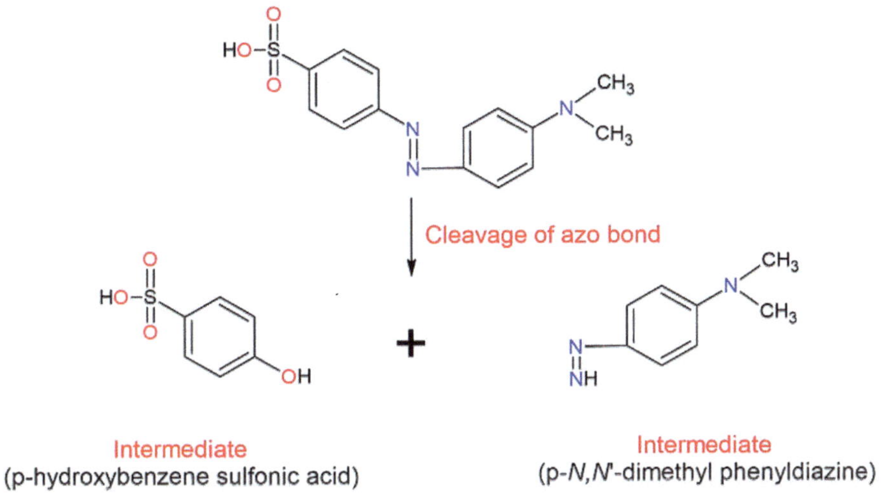

Fig. 3 Degradation pathway of the azo dye Methyl Orange by laccase [6]

decolorization. The laccase-mediated degradation process involves the breakdown of the dye chromophore into smaller molecules, which may lead to a reduction in toxicity. Osma et al. [55] looked into how *T. pubescens* laccase breaks down RBBR and came up with a reaction route that includes known intermediates and degradation products, as demonstrated in Fig. 4a. Additionally, the observed reactions comprised oxidation, deamination, hydroxylation, and reduction. Hadibarata et al. [32] demonstrated the decomposition of RBBR by laccase derived from *Polyporus* sp. S133, as illustrated in Fig. 4b. In contrast to *T. pubescens*, *Polyporus laccase* exhibited no deamination, yielding products containing amino groups. Even though Hadibarata et al. found low-molecular-weight products, they found that these products were still harmful. In contrast, Osma et al. [55] assessed phytotoxicity with ryegrass and found that the degradation products of RBBR, while still toxic, were less detrimental than the original dye.

Peroxidases

Peroxidases are oxido-reductive hemoprotein enzymes with a heme group, specifically a compound of Fe^{3+} ion with protoporphyrin IX, and both bacterial and fungal organisms are capable of producing peroxidases. Since these enzymes don't pick a specific target, they are widely used to get rid of the dye in textile dye waste [78]. Peroxidases, in contrast to laccases, utilize H_2O_2 as the terminal electron acceptor for the oxidation of substrates, including azo dye molecules, through a comparable degradation mechanism [25]. The two main ligninolytic peroxidases involved in dye degradation are manganese peroxidase (MnP) lignin peroxidase (LiP). Originally

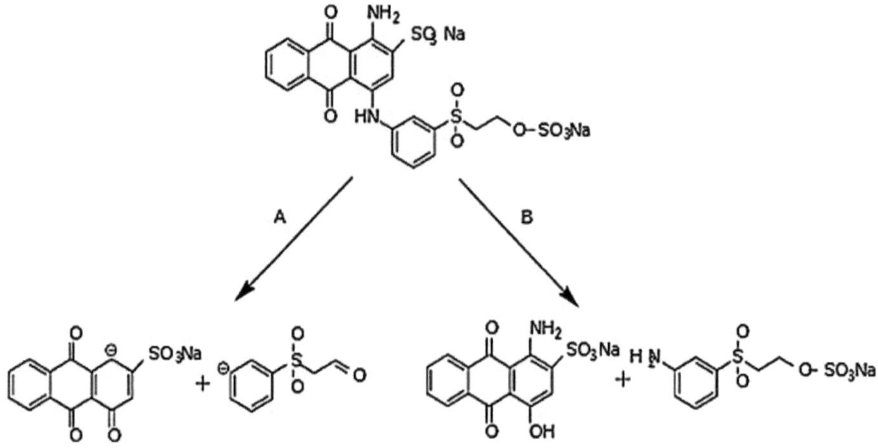

Fig. 4 Degradation pathway of Remazol Brilliant Blue R by laccase from **a** *Trametes pubescens* [55] and **b** *Polyporus* sp. S133 [32]

extracted from the fungal species *P. chrysosporium* [28], these enzymes have shown significant redox potential activity in the decolorization of dyes in vitro using purified enzymes. The process starts with the oxidation of the peroxidase enzyme through H_2O_2, enabling it to oxidize specific xenobiotic compounds, including sulfonated azo dyes, as elucidated in the degradation pathway for these dyes [49], LiP enables the direct oxidation of the dyes, while MnP initially oxidizes Mn^{2+} to Mn^{3+}, which then interacts with the dye molecules. For example, asymmetric or symmetric cleavage occurs at two distinct places within the dye's chromophore structure when LiP or MnP degrades Methyl Orange, a sulfonated monoazo dye, in the presence of H_2O_2, as shown in Fig. 5 [49].

4 Enzymatic Treatment of Textile Effluents

The enzymatic treatment of colored textile effluent presents a cost-effective and environmentally sustainable solution, as enzymes provide specificity and function under mild conditions. Biotic organisms use cellular enzymatic systems for dye degradation in labs and factories, indicating their potential for broad use. Table 2 summarizes studies on the enzymatic decolorization of industrial effluent, while Fig. 6 illustrates laboratory-scale treatment procedures. The increased enzyme activity noted in bioremediation agents underscores their effectiveness in addressing dye-laden textile effluent at a large scale.

Watharkar and Jadhav [89] showed that *Petunia grandiflora* and *Gaillardia grandiflora* can remove dyes, with 76% and 62% ADMI drops for each plant alone and 94% for the group as a hole. Roots of *G. grandiflora* demonstrated increases in

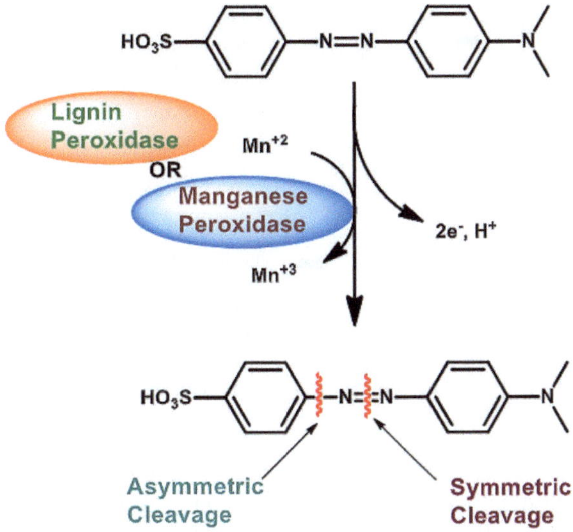

Fig. 5 Degradation mechanism of methyl orange color by peroxidase enzymes [49]

Table 2 Studies on enzyme-assisted degradation of actual dye-contaminated industrial effluents

Source species	Enzyme type	Enzyme activity level	Results	References
Asparagus densiflorus	Tyrosinase, laccase, and veratryl oxidase	205%, 178%, and 52%, respectively	COD (66%), TSS (66%), ADMI (65%), TDS (48%), BOD (61%)	[90]
Chlorella pyrenoidosa	Laccase, tyrosinase, veratryl alcohol oxidase, azoreductase, and DCIP reductase	350%, 170%, 77%, 850%, and 376%, respectively	COD (83%), ADMI (92%), TOC (79%), BOD (78%)	[45]
Fimbristylis dichotoma	NADH–DCIP reductase, lignin peroxidase and azoreductase	52.08%, 76.39%, and 57.37%, respectively	COD (72%), TSS (56%), ADMI (79%), TDS (66%), BOD (77%)	[38]
Bacillus aryabhattai DC100	Laccases and peroxidases	Not specified	COD (50%)	[58]
Lysini bacillus sp. RGS	Lignin peroxidase, azoreductase, and riboflavin reductase	576%, 845%, and 760%, respectively	COD (66%), TOC (64%)	[68]
Salvinia molesta	Veratryl alcohol oxidase, lignin peroxidase, and catalase	193%, 716%, and 194%, respectively	COD (76%), ADMI (81%), BOD (82%)	[17]

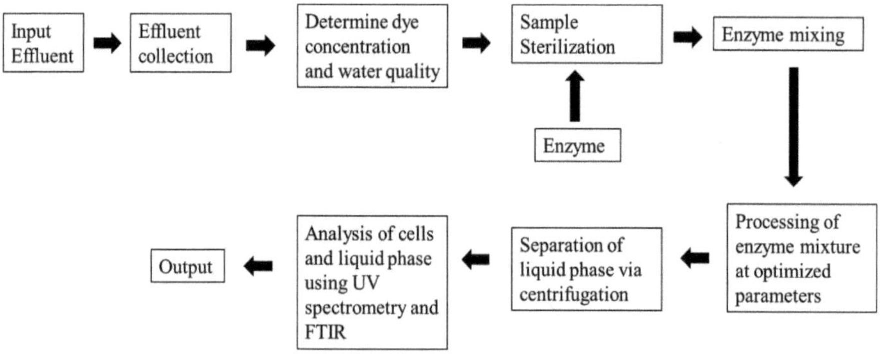

Fig. 6 Procedure for analyzing dye effluent treatment

tyrosinase activity of 379%, riboflavin reductase activity of 142%, and lignin peroxidase activity of 77%, whereas veratryl alcohol oxidase and laccase activities of *P. grandiflora* were 74% and 24% higher, respectively. The results highlighted the significance of the collective enzymatic machinery in dye degradation. Lade et al. [45] used an immobilized microbial consortium for the treatment of textile effluent, successfully achieving a 92% reduction in ADMI, along with decreases in COD (83%), TOC (79%), and BOD (78%).Moreover, the induction of enzyme activity was significant: 170%, 77%, 350%, 376% and 850% for tyrosinase, veratryl, laccase, alcohol oxidase, DCIP reductase, and azoreductase respectively. This demonstrated the substantial role of these enzymes in the dye degradation process. A separate study used *Lysinibacillus* sp. RGS for the treatment of textile effluent, achieving significant reductions in chemical oxygen demand (COD) by 90% and total organic carbon (TOC) by 93%. The activities of riboflavin reductase, lignin peroxidase, and azoreductase enzymes went up by 576%, 760%, and 845%, respectively. This shows that pollutants can be used as substrates for enzymes and that they can break down pollutants effectively [68]. Chandanshive et al. [17] examined large-scale wastewater treatment with the aquatic fern *Salvinia molesta*, resulting in reductions of 76% COD, 82% BOD, and 81% ADMI. Significant increases in enzyme activities were observed, with catalase, lignin peroxidase, and veratryl alcohol oxidase showing improvements of 716%, 194%, and 193%, respectively, confirming their usefulness in the breakdown of pollutants. Paz et al. [58] studied the biological treatment of textile effluent utilizing Bacillus aryabhattai DC100, known for its production of laccases and peroxidases. Their optimized treatment achieved a 50% reduction in COD in liquid effluent and a 14% reduction in semi-liquid sludge. Kadam et al. [38] similarly treated textile wastewater using *Ammannia baccifera*, *Fimbristylis dichotoma*, and their co-plantation consortia. Significant reductions were observed in COD (72%), TDS (66%), ADMI (79%), BOD (77%), and TSS (56%), coupled with a remarkable 92.47% increase in lignin peroxidase activity from the plant consortium.

5 Factors Influencing Enzymatic Degradation

The biodegradation of pollutants in textile wastewater treatment is contingent upon abiotic and biotic factors that affect microbial communities and their effectiveness, as illustrated in Fig. 7. These parameters influence the speed of biochemical reactions, maximizing the microbe-to-substrate ratio and maintaining homogeneous microbial distribution.

The efficacy of biological treatment is affected by factors such as organic loading, temperature, oxygen concentrations, oxygen diffusion, and dye persistence [50]. Additionally, the process of chemical degradation as a pretreatment offers simple carbon and nitrogen sources, thus enhancing the efficiency of microbial decolorization [67].

Biotic Factors

Typical biological factors in the treatment of effluents encompass community development, competition for nutrients, enzyme activity, composition, horizontal gene transfer mutation, and biomass. The competition for scarce nitrogen and carbon sources, coupled with antagonistic interactions among bacterial species and their

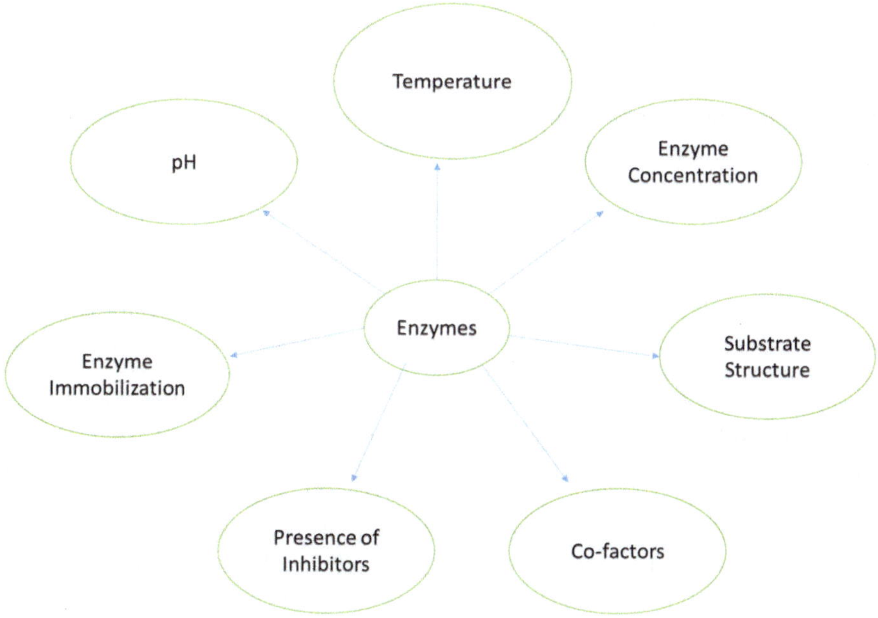

Fig. 7 Key factors affecting enzymatic dye degradation

predators, significantly influences the effectiveness of biodegradation in textile effluent treatment [2]. The efficiency of biological treatment is affected by the organic load, along with the diversity and quantity of microbial communities that can break down contaminants. The rate of degradation is also affected by the ability of enzymes to work and the ratio of microbes to food [77]. Enzyme blockage or less microbial growth can slow down the biodegradation process [52]. Additionally, other microorganisms, such as bacteriophages and protozoa, can disrupt bacterial breakdown by blocking enzymatic activity [60]. Therefore, for biodegradation to work well, there must be enough enzyme activity and microbial cells. The specific types of dyes found in the effluent are influenced by the uniqueness of enzymes and the availability of food supplies [62]. This indicates that the degree of degradation is linked to the enzymes' capacity to identify certain pollutants and their accessibility.

Abiotic Factors

The biodegradation process of dyes is affected by various environmental factors, including oxygen levels, temperature, retention time, types of dyes, dye concentration, mixing conditions, pH, and the availability of nitrogen and carbon sources [4]. Also, these elements influence metabolic turnover. Moreover, geographical location, process architecture, reaction circumstances, bioreactor types, and effluent characteristics are essential for optimizing treatment facilities. Research also indicates that the treatment methods for textile wastewater are influenced by elements such as aerobic or anaerobic conditions, dye concentration, and dye structure [40]. There are different amounts and kinds of dyes in textile effluent, and more dyes make it harder for microbes to remove the dye [27, 31]. The evaluation of microbial potential for various dye structures in textile effluent is essential. Guadie et al. [31] observed significant decolorization within 24 h, with no substantial differences noted after 96 h. This indicates that the dye structure influences degradation efficiency in the initial phase but not over an extended period. The structural complexity of dye molecules, particularly the substituent groups on the aromatic ring, influences microbial degradation [57]. Additionally, higher dye concentrations can reduce decolorization by inhibiting microbial activity [22]. Furthermore, for monoazo, diazo, and triazo dyes, *Streptomyces albidoflavus* demonstrated decolorization rates of 65%, 63%, and 54%, respectively. Biodegradation is also influenced by pH, with optimal ranges typically between 6 and 10 for the majority of microorganisms [70]. Compared to neutrophilic microbes, extremeophiles are more resilient to pH changes. For instance, Jamee and Siddique [36] discovered that *A. baumannii* was less effective at removing color when the pH was acidic (3) but more effective when the pH was between 5 and 9. Decolorization of various dyes worked best between pH 7 and pH 9, respectively. This highlights the significance of pH optimization for efficient microbial decolorization. Temperature influences microbial degradation through its effects on enzyme activity and microbial growth [46]. Too hot or too cold can make things less efficient. For example, *Bacillus albus* was able to eliminate 98% of dye at 30 °C but less at 40 °C

[43]. In a similar manner, *Klebsiella pneumoniae* exhibited a 95% decolorization rate at 35 °C, but this rate diminished at 60 °C [54].

The procedure of dye decolorization is influenced by anaerobic, facultative, or aerobic conditions [36]. Azoreductase plays a crucial role in the reduction of azo dyes, operating effectively in anaerobic environments and utilizing carbon sources such as glucose [7]. Aerobic microbes face difficulties with utilizing dyes as carbon sources in textile effluent because of competition for dissolved oxygen [60]. Anaerobic conditions are essential for the initial treatment of dyes. Redox mediators are crucial for improving reductive processes, such as the reduction of azo dyes [13]. The effectiveness of biodegradation is also influenced by the electron density of the dyes' chromophore groups [5].

6 Conclusion

Enzymes are essential for the degradation of dyes found in industrial wastewater. Biological agents like bacteria, fungi, algae, and plants make enzymes that break down dye compounds quickly and cheaply, making them an environmental friendly alternative to traditional ways of treatment. Among these enzymes, such as laccase, azoreductase, and peroxidase, extensive research has been conducted regarding their effectiveness in the degradation of dyes. Peroxidase enzymes can work well with a wide range of industrial dyes because they can handle high temperatures and changing pH levels. The decolorization process by enzymes is notably affected by operational factors including pH, temperature, electron donors, and co-substrates, which require optimization for effective treatment. Immobilized enzymes exhibit superior decolorization effectiveness relative to their soluble equivalents. However, some enzymes may generate intermediate chemicals, such as aromatic amines and phenols, which exhibit greater toxicity than the parent dyes. To fix this, adding several oxidoreductase enzymes could help change these harmful intermediates into harmless, insoluble products, which would make the treatment process better overall. More study is needed to look into how immobilized bacterial enzymes like azoreductase and laccase can be used with other enzymes for more effective treatment of mixed-dye industrial effluents. Additionally, molecular studies that focus on the active sites and structural features of these enzymes will help us learn more about how they react with different dyes. This makes enzymatic dye degradation a hopeful and long-term method for dealing with dye-contaminated wastewater.

Acknowledgements The authors extend their appreciation to the editors and reviewers for their insightful feedback, recognizing their contributions in improving the quality of the manuscript.

Conflict Interests No potential conflict of interest was reported by the author(s).

References

1. Adnan, L. A., Sathishkumar, P., Mohd Yusoff, A. R., & Hadibarata, T. (2015). Metabolites characterisation of laccase mediated Reactive Black 5 biodegradation by fast growing ascomycete fungus Trichoderma atroviride F03. *International Biodeterioration and Biodegradation, 104*, 274–282. https://doi.org/10.1016/j.ibiod.2015.05.019
2. Ajaz, M., Shakeel, S., & Rehman, A. (2020). Microbial use for azo dye degradation—A strategy for dye bioremediation. *International Microbiology, 23*(2), 149–159. https://doi.org/10.1007/s10123-019-00103-2
3. Akan, J. C., Abdulrahman, F. I., Ayodele, J. T., & Ogugbuaja, V. O. (2009). Impact of tannery and textile effluent on the chemical characteristics of Challawa River, Kano State, Nigeria. *Australian Journal of Basic and Applied Sciences, 3*(3), 1933–1947.
4. Aragaw, T. A. (2021). Functions of various bacteria for specific pollutants degradation and their application in wastewater treatment: A review. *International Journal of Environmental Science and Technology, 18*(7), 2063–2076. https://doi.org/10.1007/s13762-020-03022-2
5. Araghi, M. S., Olya, M. E., Marandi, R., & Siadat, S. D. (2016). Investigation of enhanced biological dye removal of colored wastewater in a lab-scale biological activated carbon process. *Applied Biological Chemistry, 59*(3), 463–470. https://doi.org/10.1007/s13765-016-0177-4
6. Ardila-Leal, L. D., Poutou-Piñales, R. A., Pedroza-Rodríguez, A. M., & Quevedo-Hidalgo, B. E. (2021). A brief history of colour, the environmental impact of synthetic dyes and removal by using laccases. *Molecules, 26*(13). https://doi.org/10.3390/molecules26133813
7. Arora, P. K. (2015). Bacterial degradation of monocyclic aromatic amine. *Frontiers in Microbiology, 6*(July), 1–14. https://doi.org/10.3389/fmicb.2015.00820
8. Baldev, E., MubarakAli, D., Ilavarasi, A., Pandiaraj, D., Ishack, K. A. S. S., & Thajuddin, N. (2013). Degradation of synthetic dye, Rhodamine B to environmentally non-toxic products using microalgae. *Colloids and Surfaces B: Biointerfaces, 105*, 207–214. https://doi.org/10.1016/j.colsurfb.2013.01.008
9. Banat, I. M., Nigam, P., Singh, D., & Marchant, R. (1996). Microbial decolorization of textile-dye containing effluents: A review. *Bioresource Technology, 58*(3), 217–227.
10. Baughman, G. L., & Weber, E. J. (1994). Transformation of dyes and related compounds in anoxic sediment: Kinetics and products. *Environmental Science & Technology, 28*(2), 267–276. https://doi.org/10.1021/es00051a013
11. Ben Younes, S., & Sayadi, S. (2013). Detoxification of Indigo carmine using a combined treatment via a novel trimeric thermostable laccase and microbial consortium. *Journal of Molecular Catalysis B: Enzymatic, 87*(September), 62–68. https://doi.org/10.1016/j.molcatb.2012.10.007
12. Benkhaya, S., M'rabet, S., & El Harfi, A. (2020). A review on classifications, recent synthesis and applications of textile dyes. *Inorganic Chemistry Communications, 115*, Article 107891.
13. Bharagava, R. N., & Chowdhary, P. (2018). Emerging and eco-friendly approaches for waste management. In *Emerging and eco-friendly approaches for waste management* (March, pp. 1–435). https://doi.org/10.1007/978-981-10-8669-4
14. Call, H. P., & Mücke, I. (1997). History, overview and applications of mediated lignolytic systems, especially laccase-mediator-systems (Lignozym®-process). *Journal of Biotechnology, 53*(2–3), 163–202. https://doi.org/10.1016/S0168-1656(97)01683-0
15. Campos, R., Kandelbauer, A., Robra, K. H., Cavaco-Paulo, A., & Gübitz, G. M. (2001). Indigo degradation with purified laccases from Trametes hirsuta and Sclerotium rolfsii. *Journal of Biotechnology, 89*(2–3), 131–139. https://doi.org/10.1016/s0168-1656(01)00303-0
16. Chacko, J. T., & Subramaniam, K. (2011). Enzymatic degradation of azo dyes—A review. *International Journal of Environmetal Sciences, 1*(6), 1250–1260. https://doi.org/10.6088/ijes.00106020018
17. Chandanshive, V. V., Rane, N. R., Gholave, A. R., Patil, S. M., Jeon, B. H., & Govindwar, S. P. (2016). Efficient decolorization and detoxification of textile industry effluent by Salvinia

molesta in lagoon treatment. *Environmental Research, 150*, 88–96. https://doi.org/10.1016/j.envres.2016.05.047
18. Chavan, R., Patil, R., & Chougale, A. (2024). *Efficient dye removal strategies: Exploring the role of biochar*.
19. Chen, H. (2006). Recent advances in azo dye degrading enzyme research. *Current Protein & Peptide Science, 7*(2), 101–111. https://doi.org/10.2174/138920306776359786
20. Chen, Z., Feng, M., Wang, Y., & Ling, X. (2024). Comparison of treatment performance and microbial community evolution of typical dye wastewater by different combined processes. *Ecotoxicology and Environmental Safety, 275*, Article 116226.
21. Chhabra, M., Mishra, S., & Sreekrishnan, T. R. (2009). Laccase/mediator assisted degradation of triarylmethane dyes in a continuous membrane reactor. *Journal of Biotechnology, 143*(1), 69–78. https://doi.org/10.1016/j.jbiotec.2009.06.011
22. Chittal, V., Gracias, M., Anu, A., Saha, P., & Bhaskara Rao, K. V. (2019). Biodecolorization and biodegradation of azo dye Reactive Orange-16 by marine nocardiopsis sp. *Iranian Journal of Biotechnology, 17*(3), e1551. https://doi.org/10.29252/ijb.1551
23. Das, S. C., Khan, O., Khadem, A. H., Rahman, M. A., Bedoura, S., Uddin, M. A., & Islam, M. S. (2024). Evaluating the biocatalytic potential of fruit peel-derived eco-enzymes for sustainable textile wastewater treatment. *Results in Engineering, 21*, Article 101898.
24. Delclos, K. B., Tarpley, W. G., Miller, E. C., & Miller, J. A. (1984). 4-Aminoazobenzene and N, N-dimethyl-4-aminoazobenzene as equipotent hepatic carcinogens in male C57BL/6× C3H/He F1 mice and characterization of N-(deoxyguanosin-8-yl)-4-aminoazobenzene as the major persistent hepatic DNA-bound dye in these mice. *Cancer Research, 44*(6), 2540–2550.
25. Durán, N., Rosa, M. A., D'annibale, A., & Gianfreda, L. (2002). Applications of laccases and tyrosinases (phenoloxidases) immobilized on different supports: A review. *Enzyme and Microbial Technology, 31*, 907–931. http://bioinf.leeds
26. Durão, P., Bento, I., Fernandes, A. T., Melo, E. P., Lindley, P. F., & Martins, L. O. (2006). Perturbations of the T1 copper site in the CotA laccase from Bacillus subtilis: Structural, biochemical, enzymatic and stability studies. *Journal of Biological Inorganic Chemistry, 11*(4), 514–526. https://doi.org/10.1007/s00775-006-0102-0
27. El Awady, M. E., El-Shall, F. N., Mohamed, G. E., Abd-Elaziz, A. M., Abdel-Monem, M. O., & Hassan, M. G. (2024). Exploring the decolorization efficiency and biodegradation mechanisms of different functional textile azo dyes by Streptomyces albidoflavus 3MGH. *BMC Microbiology, 24*(1), 1–22. https://doi.org/10.1186/s12866-024-03347-9
28. Gomare, S. S., Jadhav, J. P., & Govindwar, S. P. (2008). Degradation of sulfonated azo dyes by the purified lignin peroxidase from Brevibacillus laterosporus MTCC 2298. *Biotechnology and Bioprocess Engineering, 13*(2), 136–143. https://doi.org/10.1007/s12257-008-0008-5
29. Gomare, S. S., Tamboli, D. P., Kagalkar, A. N., & Govindwar, S. P. (2009). Eco-friendly biodegradation of a reactive textile dye Golden Yellow HER by Brevibacillus laterosporus MTCC 2298. *International Biodeterioration & Biodegradation, 63*(5), 582–586. https://doi.org/10.1016/j.ibiod.2009.03.005
30. Goszczynski, S., Paszczynski, A., Pasti-Grigsby, M. B., Crawford, R. L., & Crawford, D. L. (1994). New pathway for degradation of sulfonated azo dyes by microbial peroxidases of Phanerochaete chrysosporium and Streptomyces chromofuscus. *Journal of Bacteriology, 176*(5), 1339–1347. https://doi.org/10.1128/jb.176.5.1339-1347.1994
31. Guadie, A., Gessesse, A., & Xia, S. (2018). Halomonas sp. strain A55, a novel dye decolorizing bacterium from dye-uncontaminated Rift Valley Soda lake. *Chemosphere, 206*, 59–69. https://doi.org/10.1016/j.chemosphere.2018.04.134
32. Hadibarata, T., Yusoff, A. R. M., & Kristanti, R. A. (2012). Decolorization and metabolism of anthraquionone-type dye by laccase of white-rot fungi polyporus sp. S133. *Water, Air, and Soil Pollution, 223*(2), 933–941. https://doi.org/10.1007/s11270-011-0914-6
33. Heinfling, A., Martínez, M. J., Martínez, A. T., Bergbauer, M., & Szewzyk, U. (1998). Transformation of industrial dyes by manganese peroxidases from Bjerkandera adusta and Pleurotus eryngii in a manganese-independent reaction. *Applied and Environmental Microbiology, 64*(8), 2788–2793. https://doi.org/10.1128/aem.64.8.2788-2793.1998

34. Husain, Q. (2006). Potential applications of the oxidoreductive enzymes in the decolorization and detoxification of textile and other synthetic dyes from polluted water: A review. *Critical Reviews in Biotechnology, 26*(4), 201–221. https://doi.org/10.1080/07388550600969936
35. Inamdar, S., Vanarase, K., & Inamdar, S. (2023). Enzyme engineering for advanced treatment of textile wastewater. In *Current developments in bioengineering and biotechnology* (pp. 679–710). Elsevier.
36. Jamee, R., & Siddique, R. (2019). Biodegradation of synthetic dyes of textile effluent by microorganisms: An environmentally and economically sustainable approach. *European Journal of Microbiology and Immunology, 9*(4), 114–118. https://doi.org/10.1556/1886.2019.00018
37. Johannes, C., & Majcherczyk, A. (2000). Laccase activity tests and laccase inhibitors. *Journal of Biotechnology, 78*(2), 193–199. https://doi.org/10.1016/S0168-1656(00)00208-X
38. Kadam, S. K., Chandanshive, V. V., Rane, N. R., Patil, S. M., Gholave, A. R., Khandare, R. V., Bhosale, A. R., Jeon, B. H., & Govindwar, S. P. (2018). Phytobeds with Fimbristylis dichotoma and Ammannia baccifera for treatment of real textile effluent: An in situ treatment, anatomical studies and toxicity evaluation. *Environmental Research, 160*(September), 1–11. https://doi.org/10.1016/j.envres.2017.09.009
39. Kamati, S. N., Yan, J., & Fan, J. (2024). A review on progresses in reactive dye-containing wastewater treatment. *Water Practice & Technology,* wpt2024142.
40. Kapoor, R. T., Danish, M., Singh, R. S., Rafatullah, M., & Abdul, A. K. (2021). Exploiting microbial biomass in treating azo dyes contaminated wastewater: Mechanism of degradation and factors affecting microbial efficiency. *Journal of Water Process Engineering, 43*(August), Article 102255. https://doi.org/10.1016/j.jwpe.2021.102255
41. Keck, A., Klein, J., Kudlich, M., Stolz, A., Knackmuss, H. J., & Mattes, R. (1997). Reduction of azo dyes by redox mediators originating in the naphthalenesulfonic acid degradation pathway of Sphingomonas sp. strain BN6. *Applied and Environmental Microbiology, 63*(9), 3684–3690. https://doi.org/10.1128/aem.63.9.3684-3690.1997
42. Khataee, A. R., Movafeghi, A., Torbati, S., Salehi Lisar, S. Y., & Zarei, M. (2012). Phytoremediation potential of duckweed (Lemna minor L.) in degradation of C.I. Acid Blue 92: Artificial neural network modeling. *Ecotoxicology and Environmental Safety, 80*, 291–298. https://doi.org/10.1016/j.ecoenv.2012.03.021
43. Kishor, R., Purchase, D., Saratale, G. D., Saratale, R. G., Ferreira, L. F. R., Bilal, M., Chandra, R., & Bharagava, R. N. (2021). Ecotoxicological and health concerns of persistent coloring pollutants of textile industry wastewater and treatment approaches for environmental safety. *Journal of Environmental Chemical Engineering, 9*(2). https://doi.org/10.1016/j.jece.2020.105012
44. Kumar, J. E., Mulai, T., Kharmawphlang, W., Sharan, R. N., & Sahoo, M. K. (2023). The efficiency of Fenton, Fenton/MW and UV/oxidant processes in the treatment of a mixture of higher concentrations of azo dyes. *Chemical Engineering Journal Advances, 15*, Article 100515.
45. Lade, H., Govindwar, S., & Paul, D. (2015). Mineralization and detoxification of the carcinogenic azo dye Congo red and real textile effluent by a polyurethane foam immobilized microbial consortium in an upflow column bioreactor. *International Journal of Environmental Research and Public Health, 12*(6), 6894–6918. https://doi.org/10.3390/ijerph120606894
46. Li, H.-H., Wang, Y.-T., Wang, Y., Wang, H.-X., Sun, K.-K., & Lu, Z.-M. (2019). Bacterial degradation of anthraquinone dyes. *Journal of Zhejiang University, Science B, 20*(6), 528–540. https://doi.org/10.1631/jzus.B1900165
47. López, C., Valade, A. G., Combourieu, B., Mielgo, I., Bouchon, B., & Lema, J. M. (2004). Mechanism of enzymatic degradation of the azo dye Orange II determined by ex situ 1H nuclear magnetic resonance and electrospray ionization-ion trap mass spectrometry. *Analytical Biochemistry, 335*(1), 135–149. https://doi.org/10.1016/j.ab.2004.08.037
48. Mate, D. M., & Alcalde, M. (2015). Laccase engineering: From rational design to directed evolution. *Biotechnology Advances, 33*(1), 25–40. https://doi.org/10.1016/j.biotechadv.2014.12.007

49. McMullan, G., Meehan, C., Conneely, A., Kirby, N., Robinson, T., Nigam, P., Banat, I. M., Marchant, R., & Smyth, W. F. (2001). Microbial decolourisation and degradation of textile dyes. *Applied Microbiology and Biotechnology, 56*(1–2), 81–87. https://doi.org/10.1007/s002530000587
50. Mishra, S., & Maiti, A. (2018). The efficacy of bacterial species to decolourise reactive azo, anthroquinone and triphenylmethane dyes from wastewater: A review. *Environmental Science and Pollution Research, 25*(9), 8286–8314. https://doi.org/10.1007/s11356-018-1273-2
51. Mohan, S. V., Prasad, K. K., Rao, N. C., & Sarma, P. N. (2005). Acid azo dye degradation by free and immobilized horseradish peroxidase (HRP) catalyzed process. *Chemosphere, 58*(8), 1097–1105. https://doi.org/10.1016/j.chemosphere.2004.09.070
52. Mohanty, S. S., & Kumar, A. (2021). Enhanced degradation of anthraquinone dyes by microbial monoculture and developed consortium through the production of specific enzymes. *Scientific Reports, 11*(1), 1–15. https://doi.org/10.1038/s41598-021-87227-6
53. Mugdha, A., & Usha, M. (2012). Enzymatic treatment of wastewater containing dyestuffs using different delivery systems. *Scientific Reviews & Chemical Communications, 2*(1), 31–40. https://www.academia.edu/download/107742822/enzymatic-treatment-of-wastewater-containing-dyestuffs-using-different-delivery-systems.pdf
54. Mustafa, G., Tariq Zahid, M., Ali, S., Zaghum Abbas, S., & Rafatullah, M. (2021). Biodegradation and discoloration of disperse blue-284 textile dye by Klebsiella pneumoniae GM-04 bacterial isolate. *Journal of King Saud University—Science, 33*(4), Article 101442. https://doi.org/10.1016/j.jksus.2021.101442
55. Osma, J. F., Toca-Herrera, J. L., & Rodríguez-Couto, S. (2010). Transformation pathway of Remazol Brilliant Blue R by immobilised laccase. *Bioresource Technology, 101*(22), 8509–8514. https://doi.org/10.1016/j.biortech.2010.06.074
56. Oturkar, C. C., Nemade, H. N., Mulik, P. M., Patole, M. S., Hawaldar, R. R., & Gawai, K. R. (2011). Mechanistic investigation of decolorization and degradation of Reactive Red 120 by Bacillus lentus BI377. *Bioresource Technology, 102*(2), 758–764. https://doi.org/10.1016/j.biortech.2010.08.094
57. Patel, V. R., Bhatt, N. S., & BỳBhatt, H. (2013). Involvement of ligninolytic enzymes of Myceliophthora vellerea HQ871747 in decolorization and complete mineralization of Reactive Blue 220. *Chemical Engineering Journal, 233*, 98–108. https://doi.org/10.1016/j.cej.2013.07.110
58. Paz, A., Carballo, J., Pérez, M. J., & Domínguez, J. M. (2017). Biological treatment of model dyes and textile wastewaters. *Chemosphere, 181*, 168–177. https://doi.org/10.1016/j.chemosphere.2017.04.046
59. Pereira, L., Coelho, A. V., Viegas, C. A., Dos Santos, M. M. C., Robalo, M. P., & Martins, L. O. (2009). Enzymatic biotransformation of the azo dye Sudan Orange G with bacterial CotA-laccase. *Journal of Biotechnology, 139*(1), 68–77. https://doi.org/10.1016/j.jbiotec.2008.09.001
60. Pinheiro, L. R. S., Gradíssimo, D. G., Xavier, L. P., & Santos, A. V. (2022). Degradation of azo dyes: Bacterial potential for bioremediation. *Sustainability (Switzerland), 14*(3), 1–23. https://doi.org/10.3390/su14031510
61. Rahman, M., & Tabassum, Z. (2024). Biotechnological approach to treat textile dyeing effluents: A critical review analysing the practical applications. *Textile and Leather Review, 7*, 124–152.
62. Robinson, P. K. (2015). Enzymes: Principles and biotechnological applications. *Essays in Biochemistry, 59*, 1–41. https://doi.org/10.1042/bse0590001
63. Roriz, M. S., Osma, J. F., Teixeira, J. A., & Rodríguez Couto, S. (2009). Application of response surface methodological approach to optimise Reactive Black 5 decolouration by crude laccase from Trametes pubescens. *Journal of Hazardous Materials, 169*(1–3), 691–696. https://doi.org/10.1016/j.jhazmat.2009.03.150
64. Sakthishabarish, K., Karthika, T., Sandhya, S. N., Yazhini, A., & Krishnan, K. (2023). Bacterial bioremediation of textile effluent dyes contaminated sites. *Research Journal of Biotechnology, 18*(10), 127–138. https://doi.org/10.25303/1810rjbt1270138

65. Sanmathi, K. R. (2023). Biodegradation on the removal of dyes from textile effluent. *International Journal of Advanced Research in Science, Communication and Technology,* 4–8. https://doi.org/10.48175/ijarsct-13102
66. Sarafpour, M., Alihosseini, F., & Bayat, M. (2022). New laccase-mediated system utilized for bio-discoloration of indigo-dyed denim fabrics. *Applied Biochemistry and Biotechnology, 194*(12), 5848–5861. https://doi.org/10.1007/s12010-022-04066-5
67. Saratale, R. G., Saratale, G. D., Chang, J. S., & Govindwar, S. P. (2011). Bacterial decolorization and degradation of azo dyes: A review. *Journal of the Taiwan Institute of Chemical Engineers, 42*(1), 138–157. https://doi.org/10.1016/j.jtice.2010.06.006
68. Saratale, R. G., Saratale, G. D., Govindwar, S. P., & Kim, D. S. (2015). Exploiting the efficacy of Lysinibacillus sp. RGS for decolorization and detoxification of industrial dyes, textile effluent and bioreactor studies. *Journal of Environmental Science and Health. Part A, Toxic/Hazardous Substances & Environmental Engineering, 50*(2), 176–192. https://doi.org/10.1080/10934529.2014.975536
69. Šekuljica, N. Ž., Prlainović, N. Ž., Stefanović, A. B., Žuža, M. G., Čičkarić, D. Z., Mijin, D. Ž., & Knežević-Jugović, Z. D. (2015). Decolorization of anthraquinonic dyes from textile effluent using horseradish peroxidase: Optimization and kinetic study. *The Scientific World Journal, 2015*, 371–625. https://doi.org/10.1155/2015/371625
70. Selvakumar, S., Manivasagan, R., & Chinnappan, K. (2013). Biodegradation and decolourization of textile dye wastewater using Ganoderma lucidum. *3 Biotech, 3*(1), 71–79. https://doi.org/10.1007/s13205-012-0073-5
71. Semião, M. A., Haminiuk, C. W. I., Brugnari, T., Mannes, Y., Nagata, W. A. R., & Maciel, G. M. (2023). Synergistic treatment of textile wastewaters using spent diatomaceous earth loaded with laccases: A cost-effective and eco-friendly approach. *Journal of Water Process Engineering, 56*, Article 104552.
72. Sharma, J., Sharma, S., & Soni, V. (2021). Classification and impact of synthetic textile dyes on Aquatic Flora: A review. *Regional Studies in Marine Science, 45*, Article 101802.
73. Shi, X., Liu, Q., Ma, J., Liao, H., Xiong, X., Zhang, K., Wang, T., Liu, X., Xu, T., Yuan, S., Zhang, X., & Zhu, Y. (2015). An acid-stable bacterial laccase identified from the endophyte Pantoea ananatis Sd-1 genome exhibiting lignin degradation and dye decolorization abilities. *Biotechnology Letters, 37*(11), 2279–2288. https://doi.org/10.1007/s10529-015-1914-1
74. Sim, C. S. F., Chen, S. H., & Ting, A. S. Y. (2018). Endophytes: Emerging tools for the bioremediation of pollutants. In *Emerging and eco-friendly approaches for waste management* (pp. 189–217). https://doi.org/10.1007/978-981-10-8669-4_10
75. Singh, G., Capalash, N., Goel, R., & Sharma, P. (2007). A pH-stable laccase from alkali-tolerant γ-proteobacterium JB: Purification, characterization and indigo carmine degradation. *Enzyme and Microbial Technology, 41*(6–7), 794–799. https://doi.org/10.1016/j.enzmictec.2007.07.001
76. Singh, L., & Singh, V. P. (2017). Decolourization of azo (acid red) and anthraquinonic (basic blue) dyes by the fungus Aspergillus flavus. *International Journal of Biomedical Engineering and Clinical Science, 3*(1), 1–5.
77. Singh, R. L., Singh, P. K., & Singh, R. P. (2015). Enzymatic decolorization and degradation of azo dyes–A review. *International Biodeterioration & Biodegradation, 104*, 21–31. https://doi.org/10.1016/j.ibiod.2015.04.027
78. Singh, S. N. (2015). *Microbial degradation of synthetic dyes in wastewaters. Environmental science and engineering* (Subseries: *Environmental science*), 9783319109411, iii–iv. https://doi.org/10.1007/978-3-319-10942-8
79. Stolz, A. (2001). Basic and applied aspects in the microbial degradation of azo dyes. *Applied Microbiology and Biotechnology, 56*(1–2), 69–80. https://doi.org/10.1007/s002530100686
80. Sudarshan, S., Harikrishnan, S., RathiBhuvaneswari, G., Alamelu, V., Aanand, S., Rajasekar, A., & Govarthanan, M. (2023). Impact of textile dyes on human health and bioremediation of textile industry effluent using microorganisms: Current status and future prospects. *Journal of Applied Microbiology, 134*(2), lxac064.

81. Telke, A. A., Ghodake, G. S., Kalyani, D. C., Dhanve, R. S., & Govindwar, S. P. (2011). Biochemical characteristics of a textile dye degrading extracellular laccase from a Bacillus sp. *ADR. Bioresource Technology, 102*(2), 1752–1756. https://doi.org/10.1016/j.biortech.2010.08.086
82. Telke, A. A., Kadam, A. A., Jagtap, S. S., Jadhav, J. P., & Govindwar, S. P. (2010). Biochemical characterization and potential for textile dye degradation of blue laccase from Aspergillus ochraceus NCIM-1146. *Biotechnology and Bioprocess Engineering, 15*(4), 696–703. https://doi.org/10.1007/s12257-009-3126-9
83. Telke, A. A., Kalyani, D. C., Dawkar, V. V., & Govindwar, S. P. (2009). Influence of organic and inorganic compounds on oxidoreductive decolorization of sulfonated azo dye C.I. Reactive Orange 16. *Journal of Hazardous Materials, 172*(1), 298–309. https://doi.org/10.1016/j.jhazmat.2009.07.008
84. Umamaheswari, K., & Sumathi, R. K. (2024). Integrated enzymatic bioremediation: Efficient decolorization of textile DYE effluents and assessing environmental safety. *Uttar Pradesh Journal of Zoology, 45*(16), 189–197.
85. Van der Zee, F. P., & Villaverde, S. (2005). Combined anaerobic–aerobic treatment of azo dyes—a short review of bioreactor studies. *Water Research, 39*(8), 1425–1440.
86. Verma, A. K., Dash, R. R., & Bhunia, P. (2012). A review on chemical coagulation/flocculation technologies for removal of colour from textile wastewaters. *Journal of Environmental Management, 93*(1), 154–168.
87. Vikrant, K., Giri, B. S., Raza, N., Roy, K., Kim, K.-H., Rai, B. N., & Singh, R. S. (2018). Recent advancements in bioremediation of dye: Current status and challenges. *Bioresource Technology, 253*, 355–367. https://doi.org/10.1016/j.biortech.2018.01.029
88. Wagaw, T., & Babu, K. M. (2023). Textile waste recycling: A need for a stringent paradigm shift. *AATCC Journal of Research, 10*(6), 376–385.
89. Watharkar, A. D., & Jadhav, J. P. (2014). Detoxification and decolorization of a simulated textile dye mixture by phytoremediation using Petunia grandiflora and Gailardia grandiflora: A plant-plant consortial strategy. *Ecotoxicology and Environmental Safety, 103*(1), 1–8. https://doi.org/10.1016/j.ecoenv.2014.01.033
90. Watharkar, A. D., Kadam, S. K., Khandare, R. V., Kolekar, P. D., Jeon, B. H., Jadhav, J. P., & Govindwar, S. P. (2018). Asparagus densiflorus in a vertical subsurface flow phytoreactor for treatment of real textile effluent: A lab to land approach for in situ soil remediation. *Ecotoxicology and Environmental Safety, 161*(February), 70–77. https://doi.org/10.1016/j.ecoenv.2018.05.078
91. Wong, Y., & Yu, J. (1999). Laccase-catalyzed decolorization of synthetic dyes. *Water Research, 33*(16), 3512–3520. https://doi.org/10.1016/S0043-1354(99)00066-4
92. Xie, Z. (2024). Progress of dye wastewater treatment methods. *World, 2*(2), 35–44.
93. Yaghoot-Nezhad, A., Saebnoori, E., Danaee, I., Elahi, S., Panah, N. B., & Khosravi-Nikou, M. R. (2022). Evaluation of the oxidative degradation of aromatic dyes by synthesized nano ferrate (VI) as a simple and effective treatment method. *Journal of Water Process Engineering, 49*, Article 103017.
94. Yang, J., Yang, X., Lin, Y., Ng, T. B., Lin, J., & Ye, X. (2015). Laccase-catalyzed decolorization of malachite green: Performance optimization and degradation mechanism. *PLoS ONE, 10*(5), Article e0127714. https://doi.org/10.1371/journal.pone.0127714
95. Yang, J., Yang, X., Lin, Y., Ng, T. B., Lin, J., & Ye, X. (2015). Laccase-catalyzed decolorization of malachite green: Performance optimization and degradation mechanism. *PLoS ONE, 10*(5), 1–14. https://doi.org/10.1371/journal.pone.0127714
96. Yin, Q., Wang, Y., Li, W., Wei, C., Chen, Y., Liu, C., Ling, X., & Chen, Z. (2024). Highly efficient removal of contaminant from typical dye wastewater by using individual AOPs-A/O processes: Design, performance, and mechanism. *Journal of Environmental Chemical Engineering, 12*(4), Article 113103.
97. Zeng, X., Cai, Y., Liao, X., Zeng, X., Li, W., & Zhang, D. (2011). Decolorization of synthetic dyes by crude laccase from a newly isolated Trametes trogii strain cultivated on solid agro-industrial residue. *Journal of Hazardous Materials, 187*(1–3), 517–525. https://doi.org/10.1016/j.jhazmat.2011.01.068

98. Zeng, X., Cai, Y., Liao, X., Zeng, X., Luo, S., & Zhang, D. (2012). Anthraquinone dye assisted the decolorization of azo dyes by a novel Trametes trogii laccase. *Process Biochemistry, 47*(1), 160–163. https://doi.org/10.1016/j.procbio.2011.10.019
99. Zhang, Z., Xin, Z. Z., Ma HongRui, M. H., Luo YiChao, L. Y., Zhu Chao, Z. C., & Fan Yang, F. Y. (2015). Efficiency of different treatment process of fur dyeing wastewater and impact of bacterial community in activated sludge.
100. Zheng, F., Cui, B. K., Wu, X. J., Meng, G., Liu, H. X., & Si, J. (2016). Immobilization of laccase onto chitosan beads to enhance its capability to degrade synthetic dyes. *International Biodeterioration and Biodegradation, 110*(4), 69–78. https://doi.org/10.1016/j.ibiod.2016.03.004
101. Zille, A., Gornacka, B., Rehorek, A., & Cavaco-Paulo, A. (2005). Degradation of azo dyes by Trametes villosa laccase over long periods of oxidative conditions. *Applied and Environmental Microbiology, 71*(11), 6711–6718. https://doi.org/10.1128/AEM.71.11.6711-6718.2005

Bioremediation of Heavy Metals in Textile Industry Effluents for Sustainable Future

Sana Farooq, Humera Aziz, Muhammad Rizwan, Umair Riaz, Fernanda Maria Policarpo Tonelli, Tariq Mehmood, and Tabinda Athar

1 Introduction

Textile processing relies on chemical-intensive methods that involve non-biodegradable and environmentally harmful compounds. However, a considerable concentration of residuals is still present in treated water after the tertiary treatment. Even though the accumulated hazardous, non-recyclable chemicals in the bilge waters are not easily separable from the water and require tertiary, or additional, treatments due to their peculiarities under the processes. If such waste water is not disposed of in the right manner it pollutes the environment when discharged [3]. Most textile wet processing facilities are located in the world's developing and underdeveloped countries. Another reason is that most of these plants are small with limited capacity, they do not possess necessary equipment to handle wastewater. Consequently, there is often a direct discharge to the environment with little or no treatment to make the discharge safe. Most of these facilities are situated near the

S. Farooq · H. Aziz (✉) · M. Rizwan
Department of Environmental Sciences, Government College University, Faisalabad, Pakistan
e-mail: humeraaziz.uaf@gmail.com

U. Riaz
Department of Soil and Environmental Sciences, MNS-University of Agriculture, Multan, Pakistan

F. M. P. Tonelli
Federal University of São João del-Rei, São João del-Rei, Brazil

T. Mehmood
Department Sensors and Modeling, Leibniz Institute for Agricultural Engineering and Bioeconomy (ATB), Potsdam, Germany

T. Athar
Microbial Biogeochemistry Lab, Department of Geosciences, College of Arts and Sciences, Baylor University, Waco, TX, USA

R. A. Bhat et al. (eds.), *Biocentric Approaches in Textile Waste Management*, Textile Science and Clothing Technology, https://doi.org/10.1007/978-3-032-04974-2_6

regions that have had a direct contact with the ecological impact of this developmental exercise [37].

The usage of fresh water is high because it is utilized as a medium for a number of processes, such as a solvent for chemical dissolution, a wash-off agent, and to generate vapor for process bath heating. The type of raw material used, the colors and finishing compounds used, and the handling method employed all affect how much water is utilized. 50 to 100 L of water are needed to process 1 kg of textile materials. Wet processing discharges textile effluent which is water containing unfixed dyes, chemical and auxiliary matters. The components of cloth dyes as the textile effluents involve mixtures of both the organic and inorganic particles in different concentrations and with different properties [31].

Multiple studies have revealed that the textile wastewater contains color, heavy metals, salt, sulphide compounds, silicate, and caustic. Textile dyes are typically the cause of wastewater's color, pH, and chemical oxygen demand. They can be broadly classified into two categories: aqueous and nonaqueous, since water is one of the most crucial elements of textile processes. The reagents and water are utilized by numerous stages in the wet system and are discharged in the effluent. Thus, these processes are the main culprits of toxins which are emitted by the textile industry. Additionally, its effluent's chemical burden is indicated by the values of pH, total solids, chemical oxygen demand, and biochemical oxygen demand. Especially during the wet manufacturing stages, the textile dyes have a high chemical content [34]. When textile wastewater samples were gathered from multiple dyeing companies, observation surveys showed the true variations in TSS, COD, and BOD. These variances illustrate the type of technique that has an impact on the character of the effluent [23].

This chapter also found that the type of process, as well as the equipment and raw materials used, affect the properties of textile effluents. The type of equipment used, the variety of unit activities, the quantity of raw materials used, and the process type all affect the effluent's characteristics. Conventional methods are favored in most textile industries because they are known and economical. The process, equipment, chemicals, and amount of water determine the effluent characteristics. Wastewater production is significantly impacted by any major changes made to one of these factors. Consequently, the process is identified as the polluting source and thus becomes the reason behind the ongoing pollution load. Wet processing processes are thought to have contributed significantly to the pollutant burden. The research described above make it clear that the pollution load is greatly influenced by the sort of textile, the size of the manufacturing facility, its operation, the tools, chemicals used, and the amount that is produced [25].

Heavy Metals in Textile Effluents and Their Effects on Human Health or Ecosystem

Metals are used to give color to many dyes; typical heavy metals including copper, zinc, chromium, lead, cobalt, nickel, and manganese are often linked to these colors. These metals' release into soil and water can have serious negative effects on the environment and human health. Numerous health problems, such as headaches, congenital abnormalities, nausea, and irritation of the skin and lungs, have been connected to dye exposure. Furthermore, mutagenic chemicals are present in several colors. Because organic dyes contain azo and nitro groups, incomplete chemical degradation may lead to the creation of aromatic amines. These aromatic amines can cause human tumors and are carcinogenic. Organic dye-derived non-biodegradable chemicals linger in the environment and have the ability to enter and upset the food chain. This bioaccumulation of heavy metals effect both animals and humans [31].

Heavy metals naturally accumulated in soil, water and human body. They are extremely dangerous, especially in high concentrations. Certain heavy metals are harmful to human and other species even at trace concentrations, as they cannot be broken down or eliminated by living beings. Consuming tainted water, eating food that has been contaminated by heavy metals throughout the food chain, or breathing in air that has higher concentrations of heavy metals close to emission sources can all result in heavy metal toxicity [41]. High levels of COD, BOD, TDS, and TSS, combined with low DO and vibrant colors, are characteristics of textile effluents. Color's aesthetic value at the discharge point is not the main problem. Additionally, it reduces light penetration in water bodies that receive it and hence limiting the photosynthetic activity in aquatic plants. The main types of inorganic pollutants are metallic salts and basic and acidic substances. Such inorganic substances cause the receiving water bodies to undergo a variety of metabolic reactions that deteriorate the water's quality. Textile effluents with excessive amounts of salt deteriorate soil quality and render it unsuitable for agricultural use [35].

In addition to having detrimental effects on several vital human organs, the majority of heavy metals have been shown to hinder physiological functions like photosynthesis and transpiration in plants and other organisms. Intracellular concentrations of heavy metals can vary from nanomolar to femtomolar due to interspecies differences in metallic stoichiometry [5]. Nonetheless, trace levels of a number of heavy metals are necessary for many metabolic processes, especially as cofactors for enzymes. Excessive amounts of heavy metals bind to biomolecules' functional groups and disrupt their operation, which impacts fundamental metabolism. Examples of these effects includes direct DNA damage, the formation of toxic free radicals and reactive oxygen species, alterations in cell redox state, loss of cell membrane permeability, and the manufacture of nonfunctional protein-metal adducts. Remediation of HM-contaminated terrestrial and aquatic environments is essential to re-establishing the disrupted biological balance of our planet [1].

Table 1 Common heavy metals in textile wastewater along with its effects on human health with its permissible limits

Heavy metals	Effects on human health	WHO permissible limit (μg)	References
Chromium	Acute gastrointestinal pain, liver and kidney damage, necrosis, diarrhea, nausea, rhinitis, laryngitis, pulmonary fibrosis, prostate, stomach, kidney, urinary system, and bone cancer, carcinogenic and genotoxic conditions	50	[26]
Zinc	Gastrointestinal distress, nausea, cramps, skin rashes, vomiting, anemia, neurological symptoms, fatigue, increased thirst, and feeling dizzy	3000	[10]
Cadmium	The immune system, circulatory system, lungs, tests, brain, liver, kidneys, and bones	3	[19]
Mercury	Immune system, cardiovascular system, endocrine system, brain, lungs, kidneys, liver, and reproductive system	6	[47]
Nickel	Skin, renal, pulmonary fibrosis, gastrointestinal issues, and lungs	70	[33]

Need for Heavy Metals Effective Remediation

Like other organic molecules, heavy metals are typically hazardous by nature or difficult to break down through basic biological processes. Their tendency to bioaccumulate and their gradual discharge into water bodies pose significant risks to living beings. As a result, treating wastewater that contains metals before releasing it into the environment is crucial [9]. Table 1 lists some prevalent heavy metals, along with their potential health risks.

Treatment Methods for Heavy Metals in Textile Effluents

Numerous methods including coagulation, membrane separation, electrocoagulation, precipitation, Fenton's oxidation, sonication, irradiation, adsorption [20] and photocatalysis have been explored for treating both synthetic and real textile effluents [17]. The selection of removal methods depends on different factors including industry applicability, convenience of use, economics, efficiency, use of energy, and effects on the environment. Some of the conventional technologies that have developed over time are weaving, which is also a component of the textile industry's

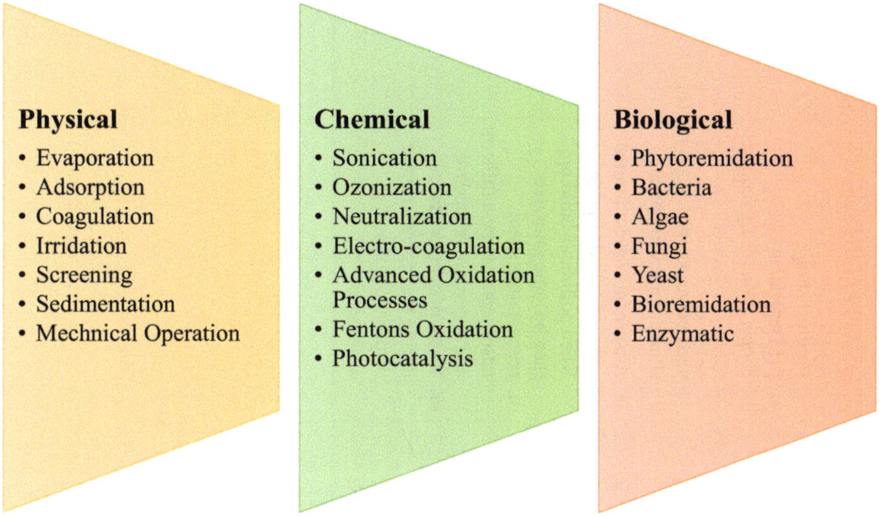

Fig. 1 Textile effluents treatment methods

growth. But many manufacturing operations are under pressure from the need for high-quality textiles and operating costs. Based on the procedures performed within the textile unit, different types of effluent are produced. Therefore, creating a treatment approach that considers the main elements affecting textile functioning is difficult. Textile processes are often divided into three categories: physical, chemical, and biological which is described in Fig. 1 based on the information from [25].

Limitations of These Methods

Several traditional and innovative techniques for treating textile wastewater are covered in the section under "Treatment Methods for Heavy Metals in Textile Effluents." Coagulation, Fenton's method, ozonation, electrochemical method, membrane technology, photocatalysis, bioremediation, adsorption, and irradiation are the methods used to treat textile wastewater effluent. The approaches are chosen based on efficacy, affordability, and convenience of use. Sludge production is one of the main obstacles to coagulation. Furthermore, the sludge load is greatly increased by coagulation, which employs extra material as a coagulant or additive [46]. Although a specially made electrochemical cell is needed, electrochemical technologies greatly lessen the silt generation issue related to the coagulation process. The electrochemical process necessitates, at the very least, an electrochemical cell that is powered continuously. One major challenge for electrochemical operations is the degradation and upkeep of an electrode [8]. Pros and cons of heavy metals removal methods are described in Table 2.

Table 2 Heavy metals removal conventional methods and their characteristics along with their pros and cons

Methods	Characteristics	Pros	Cons	References
Coagulation	i. Needs additional chemicals ii. Time consuming iii. Require a mixing device	• Effective in removing SS, color and natural organic matter	i. Produce sludge ii. pH Dependent iii. Proper monitoring requires	[36]
Electrocoagulation	i. An electromagnetic cell is required to carry treatment	• Extremely efficient at eliminating contaminants, including heavy metals	i. Need power supply ii. More complex to operate	[8]
Adsorption	i. Molecules frequently migrate from high concentration areas to low concentration areas ii. In many procedures, recovering the adsorbent and regenerating it can be difficult iii. Time, pH, temperature, adsorbent quantity, and dye starting concentration are some of the factors that affect adsorption	• Selective processes allowing specific pollutants to adsorb • No need of additional chemicals • Many adsorbents can be reused again • Relatively simple to operate • It is applicable to both the liquid and gas phases	i. Adsorbents have a limited capacity for adsorption, and when pollutants build up, they become saturated ii. regenerating spent adsorbents can be a difficult and expensive procedure iii. Waste may be produced during the regeneration or disposal of exhausted adsorbents	[4]

(continued)

Table 2 (continued)

Methods	Characteristics	Pros	Cons	References
Photocatalysis	i. Precise nanomaterial choice is required to customize a photocatalyst ii. It is a method that effectively removes a specific contaminant on a laboratory size iii. A variety of contaminants could impede the procedure	• It is possible to effectively degrade a wide range of organic and inorganic pollutants, including heavy metals, dyes, insecticides, and volatile organic compounds (VOCs) and eliminated from soil, water, and air by photocatalysis. This helps to protect and clean up the environment	i. Its implication can be complex ii. Costly	[39]
Fenton's oxidation	i. Sludge is the major drawback of this method	• Fast reaction rates • Efficient pollutant removal	i. To achieve competent removal proper design important ii. Some cases produce iron sludge which need disposal	[23]
Membrane separation	i. Numerous membranes have been developed for the treatment of textile effluents, demonstrating the technique's effectiveness on a broad scale	• Membranes can be designed to separate particular components according to their size, shape, charge, or other characteristic • Target chemicals can be separated from complex combinations thanks to this selectivity	i. The accumulation of particles, germs, or substances in the feed stream on the membrane's surface or inside its pores is known as membrane fouling ii. This may result in decreased separation efficiency and the need for regular membrane cleaning or replacement	[40]

2 Bioremediation for Heavy Metals Treatment

By using biological organisms and their parts to accumulate, adhere, trap, breakdown, transform, mobilize, or confine environmental poisons prevalent in various environmental fields, bioremediation is a technique for stopping this kind of waste. Numerous plant and microbial species, such as bacteria, fungi, yeasts, and microalgae, have demonstrated a great deal of promise for bioremediation of heavy metal-derived pollution in aquatic environments by attaching to and absorbing metals [16]. Additionally, the bioremediation process uses bacteria to eliminate effluent from textiles. Numerous species are used in wastewater treatment processes, such as bacteria, fungi, and algae, which have been demonstrated to be helpful in the breakdown of organic contaminants. Once more, the kind of biomaterial and contaminant causing the process determines how these organisms behave. The deterioration is caused by distinct biological mechanisms that need more research. Additionally, another irradiation-related technique breaks the bindings by irradiating the effluent using gamma rays. Because the irradiation experiment requires a particular facility to shield people from radiation exposure, the method will cost more to develop [22]. In contrast to conventional chemical and physical cleanup techniques, bioremediation is less intrusive, simpler to use, and more environmentally friendly. The procedures used in bioremediation can remove, detoxify, and lessen the toxic effects of harmful chemicals such as pesticides, heavy metals, xenobiotics, organic halogens, industrial waste water, and chlorine-based compounds [38].

Although genetically engineered organisms and populations of microbes have been used in molecular research to speed up decomposition and combat a greater range of toxins, new knowledge has expanded the application of bioremediation. Recent studies have detailed the potential for developing novel microbial strains with improved mechanisms for pollutant degradation, which could lead to new developments in this area. Furthermore, it is feasible to reveal intricate pollution scenarios and offer an extra path for environmental restoration by combining bioremediation with phytoremediation, which uses plants as a medium to support microbial operations. Further study has investigated the potential applications of biomathematics and machine learning in both anticipating and improving the bioremediation processes. By the help of big data on environmental conditions and microbial genes, research institutes can pinpoint important elements that influence the effects of bioremediation and even optimize treatments. An important step towards improving the efficiency and economics of bioremediation is the data-driven approach. Bioremediation develops in parallel with technology and research recognized for improving its accessibility and effectiveness in cleaning contaminated areas. To this aim, harnessing the skills of microbes to clean and safeguard our environment is still one of the most essential ways for solving the myriad difficulties stemming from pollution [27].

3 Bioremediation Types

Based on the location where the cleaning processes take place, bioremediation can be divided into two categories (Fig. 2).

Insitu Bioremediation

An essential technique for purging the environment of hazardous substances is in-situ bioremediation. This method involves conducting remediation that takes place right at the site of the pollution. The process is known as bioremediation if it uses plants and animals. In situ bio-processing, also known as directly bio-processing, treats polluted groundwater or soil while requiring time-consuming and expensive cleanup procedures. This method efficiently eliminates chemical poisons by using innocuous bacteria, plants, and animals. Choosing local bioremediation has several benefits, such as using natural processes, lowering shipping costs, and requiring less physical effort. There are a number of solutions for on-site bioremediation, including as composting, slurping, bioventing, sparking, natural attenuation, and microorganisms that support phytoremediation. The effectiveness of in-situ bioremediation may be restricted by several reasons, notwithstanding its advantages. Various organic chemicals frequently contaminate the soil, especially in industrialized areas. Moreover, low temperatures, anaerobic conditions, deficits in nutrients and co-substrates, a lack of degradation capacity, and the bioavailability of organic compounds can all impede the breakdown of organic compounds. Environmental factors have a big impact on how well in-situ bioremediation works. Significant contributions are played by elements such soil type, neutral pH, seasonal temperature fluctuations, dietary supplements, oxygen levels, and gene expression. In-situ bioremediation can become a potent instrument for reducing environmental contamination and encouraging ecosystem restoration if these issues are successfully resolved [42]. Here some major types of in-situ bioremediation are given [27].

Fig. 2 Bioremediation types

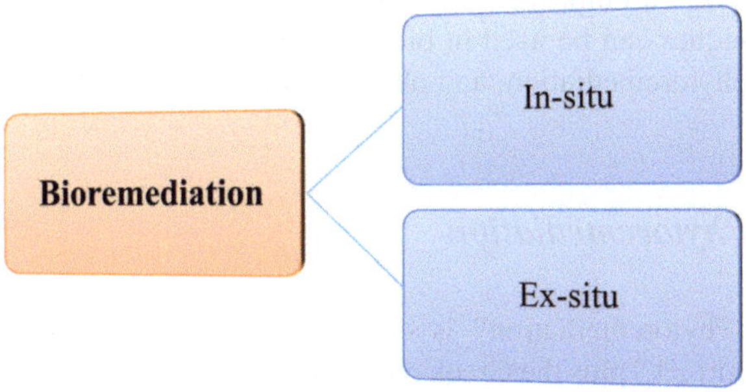

(i) Natural attenuation
(ii) Biostimulation
(iii) Bioaugmentation
(iv) Phytoremediation
(v) Bioventing
(vi) Electrobioremediation.

Ex-situ

Pollutants are actually exported from the contaminated site and treated at a different location as part of ex-situ biodegradation. According to [2], efficiency metrics are used to choose ex-situ biodegradation methods [2]. The nature and number of pollutants, the expense of treatment, the polluted site's location, or its topography are some of the variables that affect the decision to use ex-situ bioremediation techniques. In wastewater treatment microorganisms and plants are involved to uptake or convert heavy metals into less toxic forms. Major types of ex-situ bioremediation are given below [27].

(i) Bioreactors
(ii) Landfarming
(iii) Composting
(iv) Bioaugmentation
(v) Bioleaching
(vi) Phytoremediation
(vii) Biosorption.

4 Mechanism of Bioremediations

A type of biotechnology called bioremediation uses biological sources of plants, microorganisms, algae, etc. to remove pollutants from the environment. Furthermore, it employs straightforward, reasonably priced methods. The following techniques can be used in bioremediation: mycoremediation, bacterial bioremediation, phytoremediation, and phycoremediation (Fig. 3).

Phytoremediation

"Phytoremediation" is derived from the Latin word "remedium," meaning "to correct," plus the Greek prefix "phyton," meaning "plant." Utilizing particular plants, like vegetation, aquatic plants, grasses, and shrubs, phytoremediation is a method

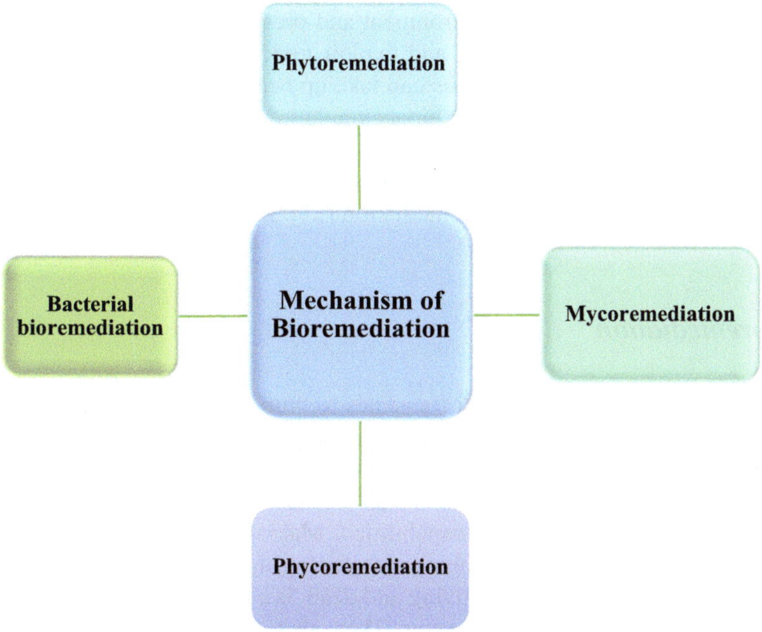

Fig. 3 Mechanism of bioremediation

that eliminates toxins from the environment. When comparison with other conventional remediation techniques, phytoremediation has some advantages. It can reduce environmental harm and is reasonably priced for huge amounts of water containing a variety of toxins. The following compounds can be subjected to phytoremediation: petroleum hydrocarbon insecticides and pesticides, heavy metals, explosives, metalloids, inorganic materials, radioactive elements, organic manufacturing wastes, and chlorinated solvents. Extraction, volatilization, stabilization, rhizo-filtration, and degradation are plant-based methods for heavy metal decontamination [24].

Microbial Bioremediation

Microorganisms are seen as a significant alternative solution to problems. Metals are crucial for the proliferation, differentiation, and metabolism of microorganisms. Even if microbes are more at risk due to the increased pollution toxicity, microorganisms have evolved several strategies to cope with heavy metal exposure. A number of methods used in bioremediation, including biotransformation, bioaccumulation, biosorption, bio-separation, biodegradation, bio-bleaching, and biodetoxification, underlie the associations among metals and microbes which extract metals from wastewater [6]. To restore the environment and prevent further pollution, microorganisms could change, transform, and exploit toxic pollutants to produce energy and biomass. Interestingly, microbes can take up heavy metals through a proactive procedure via intracellular storage (bioaccumulation) within its living biomass and a passive process from external adherence to both living and dead biomass. Positive results have been obtained when single, consortium, and immobilized metal-resistant strains are used for heavy metals remediation [21].

Mycoremediation

The term "mycoremediation," that translates to "fungi removing toxic chemicals." Fungi are varied microorganisms that are eukaryotic with a wide range of applications, including molds, yeasts, and mushrooms [32]. Fungi have recently gained popularity for the removal of heavy metals because of their enormous size, abundance, and excellent adsorption capabilities. Making a large number of them is comparatively simple and affordable, and altering their genes is similarly simple. Additionally, their biomass, both living and dead, is a natural supply of high-quality feed for treating industrial wastewater. They also emit a variety of extracellular metabolites, such as proteins, organic acids, and enzymes [15]. Fungi have become the preferred material for heavy metal removal because of their high energy content and adsorption properties. From a genetic perspective, they are highly adaptable, and they can be produced in large quantities at affordable prices. Industrial wastewater can be effectively treated with a variety of extracellular metabolites, such as proteins, organic acids, enzymes, and their biomass (dead or alive) [11].

5 Processes Involved in Mycoremediation

Since fungi are ubiquitous, they have been able to change to the majority of waste types. The methods by which fungi purify the heavy metal-filled environment can be classified as metabolically passive and metabolically active which is further described in following flow chart (Fig. 4).

Biosorption and Bioaccumulation

One of the crucial procedures that relies on a higher affinity within the biosorbent (biological substance) and sorbate (metal ions) is biosorption. It carried on till both elements were balanced. Heavy metals are eliminated by fungi by adhering them on its surfaces by attaching them to functional molecules that are present on the

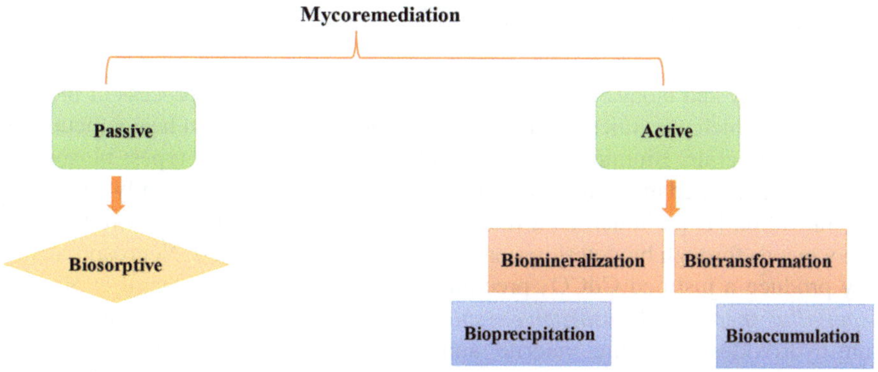

Fig. 4 Flowchart mycoremediation and its mechanisms

outermost layer of the cell, which include amine, carboxyl, phosphate, hydroxyl, and sulfhydryl groups [24].

Anionic compounds found in the outermost fungal cell barrier give cationic heavy metals to attach to. Various chemical compounds on the exterior structures of fungal cells which contribute to metal binding distinguish them from those of other evolutionary units. The terms "bioaccumulation" and "biosorption," both rely on the biomass's chemical state, have proven rather ambiguous in the literature. The majority of living cells are employed in mechanisms classified as "bioaccumulation," whereas "biosorption" activities use dead biomass [13].

Bio-immobilization

Microorganisms are used in the bio-immobilization process to decrease metals mobility and convert soluble metals into solids that are insoluble. Numerous investigations have demonstrated that fungi can precipitate heavy metals and then release them [45]. Table 3 displays a list of fungal species that have been immobilized heavy metals.

Table 3 Different fungi spp. that immobilized heavy metals

Fungi	Uptake metal	References
Rhizopus delemar	Copper	[7]
Penicillium janthinillum	Lead and copper	[44]
Fusarium oxysporum	Mercury	[43]
Penicillium chrysogenum	Cadmium	[28]
Aspergillus niger	Cadmium	[14]

Biomineralization

Metal mobility and bioavailability are reduced by the biological process of biomineralization, which uses microorganisms to immobilize and convert heavy metal ions into solid minerals. Biomineralization can be classified into two types: biosourced bioprocess regulated mineralization and source-triggered physiologically stimulated mineralization [18]. Numerous investigations have shown that fungi could oxidize heavy metals through biodeterioration. According to Li et al. [29], Neurospora crass may produce a justified $CdCO_3$ precipitate [29]. According to Dhami et al. [12], *Aspergillus species* and *Fusarium oxysporum* can produce carbonates of lead, strontium hydroxides, aragonite, vaterite, and calcite. Furthermore, there is proof that *Myrothecium gramineum* and *Pestalotiopsis* sp. can produce the minerals $SrCO_3$ and $CaCO_3$ [12]. Furthermore, Li et al. [30] noted that *Myrothecium gramineum* and *Pestalotiopsis* sp. are capable of mineralizing $CaCO_3$ and $SrCO_3$ [30].

6 Conclusion and Future Prospects

Textile industry is one of the largest freshwater users and a major wastewater contributor in the world, which has adverse effects on aquatic life and human health because it releases many harmful chemicals like heavy metals and dyes in its effluent without any treatment. In the past physical, chemical and biological treatments were used to remove heavy metals. In this era, interest toward green technology (bioremediation) is increasing due to its cost effectiveness, environment friendly and ecological friendly. This study highlights the importance of adopting bioremediation to treat effluent effectually. These techniques can decrease the risk toward human health due to contact with heavy metals and promoting sustainability especially in agriculture sectors where wastewater can be reused. Future research should look toward new sp. of microorganisms that have a high potential for heavy metals removal.

References

1. Ali, H., Khan, E., & Ilahi, I. (2019). Environmental chemistry and ecotoxicology of hazardous heavy metals: Environmental persistence, toxicity, and bioaccumulation. *Journal of Chemistry, 2019*(1), 6730305.
2. Atlas, R. M., & Philp, J. (2005). Bioremediation: Applied microbial solutions for real-world environmental cleanup.
3. Azanaw, A., Birlie, B., Teshome, B., & Jemberie, M. (2022). Textile effluent treatment methods and eco-friendly resolution of textile wastewater. *Case Studies in Chemical and Environmental Engineering, 6*, Article 100230.
4. Bakar, A. A., Mazlan, W. W., Akbar, N., Badrealam, S., & Ali, K. M. (2019). Agriculture waste from banana peel as low cost adsorbent in treating methylene blue from batik textile waste water effluents. *Journal of Physics: Conference Series.*

5. Balzano, S., Sardo, A., Blasio, M., Chahine, T. B., Dell'Anno, F., Sansone, C., & Brunet, C. (2020). Microalgal metallothioneins and phytochelatins and their potential use in bioremediation. *Frontiers in Microbiology, 11*, 517.
6. Barrech, D., Ali, I., & Tareen, M. (2018). 1. A review on mycoremediation—The fungal bioremediation. *Pure and Applied Biology (PAB), 7*(1), 343–348.
7. Başkan, G., & Açıkel, Ü. (2024). Comparison of the bioaccumulation and biosorption of copper ions by Rhizopus delemar and Candida lipolytica. *Bioremediation Journal,* 1–17.
8. Bener, S., Bulca, Ö., Palas, B., Tekin, G., Atalay, S., & Ersöz, G. (2019). Electrocoagulation process for the treatment of real textile wastewater: Effect of operative conditions on the organic carbon removal and kinetic study. *Process Safety and Environmental Protection, 129*, 47–54.
9. Bilal, M., Ihsanullah, I., Younas, M., & Shah, M. U. H. (2021). Recent advances in applications of low-cost adsorbents for the removal of heavy metals from water: A critical review. *Separation and Purification Technology.*
10. Chasapis, C. T., Ntoupa, P.-S.A., Spiliopoulou, C. A., & Stefanidou, M. E. (2020). Recent aspects of the effects of zinc on human health. *Archives of Toxicology, 94*, 1443–1460.
11. Dell'Anno, F., Rastelli, E., Buschi, E., Barone, G., Beolchini, F., & Dell'Anno, A. (2022). Fungi can be more effective than bacteria for the bioremediation of marine sediments highly contaminated with heavy metals. *Microorganisms, 10*(5), 993.
12. Dhami, N. K., Quirin, M. E. C., & Mukherjee, A. (2017). Carbonate biomineralization and heavy metal remediation by calcifying fungi isolated from karstic caves. *Ecological Engineering, 103*, 106–117.
13. Dhankhar, R., & Hooda, A. (2011). Fungal biosorption–An alternative to meet the challenges of heavy metal pollution in aqueous solutions. *Environmental technology, 32*(5), 467–491.
14. Dipannita, D., Pinky, B., Devi, N. N., & Mayuri, C. (2021). Biosorption of cadmium by fungi isolated from Bharalu River, Assam. *Journal of Tropical Life Science, 11*(3).
15. El-Bondkly, A. M. A., & El-Gendy, M. M. A. A. (2022). Bioremoval of some heavy metals from aqueous solutions by two different indigenous fungi Aspergillus sp. AHM69 and Penicillium sp. AHM96 isolated from petroleum refining wastewater. *Heliyon, 8*(7).
16. Fulke, A. B., Ratanpal, S., & Sonker, S. (2024). Understanding heavy metal toxicity: Implications on human health, marine ecosystems and bioremediation strategies. *Marine Pollution Bulletin, 206*, Article 116707.
17. Fulzele, N. N., Bhanvase, B. A., & Pandharipande, S. L. (2022). Sonochemically prepared rGO/Ag_3PO_4/CeO_2 nanocomposite photocatalyst for effective visible light photocatalytic degradation of methylene dye and its prediction with ANN modeling. *Materials Chemistry and Physics, 292*, Article 126809.
18. Gadd, G. M. (2010). Metals, minerals and microbes: Geomicrobiology and bioremediation. *Microbiology, 156*(3), 609–643.
19. Genchi, G., Sinicropi, M. S., Lauria, G., Carocci, A., & Catalano, A. (2020). The effects of cadmium toxicity. *International Journal of Environmental Research and Public Health, 17*(11), 3782.
20. Girhe, P., Barai, D., & Bhanvase, B. (2021). Adsorption of metals using activated carbon derived from coal. In *Clean coal technologies: Beneficiation, utilization, transport phenomena and prospective* (pp. 233–265). Springer.
21. Goutam, J., Sharma, J., Singh, R., & Sharma, D. (2021). Fungal-mediated bioremediation of heavy metal–polluted environment. *Microbial Rejuvenation of Polluted Environment:, 2*, 51–76.
22. Haddaji, D., Ghrabi-Gammar, Z., Hamed, K. B., & Bousselmi, L. (2019). A re-circulating horizontal flow constructed wetland for the treatment of synthetic azo dye at high concentrations. *Environmental Science and Pollution Research, 26*(13), 13489–13501.
23. Islam, M., & Mostafa, M. (2018). Textile dyeing effluents and environment concerns-A review. *Journal of Environmental Science and Natural Resources, 11*(1–2), 131–144.
24. Jeyakumar, P., Debnath, C., Vijayaraghavan, R., & Muthuraj, M. (2023). Trends in bioremediation of heavy metal contaminations. *Environmental Engineering Research, 28*(4).

25. Kallawar, G. A., & Bhanvase, B. A. (2024). A review on existing and emerging approaches for textile wastewater treatments: Challenges and future perspectives. *Environmental Science and Pollution Research, 31*(2), 1748–1789.
26. Kumar, A. S. K., Jiang, S.-J., & Tseng, W.-L. (2015). Effective adsorption of chromium (VI)/Cr (III) from aqueous solution using ionic liquid functionalized multiwalled carbon nanotubes as a super sorbent. *Journal of Materials Chemistry A, 3*(13), 7044–7057.
27. Kuppan, N., Padman, M., Mahadeva, M., Srinivasan, S., & Devarajan, R. (2024). A comprehensive review of sustainable bioremediation techniques: Eco friendly solutions for waste and pollution management. *Waste Management Bulletin.*
28. Li, J., Zhang, W., Lin, Z., Huang, Y., Bhatt, P., & Chen, S. (2021). Emerging strategies for the bioremediation of the phenylurea herbicide diuron. *Frontiers in Microbiology, 12*, Article 686509.
29. Li, Q., Csetenyi, L., & Gadd, G. M. (2014). Biomineralization of metal carbonates by Neurospora crassa. *Environmental Science & Technology, 48*(24), 14409–14416.
30. Li, Q., Csetenyi, L., Paton, G. I., & Gadd, G. M. (2015). $CaCO_3$ and $SrCO_3$ bioprecipitation by fungi isolated from calcareous soil. *Environmental Microbiology, 17*(8), 3082–3097.
31. Madhav, S., Ahamad, A., Singh, P., & Mishra, P. K. (2018). A review of textile industry: Wet processing, environmental impacts, and effluent treatment methods. *Environmental Quality Management, 27*(3), 31–41.
32. Mohmand, A. Q. K., Kousar, M. W., Zafar, H., Bukhari, K. T., & Khan, M. Z. (2011). Medical importance of fungi with special emphasis on mushrooms. *ISRA Medical Journal, 3*(1), 1–44.
33. Nieboer, E., Rossetto, F. E., & Menon, C. R. (2024). Toxicology of nickel compounds. In *Metal ions in biological systems* (pp. 359–402). CRC Press.
34. Paździor, K., Wrębiak, J., & Ledakowicz, S. (2020). Treatment of industrial textile wastewater in biological aerated filters–Microbial diversity analysis. *Fibres & Textiles in Eastern Europe, 1*(139), 106–114.
35. Periyasamy, A. P. (2024). Recent advances in the remediation of textile-dye-containing wastewater: Prioritizing human health and sustainable wastewater treatment. *Sustainability, 16*(2), 495.
36. Reck, I. M., Baptista, A. T. A., Paixão, R. M., Bergamasco, R., Vieira, M. F., & Vieira, A. M. S. (2020). Application of magnetic coagulant based on fractionated protein of Moringa oleifera Lam. seeds for aqueous solutions treatment containing synthetic dyes. *Environmental Science and Pollution Research, 27*, 12192–12201.
37. Saxena, G., Chandra, R., & Bharagava, R. N. (2017). Environmental pollution, toxicity profile and treatment approaches for tannery wastewater and its chemical pollutants. *Reviews of Environmental Contamination and Toxicology, 240*, 31–69.
38. Sharma, I. (2020). Bioremediation techniques for polluted environment: Concept, advantages, limitations, and prospects. In *Trace metals in the environment-new approaches and recent advances.* IntechOpen.
39. Sharma, M., Sondhi, H., Krishna, R., Srivastava, S. K., Rajput, P., Nigam, S., & Joshi, M. (2020). Assessment of GO/ZnO nanocomposite for solar-assisted photocatalytic degradation of industrial dye and textile effluent. *Environmental Science and Pollution Research, 27*, 32076–32087.
40. Sonawane, A. V., & Murthy, Z. (2023). A metal organic framework decorated 2-dimensional nanomaterial based nanocomposite photocatalyst for photocatalytic degradation of dyes from textile industry wastewater. *Environmental Science: Water Research & Technology, 9*(10), 2515–2537.
41. Türksoy, R., Terzioğlu, G., Yalçın, İE., Türksoy, Ö., & Demir, G. (2021). Removal of heavy metals from textile industry wastewater. *Frontiers in Life Sciences and Related Technologies, 2*(2), 44–50.
42. Urionabarrenetxea, E., Garcia-Velasco, N., Anza, M., Artetxe, U., Lacalle, R., Garbisu, C., Becerril, T., & Soto, M. (2021). Application of in situ bioremediation strategies in soils amended with sewage sludges. *Science of the Total Environment, 766*, Article 144099.

43. Văcar, C. L., Covaci, E., Chakraborty, S., Li, B., Weindorf, D. C., Frențiu, T., Pârvu, M., & Podar, D. (2021). Heavy metal-resistant filamentous fungi as potential mercury bioremediators. *Journal of Fungi, 7*(5), 386.
44. Wang, R., Fan, X.-W., & Li, Y.-Z. (2022). Efficient removal of a low concentration of Pb (II), Fe (III) and Cu (II) from simulated drinking water by co-immobilization between low-dosages of metal-resistant/adapted fungus Penicillium janthinillum and graphene oxide and activated carbon. *Chemosphere, 286*, Article 131591.
45. Yahaya, Y. A., Don, M. M., & Bhatia, S. (2009). Biosorption of copper (II) onto immobilized cells of Pycnoporus sanguineus from aqueous solution: Equilibrium and kinetic studies. *Journal of Hazardous Materials, 161*(1), 189–195.
46. Yu, R., Chen, H., Cheng, W., & Huang, H. (2017). Applying online image analysis to simultaneously evaluate the removals of suspended solids and color from textile wastewater in chemical flocculated sedimentation. *Journal of Environmental Informatics, 29*(1).
47. Zulaikhah, S. T., Wahyuwibowo, J., & Pratama, A. A. (2020). Mercury and its effect on human health: A review of the literature. *International Journal of Public Health Science, 9*(2), 103–114.

Biological Decolorization of Textile Dye Wastewater Using Microbial Consortia

Rubiya Dar and Tanvir-ul-Hassan Dar

1 Introduction

Water resources have become increasingly scarce for human survival over the last few decades, with one in every three people worldwide lacking access to clean water. The main contributing factors to the lack of clean water are the exponential increase in population, agriculture, urbanisation, and industrialization [20, 56, 119, 179, 197, 257]. Different industrial sectors raise serious environmental and public health concerns. Textile dyeing is one of the most water-intensive industries, releasing substances such as recalcitrant organic molecules, high concentrations of Biological Oxygen Demand (BOD), Chemical Oxygen Demand (COD), fibres, surfactants, detergents and solvents [185]. The textile industry is one of the oldest and important manufacturing industries in the world that has contributed immensely to the development of many economies, employing approximately 35 million people around the world [69]. Despite its undeniable importance the textile industry is unfortunately one of the highly polluting industries. Textile industries consume huge volumes of freshwater for its various wet processes like desizing, scouring, bleaching, dyeing as well as finishing of textiles and release equal amounts of wastewaters [267]. These processes consume large amounts of chemicals in the form of alkalis, salts, surfactants, dyes, pigments and water [193]. An average textile mill producing 8000 kg of fabric per day uses around 1.6 million litres of water, with dyeing and printing accounting for 24% of this total [119]. The textile and dyeing industry is projected to

R. Dar (✉)
Centre of Research for Development (CORD), University of Kashmir, Jammu and Kashmir, India
e-mail: rubiyadar@uok.edu.in

Present Address:
Tanvir-ul-Hassan Dar
Department of Biotechnology, School of Biosciences and Biotechnology, BGSB University, Rajouri, Jammu and Kashmir, India
e-mail: tanvirulhasan@gmail.com

R. A. Bhat et al. (eds.), *Biocentric Approaches in Textile Waste Management*, Textile Science and Clothing Technology, https://doi.org/10.1007/978-3-032-04974-2_7

consume approximately 700,000 tonnes of dyes each year, with 10–15% remaining unfixed and eventually finding their way into water bodies [196]. Because textile dyes block light from passing through the water, their disposal into bodies of water has an impact not only on the water's look but also on the stability of aquatic habitats. According to Lambert and Davy [137], this results in lower oxygen levels in the aquatic environment, which has an adverse effect on the aquatic organisms that live there. Because of the risks that these dye pollutants bring to aquatic life as well as humans, the disposal of untreated textile dye waste water has thus become a major concern in many nations. To eliminate dyes from textile effluents, a variety of physicochemical [13, 228], electrochemical, and biological treatment techniques have been employed, including adsorption, coagulation, flocculation, membrane technologies, fenton-oxidation, ozonation, and electrolysis [123, 231]. These techniques each have pros and cons in terms of effectiveness, cost-effectiveness, and the production of hazardous byproduct waste, which may help to explain why, in order to achieve optimal efficiency, a variety of techniques are typically employed to treat textile dye waste water. Due to its affordability, environmental friendliness, and low sludge production, bioremediation has emerged as a unique issue [63]. The use of pure culture to decolorise synthetic dyes has been the subject of numerous investigations. The development of more effective microbial decolorisation techniques is required because of the chemical complexity of these dyes [228]. The use of microbial agents to remediate dye effluents has become more popular recently. For effective dye removal, a number of microbial consortia have been documented [31, 187, 205]. This has mostly been brought about by the benefits that biological waste water treatment techniques offer over conventional techniques. Numerous microbiological species, including bacteria, fungi, yeast, and algae, have been employed to cleanse wastewater containing dyes. For the treatment of dye wastewater, algae, yeast, and fungi have historically been the favoured microbial agents. However, bacterial organisms have recently drawn increased attention because of their prolonged growth cycles and poor decolorising dye effectiveness [111, 180]. Shorter growing time and the capacity to extract and breakdown alizedyes are two advantages of bacterial isolates. Numerous studies on the optimisation of bacterial isolates for wastewater treatment have been published. Finding bacterial strains that can be used to break down and decolorise dye waste water and attempting to comprehend their mechanism of action have been the main goals of several of the investigations.

Classification of Textile Dyes

These days, the textile business, along with other industries like paper, plastics, leathers, and cosmetics, uses a variety of dyes and pigments because of their colouring qualities. Textile processing uses a lot of water, which is then discharged into the environment untreated. One of the primary pollutants released into the environment by the textile industry is dye effluent. Dye effluents are one of the main pollutants carried to the environment, mainly by textile industries. These dyes are most commonly classified according to their source, chemical structure and industrial application [73]. Natural and synthetic dyes are the two types of dyes. While synthetic textile dyes are made from organic and inorganic chemicals, typically from petroleum byproducts or earth mineral types of synthetic resources, natural textile colours are derived from plants, insects, and minerals [94]. Compared to natural textile colours, synthetic dyes are used in the textile industry the most. Their solubility in water, ease of absorption, and extremely quick colouration in comparison to natural textile dyes are the main reasons for their appeal [111, 129]. A limited shade range, poor colour fastness, and a low degree of fixing are just a few of the many issues that natural textile dyes face [263]. However, due to their advantageous functional qualities, which include anti-microbial, anti-bacterial, anti-inflammatory, anti-insect, UV protection and deodorising qualities, natural textile dyes can still be used [80]. Classification of dyes by chemical structure focuses on the two main components. The first component consists of a set of atoms known as chromophores, which lend colour to the dye and the second component consists of electron withdrawing chromophores known as auxochromes. Chromophores include azo (–N=N–), carbonyl (–C=O), methine (–CH=), nitro (NO_2), quinoid groups, triphenylmethane, phthalein, triphrnylmethyl, indigo, azine, xathene, oxazine, phthalocyanine and triarylenethane, anthraquinone, nitrated and nitrosated, diphenylmethane, diarylmethane and polymethic dyes [94]. The examples of main auxochromes are amine (–NH_3), carboxyl (–COOH), sulfonate (–SO_3H) and hydroxyl (–OH) groups [182]. On the basis of their chemical properties, auxochromes can fit into the following dye classes, reactive, direct, acid, basic, pigment, vat, anionic, mordant, Sulphur, solvent and disperse dyes [107, 188, 209, 256]. The most extensively used dye classes are azo, anthraquinone and triphenylmethane dyes. Azo dyes have the greatest colour variation and are widely regarded as the largest dye class due to their colouring capabilities [22]. Azo dyes account for roughly 80% of global commercial dye production due to their ease of synthesis [102]. Griss reported the first azo dye, Aniline Yellow, in 1858. Every year, over 7×105 tonnes of dyestuffs are manufactured, and over 10,000 dyes are commercially accessible [189, 195]. One or more azo groups (R1-N = N-R2), as well as their stability and resistance to microbial attack, are characteristics of azo dyes. It is extremely difficult to remove them from waste water using standard treatments due to their poor absorbability to the fibres and their persistent chemical nature [14]. The primary structural issue with these dyes is their stability because of their aromatic character, which makes degradation challenging. Currently, practice of using green technologies mainly for the purpose of protecting environment and for reducing environmental pollution is gaining wide attention.

Environmental and Health Impacts of Textile Wastewaters

Textile effluent affects the environment in a number of ways. The release of textile waste waters causes dark coloration of surface waters that captures the attention of both the public and the authorities. Furthermore, dyes and pigments in textile effluent alter pH, create high levels of biological and chemical oxygen demand (BOD, COD) and increase total organic carbon and suspended particles [65]. All of these consequences disrupt and restrict aquatic life's biological processes [153], degrade the aesthetic quality and transparency of water bodies and hinder light penetration into deeper layers of water bodies, lowering photosynthesis [73]. The textile sector is a major economic driver, but it is also one of the top consumers of chemicals, energy and water [139]. The textile industry employs about 8000 chemicals, the majority of which are dangerous to the environment and human health. The negative environmental impact of the textile industry is mostly as a result of untreated effluent discarded into the water bodies which accounts for 80% of the total industry emission [114]. What further complicates the situation is the presence of non-biodegradable organic compounds from textile dyes in the effluent. According to Yaseen and Scholz [262], textile wastewater contains colours, suspended particles, raw materials, inorganic salts and other compounds. Without suitable and effective wastewater treatment techniques, dyes and other textile wastewater ingredients leak into the environment [73]. The colouring in waste water is caused by inadequate dye exhaustion during the dyeing process, which results in excess dye being discharged into the environment [36]. In addition to colours, wastewater contains other harmful substances such as sulphur, naphthol, nitrates, acetic acid, soaps, chromium compounds and heavy metals such as copper, nickel, arsenic, lead, cadmium, mercury and cobalt, as well as formaldehyde-based fixing agents and hydrocarbon softeners. These contaminants, when washed into water bodies, hurt the environment, disrupt the natural balance and endanger humanity [153]. Other undesirable effects include bacterial overgrowth and pestilential odours. Dye contamination in the environment can accumulate naturally across the food chain in aquatic organisms, posing a risk to species at the top of the food chain, such as humans [98]. These harmful compounds from dyeing effluent in wastewater survive in water bodies because they are resistant to biodegradation, light, temperature fluctuations, detergents, chemicals, soaps, and bleaches. The toxicity of dyes in wastewater can also modify the soil by eliminating microorganisms, affecting agricultural production. Irrigating with liquid textile effluent may clog soil pores and harden the texture, impeding root penetration [150].

Textile dyes, synthetic dyes, textile industry pollutants and their metabolic intermediate products are found to be highly toxic, carcinogenic and mutagenic posing high risk to human health [139, 154, 259]. Exposure to these dyes and pigments in water causes respiratory sharpness, which in severe cases might impair a person's immune system. Itchy and watery eyes, sneezing and general asthma symptoms including coughing and wheezing are among the respiratory difficulties' symptoms. Other compounds in dyes may potentially produce these symptoms, as well as skin irritation [150]. Other synthetic substances used in the textile business, such as optical whiteners, soda ash, caustic soda and bleach, as well as formaldehyde-based resins, ammonia, acetic acid, and shrink-resistant chemicals, have health consequences. When absorbed via the skin, they may cause allergic symptoms such as itchy and blocked noses, sniffling and sore eyes [153]. Furthermore, dye effluents contain trace metals, which can cause severe skin irritation, dermatitis, ulceration and nausea if exposed. There are wastewater regulating bodies around the world that work to keep the number of harmful compounds released into the environment under control. Methods for measuring wastewater contents differ according to standards and industry recommendations. The source, properties and planned use of treated wastewater all have an impact on effluent discharge parameters. The wastewater quality parameters include pH between 5.5 and 9, Carbon Oxygen Demand of 20 mg/l, Biological Oxygen Demand of less than 60 mg/l, oil and grease of 10 mg/l and total suspended matter of no more than 100 mg/l. The goal of wastewater treatment is to reduce hazardous substances such as Nonylphenol Ethoxylates (NPEOs and NPEs), azo dyes, brominate and chlorinated flame retardants, chlorobenzenes, chlorophenols, chlorinated solvents, heavy metals, organotin compounds, perfluorinated chemicals, phthalates and short-chained paraffins to acceptable levels. The major hazardous effects of azo dyes are induced by aromatic amines produced during biodegradation. In addition, the standard Salmonella mutagenicity assay identifies a substantial number of aromatic amines as either cancer-causing agents or established mutagens [61]. Some azo dyes, even when not cleaved into aromatic amines, can cause cancer. However, the carcinogenicity of many azo dyes is attributed to their cleavage product, benzidine. Benzidine causes several human and animal tumours. Another azo dye component, p-phenylenediamine (p-PDA), causes contact allergies [232]. Other prominent dyes, such as triphenylmethane and anthraquinone, have also been proven to have deleterious affects [59, 63, 221]. Some dyes from the triphenylmethane group have been shown to be poisonous, mutagenic and carcinogenic [59, 221]. The known negative consequences of the textile industry on the environment and humanity highlight the importance of using sustainable dyeing processes. These may include optimising the dyeing process, enhancing wash fastness and identifying and implementing novel, practical and cost-effective strategies for treating textile wastewater before it is discarded into the environment. Reducing the process temperature may also save energy.

Characteristics of Textile Wastewaters and Textile Effluents

Textile dyeing industries are facing problems to meet the green practices standards for safe discharge of wastewater due to its complex nature. Depending on the process, the machinery used in the factory, the type of fabric produced, the chemicals used, the fabric's weight, the season and fashion trends, the wastewater composition of the textile industry differs from mill to mill and from nation to nation [43, 124]. Textile waste waters are a complex mixture of salts, acids, heavy metals, organ-chlorine-based pesticides, pigments and dyes [185, 228]. Textile waste waters generated from the different wet processes are characterized by high pH, temperature [207, 216], COD, BOD [261], detergents, surfactants, suspended and dissolved solids, dispersants, leveling agents, toxic organics, chlorinated compounds, sulphide and formaldehyde, may be added to improve dye adsorption onto the fibers [189]. Such effluents are also characterized by the presence of heavy metals, such as Cr, Zn,

Cu, Fe, Pb, Hg, Cu, Co and Al due to metal-based complexes dyes [6, 102, 108, 214, 222, 228]. The textile industry is one of the most water-intensive sectors (using around 100–200 L of high-quality water per kg of textile product) and the most polluting of all industrial operations. The yearly global textile manufacturing is around 30 million tonnes, requiring 700,000 tonnes of various dyes, which pose significant environmental pollution problems. Textile manufacturing requires a wide range of machines and procedures, the most popular of which are desizing, scouring, bleaching, mercerising, dyeing, printing and finishing (Fig. 1) to produce the desired shape and qualities of the product [147]. The process determines the characteristics and quantity of textile waste waters. Dyeing is the one that uses the most water, both during the dye bath and the washing phase as well as when imparting colour to the fibres. Another important source of textile waste liquids is the finishing and mercerising processes. Additionally, the character of textile waste fluids is drastically changed by machinery, chemicals and equipment including stabilisers and detergents [185]. The fact that a variety of dyestuffs are needed for the dyeing and finishing operations in particular is another significant element that raises the ecotoxicity and volume of textile waste fluids. The range of dyes relies to the fabric used. For example, cellulose fiber necessitates the application of direct, reactive, vat, azo or sulfide dyes. Acid dyes are used mostly for wool and silk. Azo and dispersion dyes are added to the polyester fiber. A high number of these colours is released in the textile waste fluids due to their degree of attachment to fibers. Dye concentrations in textile effluent are reported throughout a wide range of values. An estimated 200,000 tonnes of these resistant compounds are wasted annually as effluent due to defective dying procedures, out of the 0.7 million tonnes of dyes generated in the textile industry worldwide each year [112]. Unfortunately, because to their strong resistance to microbiological, chemical and physical degradation, the majority of these continue to pose a threat to the environment, as well as to flora and fauna. According to [134], the textile effluent has a dye content of 10–50 mg/l. Nonetheless, it has been observed that reactive dyes in cotton factories are released in concentrations of 60 mg/l [223] and 100–200 mg/l [84]. The dye effluents mentioned by Vandevivere et al. [251] ranged from 600 to 800 mg/l. Koprivanac et al. [130] reported a reactive dye effluent concentration of 7000 mg/l. This concentration may be related to wastewater discharge from a particular textile company and is incredibly high when compared to other benchmarks. According to Abid et al. [1], 14 Ramadan textile businesses in Iraq had outflow dye concentrations ranging from 20 to 50 mg/l. According to Sivakumar [229], an Indian textile factory's final clarifier has an outflow concentration of 45 mg/l of the dye Acid Orange 10. The dye quantities released from dye houses, however, varied from 10 to 250 mg/l, according to Ghaly et al. [89]. The dye intensity in the coloured effluents was observed to range from 1000 to 1500 ADMI units, depending on the American Dye Manufactures Institute (ADMI).

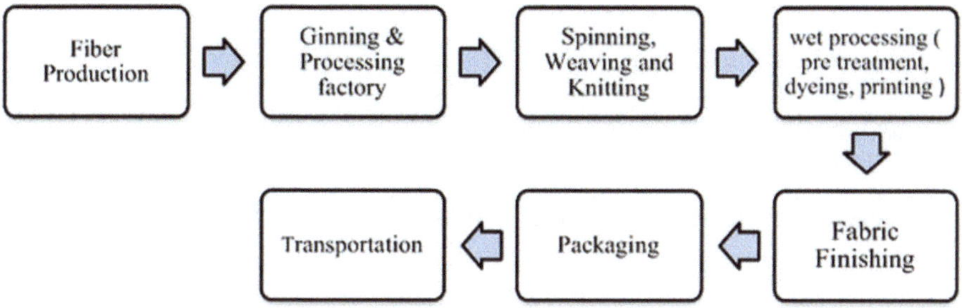

Fig. 1 Textile production supply chain. *Source* Adapted with changes from Gardetti and Torres [85], UNDP Report [249]

Different Treatments of Textile Wastewaters

Given the various risks associated with dyes, it is imperative that they be removed before being released into the environment. As a result, numerous technologies have been created to provide effective and affordable treatment. They may be chemical, biological or physical.

Physico-Chemical Methods

During the past two decades, several physicochemical decolorization techniques, such as sedimentation, coagulation, flocculation, filtration, reverse osmosis, electrocoagulation, adsorption, oxidation and membrane techniques (Fig. 2) have been reported [9]. Physical methods are mainly used for the primary treatment of textile waste waters. Based on dye coagulation-flocculation, they are successful for removing primarily sulphur and dispersion dyes, but have extremely low capacity for acid, direct, reactive and vat dyes [234]. Despite being a very efficient adsorbent for a variety of dyes, activated carbon is not frequently utilised because of its high cost and pH-dependent removal, nonetheless, it can only treat a small volume of wastewater and operates slowly [201]. In a study conducted in Northern Taiwan, [142] treated actual textile wastewater from several dyeing and finishing industries utilising ozonation, coagulation and the activated sludge process in multiple reactors. The results demonstrated the effectiveness of combining these methods for the removal of COD and dye. Georgiou et al. [88] used an ultraviolet reactor with hydrogen peroxide to treat actual textile effluents that were collected from the equalisation tank close to the cotton textile factory in Komotini, Greece. The Levafix and Remazol dyes were successfully completely degraded and mineralised by the system. Brik et al. [43] assessed the effectiveness of membrane bioreactor technology in treating mixed textile effluents from the polyester manufacturing finishing sector. The authors proposed that a polishing phase is required when a treatment target is also for reuse purposes, as the technique was able to remove about 87% of the colour. Similarly

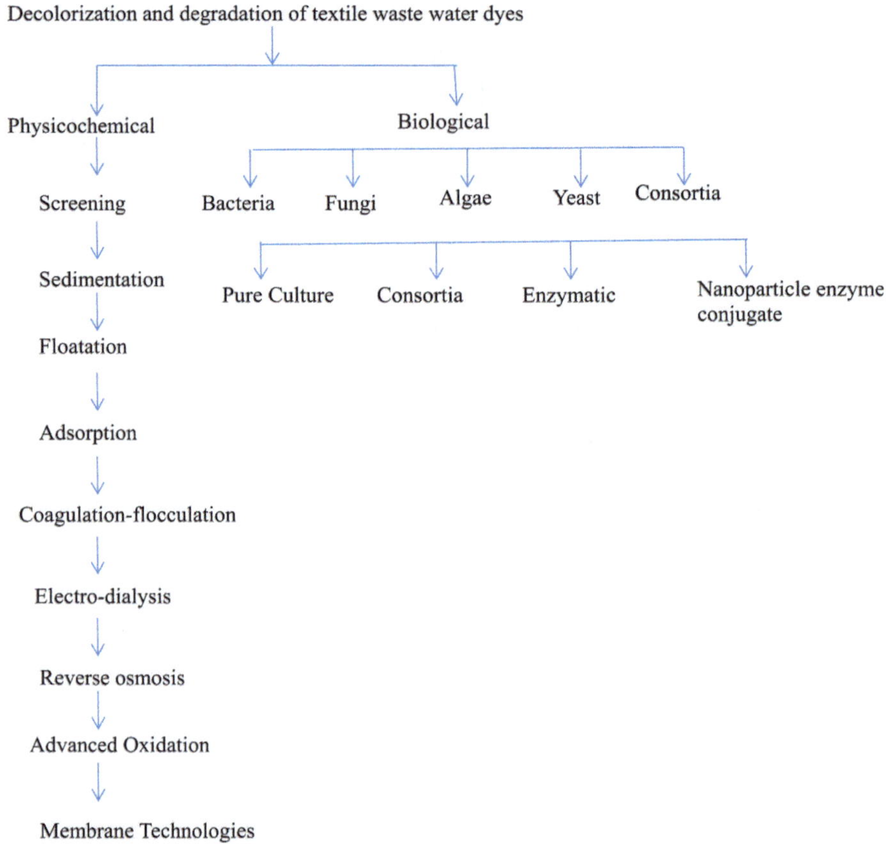

Fig. 2 Different treatment methods of textile wastewaters

numerous studies have been carried out on physico-chemical parameters/methods like electrochemical oxidation method, nanofiltration membrane technique, ultrafiltration and nanofiltration technique, adsorption, photocatalytic technique, coagulation and intermediate ozonation, electrocagulation, etc. to treat textile waste water effluents [12, 15, 23, 25, 40, 44, 55, 108, 149, 192, 194, 239, 240]. Excessive chemical treatment is costly and causes secondary contamination [109]. These approaches can merely move the dyestuff from one phase to another, leaving the problem mostly unresolved. Briefly, their continued application demonstrated their limitations, paving the way for biological treatment.

Biological Treatment

For many uses, biological processes offer an alternative to current technologies (Fig. 2). Numerous microbiological sources, such as bacteria, yeasts, fungi, algae and actinomycetes, have been documented [234]. Because of their capacity to last in extremely severe conditions, which presents a barrier for mesophilic microbes, extremophilic bacteria are frequently used in the treatment of textile waste waters [170, 172]. The capacity of plant growth-promoting rhizobacteria to decolorise different textile dyes is also intriguing [117, 147]. According to their oxygen requirements, these bioprocesses can be categorised as facultative, anaerobic, aerobic or a combination of these [264]. Additionally, they rely on the enzymes produced by whole cells or microbes [146]. The adaptability and activity of specific bacteria improved by their application under appropriate conditions determine the efficiency of microbial decolorisation [126]. Microorganisms that are co-cultivated or of a particular strain can accomplish this decolorisation.

Role of Bacteria in Biodegradation

Numerous scientists have investigated various bacterial species for the bioremediation of metals, dyes, soils and waterways affected by textile effluents. Among the primary contaminants of textile wastewater are dyes that contain formaldehyde, chlorine, heavy metals, halogen-containing compounds and other carcinogenic substances [110]. Because of their hazardous nature and high industry consumption rates, bacterial strains have been successfully used to degrade and decolorise textile effluents that contain dyes. These strains have been specifically targeted at azo dyes [215, 234, 266], anthraquinone [48, 178, 253], and triphenylmethane-based dyes [26, 27, 59], largely because of their hazardous nature and their large consumption rates within the industry. The majority of the bacterial species used to degrade and decolorise dyes have been isolated from soil collected near textile manufacturing plants [127], textile effluent sludge waste [158] and dye-contaminated textile wastewater [41]. Because they are exposed to the harsh circumstances produced by the textile effluent waste, bacteria from these areas develop accustomed to them and may eventually degrade dyes. Numerous bacterial strains have been recognised as viable contenders for the decolorisation and degradation of various dyes. Table 1 lists a few strains that have been effectively employed to break down various dye kinds. It has been demonstrated that under ideal circumstances, the majority of the strains that have been found can effectively degrade dyes. The efficient breakdown of dyes depends on a number of environmental and dietary parameters, as dye decolorisation and degradation are influenced by these factors. It is also crucial to remember that ideal simulated conditions might not apply to actual industrial textile effluent, which is more complicated due to the presence of many chemicals that alter how the bacteria and dye compounds interact. Researchers have turned their attention from employing individual bacterial isolates to using bacterial consortia in order to increase the effectiveness of bacteria in degrading colours. It has been demonstrated that employing a bacterial consortium rather than individual bacterial cultures results in higher dye degradation efficiency [245]. Compared to pure single cultures, this might be the outcome of the synergistic metabolic networks produced in consortiums. Utilising a consortium also has the benefit of enabling a one-pot treatment system. According to Junnarkar (2006), the majority of dyes can only be fully broken down in two steps: anaerobic and then either aerobic or anaerobic reduction. A class of aerobic and anaerobic bacteria has been demonstrated to degrade various azo dyes, however aerobic bacterial degradation of azo dye is ineffective. Among the bacteria utilised for anaerobic degradation are *Bacillus subtilis* and *Pseudomonas leuteola.* By allowing two or more bacteria to live together in the same culture media, mixed cultures increase the likelihood that dyes will degrade because of the microbial community's metabolic processes working in concert. As a result, the majority of research now focusses on treating wastewater containing textile dyes utilising consortiums rather than pure single cultures. In a single step, Dissanayake et al. [72] converted Congo red and methyl orange into straightforward, non-toxic chemicals using a microaerophilic mixed bacterial consortia. Within 50 h, the consortium was able to decolorise and break down the dyes with an efficiency of more than 95%, according to the biodegradation investigations. The synergistic effect of the bacteria's co-metabolic activities within the consortium was credited with the colours' successful biodegradation. In a similar vein, Das and Mishra [67] discovered that a bacterium consortium comprising Bacillus pumilus HKG212 and Zobellella taiwanensis AT 1–3 was able to decolorise Reactive Green 19 with a degradation efficiency above 97% under ideal conditions, as opposed to 86.80 and 89.50% degradation efficiency noted for the pure cultures of *Bacillus pumilus* HKG212 and Zobellella taiwanensis AT 1–3 under comparable circumstances. Mixed cultures are effective for degrading azoic colours because they avoid the need for a two-stage anaerobic–aerobic sequence, which is typically required to degrade azoic dyes. Mohanty et al. [164] discovered that a bacterial consortium-BP consisting of *Bacillus fexus* TS8 (BF), *Proteus mirabilis* PMS (PM), and *Pseudomonas aeruginosa* NCH (PA) was able to decolorise an anthraquinone dye Indanthrene Blue RS with a decolorisation efficiency of 100% within 9 h, as opposed to pure cultures of the bacterial strains, which degraded the dyes in an average of 19 h. The researchers attributed the dye's rapid breakdown by the bacteria in the consortium to the simultaneous action of the oxidoreductive enzymes in the reaction media. Rathour et al. [198] processed raw textile wastewater including 27 different dyes using a new microaerophilic bacterial community, DR4. The authors concluded that the leading genus in the bacterial population was *Comamonas*. Nitrogen supplies and optimised electron donors are two critical aspects in bacterial growth for maximising dye breakdown efficiency. The results showed that a 32-h reaction period at 37 °C reduced COD by 98% and removed dye by 88%. The scientists found that the DR4 Bacterial community offers a feasible treatment option for soil contaminated with textile wastewater. Chanwala et al. [52] evaluated the predominant bacteria species responsible for textile wastewater decolorisation using bacterial strain isolation. The 16S rDNA sequencing study was used and the scientists concluded that bacterial isolate TS20 was the most common bacteria species for colour removal. Hossen et al. [105] evaluated the use of three bacteria isolates (*Bacillus sp.* AZ28, *Bacillus cereus* AZ27 and *Alcaligenes faecalis* AZ26) from textile effluent to remediate Novacron Super Black G dye (NSB-G). After 96 h, the study successfully removed 90% of the NSB-G dye at 37 °C with a pH of 8. The results also showed that adding 500 mg/L of NSB-G dye resulted in optimal bacteria growth.

Table 1 List of bacterial strains that have been used to successfully decolourise and degrade textile dyes

Bacterial strain	Target dye (s)	Treatment conditions Temp, pH, Initial conc, time	Efficiency	Reference
Consortium: Neisseria sp., Vibrio sp., Bacillus sp Bacillus sp. and., *Aeromonas sp*	Reactive dyes: Novacron Orange FN-R, Novacron Brilliant Blue FN R, Novacron Super Black G, Bezema Yellow S8-Gand Bezema RedS2-B	37° C, pH 7, static conditions	Monoculture: 0–90% Consortium 65–90%	[122]
Alcaligenes faecalis AZ26, Bacillus cereus AZ27 and *Bacillus sp*	Novacron Super Black G (NSB-G)	37 °C, pH 8.0, 200 mg/ L 96 h, static conditions	90%	[105]
Pseudomonas aeruginosa (RS1) and *Thiosphaera pantotropha ATCC 35512*	Reactive Yellow 145 (RY145)		96 and 72 h	Garg et al. (2020)
Staphylococcus sp. K2204	Remazol Brilliant Blue R (RBBR)	100 mg/ L37 °C, 12 h, static condition	100%	[253]
Bacillus megaterium KY848339.1	Acid Red (AR337)	pH 7, 30 °C, 24 h, 500 mg/L	91%	[82]
M. yunnaenensis	Methyl Orange	pH 7, 30 °C, 100 mg/L	98%	Carolin et al. (2021)
Acinetobacter baumannii JC359	Reactive Black 5 (B-GDN), Reactive Red 120 (RP) and Reactive Blue 19 (RNB)	PH 7, 7, 9 (RP,B-GDN and RNB) 37 °C, 48 h, 500 mg/L	98.8% for B-GDN, 96% for RP and 96.2% for RNB	Ameenudeen et al. (2021)
Acinetobacter baumannii	Reactive Red 198	pH 7, 37 °C, 72 h, 200 mg/L	96.20%	Unnikrishnan et al. (2018)
Bacillus stratosphericus SCA1007	Methyl Orange	pH 7, 35 °C,12 h, 150 mg/L static conditions	100%	Akansha et al. (2019)

(continued)

Table 1 (continued)

Bacterial strain	Target dye (s)	Treatment conditions Temp, pH, Initial conc, time	Efficiency	Reference
Streptomyces DJP15	Reactive Blue 222	pH 7, 35 °C, 48 h, 50 mg/L, static conditions		Pillai (2017)
Bacillus pseudomycoides	Acid Black 24	pH7, 37 °C, 24 h, 40 mg/L	96%	[132]
Escherichia coli NG188	Remazol Brilliant Blue	pH 8, 40 °C, 72 h, 50 mg/L, static conditions	72%	El-Badan et al. (2020)

Algal Degradation and Decolorization of Textile Dyes (Phycoremediation)

Phycoremediation is the employment of macroalgae or microalgae to remove or biodegrade contaminants from wastewater such as textile dyes and xenobiotics, while simultaneously propagating biomass. There are numerous procedures for treating polluted water, industrial effluents, dyes, various colourants, bleaching and tanning agents with microalgae both aerobically and anaerobically [148]. Although bacteria play an important role in the degradation of environmental pollutants, microalgae have been proven to have the ability to breakdown organic pollutants in addition to providing oxygen for aerobic bacteria, which can only destroy pollutants when oxygen is present. Indeed, numerous studies have demonstrated that existing environmental concerns, such as textile industry wastewater treatment challenges, can be conquer by employing microalgae to degrade contaminants into eco-friendly compounds [131]. Many health problems are caused by mutagenic and carcinogenic azo dyes and their intermediate chemicals, which also cause mutagenic activity in surface waters that are mostly contaminated by textile waste water. Their discharge into surface water also causes serious problems, such as blocking light penetration and impeding the entry of oxygen into water bodies [125, 176]. Since they are photoautotrophs, the algae don't need a carbon source. Using algae to treat wastewater is a cost-effective and efficient method. Photosynthetic algae flourish even in textile industry effluents, making them a potential choice for waste water treatment [75]. Algal azo reductase enzymes, which can break down azo dyes, can provide effective colour removal. *Chlorella* and *Oscillatoria* species have been found to make aromatic amines by degrading azo dyes, which then lead to the creation of simple organic compounds by further breakdown. Only a few number of other algae species could use azo compounds as a source of carbon and nitrogen. More than 30 azo dyes have been demonstrated to be broken down and decolored into less complex aromatic amines by *Chlorella* and *Oscillatoria* species [265]. Oyebamiji et al. [175] used six distinct microalgae strains (*Chlorellaceae* family) to investigate the treatment of textile wastewater in terms of biomass generation, heavy metal reduction and decolorisation. The treated wastewater effluent showed a remarkable reduction in heavy metals (Cu, Pb, Al, Se, and V) as well as decolorisation. The scientists came to the conclusion that recovering microalgae biomass in conjunction with treating wastewater from textiles is a viable and promising way to produce biofuel and counteract the harmful effects of wastewater that is high in nutrients. Behl et al. [34] used electrogenic microalgae, *Desmodesmus spp.*, to break down azo dyes. By combining graphene oxide nanosheets with the green algae *Desmodesmus spp.*, the scientists created a new graphene oxide (GO) and algae-based sustainable nanocomposite. The bionanocomposite effectively eliminated 90% of the DR 31 dye in 150 min when exposed to visible light, according to the results. The authors draw the conclusion that using GO sheets in conjunction with the algae resulted in improved lipid production and increased decolorisation efficiency. Additionally, this bionanocomposite provides a recyclable, cost-effective and environmentally responsible alternative for treating textile water. In an ongoing effort to improve the efficiency of bacteria dye wastewater remediation, algal–bacterial consortiums have recently emerged as a cost-effective wastewater treatment method capable of removing nutrients, heavy metals, minerals and dyes from wastewater [70, 140]. Algae have previously been utilised to treat wastewater [181, 191], therefore combining them with bacteria takes advantage of the benefits provided by both microbes to provide better dye wastewater treatment systems. Dye decolorisation in algal bioremediation occurs through three mechanisms: assimilative chromophores for algal biomass production, CO_2 and H_2O transformation of coloured molecules to non-coloured compounds and chromophores adsorption on algal biomass [180]. According to studies, algal–bacterial consortiums offer an environment that can withstand temperature, pH and dissolved inorganic compound concentration extremes—all of which can be limiting factors in the bacterial biodegradation of wastewater containing dyes [141, 177]. Ayed et al. [29] treated real textile effluent containing CI Reactive Blue 40 (CI RB 40) using a group of algal–bacterial probiotic strains (*Pseudomonas putida*, *Chlorella*, and *Lactobacillus plantarum*). According to the study, the consortium removed COD and colour from wastewater with 89 and 99% efficiency respectively. Phototoxicity experiments validated the consortium's ability to convert the dye into harmless chemicals.

Role of Fungi in Degradation and Decolorization of Textile Dyes (Mycoremediation)

Mycoremediation refers to the practical application of fungi for bioremediation of contaminated soils and textile effluents. Mycoremediation has been shown to breakdown a variety of hazardous pollutants, including petroleum hydrocarbon compounds, polychlorinated biphenyls, heavy metals (by biosorption), phenolic-derived compounds, degradation-resistant insecticides, and so on. In this mycoremediation process, fungi exploit these dangerous substances as a source of nutrients, converting them into simpler forms and decolorising them in the process. Fungal degradation efficiency can be used to categorise fungi, such as mycorrhizal fungal degradation, fungal biosorption and ligninolytic fungal degradation [96]. According to Dos Santos et al. [207], a number of fungus, including *Bjerkandera adusta*, *Trametes versicolour*, and *Phanerochaete chrysosporium*, have been authorised for use in textile dye treatment due to their ligninolytic properties. An environmentally favourable biosorbent was *Aspergillus niger*. Both in vitro and in vivo research have been conducted by the authors. The findings of the batch absorption showed that the fungi had considerable decolorisation ability in addition to a high removal efficiency of 98% for acid dye. As a result, the scientists came to the conclusion that the fungal strain *Aspergillus* can be used as a promising biosorbent for wastewater treatment and that modifying the dye solution's pH can improve performance even more. Other strains that were employed to achieve the goals of decolorisation and organics removal of actual textile industry effluent were the filamentous fungus *Aspergillus carbonarius* M333 (AC) and *Penicillium glabrum* Pg1 (PG) [24]. The COD removal and decolorisation efficiency were examined in both batch and continuously fed packed bed bioreactors. The continuous flow bioreactor achieved 79% decolorisation and 68% COD removal efficiency, according to the data. Additionally, the batch reactor achieved 70% COD removal and 98% decolorisation respectively. The authors came to the conclusion that, in terms of COD and colour removal efficiency, PG fungi are the most effective treatment when compared to AC. In order to create an ultrafiltration membrane, Pazdzior et al. [183] used the filamentous fungus *Aspergillus carbonarius* M333 to handle actual textile wastewater. The growth duration of 3, 5, and 9 days was used by the authors to measure the fungal growth. Using the fungus-derived ultrafiltration membrane, the study results in a 73% reduction in COD and a 91% decolorisation. White rot fungi have been demonstrated to have exceptional dye decolorising capabilities in textile effluent treatment. They can effectively mineralise a variety of pollutants, whereas bacteria are substrate-specific. White rot fungi's lignin-degrading enzymes, such as lignin peroxidase (LiP), manganese peroxidase (MnP) and laccase, which are not substrate-specific, are the primary cause of their degradative efficiency in breaking down substances like pollutants and aromatic compounds. According to Toh et al. [246], lignin peroxidase, which is secreted by *Phanerochaete Chrysosporium*, effectively breaks down azo dye into simpler molecules. White rot fungus break down lignin, a polymer found in woody plants. A white-rot fungus called *Phanerochaete chrysosporium* can break down a range of xenobiotic contaminants, including dioxins, polychlorinated biphenyls (PCBs) and other chloro-organics. The most common, resilient and model white-rot fungus known to decolorise a variety of dyes from textile waste fluids is *Phanerochaete chrysosporium* [226, 236]. Other known white rot fungi that exhibit the capacity to eliminate colour from dyes are *Coriolus versicolo*, *Trametes versicolour*, *Pleurotus ostreatus* and *Coriolopsis polysona*. The primary enzymes that white rot fungi use to remove colour are LiP and Mn-dependent peroxidase (MnP) or laccase.

The ability of *Trametes versicolour* to decolorise various coloured chemicals emitted by textile and other industries makes it a species of interest. Furthermore, according to Eichlerova et al. [77], three promising strains—Dichomitus squalens, Ischnoderma resinosum, and Pleurotus calyptratus—effectively decolorise both Orange G and Remazol Brilliant Blue R. The chromium metal complex dye Isolan Dark Blue 2SGL-01 can be decolored by the white rot fungus *Irpex lacteus* [118]. Furthermore, it was discovered that, in submerged culture, *Trametes pubescens* was the strain that performed the best at decolorising the azo dye Congo Red [224]. Remazol Brilliant Blue R dye can be eliminated by *Trametes maxima* CPB30 [136] and *Fomes fomentarius* [171]. However, the degradation of dyes in textile waste waters by white-rot fungi has some inherent disadvantages, such as a long growth phase and the need for nitrogen-restricted environments, unreliable enzyme production and a large reactor size due to the long holding time for complete degradation [19, 102], Multiple fungal species, particularly the genus *Aspergillus* identified by El-Rahim et al. [79], may decolorise a wide range of dyes. Jin et al. [113] also reported this genus.

Yeast Degradation Ability

Significant benefits of yeasts are their affordability, ease of access to biomass and resilience to harsh environmental conditions. The yeast is an excellent microorganism for azo dye bioremediation because of its significant absorption of dyes, heavy metals and colourants. Additionally, it has been discovered that a number of yeasts can handle textile waste effients [53]. The primary mechanisms by which yeast breaks down azo dyes are enzymatic degradation, adsorption or a combination of the two. *Candida tropicalis* effectively removes up to 94% and 44% of the Remazol Blue and Reactive Red dyes, respectively. The genus *Candida* is frequently employed in dye decolorisation by biodegradation [4, 155, 242], biosorption [71, 255] and bioaugmentation [100]. *Kluyveromyces marxianus* IMB3C, a thermotolerant yeast, has been proven to eliminate Remazol Black-B [157]. *Saccharomyces cerevisiae* successfully decolorises methyl red at various pH levels via azoreductase. There is an increasing interest in biological treatment of waste water components such as textile effluents, particularly dyes, metals, inorganic nutrients and organic compounds, via biosorption, biodegradation, bioaccumulation and enzymatic mineralisation [3].

Enzymatic Degradation of Textile Effluents

Enzymes have various positive properties connected to degradation. Textile wastewater is the most harmful effluent released by the industrial sector in terms of quantity and composition. The limitations of using microbes for degrading activities can be easily overcome by using enzymes as an alternative. Enzymes are capable of efficiently breaking down a particular contaminant while having no influence on the other elements in wastewater. As a result, contaminants that are difficult to break down can be readily broken down by enzymes. The most significant enzyme classes actively engaged in bioremediation processes include hydrolases, dehalogenases, transferases and oxidoreductases. Laccase is an example of an enzyme that is active in the elimination of contaminants. When it comes to producing enzymes, fungi are the best producers. By generating extracellular enzymes such as phenol oxidases and peroxidases, the fungus decrease azo dyes (Duran et al., 2002). Commercial fungal enzymes are produced using the surface culture approach. According to Heinfling et al. (1998), white rot fungi are incredibly effective organisms with remarkable lignin degrading capabilities. White rot fungi that carry out biodegradation produce extracellular oxidoreductases including laccases, peroxidases and oxidases. These enzymes's primary function is to break down lignin but because they are not substrate-specific, they can also break down several aromatic persistent chemicals that contribute to environmental problems. White rot fungi are more beneficial than bacteria for decomposition. Extracellular enzymes like lignin peroxidase (LiP), manganese peroxidases (MnP), laccase and Mn-independent versatile peroxidases (VP) are non-substrate specific, which allows white rot fungi to break down resistant aromatic colour compounds with remarkable efficiency [113]. Fungi must be able to modify their metabolism in response to shifting carbon and nitrogen levels in order to survive. To accomplish this metabolic activity, fungi create both intracellular and extracellular enzymes, which then perform nonspecific degradation of various refractory effluents, including organic waste, aromatic hydrocarbons, dye chemicals etc. The non-substrate specific nature of fungal enzyme systems accounts for their ability to breakdown coloured and metallic effluents [209]. Similar enzymes, due to their limited substrate specificity (lignin), have been successfully used in dye degradation. Only a few studies have shown that brown-rot fungi can remove various colours [265]. Compared to enzymes made by plants and animals, microbial enzymes have numerous attractive qualities, including the potential to create the highest possible yields, simple genetic manipulation, convenient availability due to the lack of seasonal changes and higher growth rates of microorganisms. Microbial enzymes are particularly significant since they are simple to make and have a high stability. White rot fungus have been shown to decolorise a number of dyes, indicating that these organisms constitute a significant option for the treatment of industrial dye-containing effluents. Another study found that Crystal Violet dye is broken down by white rot fungi by *N* demethylation. The primary process by which bacteria break down azo dyes involves the azoreductase enzyme and an electron source breaking the azo bonds. Effective biotreatment of azo dyes involving textile waste water can be achieved by creating treatment procedures that employ microorganisms that produce azoreductase. According to Sarayu and Sandhya (2010), these azoreductase enzymes have the ability to catalyse the reduction-based cleavage of azo compounds. Another study discovered that Crystal Violet dye is broken down by white rot fungi via *N* demethylation. The fundamental method by which bacteria break down azo dyes requires the azoreductase enzyme and an electron donor that break the azo bond. Effective biotreatment of azo dyes using textile waste water can be accomplished by developing treatment processes that use microorganisms that produce azoreductase. According to Sarayu and Sandhya (2010), these azoreductase enzymes can catalyse the reduction-based cleavage of azo compounds. Azoreductases are the most significant bacterial enzymes engaged in degradation processes (Table 1). According to Chang and Kuo (2000), azoreductases break azo bonds (–N = N–) to produce aromatic amines, which are then further transformed into CO_2 and H_2O in aerobic conditions. Darwesh et al. [66] used the immobilised peroxidase enzyme embedded on modified Fe_3O_4 magnetic nanoparticles to co-precipitate glutaraldehyde in order to detoxify dyes found in wastewater from industrial textiles. Additionally, in a variety of environmental conditions, the study discovered that the immobilised peroxidase enzyme was far more stable than the free form. The immobilised enzyme has not been found to be significantly affected by temperature or pH, according to the authors. All of this points to the immobilised peroxidase enzyme's potential utility in industrial remediation applications, given its high stability and resistance to environmental changes. Srinivasan et al. [235] used molecular docking analysis to investigate the interaction between textile dyes and enzymes. Joyfix Red RB, Reactive Red F3B, Tur-quoise CL-5, Joyfix Yellow MR, Remazol Blue RGB and Remazol RedRGB are the six textile dyes that have been utilised. Understanding different affinities, active sites and enzyme/dye interactions is accomplished through the use of docking studies. According to the docking analysis, phthalocyanine and anthraquinone were broken down by the oxidoreductive enzymes laccase and azoreductase. Navada and Kulal [169] investigated the degradation of aniline blue and textile dye effluent with laccase derived from gamma-irradiated endophytic fungus (*Phomopsis spp.).* The scientists discovered a twofold increase in laccase synthesis in 1.2 kGy gamma irradiated endophytic fungus. As a result, higher laccase led to faster dye degradation. The maximum dye removal (COD of 67% and BOD of 47%) and laccase activity were found at 30 °C and pH of 5.

Microbial Mechanism to Degrade and Decolorize Azo Dye

Biological dye management in the textile industry is an effective and environmentally beneficial approach. In general, biodegradation of azo dyes occurs via two mechanisms: direct decolorisation and indirect/mediated decolorisation. Both processes rely heavily on enzymes to breakdown refractory chemicals in the microbial system [227]. Several studies show that laccase [144, 254], azoreductase [58, 121], lignin peroxidase [267], NADH-DCIP reductase [133] and hexane oxidase [65] degrade complex organic substances. Laccases and azoreductases are two of these enzyme groups that have demonstrated a high capacity to decolorise a wide variety of well-known industrial dyes. The breaking of azo bonds in azo compounds, which results in the colour removal of azo compounds is the microbial process for the degradation and decolorisation of the majority of problematic textile dyes that are azo dyes. The reduction of azo bonds (–N=N–) by azo reductase enzymes in anaerobic conditions is the primary cause of the biodegradation of azo dyes [102]. This technique decolorises the dye and produces colourless amines by transferring four electrons from electron donors to the electron acceptor (azo dye) in two steps at the azo linkage [228]. After that, the produced metabolites or aromatic amines undergo additional aerobic or anaerobic degradation. In the biodegradation of anthroquinone and azo dyes by a fungus strain *Lentinus spp.*, the function of laccase in asymmetric cleavage of azo dye was extensively documented [106]. The laccases used to remove several azo dyes from *Brevibacillus spp.* (Z1) [42] and *Anoxybacillus Gonensis* P39 [260] were purified and described. Under aerobic conditions, respiration may dominate NADH utilisation, preventing electron transport from NADH to the azo linkage. Thus, decolorisation could be linked to non-specific extracellular processes involving reduced chemicals produced by anaerobic biomass [228]. In anaerobic settings, a modest redox potential ($\leq$ 50 mV) effectively decolorises azo dyes. Azoreductases are oxygen sensitive and released extracellularly [190]. Researchers employ redox mediators to speed up the dye removal process. Several investigations have demonstrated that the biodegradation of azo dyes is greatly enhanced by the inclusion of redox mediator such as synthetic electron carriers, anthraquinone-2-sulfonate or quinone compounds. However, quinone addition's expensive cost restricted its use. Other redox mediators such as activated carbon, carbon black and carbon nanotube were utilised to make the application of bacteria-mediated decolorisation feasible. The majority of dyes are mineralised into simple inorganic chemicals that pose little hazard to living things [225]. Reduced flavin nucleotides are present in the microbial electron transport chain. These nucleotides are oxidised using an azo dye, which is also reduced throughout the process. The reduced flavin nucleotides are then reoxygenated. For decolorisation and degradation to occur at a practical rate, an additional carbon source is often required. Some aerobic microbes rely only on azo dyes for carbon and nitrogen, while in others, azo group reduction is only catalysed by oxygen-tolerant azo reductases. This carbon is transformed into methane and CO_2, releasing electrons. These electrons are then sent down the electron transport chain to the azo dye, which acts as an electron acceptor. Then electrons interact with the azo dye, reducing the azo bonds and causing decolorisation. Several microorganisms, including bacterial strains, fungi, actinomycetes and algae, have been extensively investigated for their potential to decolonise azo dyes.

Recent Developments in the Breakdown of Bacterial Dyes

Utilising Immobilised Microorganisms to Clean Wastewater Containing Dyes

The employment of pure and bacterial consortiums in bioremediation applications is a problem because the bacteria are quickly affected by environmental conditions such as pH, temperature, and starting dye concentration, among others, resulting in reduced dye decolorisation. To overcome these problems and improve the bioremediation process, immobilisation techniques have gained popularity in the wastewater treatment area. Immobilisation techniques offer several advantages over free bacterial cells, including improved bacterial stability, survivability and reusability of the bacteria without loss of enzymatic activity [159], resulting in improved dye decolorisation and degradation. Bacterial organisms are cultivated and immobilised onto a suitable support system using one of several methods, including microencapsulation, matrix entrapment, covalent binding and adsorption, before being exposed to dye-contaminated water. To achieve maximum biodegradation efficiency, the support material used for bacterial immobilisation must be capable of protecting the bacterial organisms from toxic environmental conditions, providing operational stability, being lightweight, cost-effective, mechanically and chemically stable, inert and having good diffusivity. Support materials for bacterial immobilisation have included natural and synthetic polymers such as alginate, chitosan, cellulose, polyurethane, polyvinyl alcohol, polyacrylamide, polyethylene and polyvinyl chloride. In order to better understand the differences in degradation capacity between immobilised bacteria and loose cells, Reddy and Osborne [199] isolated *Pseudomonas guariconensis* from paddy rhizosphere and tested its ability to digest reactive dye Reactive red 190. The study evaluated the ability of free P*seudomonas guariconensis* bacterial cells to calcium alginate immobilised bacterial cells. The study found that immobilised cells degraded the dye with an efficiency of 91%, compared to 86% for free cells. Immobilisation of bacterial cells has been demonstrated to boost enzyme activity and bacterial cell stability, which may explain the differences in dye degradation efficiency amongst bacterial strains. Similarly, Cai et al. [45] immobilised *Bacillus sp.* JF4 in polyvinyl alcohol-calcium alginate-activated carbon beads to decolorise Reactive Blue 19 (RB 19). The immobilisation of the bacterial strain proved efficient, with a degradation efficiency of 100% compared to 92.1% for free bacteria. The immobilised bacteria also demonstrated excellent resistance to RB 19 dye concentrations. Increased dye concentration is one of the limiting factors in bioremediation, as it inhibits bacterial enzyme activity due to dye toxicity. Immobilisation has also been observed in naturally occurring bacterial populations capable of forming biofilms. Bacterial biofilms are surface-associated cells that are immobilised in extracellular polymeric compounds that they create themselves [166]. Depending on the type of bacteria, some biofilms can destroy various colours. Biofilms are appealing because they are inexpensive, reusable and can adapt to harmful settings as compared to pure and mixed bacterial cultures [162]. The utilisation of biofilms as a bioremediation alternative to typical wastewater treatment procedures has lately sparked interest among academics looking for more environmentally friendly wastewater treatment systems. Various studies have found that biofilm-forming bacteria can effectively remediate textile wastewater [32, 167]. Tahir et al. [241] used biofilm-forming *Staphylococcus* and *Bacillus* bacteria to decolorise and convert Mordant black 11 dye. They discovered that both bacterial strains could breakdown the dye with efficiencies more than 50%, with *Bacillus spp.* achieving a decolorisation efficiency of 75.2% in a glucose-supplemented medium. However, the study discovered that high Mordant black 11 concentrations influenced the bacterial strains, with dye breakdown decreasing as the dye concentration increased from 50 to 150 mg/l, with maximal degradation reported at 50 mg/l. Similar research [7, 128] have shown that increased dye concentration in the culture medium creates a hazardous environment, which inhibits enzymes and reduces dye breakdown efficiency. Other research [97, 132, 258] have found that biofilms perform better in dye degradation than free forms under identical conditions.

Microbial Fuel Cells

Microbial fuel cells (MFC) are a potential bio-electrochemical technique that may degrade both organic and inorganic contaminants while also producing electric energy. MFCs use microbes as catalysts to oxidise organic molecules. The oxidation of organic substances generates electrons and protons at the anode, which travel to the cathode via an external circuit, complete the electrochemical cycle and generating electric energy [145]. Electron transfer is critical in dye decolorisation because a portion of the generated electrons is donated to the dye for reductive decolorisation and the remainder to the anode for electricity generation. MFCs have been found to have significant potential for degrading azo dyes [233, 237]. Reyes et al. recently used an indigenous bacterial consortia to breakdown four anthraquinone dyes, Acid blue 62, Acid blue 25, Acid blue 40 and Reactive blue 19, at an initial dye concentration of 50 mg L^{-1} using a microbial fuel cell. The study found that the consortium was able to degrade the dyes with removal efficiencies of 48%, 17%, 9% and 5% respectively. The findings revealed a decline in dye removal effectiveness as the complexity of the dye complexes grew. Reactive blue 19 exhibited the lowest dye removal efficiency at 5% due to the presence of a sodium 2-((3 aminophenyl) sulfonyl) ethyl sulphate substituent, which resulted in a more efficient charge transfer across the dye's chemical structure, enhancing conjugation duration and hence, stability. This demonstrates that Reactive blue 19 was the most refractory of the dyes examined and the bacterial consortia failed to successfully breakdown it [200]. Similarly, Saba et al. studied the decolorisation of an azo dye Reactive black 5 (RB 5) and an anthraquinone dye Reactive blue 4 (RB4) using an open circuit microbial fuel cell and a microbial consortia derived from bovine rumen fluid. The study found that RB5 was decoloured with higher than 90% efficiency in 120, 165 and 225 min at concentrations of 50, 100 and 200 mg L^{-1}, respectively. RB4 dye at 50 and 100 mg L^{-1} took 225 and 300 min to decolorise respectively, whilst 200 mg L^{-1} dye did not decolorise at all. The variations in dye decolorisation efficiency by the bacterial community suggest that dye breakdown occurred faster in RB5 than in RB4. This could have been caused to differences in dye structures. Reactive Blue 4's dichlorotriazine group, along with the C¼O chromophore group, enhances the dye's stability and renders it resistant to degradation. The study also discovered that in closed-circuit conditions, dye decolorisation increased as the external load decreased [204]. Anthraquinone dyes contain anthraquinone chromophore groups, which have been proven to be more stable than other chromophores, such as azo, due to their complex structures, making biodegradation difficult [103].

Membrane Bioreactors

Membrane bioreactors (MBR) integrate membrane technology with biodegradation technology. Compared to standard wastewater treatment systems, this has several benefits, such as minimal maintenance, high biocatalyst retention, high effluent quality and less sludge production [62, 68, 138]. The primary issue with membrane bioreactors, however, is membrane fouling brought on by the accumulation of microorganisms, colloids and solutes on the membrane surface. This impairs the transmembrane pressure and the membrane permeate flux [17], requiring constant membrane cleaning and resulting in a shorter membrane lifespan [50]. The focus of membrane bioreactor research has recently switched to additional membrane bioreactor derivatives, including membrane aerated biofilm reactors (MABR), biofilm membrane bioreactors (BFMBR) and anaerobic membrane bioreactors (AnMBR). These derivatives have been demonstrated to lessen the environmental impact of energy-related emissions, freshwater resource eutrophication and marine environment eutrophication when compared to conventional aerobic membrane bioreactors. An anaerobic baffled membrane bioreactor, which combines an anaerobic baffled reactor and a membrane bioreactor, was used by Liu et al., to detoxify methyl orange. As a result, a baffled membrane bioreactor with low membrane fouling was able to degrade dyes. After undergoing four anaerobic treatment cells prior to being fed into the membrane bioreactor, the study discovered that the methyl orange had totally lost its hue. Demethylation and desulfonation were two of the two distinct pathways that the dye degradation followed, according to an analysis of the seven distinct degradation intermediates from the anaerobic and aerobic treatments. The membrane accumulated 0.19–0.32 gMLSS/m^2 of filter cake during the system's 110-day operation. This could have been caused by the pretreatment in the anaerobic baffled reactor, which broke down the majority of potential membrane foulants [143]. In a different investigation, Berkessa et al., treated wastewater containing the anthraquinone Remazol brilliant blue R (RBBR) dye using an anaerobic dynamic membrane bioreactor. The study's findings demonstrated that the reactor successfully eliminated over 97% of the colour and COD. With a total suspended solid rejection of > 98.8%, the membrane also demonstrated good rejection of substances with a high molecular weight. It is important to note that the ozonation oxidation method was used to pretreat the dye effluent before it was treated in the bioreactor. This made it easier for the dye molecules to break down, particularly the aromatic chemicals that are hard to biodegrade in anaerobic environments [38]. Given that certain dye molecules have been demonstrated to be dangerous and resistant to biodegradation [174], pretreatment of dye wastewater is crucial to guaranteeing the production of high-quality effluent following the last biological post-treatment procedure. Therefore, the pretreatment procedure might be the cause of the great dye removal efficiency the researchers were able to get. Researchers have developed a system known as electrically enhanced membrane bioreactors (eMBR) that combines MBRs with electrochemical devices to improve membrane performance. These methods have been demonstrated to enhance phosphorus and micropollutant removal [90], decrease membrane fouling in MBRs [104, 160] and more. MBRs have employed electrophoresis, electroosmosis and electrocoagulation to enhance pollutant removal and lessen membrane fouling [33, 95]. Hawari et al. showed that membrane fouling in submerged MBRs might be inhibited by the dielectrophoretic (DEP) mobility of particles in an inhomogeneous electrical field. The study discovered a strong correlation between the electrical field's frequency and intensity and fouling-suppression performance. Because the biomass particles closer to the membrane's surface experience a larger DEP force, a stronger electrical field was found to recover the filtrate flux more effectively. Due to an electrothermal impact, the permeate flux decreased above an intensity and frequency value of 130 V and 1 kHz, respectively [99]. The effectiveness and membrane fouling of a laboratory-scale eMBR for the treatment of wastewater gathered from the textile industry were documented by Belli et al. The researchers discovered that, when employing an electrical current density of 10 A m^2 and 15 A m^2, respectively, the electrocoagulation process greatly increased the reactor's colour removal efficiency, achieving efficiencies of 50 and 70%. Furthermore, the study discovered that the bioreactor had an efficiency of over 90% in removing NH4þ-N. Throughout the experiment, COD elimination remained consistent at 70–75%. Better-mixed filterability conditions were produced by applying the electric current, which decreased the rate of membrane fouling and the need for membrane chemical cleaning [35].

Bio-advanced Oxidation Processes

Advanced oxidation processes (AOPS) use in-situ produced hydroxyl or sulphate radicals to destroy organic and inorganic contaminants [21]. Fenton oxidation and ozonation are two AOPs that have been utilised in conjunction with biological therapy to degrade dyes [183]. Each of these advanced oxidation techniques has previously been proven to be effective in dye removal [93, 173]. However, it has been noted that using AOPs can be costly, complicated and in some situations, ineffective [213]. Biodegradation has some limitations, particularly on a large scale, because the process is time-consuming due to its reliance on microbial metabolic processes, not all microorganisms can degrade all types of dyes due to enzyme specificity and biodegradation is only possible for biodegradable pollutants [54]. To address the problems of both AOPs and biodegradation researchers have developed wastewater treatment technologies that mix the two in a single system. Each of these systems can be utilised as a pretreatment or post-treatment step depending on the dye's degradability or the byproducts of each treatment procedure. For example, if a dye is not biodegradable but its byproducts are, biodegradation might be utilised as a post-treatment step, with one of the AOPs capable of degrading the pretreatment dye and vice versa. Castro et al. [47] treated an aqueous solution containing Reactive orange 16 (RO16) with ozonation as a pretreatment before proceeding to a biological treatment in a moving-bed biofilm reactor (MBBR). The study discovered that 97% of the dye was removed after 5 min of ozone exposure but the procedure had inadequate COD removal, with just 48% removal, indicating incomplete dye mineralisation. Further treatment in the MBBR resulted in COD and ammonium removal rates of

93.1 and 97.2%, respectively. Goswami et al. used a packed bed bioreactor (PBBR) containing Arjuna (*Terminalia arjuna*) seeds charcoal immobilised with *Providencia stuartii* and ozonation to decolorise Congo red (CR) dye. The biological treatment was performed as a pre-treatment step, resulting in 92% dye removal efficiency. Follow-up ozonation of the biologically treated sample removed all leftover dye molecules in 30 min [91]. [243] conducted a study to remove Remazol yellow RR dye from textile waste water using a bacterial strain *Aeromonas hydrophila* SK16 (A. hydrophilia) and AOP-H_2O_2. The researchers discovered that when used alone, *A. hydrophilia* had a 90% dye removal effectiveness, whereas AOP-H_2O_2 had a 63.07% under ambient circumstances within 9 h at pH 6. In comparison, the researchers discovered that the bio-AOP combination approach, which subjected the biologically treated dye to AOP with 4% H_2O_2, achieved maximum dye removal within 3 h while also eliminating 84.88% and 82.76% of BOD and COD respectively. Shanmugam et al. [217] investigated the degradation of Acid blue 113 (AC113) using Fenton oxidation in conjunction with biodegradation via a bacterial consortia. The investigation included both simulated and actual dye baths. Pre-treatment with Fenton oxidation in the simulated dye solution resulted in a maximum dye degradation efficiency of 85%. Analysis of the treated sample revealed the existence of lower molecular weight aromatic amine molecules. This indicated that the oxidation was capable of breaking down the dye into smaller metabolites, however, the identified compounds: benzene acetic acid, diethyl phthalate and n-hexadecanoic acid are not environmentally friendly and thus necessitated additional treatment to break them down into smaller nontoxic intermediates. This was accomplished by using biological deterioration as a post-treatment step. The biological degradation process was able to further convert these chemicals into more ecologically friendly metabolites, which included benzoic acid, 4-ethoxy-, ethyl ester, pyrrolo [1, 2-a]pyrazine1,4-dione, pyrazine1,4-dione, hexahydro-3 (2 methylpropyl), and hexadecanoic acid. The Fenton oxidation of the dye bath resulted in a decreased dye removal efficiency (89.5%), and analysis of the oxidised material indicated the presence of naphthalene phthalic anhydride, phenol, 3, 5-bis (1, 1-dimethyl ethyl), phthalic acid, butyl hept-4-yl ester and phenol, 4, 40-methylenebis. Following the final biological treatment, the sample revealed the presence of biodegradable alkenes and hydrocarbons. As a result, the collective process was able to transform AC113 into smaller, more biodegradable molecules [219]

Degradation of Textile Effluent Through Biosorption

Biosorption is the absorption or uptake of chemicals by microbial mass, or more accurately, "the capability of microbes to amass heavy metals from wastewater through metabolic or physio-chemical paths of absorption" (Ahalya et al., 2003). It primarily happens through the cell wall, with adsorption and absorption dependent on biomass type. Algae, yeast, filamentous fungi and bacteria can remove colours from wastewater through biosorption using their biomass. Heteropolysaccharide and lipid are essential components of the cell wall, with various functional groups such as amino, carboxyl and hydroxyl groups, as well as other charged components that have biosorption capabilities. Strong attraction forces form between the azo dye and the functional groups of the cell walls of several microorganisms, including yeast, algae bacterium and filamentous fungi. Thus, these cell wall components contribute to bacteria' biosorption capacity. Biosorption is the ability of a specific type of microbial biomass to interact with and accumulate trash and heavy metals found in effluents. Biomass has the ability to operate like a chemical element or ion exchanger derived from a biological source. The primary component of fungi, bacteria and algae responsible for biosorption is the cell wall. Until date, it has been extensively documented that biosorption is an ideal option for cleaning metal-containing effluents. Biosorption using microorganisms, particularly fungi, for colourant removal from textile waste water has received a lot of interest. The decolorisation of synthetic dyes and dye effluents has been investigated utilising several fungus. *Aeromonas sp.*, *Pseudomonas luteola*, *E. coli*, *Bacillus subtilis* and *Staphylococcus aureus* are among the bacterial strains used for their biosorbent capabilities in the decontamination of reactive colours. Biosorption has several advantages over traditional treatment procedures, including high effectiveness, low cost, sludge minimisation, metal recovery, biosorbent regeneration, ease of metal recovery and so on. Bacteria decolourize textile dyes via biosorption, biodegradation or a combination of the two methods. During biosorption, the dye is adsorbed onto the bacteria's cells, affecting the dye's hue. This is characterised by a reduction in the dye's absorption spectrum in proportion to the dye adsorbed by the bacterial cells. Alternatively, after biodegradation, the dye structure is broken down into its simplest compounds, as evidenced by the disappearance of dye absorption peaks from the dye spectrum or the formation of new peaks if new metabolites are formed during the biodegradation process [151]. Other research has linked the loss of absorption peaks in the spectra of treated dyes to the dye degradation mechanism [28, 57]. Degradation is consequently a more advantageous mechanism since it produces less harmful metabolites. Adenan et al. [5] investigated the effect of both dead and live cells from the Streptomyces bacillaris bacterial species on the degradation of triphenylmethane dyes [(Malachite green (MG), Methyl violet (MV), Crystal violet (CV) and Cotton blue (CB)] and concluded that both mechanisms were responsible for dye removal. The intensity of the adsorption peaks of triphenylmethane dyes treated with live cells was greatly reduced (MG, MV, and CV) or totally abolished (CB), indicating a dye degradation mechanism. In contrast, the study discovered that triphenylmethane dyes treated with dead cells were removed through adsorption because all dye absorption peaks of the treated dye remained visible due to the dye structure remaining unchanged during the dye removal process and the bacterial cells changed colour to match the target dye. Similarly, Mishra and Maiti [161] found in their investigation on the decolorisation of an anthraquinone-based dye reactive blue 19 by the strain *B. cohnii* LAP217 that the dye removal mechanism was due to biosorption of the dye on the surface of the bacterial cells.

Analytical Methods for Evaluation of Dye Degradation Mechanism

Several analytical techniques are used to detect the intermediates and aromatic amines produced during azo dye decolorisation. UV Vis analysis is a preliminary approach for studying products formed during biodegradation after incubation under anaerobic circumstances by monitoring changes in the ultraviolet visible (UV–Vis) spectra. To quantify dye decolorisation, a UV–Vis spectrophotometer was used to measure the dye's absorbance at λ max [64]. Thin Layer Chromatography (TLC) and High Performance Thin Layer Chromatography (HPTLC) methods are employed to detect metabolites further [163]. These techniques provide information on the kinds and quantities of metabolites produced. However, HPLC has been widely employed to confirm dye degradation [228]. Additionally, Fourier Transform Infrared Spectroscopy (FTIR) is frequently used to verify that the azo group has been removed from the azo dye and to create new kinds of compounds [217]. Gas Chromatography-Mass Spectrometry (GC–MS) and Liquid Chromatography-Mass Spectrometry (LC–MS) are two more potent methods for qualitatively determining metabolites. The assessment is carried out using molecular weights and structural data, which are then utilised to suggest the dye's degradation process. In addition to these approaches, sophisticated techniques such as Nuclear Magnetic Resonance (NMR) are currently used to get comprehensive quantitative studies [30].

Factors Influencing the Degradation of Dyes by Bacteria

Environmental elements including pH, temperature, oxygen and agitation, initial dye concentration and nutrition content all affect how successfully bacteria biodegrade textile dyes.

Effect of pH and Temperature

Because of the wide range of chemicals employed in textile operations, textile effluents have an extremely complicated nature. High pH, colour molecules and additives like detergents and salts are the main characteristics of the effluent. It has been demonstrated that effluent pH affects dye biodegradation; most dyes degrade well at neutral to slightly alkaline pH [83, 187]. Table 1 shows the pH range needed for various bacterial strains to remove the most dye. This demonstrates that pH plays a crucial role in bacterial growth; if the pH of the effluent does not promote bacterial growth, dye degradation will not be feasible since an incorrect pH would make the bacteria dormant. The pH of the effluent must be changed to accommodate the bacteria being utilised or appropriate bacteria that can tolerate the pH of the effluent must be selected for the degradation process, in order to guarantee that dye degradation occurs. The bacterial strain *Bacillus megaterium* KY848339.1 was discovered by Ewida et al. [82] to be able to decolorise the azo dye Acid red 337 at an optimal pH of 7 with an efficiency of 91% at an initial concentration of 500 mg/l in under 24 h. In a similar vein, Ezhilarasu [83] investigated how *Bacillus* species decolorised the azo dyes Navy blue and Methyl red at varying pH levels (5–8). It was discovered that an ideal pH of 7 produced the highest dye removal effectiveness for both colours. In a different investigation to ascertain the *Enterobacter sp.* CV–S1 bacterial strain's biodegradation effectiveness on a triphenylmethane-based dye Crystal violet, Roy et al. [202] discovered that the bacterial strain could totally decolorise Crystal violet at a pH of 6.5. Another factor that is crucial to the decolorisation and deterioration of dyes is temperature. For bacterial organisms to maintain their development, survival and metabolic processes, the ideal temperature is required. According to Çetin and De Onmez [49], bacteria may experience cell deactivation or inactivation as a result of low or high temperatures, which will impair their capacity for biodegradation. According to studies, the ideal temperature range for bacterial dye degradation in both pure cultures and consortia is between 30 and 40 °C [86]. When Parmar and Shukla [180] evaluated the ability of a *Staphylococcus hominis Subsp. hominis* DSM 20328 bacterial strain to decolorise an anthraquinone-based dye, C.I. Reactive blue 4, they discovered that the highest decolorisation efficiency was attained at 37 °C. They found that when the temperature rose over 37 °C, the decolorisation efficiency decreased. They ascribed this to the bacteria's loss of cell viability as a result of the high temperatures. According to Naseer et al., a bacterial consortium used to break down and detoxify Navy blue CBF dye was temperature dependent. The decolorisation efficiency increased as the temperature rose up to an optimal of 35 °C, after which it was observed to drastically decline [168]. In a similar line, Bharagava et al. [39] explored how a bacterial strain A. hydrophila decolorised the triphenylmethane dye Crystal Violet. They discovered that the bacteria's maximum decolorisation effectiveness was at 35 °C, and that it dropped as the temperature rose above that point.

Initial Dye Concentration

The efficiency of the dye degradation or decolorisation process is influenced by the concentration of dyes inside the reaction medium. According to some studies, the dye concentration in textile waste water ranges from 10 to 50 mg/L [134], while other studies report concentrations ranging from 10 to 250 mg/L [18]. Therefore, for the bacteria to be useful in industrial applications, it is crucial that any possible organisms be able to efficiently breakdown or decolorise the target dyes without any hindrance at these dye concentrations. Due to the toxicity of dyes at high concentrations, which prevent the bacteria's metabolic processes, several studies have demonstrated that an increase in dye concentration results in a decrease in the bacteria's efficiency of degradation [120, 220]. According to reports, the insufficient biomass to dye ratio and the excess dye molecules' obstruction of enzyme active sites are the additional causes of the decreased degradation efficiency at high dye concentrations [64]. Parmar and Shukla [180] investigated the impact of varying dye concentrations on the degradation of C.I. Reactive blue 4, an anthraquinone-based dye, by *Staphylococcus hominis subsp. hominis* DSM 20328. They discovered that a higher dye concentration resulted in lower dye degradation efficiency. The results demonstrated that a concentration of 50 mg/L produced the highest degradation efficiency of 97%, while a concentration of 1000 mg/L produced the lowest. Similarly, in their investigation into the impact of bacteria isolated from a textile waste dump on the degradation of an azo-based dye, Procion red H-3B, when an initial dye concentration of 50 mg/L was utilised, Bera and Tank [37] reported a maximum decolorisation effectiveness of 96%. When the dye concentration was raised over 600 mg/L, the degradation efficiency dropped precipitously to 30%. Nevertheless, it has been demonstrated that certain strains may break down dyes at high concentrations. Masarbo et al. [156] investigated the influence of methyl orange concentration on the decolorisation capacity of three bacterial strains: *Bacillus spp.* AK1, *Lysinibacillus spp.* AK2 and *Kerstersia spp.* VKY1. The results showed that the three strains could decolorise the dye at concentrations ranging from 600 to 800 mg/L, after which the decolorisation effectiveness dropped.

Carbon and Nitrogen Sources

Since most dyes lack adequate carbon to give bacteria the energy to develop, studies have shown that adding carbon and nitrogen sources to growth cultures increases the effectiveness of most bacteria's degradation [74, 206, 210, 211]. Glucose, sucrose, starch, yeast extract, beef extract, peptone, tyrosine, ammonium chloride and ammonium sulphate are among the frequently used sources of carbon and nitrogen [135]. The addition of carbon to the culture media aids in the reduction of the azo bonds and serves as both an energy source and an electron donor in the breakdown of azo-based dyes. The reducing agent nicotinamide adenine dinucleotide (NADH), which serves as an electron donor in the reduction of the azo bonds, is also regenerated by the addition of nitrogen sources [164]. Various carbon and nitrogen sources have varying effects on different bacterial strains, some assimilate the extra carbon or nitrogen sources rather than employing them as electron donors [110]. This has a detrimental effect on the dye degradation efficiency because adding such additional carbon and nitrogen sources does not significantly alter the dyes' reduction, which results in low degradation efficiencies. Furthermore, since excessive concentrations of carbon or nitrogen supplements can impede bacterial cell growth, they have a detrimental effect on the bacteria's ability to degrade. Therefore, it's critical to determine the ideal ratio of carbon and nitrogen to improve the bacteria's capacity to break down dyes. In contrast to the carbon-free culture, which shown an efficiency of 27–51%, Guadie et al. discovered that *Bacillus spp.* strain CH12 demonstrated the maximum degradation efficiency of 95–100% when supplemented with various carbon sources. The addition of organic nitrogen sources, such as yeast and peptone, produced the highest degradation efficiency (> 90%) in the same study. In contrast, the maximum degradation efficiency attained by inorganic sources was less than 30%, with NaNo3 having the lowest efficiency at 8% [257].

Shaking and Static Conditions

Bacterial biodegradation efficiency is also impacted by static or shaking circumstances [120]. Microbial enzyme activity is supported by an increase in dissolved oxygen under shaking circumstances [252]. The elevated metabolic activity of the bacterial enzymes may result in enhanced dye breakdown capabilities. The presence of excess oxygen in the reaction medium can inhibit the activity of reductive enzymes such as azoreductase, despite the fact that shaking conditions increase degradation efficiency. This is because excess oxygen creates a more positive redox potential, which lowers the bacterial organisms' degradation efficiency by reducing the redox mediator rather than the dye [135]. Accordingly, only aerobic oxidative enzymes can favour shaking conditions, whereas anaerobic reductive enzymes can favour static settings. The ability of *B. vietnamensis spp.* MSB 17 to decolorise Mala chite green, a triphenylmethane dye, under both static and shaking conditions was examined by Kabeer et al. [116]. According to the study, static conditions yielded the best decolorisation efficiency, with an efficiency of > 99%, whereas shaking conditions produced an efficiency of 80%. It has also been demonstrated that azo dyes degrade well in static environments [10, 244]. Maniyam et al. discovered that the decolorisation and biodegradation of Methyl red by two *Rhodococcus strains* UCC 0016 and UCC 0008 was more efficient under static conditions than shaking conditions for both bacterial strains [151]. On the other hand, Saranraj et al. [208] used a consortium of bacteria to compare the degradation efficiency of three reactive dyes, Reactive orange-16, Reactive black-B and Reactive yellow-MR, under shaking and static conditions and discovered that the maximum decolorisation of all the dyes occurred during shaking. This demonstrates how various bacteria respond to similar culturing conditions.

2 Conclusion

Microorganisms are essential organisms due to their efficacy at degradation. Among these are microorganisms that are particularly useful for the breakdown of refractory azo dyes, which are widely employed in the textile sector. Algae are also commonly utilised for the breakdown and decolorisation of textile effluents. Several algal species are capable of degrading various dyes, colourants and even organic contaminants. Similarly, fungi play an important role in the bioremediation process. Different fungal enzymes have the ability to break down different colours more efficiently. The most important fungal species in terms of degradation are white rot fungi, which can digest complicated effluents employing a variety of enzymes. Because of their absorption and breakdown capabilities, enzymes can serve as an alternative to wastewater treatment methods. Enzymes or biological catalysts, may absorb a variety of dyes, including the most resistant azo dyes. The yeast's high dye and heavy metal absorption makes it an ideal choice for azo dye biodegradation. Thus, it can be concluded that microorganisms have enormous potential for wastewater remediation and can be successfully used to keep our environment clean and healthy. As a result, there is a need for more investigation into the degrading capacities of various microorganisms.

References

1. Abid, M. F., Zablouk, M. A., & Abid-Alameer, A. M. (2012). Experimental study of dye removal from industrial wastewater by membrane technologies of reverse osmosis and nanofiltration. *Iranian Journal of Environmental Health Science and Engineering, 9*, 17. https://doi.org/10.1186/1735-2746-9-17
2. Abioye, O. P., Mustapha, O. T., Aransiola, S. A. (2014). Biological treatment of textile effluent using Candida zeylanoides and Saccharomyces cerevisiae isolated from soil. *Advances in Biology (1)*, 670394.
3. Abioye O.P, Mustapha O.T., Aransiola SA. (2014). Biological Treatment of Textile Effluent zeylanoides and Using Candida Saccharomyces cerevisiae Isolated from Soil. *Advances in Biology*1–4.
4. Ademakinwa, A. N., & Agboola, F. K. (2015). Bioremediation of textile dye solutions, textile dye mixtures and textile effluents by laccase from *Aureobasidium pullulans* (de Bary) G. Arnaud (1918) (Fungi: Ascomycota). *Brazilian Journal of Biological Sciences, 2*, 253–262.
5. Adenan, N. H., Lim, Y. Y., & Ting, A. S. Y. (2021). Identification and optimization of triphenylmethane dyes removal by Streptomyces sp. from forest soil. *Sustainable Environment Research, 31*(1).
6. Adinew, B. (2012). Textile effluent treatment and decolorization techniques—A review. *Bulgarian Journal of Science Education, 21*, 434–456.
7. Afrin, S., et al. (2021). The degradation of textile industry dyes using the effective bacterial consortium. *Heliyon, 7*(10), Article e08102.
8. Ahalya, N., Kanamadi, R. D., Ramachandra, T. V. (2005). Biosorption of chromium (VI) from aqueous solutions by the husk of Bengal gram (Cicer arientinum).*Electronic Journal of Biotechnology, 8*(3), 0-0.
9. Ahmad, A., Mohd-Setapar, S. H., Chuong, C. S., et al. (2015). Recent advances in new generation dye removal technologies: Novel search for approaches to reprocess wastewater. *RSC Advances, 5*, 30801–30818.
10. Ajaz, M., Rehman, A., Khan, Z., Nisar, M. A., & Hussain, S. (2019). Degradation of azo dyes by *Alcaligenes aquatilis* 3c and its potential use in the wastewater treatment. *AMB Express, 9*(1).
11. Akansha, K., Chakraborty, D., Sachan, S. G. (2019). Decolorization and degradation of methyl orange by Bacillus stratosphericus SCA1007. Biocatalysis and Agricultural Biotechnology.*Biocatalysis and Agricultural Biotechnology,18,* 101044.
12. Al-Shuwaiki, N. M., Abid, B. A., & Brbooti, M. M. (2013). Color removal from industrial textile wastewater using chemical adsorption. *Journal of Engineering Technolgy Part (B), 31*, 471–489.
13. Ali, H. (2010). Biodegradation of synthetic dyes-a review. *Water, Air, & Soil Pollution, 213*, 251–273.
14. Ali, N., Hameed, A., & Ahmed, S. (2010). Role of brown-rot fungi in the bioremoval of azo dyes under different conditions. *Brazilian Journal of Microbiology, 41*, 907–915.
15. Almazan-sanchez, P. T., Linares-hernández, I., Solache-Ríos, M. J., & Martínez-Miranda, V. (2016). Textile wastewater treatment using iron-modified clay and copper-modified carbon in batch and column systems. *Water Air Soil Pollution, 227*, 1–14.

16. Ameenudeen, S., Unnikrishnan, S., Ramalingam, K. (2021). Statistical optimization for the efficacious degradation of reactive azo dyes using Acinetobacter baumannii JC359.*Journal of Environmental Management,279*,111512.
17. An, Y., Wang, Z., Wu, Z., Yang, D., & Zhou, Q. (2009). Characterization of membrane foulants in an anaerobic non-woven fabric membrane bioreactor for municipal wastewater treatment. *Chemical Engineering Journal, 155*(3), 709–715.
18. Ananthashankar, A. G. R. (2013). Production, characterization and treatment of textile effluents: a critical review. *Journal of Chemical Engineering and Process Technology, 5*(1).
19. Anastasi, A., Parato, B., Spina, F., et al. (2011). Decolourisation and detoxification in the fungal treatment of textile wastewaters from dyeing processes. *New Biotechnology, 29*, 38–45.
20. Anastasopoulou, A., Kolios, A., Somorin, T., et al. (2018). Conceptual environmental impact assessment of a novel self-sustained sanitation system incorporating a quantitative microbial risk assessment approach. *Science of the Total Environment, 639*, 657–672.
21. Andreozzi, R., Caprio, V., Insola, A., & Marotta, R. (1999). Advanced oxidation processes (AOP) for water purification and recovery. *Catalysis Today, 53*(1), 51–59.
22. Anliker, R. (1979). Ecotoxicology of dyestuffs—A joint effort by industry. *Ecotoxicology and Environmental Safety, 3*, 59–74.
23. Aouni, A., Fersi, C., Cuartas-Uribe, B., Bes-Pia, A., Alcaina-Mirand, M. I., & Dhahbi, M. (2012). Reactive dyes rejection and textile effluent treatment study using ultrafiltration and nanofiltration processes. *Desalination, 297*, 87–96.
24. Arikan, E. B., Isik, Z., Bouras, H. D., & Dizge, N. (2019). Investigation of immobilized filamentous fungi for treatment of real textile industry wastewater using up flow packed bed bioreactor. *Bioresource Technology Reports, 7*, Article 100197.
25. Avlonitis, S. A., Poulios, I., Sotiriou, D., Pappas, M., & Moutesidis, K. (2008). Simulated cotton dye effluents treatment and reuse by nanofiltration. *Desalination, 221*, 259–267.
26. Ayed, L., Chaieb, K., Cheref, A., & Bakhrouf, A. (2010). Biodegradation and decolorization of triphenylmethane dyes by Staphylococcus epidermidis. *Desalination, 260*(1–3), 137–146.
27. Ayed, L., Khelifi, E., Ben Jannet, H., et al. (2010). Response surface methodology for decolorization of azo dye Methyl Orange by bacterial consortium: Produced enzymes and metabolites characterization. *Chemical Engineering Journal, 165*(1), 200–208.
28. Ayed, L., Hkiri, N., & Hamdi, M. (2018). Congo Red decolorization and detoxification by *Aspergillus niger*: Removal mechanisms and dye degradation pathway Nedra Asses.
29. Ayed, L., Ladhari, N., El Mzoughi, R., & Chaieb, K. (2021). Decolorization and phytotoxicity reduction of reactive blue 40 dye in real textile wastewater by active consortium: Anaerobic/aerobic algal-bacterial-probiotic bioreactor, *Journal of Microbiological Methods, 181*, 106129. ISSN 0167-7012.
30. Ayed L, Khelifi E, Ben Jannet H, et al. (2010). Response surface methodology for decolorization of azo dye Methyl Orange by bacterial consortium: Produced enzymes and metabolites characterization. *Chemical Engineering Journal, 165*,(1):200–208.
31. Banat, I. M., Nigam, P., Singh, D., et al. (1996). Microbial decolorization of textile-dyecontaining effluents: A review. *Bioresource Technology, 58*(3), 217–227.
32. Barathi, S., Aruljothi, K. N., Karthik, C., Padikasan, I. A., & Ashokkumar, V. (2022). Biofilm mediated decolorization and degradation of reactive red 170 dye by the bacterial consortium isolated from the dyeing industry wastewater sediments. *Chemosphere, 286*(P3), Article 131914.
33. Bayar, S., Karagunduz, A., & Keskinler, B. (2016). Influences of electroosmosis and electrophoresis on permeate flux and membrane fouling in submerged membrane bioreactors (SMBRs). *Water Science and Technology, 74*(3), 766–776.
34. Behl, K., Joshi, M., Sharma, M., Tandon, S., Chaurasia, A. K., Bhatnagar, A., & Nigam, S. (2019). Performance evaluation of isolated electrogenic microalga coupled with graphene oxide for decolorization of textile dye wastewater and subsequent lipid production. *Chemical Engineering Journal, 375*, Article 121950.
35. Belli, T. J., Battistelli, A. A., Costa, R. E., Vidal, C. M. S., Schlegel, A. E., & Lapolli, F. R. (2019). Evaluating the performance and membrane fouling of an electro-membrane bioreactor

treating textile industrial wastewater. *International Journal of Environmental Science and Technology, 16*(11), 6817–6826.

36. Benkhaya, S., Rabet, S. M., & El Harfi, A. (2020). Areviewonclassifications, recentsynthesisand applications of textile dyes. *Inorganic Chemistry Communications, 115*, Article 107891.
37. Bera, S. P., & Tank, S. K. (2021). Screening and identification of newly isolated *Pseudomonas* sp. for biodegrading the textile azo dye C.I. Procion Red H-3B. *Journal of Applied Microbiology, 130*(6), 1949–1959.
38. Berkessa, Y. W., Yan, B., Li, T., Jegatheesan, V., & Zhang, Y. (2020). Treatment of anthraquinone dye textile wastewater using anaerobic dynamic membrane bioreactor: Performance and microbial dynamics. *Chemosphere, 238*, Article 124539.
39. Bharagava, R. N., Mani, S., Mulla, S. I., & Saratale, G. D. (2018). Degradation and decolourization potential of an ligninolytic enzyme producing Aeromonas hydrophila for crystal violet dye and its phytotoxicity evaluation. *Ecotoxicology and Environmental Safety, 156*, 166–175.
40. Bhuvaneswari, A., Asha, B., & Selvakumar, D. (2016). Start up and enhancement granulation in an anaerobic baffled reactor for the treatment of textile wastewater. *International Journal of Civil Engineering, 9*, 645–652.
41. El Bouraie, M., & El Din, W. S. (2016). Biodegradation of Reactive Black 5 by Aeromonas hydrophila strain isolated from dye-contaminated textile wastewater. *Sustainable Environment Research, 26*(5), 209–216.
42. Bozoglu, C., Adiguzel, A., Nadaroglu, H., et al. (2013). Purification and characterization of Laccase from newly isolated thermophilic Brevibacillus sp. (Z1) and its applications in removal of Textile Dyes. *Research Journal of Biotechnology, 8*, 56–66.
43. Brik, M., Schoeberl, P., Chamam, B., Braun, R., & Fuchs, W. (2006). Advanced treatment of textile wastewater towards reuse using a membrane bioreactor. *Process Biochemistry, 41*, 1751–1757.
44. Buscio, V., Marín, M. J., Crespi, M., & Gutierrez-Bouzan, C. (2015). Reuse of textile wastewater after homogenization-decantation treatment coupled to PVDF ultrafiltration membranes. *Chemical Engineering Journal, 265*, 122–128.
45. Cai, J., et al. (2020). Effective decolorization of anthraquinone dye reactive blue 19 using immobilized Bacillus sp. JF4 isolated by resuscitation-promoting factor strategy. *Water Science and Technology, 81*(6), 1159–1169.
46. Carolin, C. F., Kumar, P. S., Joshiba, G. J. (2021). Sustainable approach to decolourize methyl orange dye from aqueous solution using novel bacterial strain and its metabolites characterization.*Clean Technologies and Environmental Policy,23*(1), 173-181.
47. Castro, F. D., Bassin, J. P., & Dezotti, M. (2017). Treatment of a simulated textile wastewater containing the Reactive Orange 16 azo dye by a combination of ozonation and moving-bed biofilm reactor: Evaluating the performance, toxicity, and oxidation by-products. *Environmental Science and Pollution Research, 24*(7), 6307–6316.
48. Cerboneschi, M., Corsi, M., Bianchini, R., Bonanni, M., & Tegli, S. (2015). Decolorization of acid and basic dyes: Understanding the metabolic degradation and cell-induced adsorption/precipitation by Escherichia coli. *Applied Microbiology and Biotechnology, 99*(19), 8235–8245.
49. Çetin, D., & Dönmez, G. (2006). Decolorization of reactive dyes by mixed cultures isolated from textile effluent under anaerobic conditions. *Enzyme and Microbial Technology, 38*(7), 926–930.
50. Chang, I. S., Le Clech, P., Jefferson, B., & Judd, S. (2002). Membrane fouling in membrane bioreactors for wastewater treatment. *Journal of Environmental Engineering, 128*(11), 1018–1029.
51. Chang, J. S., and Kuo, T. S. (2000). Kinetics of bacterial decolorization of azo dye with Escherichia coli NO3.*Bioresource Technology, 75*(2), 107-111.
52. Chanwala, J., Kaushik, G., Dar, M. A., Upadhyay, S., & Agrawal, A. (2019). Process optimization and enhanced decolorization of textile effluent by Planococcus sp. isolated from textile sludge. *Environmental Technology & Innovation, 13*, 122–129.

53. Charumathi, D., & Das, N. (2012). Packed bed column studies for the removal of synthetic dyes from textile wastewater using immobilised dead C. tropicalis. *Desalination, 285*, 22–30.
54. Chaturvedi, A., Rai, B. N., Singh, R. S., & Jaiswal, R. P. (2021). A comprehensive review on the integration of advanced oxidation processes with biodegradation for the treatment of textile wastewater containing azo dyes. *Reviews in Chemical Engineering*.
55. Chatzisymeon, E., Xekoukoulotakis, N. P., Coz, A., Kalogerakis, N., & Mantzavinos, D. (2006). Electrochemical treatment of textile dyes and dyehouse effluents. *Journal of Hazardous Materials, 137*, 998–1007.
56. Chekli, L., Kim, Y. P., Sherub, P., et al. (2017). Evaluation of fertilizer-drawn forward osmosis for sustainable agriculture and water reuse in arid regions. *Journal of Environmental Management, 187*, 137–145.
57. Chen, S. H., & Ting, A. S. Y. (2015). Biosorption and biodegradation potential of triphenylmethane dyes by newly discovered Penicillium simplicissimum isolated from indoor wastewater sample. *International Biodeterioration and Biodegradation, 103*, 1–7.
58. Chengalroyen, M. D., & Dabbs, E. R. (2013). The microbial degradation of azo dyes: Minireview. *World Journal of Microbiology & Biotechnology, 29*(3), 389–399.
59. Cheriaa, J., Khaireddine, M., Rouabhia, M., & Bakhrouf, A. (2012). Removal of triphenylmethane dyes by bacterial consortium. *The Scientific World Journal*.
60. Christie, R. (2001). *Color chemistry*. The Royal Society of Chemistry.
61. Chung, K. T. (2016). Azo dyes and human health: A review. *Journal of Environmental Science and Health. Part C, Environmental Carcinogenesis & Ecotoxicology Reviews, 34*(4), 233–261.
62. Cui, Y., Gao, H., Yu, R., Gao, L., & Zhan, M. (2021). Biological-based control strategies for MBR membrane biofouling: A review. *Water Science and Technology, 83*(11), 2597–2614.
63. Cui, D., Zhang, H., He, R., et al. (2016). The comparative study on the rapid decolorization of azo, anthraquinone and triphenylmethane dyes by anaerobic sludge. *International Journal of Environmental Research and Public Health, 13*(11), 1053.
64. Cui, D., Li, G., Zhao, M., et al. (2014). Decolourization of azo dyes by a newly isolated Klebsiella sp. strain Y3, and effects of various factors on biodegradation. *Biotechnology & Biotechnological Equipment, 28*(3), 478–486.
65. Dafale, N., Wate, S., Meshram, S., et al. (2010). Bioremediation of wastewater containing azo dyes through sequential anaerobic–aerobic bioreactor system and its biodiversity. *Environmental Reviews, 18*, 21–36.
66. Darwesh, O. M., Matter, I. A., & Eida, M. F. (2019). Development of peroxidase enzyme immobilized magnetic nanoparticles for bioremediation of textile wastewater dye. *Journal of Environmental Chemical Engineering, 7*, Article 102805.
67. Das, A., & Mishra, S. (2017). Removal of textile dye reactive green-19 using bacterial consortium: Process optimization using response surface methodology and kinetics study. *Journal of Environmental Chemical Engineering, 5*(1), 612–627.
68. Deowan, S. A., Bouhadjar, S. I., & Hoinkis, J. (2015). Membrane bioreactors for water treatment. In *Advances in membrane technologies for water treatment* (pp. 155–184).
69. Desore, A., & Narula, S. A. (2018). An overview on corporate response towards sustainability issues in textile industry. *Environment, Development and Sustainability, 20*, 1439–1459. https://doi.org/10.1007/s10668-017-9949-1
70. Dhaouefi, Z., Lahmar, A., Khlifi, R., BenToumia, I., Elgueder, D., & Chekir-Ghedira, L. (2022). Evaluation of eventual toxicities of treated textile waste water using anoxic–aerobic algal–bacterial photobioreactor. *Environmental Geochemistry and Health, 4*
71. Dilarri, G., & Corso, C. R. (2018). Saccharomyces cerevisiae immobilized onto cross linked chitosan beads: Application of a novel material for the removal of dye toxicity. *Environmental Technology, 39*(14), 1851–1867.
72. Dissanayake, M., Liyanage, N., Herath, C., Rathnayake, S., & Fernando, E. Y. (2021). Mineralization of persistent azo dye pollutants by a microaerophilic tropical lake sediment mixed bacterial consortium. *Environmental Advances, 3*, Article 100038.
73. Drumond Chequer, F. M., de Oliveira, G. A. R., Anastacio Ferraz, E. R., Carvalho, J., Boldrin Zanoni, M. V., & de Oliveir, D. P. (2021). Textile dyes: Dyeing process and environmental impact. In *Eco-friendly text, dyeing and finishing*, 2013.

74. Du, L. N., et al. (2015). Efficient metabolism of the azo dye methyl orange by Aeromonas sp. strain DH-6: Characteristics and partial mechanism. *International Biodeterioration and Biodegradation, 105*, 66–72.
75. Dubey, S. K., Dubey, J., Viswas, A. J., & Tiwari, P. (2011). Studies on cyanobacterial biodiversity in paper mill and pharmaceutical industrial effluents. *British Biotechnology Journal, 1*, 61–67.
76. Duran, N., Rosa, M. A., D'Annibale, A., Gianfreda, L. (2002). Applications of laccases and tyrosinases (phenoloxidases) immobilized on different supports: a review. *Enzyme and microbial technology, 31*(7), 907-931.
77. Eichlerova, I., Homolka, L., Lisa, L., et al. (2005). Orange G and Remazol Brilliant Blue R decolorization by white rot fungi Dichomitus squalens, Ischnoderma resinosum and Pleurotus calyptratus. *Chemosphere, 60*(3), 398–404.
78. El-Badan, D., Naga, N., Mabrouk, M., Ghanem, K. (2020). Decolorization of remazol brilliant blue dye by Escherichia coli NG188 under optimized conditions.*Egyptian Journal of Botany, 60*(2), 461-475.
79. El-Rahim, W. M. A., Moawad, H., Abdel Azeiz, A. Z., et al. (2017). Optimization of conditions for decolorization of azo-based textile dyes by multiple fungal species. *Journal of Biotechnology, 260*, 11–17.
80. Elsahida, K., Fauzi, A. M., Sailah, I., & Siregar, I. Z. (2019). Sustainability of the use of natural dyes in the textile industry. *IOP Conference Series: Earth and Environmmental Science, 399*(1).
81. Environmental quality standards for water and effluent standards [Online]. Available: http://www.wepa-db.net/policies/law/laos/standards.htm. Accessed January 7, 2021.
82. Ewida, A. Y., El-Sesy, M. E., & Abou Zeid, A. (2019). Complete degradation of azo dye acid red 337 by *Bacillus megaterium* KY848339. 1 isolated from textile wastewater. *Water Science, 33*(1), 154–161.
83. Ezhilarasu, A. (2016). Textile industry Dye degrading by bacterial strain Bacillus sp. *International Journal of Advanced Research in Biological Sciences, 3*(3), 211–226.
84. Gahr, F., Hermanutz, F., & Oppermann, W. (1994). Ozonation-an important technique to comply with new German laws for textile wastewater treatment. *Water Science and Technology, 30*, 255–263.
85. Gardetti, M. A., & Torres, A. L. (Eds.). (2012). *Textiles, fashion and sustainability: A special theme issue of The Journal of Corporate Citizenship (Issue 45)* (No. 45). Greenleaf Publishing.
86. Garg, S. K., & Tripathi, M. (2017). Microbial strategies for discoloration and detoxification of azo dyes from textile effluents. *Research Journal of Microbiology, 12*(1), 1–19.
87. Garg, S., Kumar, P., Singh, S., Yadav, A., Dumée, L. F., Sharma, R. S., Mishra, V. (2020). Prosopis juliflora peroxidases for phenol remediation from industrial wastewater—An innovative practice for environmental sustainability.*Environmental Technology Innovation, 19*(4), 100865.
88. Georgiou, D., Melidis, P., & Aivasidis, A. (2002). Use of a microbial sensor: Inhibition effect of azo-reactive dyes on activated sludge. *Bioprocess and Biosystems Engineering, 25*, 79–83.
89. Ghaly, A., Ananthashankar, R., Alhattab, M., & Ramakrishnan, V. (2014). Production, characterization and treatment of textile effluents: A critical review. *Journal of Chemical Engineering and Process Technology, 5*, 1–18.
90. Giwa, A., Dindi, A., & Kujawa, J. (2019). Membrane bioreactors and electrochemical processes for treatment of wastewaters containing heavy metal ions, organics, micropollutants and dyes: Recent developments. *Journal of Hazardous Materials, 370*, 172–195.
91. Goswami, M., et al. (2020). Application of Arjuna (Terminalia arjuna) seed biochar in hybrid treatment system for the bioremediation of Congo red dye. *Bioresource Technology, 307*, Article 123203.
92. Guadie, A., Tizazu, S., Melese, M., Guo, W., Ngo, H. H., & Xia, S. (2017). Biodecolorization of textile azo dye using Bacillus sp. strain CH12 isolated from alkaline lake. *Biotechnology Reports, 15*, 92–100.

93. Guendy, H. R. (2007). Ozone treatment of textile wastewater relevant to toxic effect elimination in marine environment. *Egyptian Journal of Aquatic Research, 33*(1), 98–115.
94. Hagan, E., & Poulin, J. (2021). Statistics of the early synthetic dye industry. *Heritage Science, 9*(1), 1–14.
95. Hamedi, H., Ehteshami, M., Mirbagheri, S. A., Rasouli, S. A., & Zendehboudi, S. (2019). Current status and future prospects of membrane bioreactors (MBRs) and fouling phenomena: A systematic review. *Canadian Journal of Chemical Engineering, 97*(1), 32–58.
96. Hammel, K. E. (1996). Extra cellular free radical biochemistry of ligninolytic fungi. *New Journal of Chemistry, 20*, 195–198.
97. Haque, M. M., Haque, M. A., Mosharaf, M. K., & Marcus, P. K. (2021). Decolorization, degradation and detoxification of carcinogenic sulfonated azo dye methyl orange by newly developed biofilm consortia. *Saudi Journal of Biological Sciences, 28*(1), 793–804.
98. Hassaan, M. A., & El Nemr, A. (2017). Health and environmental impacts of dyes: Mini review. *American Journal of Environmental Science and Engineering, 1*(3), 64–67.
99. Hawari, A. H., Du, F., Baune, M., & Thöming, J. (2015). A fouling suppression system in submerged membrane bioreactors using dielectrophoretic forces. *Journal of Environmental Sciences (China), 29*, 139–145.
100. He, M., Tan, L., Ning, S., et al. (2017). Performance of the biological aerated filter bioaugmented by a yeast Magnusiomyces ingens LH-F1 for treatment of Acid Red B and microbial community dynamics. *World Journal of Microbiology & Biotechnology, 33*(2), 39.
101. Heinfling, A., Martínez, M. J., Martinez, A. T., Bergbauer, M., Szewzyk, U. (1998). Transformation of industrial dyes by manganese peroxidases from Bjerkandera adusta and Pleurotus eryngii in a manganese-independent reaction. *Applied and Environmental Microbiology, 64*(8), 2788-2793.
102. Holkar, C. R., Jadhav, A. J., Pinjari, D. P., et al. (2016). A critical review on textile wastewater treatments: Possible approaches. *Journal of Environmental Management, 182*, 351–366.
103. Hong Li, H., Tao Wang, Y., Wang, Y., Xia Wang, H., Kai Sun, K., & Mei Lu, Z. (2019). Bacterial degradation of anthraquinone dyes. *Journal of Zhejiang University. Science B, 20*(6), 528–540.
104. Hosseinzadeh, M., Bidhendi, G. N., Torabian, A., Mehrdadi, N., & Pourabdullah, M. (2015). A new flat sheet membrane bioreactor hybrid system for advanced treatment of effluent, reverse osmosis pretreatment and fouling mitigation. *Bioresource Technology, 192*, 177–184.
105. Hossen, M. Z., Hussain, M. E., Hakim, A., Islam, K., Uddin, M. N., & Azad, A. K. (2019). Biodegradation of reactive textile dye Novacron Super Black G by free cells of newly isolated *Alcaligenes faecalis* AZ26 and Bacillus spp obtained from textile effluents. *Heliyon, 5*, Article e02068.
106. Hsu, C. A., Wen, T. N., Su, Y. C., et al. (2012). Biological degradation of anthroquinone and azo dyes by a novel laccase from *Lentinus* sp. *Environmental Science and Technology, 46*(9), 5109–5117.
107. Hunger, K., Gregory, P., & Miederer, P. (2004). Important chemical chromophores of dye classes. *Industrial Dyes, Chemistry, Properties, Applications*, 13–112.
108. Hussein, F. H. (2013). Chemical properties of treated textile dyeing wastewater. *Asian Journal of Chemistry, 25*, 9393–9400.
109. Jadhav, J. P., Parshetti, G. K., Kalme, S. D., et al. (2007). Decolourization of azo dye methyl red by Saccharomyces cerevisiae MTCC 463. *Chemosphere, 68*(2), 394–400.
110. Jadhav, S. B., Phugare, S. S., Patil, P. S., & Jadhav, J. P. (2011). Biochemical degradation pathway of textile dye Remazol red and subsequent toxicological evaluation by cytotoxicity, genotoxicity and oxidative stress studies. *International Biodeterioration and Biodegradation, 65*(6), 733–743.
111. Jamee, R., & Siddique, R. (2019). Biodegradation of synthetic dyes of textile effluent by microorganisms: An environmentally and economically sustainable approach. *European Journal of Microbiology and Immunology, 9*(4), 114–118.
112. Ji, Y., Li, X. T., & Chen, G. Q. (2008). Interactions between a poly (3 hydroxybutyrate-co-3-hydroxyvalerate co-3-hydroxyhexanoate) terpolyester Leptosphaerulina sp. from Colombia. *Bioresource Bioprocess, 3*, 1–14.

113. Jin, X. C., Liu, G. Q., Xu, Z. H., et al. (2007). Decolorization of a dye industry effluent by Aspergillus fumigatus XC6. *Applied Microbiology and Biotechnology, 74*(1), 239–243.
114. Jose, A., & Ray, J. G. (2018). Toxic heavy metals in human blood in relation to certain food and environmental samples in Kerala, South India. *Environmental Science and Pollution Research, 25*(8), 7946–7953.
115. Junankar, A. A., Purohit, J. K., Bhende, N. V. (2019). Effective utilization of nanofluids for machining process enhancement.*In AIP Conference Proceedings, 2148*, p. 030040. AIP Publishing LLC.
116. Kabeer, F. A., John, N., & Abdulla, M. H. (2019). Biodegradation of malachite green by a newly isolated *Bacillus vietnamensis* sp. MSB17 from continental slope of the Eastern Arabian Sea: Enzyme analysis, degradation pathway and toxicity studies. *Bioremediation Journal, 23*(4), 334–342.
117. Kadam, A. A., Kulkarni, A. N., Lade, H. S., et al. (2013). Exploiting the potential of plant growth promoting bacteria in decolorization of dye Disperse Red 73 adsorbed on milled sugarcane bagasse under solid state fermentation. *International Biodeterioration & Biodegradation, 86*, 1–8.
118. Kalpana, D., Shim, J. H., Oh, B. T., et al. (2011). Bioremediation of the heavy metal complex dye Isolan Dark Blue 2SGL-01 by white rot fungus Irpex lacteus. *Journal of Hazardous Materials, 198*, 198–205.
119. Kant, R. (2012). Textile dyeing industry an environmental hazard. *Natural Science, 4*, 22–26. https://doi.org/10.4236/ns.2012.41004
120. Kapoor, R. T., Danish, M., Singh, R. S., Rafatullah, M., & Abdul, A. K. (2021). Exploiting microbial biomass in treating azo dyes contaminated wastewater: Mechanism of degradation and factors affecting microbial efficiency. *Journal of Water Process Engineering, 43*, Article 102255.
121. Karatay, S., Kiliç, N. K., & Dönmez, G. (2015). Removal of Remazol Blue by azoreductase from newly isolated bacteria. *Ecological Engineering, 84*, 301–304.
122. Karim, M. E., Dhar, K., & Hossain, M. T. (2018). Decolorization of textile reactive dyes by bacterial monoculture and consortium screened from textile dyeing effluent. *Journal of Genetic Engineering & Biotechnology, 16*(2), 375–380.
123. Katheresan, V., Kansedo, J., & Lau, S. Y. (2018). Efficiency of various recent wastewater dye removal methods: A review. *Journal of Environmental Chemical Engineering, 6*(4), 4676–4697. ISSN 2213-3437.https://doi.org/10.1016/j.jece.2018.06.060
124. Kehinde, F. O., & Aziz, H. A. (2014). Textile waste water and the advanced oxidative treatment process, an overview. *International Journal of Innovative Research in Science, Engineering and Technology, 3*, 15310–15317.
125. Khalid, A., Arshad, M., & Crowley, D. E. (2008). Accelerated decolorization of structurally different azo dyes by newly isolated bacterial strains. *Applied Microbiology and Biotechnology, 78*, 361–369.
126. Khan, R., Bhawana, P., & Fulekar, M. H. (2013). Microbial decolorization and degradation of synthetic dyes: A review. *Reviews in Environmental Science & Biotechnology, 12*(1), 75–97.
127. Khan, S., & Malik, A. (2018). Toxicity evaluation of textile effluents and role of native soil bacterium in biodegradation of a textile dye. *Environmental Science and Pollution Research, 25*(5), 4446–4458.
128. Khan, A. U., Ur Rehman, M., Zahoor, M., Shah, A. B., & Zekker, I. (2021). Biodegradation of brown 706 dye by bacterial strain pseudomonas aeruginosa. *Water* (Switzerland), *13*(21).
129. Khehra, M. S., Saini, H. S., Sharma, D. K., Chadha, B. S., & Chimni, S. S. (2006). Biodegradation of azo dye C.I. Acid red 88 by an anoxic–aerobic sequential bioreactor. *Dyes and Pigments, 70*, 1–7.
130. Koprivanac, N., Bosanac, G., Grabaric, Z., & Papic, S. (1993). Treatment of wastewaters from dye industry. *Environmental Technology, 14*, 385–390.
131. Korner, S., & Vermaat, J. E. (1998). The relative importance of *Lemna gibba* L., bacteria and algae for the nitrogen and phosphorus removal in duckweed-covered domestic wastewater. *Water Research, 32*(12), 3651–3661.

132. Kumar, R., Sekhar, S., Kumar, J., & Thatoi, H. (2020). *Potential application of bacterial biofilm for bioremediation of toxic heavy metals and dye-contaminated environments*. Elsevier B.V.
133. Lade, H., Kadam, A., Paul, D., et al. (2015). Biodegradation and detoxification of textile azo dyes by bacterial consortium under sequential microaerophilic/aerobic processes. *EXCLI Journal, 14*, 158–174.
134. Laing, I. G. (1991). The impact of effluent regulations on the dyeing industry. *Review of Progress in Coloration and Related Topics, 21*(1), 56–71.
135. Lalnunhlimi, S., & Veenagayathri, K. (2016). Decolorization of azo dyes (Direct Blue 151 and Direct Red 31) by moderately alkaliphilic bacterial consortium. *Brazilian Journal of Microbiology, 47*(1), 39–46.
136. Lam, D. M., & Thi Chien, T. (2013). Isolation, selection and identification of laccase producing fungal strains and its dye decolorization ability. *Chemical and Biological Science, 58*, 132–138.
137. Lambert, S. J., & Davy, A. J. (2011). Water quality as a threat to aquatic plants: Discriminating between the effects of nitrate, phosphate, boron and heavy metals on charophytes. *New Phytologist, 189*(4), 1051–1059.
138. Lebleu, N., Roques, C., Aimar, P., & Causserand, C. (2010). Effects of membrane alterations on bacterial retention. *Journal of Membrane Science, 348*(1–2), 56–65.
139. Lellis, B., Favaro-Polonio, C. Z., Pamphile, J. A., & Polonio, J. C. (2019). Effects of textile dyes on health and the environment and bioremediation potential of living organisms. *Biotechnology Research and Innovation, 3*(2), 275–290.
140. Li, Y., Cao, P., Wang, S., & Xu, X. (2022). Research on the treatment mechanism of anthraquinone dye wastewater by algal-bacterial symbiotic system. *Bioresource Technology, 347*, Article 126691.
141. Lian, J., Wijffels, R. H., Smidt, H., & Sipkema, D. (2018). The effect of the algal microbiome on industrial production of microalgae. *Microbial Biotechnology, 11*(5), 806–818.
142. Lin, S. H., & Lin, C. M. (1993). Treatment of textile waste effluents by ozonation and chemical coagulation. *Water Research, 27*(12), 1743–1748.
143. Liu, J., Xiong, J., Tian, C., Gao, B., Wang, L., & Jia, X. (2018). The degradation of methyl orange and membrane fouling behavior in anaerobic baffled membrane bioreactor. *Chemical Engineering Journal, 338*, 719–725.
144. Liu, H., Cheng, Y., & Du, B. (2015). Overexpression of a novel thermostable and chloride-tolerant laccase from Thermus thermophilus SG0. 5JP17–16 in Pichia pastoris and its application in synthetic dye decolorization. *PloS One, 10*, e0119833.
145. Logan, B. E. (2009). Exoelectrogenic bacteria that power microbial fuel cells. *Nature Reviews Microbiology, 7*(5), 375–381.
146. Mahmood, S., Khalid, A., Arshad, M., et al. (2015). Detoxification of azo dyes by bacterial oxidoreductase enzymes. *Critical Reviews in Biotechnology, 36*(4), 639–651.
147. Mahmood, F., Shahid, M., Hussain, S., et al. (2017). Potential plant growth-promoting strain Bacillus sp. SR-2-1/1 decolorized azo dyes through NADH ubiquinone:Oxidoreductase activity. *Bioresource Technology, 235*, 176–184.
148. Mallick, N. (2002). Biotechnological potential of immobilized algae for wastewater NP and Metal removal: A review. *BioMetals, 15*, 377–390.
149. Manekar, P., Patkar, G., Aswale, P., Mahure, M., & Nandy, T. (2014). Detoxifying of high strength textile effluent through chemical and bio-oxidation processes. *Bioresource Technology, 157*, 44–51.
150. Mani, S., & Bharagava, R. N. 2018.Textile industry wastewater. In *Environmental and health hazards and treatment approaches I*.
151. Maniyam, M. N., Ibrahim, A. L., & Cass, A. E. G. (2020). Decolourization and biodegradation of azo dye methyl red by Rhodococcus strain UCC 0016. *Environmental Technology, 41*(1), 71–85.
152. Mansour, H., Boughzala, O., Dridi, D., et al. (2011). Textiles dyes as a source of wastewater contamination: Screening of the toxicity and treatment methods. *Journal of Water Science, 24*(3), 209–238.

153. Manzoor, J., & Sharma, M. (2019). Impact of textile dyes on human health and environment. pp. 162–169.
154. Maron, D. M., & Ames, B. N. (1983). Revised methods for the Salmonella mutagenicity test. *Mutation Research, 113*(3–4), 173–215.
155. Martins, M. A. M., Cardoso, M. H., Queiroz, M. J., et al. (1999). Biodegradation of azo dyes by the yeast Candida zeylanoides in batch aerated cultures. *Chemosphere, 38*(11), 2455–2460.
156. Masarbo, R. S., Ismailsab, M., Monisha, T. R., Nayak, A. S., & Karegoudar, T. B. (2018). Enhanced decolorization of sulfonated azo dye methyl orange by single and mixed bacterial strains AK1, AK2 and VKY1. *Bioremediation Journal, 22*(3–4), 136–146.
157. Meehan, C., Banat, I. M., & Mcmullan, G. (2000). Decolorization of Remazol Black-B using a thermotolerant yeast, Kluyveromyces marxianus IMB3. *Environment International, 26*, 75–79.
158. Meerbergen, K., Willems, K. A., Dewil, R., Van Impe, J., Appels, L., & Lievens, B. (2018). Isolation and screening of bacterial isolates from wastewater treatment plants to decolorize azo dyes. *Journal of Bioscience and Bioengineering, 125*(4), 448–456.
159. Mehrotra, T., Dev, S., Banerjee, A., Chatterjee, A., Singh, R., & Aggarwal, S. (2021). Use of immobilized bacteria for environmental bioremediation: A review. *Journal of Environmental Chemical Engineering, 9*(5), Article 105920.
160. Meng, F., Zhang, S., Oh, Y., Zhou, Z., Shin, H. S., & Chae, S. R. (2017). Fouling in membrane bioreactors: An updated review. *Water Research, 114*, 151–180.
161. Mishra, S., & Maiti, A. (2020). Effectual bio-decolourization of anthraquinone dye reactive blue-19 containing wastewater by *Bacillus cohnii* LAP217: Process optimization. *Bioremediation Journal, 24*(1), 1–20.
162. Mitra, A., & Mukhopadhyay, S. (2016). Biofilm mediated decontamination of pollutants from the environment. *AIMS Bioengineering, 3*(1), 44–59.
163. Mohana, S., Shrivastava, S., Divecha, J., et al. (2008). Response surface methodology for optimization of medium for decolorization of textile dye Direct Black 22 by a novel bacterial consortium. *Bioresource Technology, 99*(3), 562–569.
164. Mohanty, S., Dafale, N., & Rao, N. N. (2006). Microbial decolorization of reactive black-5 in a two-stage anaerobic-aerobic reactor using acclimatized activated textile sludge. *Biodegradation, 17*(5), 403–413.
165. Mohanty, S. S., & Kumar, A. (2021). Biodegradation of Indanthrene Blue RS dye in immobilized continuous upflow packed bed bioreactor using corncob biochar. *Scientific Report, 11*(1).
166. Mosharaf, M. K., et al. (2018). Metal-adapted bacteria isolated from wastewaters produce biofilms by expressing proteinaceous curli fimbriae and cellulose nanofibers. *Frontiers in Microbiology, 9*, 1–17.
167. Nagraj, A. K., & Gokhale, D. (2018). Bacterial biofilm degradation using extracellular enzymes produced by *Penicillium janthinellum* EU2D-21 under submerged fermentation, 687–698.
168. Naseer, A., et al. (2016). Degradation and detoxification of Navy Blue CBF dye by native bacterial communities: An environmental bioremedial approach. *Desalination and Water Treatment, 57*(50), 24070–24082.
169. Navada, K. K., & Kulal, A. (2019). Enhanced production of laccase from gamma irradiated endophytic fungus: A study on biotransformation kinetics of aniline blue and textile effluent decolourisation. *Journal of Environmental Chemical Engineering*, 103550.
170. Neifar, M., Chouchane, H., Mahjoubi, M., et al. (2016). Pseudomonas extremorientalis BU118: A new salt-tolerant laccase-secreting bacterium with biotechnological potential in textile azo dye decolourization. *Biotech, 6*, 107.
171. Neifar, M., Jaouani, A., & Ellouze-Ghorbel, R. (2010). Purification, characterization and decolourization ability of Fomes fomentarius laccase produced in solid medium. *Journal of Molecular Catalysis B: Enzymatic, 64*(12), 68–74.
172. Neifar, M., Maktouf, S., Ghorbel, R. E., et al. (2015). Extremophiles as source of novel bioactive compounds with industrial potential. In V. K. Gupta, M. G. Tuohy, A. O'Donovan, &

M. Lohani (Eds.), *Biotechnology of bioactive compounds: Sources and applications* (pp. 245–268). Wiley.

173. Nidheesh, P. V., Gandhimathi, R., & Ramesh, S. T. (2013). Degradation of dyes from aqueous solution by Fenton processes: A review. *Environmental Science and Pollution Research, 20*(4), 2099–2132.
174. Oller, I., Malato, S., & Sanchez-Perez, J. A. (2011). Combination of advanced oxidation processes and biological treatments for wastewater decontamination—A review. *Science of The Total Environment, 409*(20), 4141–4166.
175. Oyebamiji, O. O., Boeing, W. J., Holguin, F. O., Ilori, O., & Amund, O. (2019). Green microalgae cultured in textile wastewater for biomass generation and biodetoxification of heavy metals and chromogenic substances. *Bioresource Technology Reports, 7*, Article 100247.
176. Ozturk, A., & Abdullah, M. I. (2006). Toxicological effect of indole and its azo dye derivatives on some microorganisms under aerobic conditions. *Science of the Total Environment, 358*, 137–142.
177. Padmaperuma, G., Kapoore, R. V., Gilmour, D. J., & Vaidyanathan, S. (2018). Microbial consortia: A critical look at microalgae co-cultures for enhanced biomanufacturing. *Critical Reviews in Biotechnology, 38*(5), 690–703.
178. Pan, H., et al. (2017). Decolorization pathways of anthraquinone dye Disperse Blue 2BLN by *Aspergillus* sp. XJ-2 CGMCC12963. *Bioengineered, 8*(5), 630–641.
179. Pandit, A. B., & Kumar, J. K. (2015). Clean water for developing countries. *Annual Review of Chemical and Biomolecular Engineering, 6*, 217–246.
180. Parmar, N. D., & Shukla, S. R. (2018). Decolourization of dye wastewater by microbial methods—A review. *Indian Journal of Chemical Technology, 25*(4), 315–323.
181. Parmar, N. D., & Shukla, S. R. (2018). Biodegradation of anthraquinone based dye using an isolated strain *Staphylococcus hominis* subsp. hominis DSM 20328. *Environmental Progress & Sustainable Energy, 37*(1), 203–214.
182. Pavko, A. (2011). Fungal decolourization and degradation of synthetic dyes some chemical engineering aspects. *In Tech*. https://doi.org/10.5772/16120
183. Pazdzior, K., Bilinska, L., & Ledakowicz, S. (2019). A review of the existing and emerging technologies in the combination of AOPs and biological processes in industrial textile wastewater treatment. *Chemical Engineering Journal, 376*, Article 120597.
184. Pearce, C.I., Lloyd, J.R., Guthrie, J.T. (2003). The Removal of Colour from Textile Wastewater Using Whole Bacterial Cells: A Review. *Dyes and Pigment,58*(2), 179–196.
185. Pensupa, N., Leu, S. Y., Hu, Y., et al. (2017). Recent trends in sustainable textile waste recycling methods: Current situation and future prospects. *Topics in Current Chemistry, 375*(5), 76.
186. Pillai, H. P. J. S. (2017). Optimization of process conditions for effective degradation of azo blue dye by Streptomyces DJP15.*Journal of Pure and Applied Microbiology, 11*(4), 1757-1765.
187. Pinheiro, L. R. S., Gradíssimo, D. G., Xavier, L. P., & Santos, A. V. (2022). Degradation of azodyes: Bacterial potential for bioremediation. *Sustainability Times, 14*(3), 1–23.
188. Popli, S., & Patel, U. D. (2015). Destruction of azo dyes by anaerobic–aerobic sequential biological treatment: A review. *International Journal of Environmental Science and Technology, 12*, 405–420.
189. Prabakar, D., Manimudi, V. T., Mathimani, T., et al. (2018). Pretreatment technologies for industrial effluents: Critical review on bioenergy production and environmental concerns. *Journal of Environmental Management, 218*, 165–180.
190. Prasad, A. S. A., Kumar, G., & Thomas, D. M. (2017). Microbial decolorization of azo dyes—A mini review. *Bulletin of Chemical and Pharmaceutical Research, 1*, 30–39.
191. Premaratne, M., Nishshanka, G. K. S. H., Liyanaarachchi, V. C., Nimarshana, P. H. V., & Ariyadasa, T. U. (2021). Bioremediation of textile dye wastewater using microalgae: Current trends and future perspectives. *Journal of Chemical Technology and Biotechnology, 96*(12), 3249–3258.

192. Punzi, M., Nilsson, F., Anbalagan, A., Svensson, B. M., Jönsson, K., Mattiasson, B., & Jonstrup, M. (2015). Combined anaerobic-ozonation process for treatment of textile wastewater: Removal of acute toxicity and mutagenicity. *Journal of Hazardous Materials, 292*, 52–60.
193. Qi, H., et al. (2018). Enhanced Reader.pdf. *Nature, 388*, 539–547.
194. Qian, F., Sun, X., & Liu, Y. (2013). Removal characteristics of organics in bio-treated textile wastewater reclamation by a stepwise coagulation and intermediate GAC/O_3 oxidation process. *Chemical Engineering Journal, 214*, 112–118.
195. Rai, H. S., Bhattacharyya, M. S., Singh, J., et al. (2005). Removal of dyes from the effluent of textile and dyestuff manufacturing industry: A review of emerging techniques with reference to biological treatment. *Critical Reviews in Environmental Science and Technology, 35*(3), 219–238.
196. Rajamohan, N., & Rajasimman, M. (2013). Kinetic modeling of dye effluent biodegradation by pseudomonas stutzeri. *Engineering, Technology & Applied Science Research, 3*(2), 387–390. https://doi.org/10.48084/etasr.276
197. Rajput, R. S., Pandey, S., & Bhadauria, S. (2017). Status of water pollution in relation to industrialization in Rajasthan. *Reviews on Environmental Health, 32*(3), 245–252.
198. Rathour, R., Jain, K., Madamwar, D., & Desai, C. (2019). Microaerophilic biodegradation of raw textile effluent by synergistic activity of bacterial community DR4. *Journal of Environmental Management, 250*, Article 109549.
199. Reddy, S., & Osborne, J. W. (2020). Biodegradation and biosorption of Reactive Red 120 dye by immobilized Pseudomonas guariconensis: Kinetic and toxicity study. *Water Environment Research, 92*(8), 1230–1241.
200. Reyes, K. R. E., Tsai, P. W., Tayo, L. L., Hsueh, C. C., & Chen, B. Y. (2021). Biodegradation of anthraquinone dyes: Interactive assessment upon biodecolorization, biosorption and biotoxicity using dual-chamber microbial fuel cells (MFCs). *Process Biochemistry, 101*, 111–127.
201. Robinson, T., Mcmullan, G., Marchant, R., et al. (2001). Remediation of dyes in textile effluent: A critical review on current treatment technologies with a proposed alternative. *Bioresource Technology, 77*(3), 247–255.
202. Roy, D. C., et al. (2018). Biodegradation of Crystal Violet dye by bacteria isolated from textile industry effluents. *Peer J, 6*, 1–15.
203. Roy, D. C., et al. Bioremediation of malachite green dye by two bacterial strains isolated from textile effluents. *Current Research in Microbial Science, 1*, 37–43.
204. Saba, B., Christy, A. D., Park, T., Yu, Z., Li, K., & Tuovinen, O. H. (2018). Decolorization of reactive black 5 and reactive blue 4 dyes in microbial fuel cells. *Applied Biochemistry and Biotechnology, 186*(4), 1017–1033.
205. Samsami, S., Mohamadi, M., Sarrafzadeh, M. H., Rene, E. R., & Firoozbahr, M. (2020). Recent advances in the treatment of dye-containing waste water from textile industries:Overview and perspectives. *Process Safety and Environmental Protection, 143*, 138–163.
206. Sani, R. K., & Banerjee, U. C. (1999). Decolorization of triphenylmethane dyes and textile and dye-stuff effluent by Kurthia sp. *Enzyme and Microbial Technology, 24*(7), 433–437.
207. Dos Santos, A. B., Cervantes, F. J., & Van Lier, J. B. (2007). Review paper on current technologies for decolourisation of textile wastewaters: Perspectives for anaerobic biotechnology. *Bioresource Technology, 98*(12), 2369–2385.
208. Saranraj, P., Sivasakthivelan, P., & Arul, J. (2018). Comparison of bacterial decolourization of reactive textile dyes under static and shaking conditions comparison of bacterial decolourization of reactive textile dyes under static and shaking conditions. *Advances in Biology Research, 12*(6), 199–203.
209. Saratale, R. G., Saratale, G. D., Chang, J. S., et al. (2011). Bacterial decolorization and degradation of azo dyes: A review. *Journal of the Taiwan Institute of Chemical Engineers, 42*, 138–157.
210. Saratale, R. G., Saratale, G. D., Kalyani, D. C., et al. (2009). Enhanced decolorization and biodegradation of textile azo dye Scarlet R by using developed microbial consortium-GR. *Bioresource Technology, 100*(9), 2493–2500.

211. Saratale, R. G., Saratale, G. D., Chang, J. S., & Govindwar, S. P. (2009). Ecofriendly degradation of sulfonated diazo dye C.I. Reactive Green 19A using Micrococcus glutamicus NCIM-2168. *Bioresource Technol*ogy, *100*(17), 3897–3905.
212. Sarayu, K., Sandhya, S. (2010). Aerobic biodegradation pathway for Remazol Orange by Pseudomonas aeruginosa. *Applied biochemistry and biotechnology, 160,*(4), 1241-1253.
213. Sarria, V., Parra, S., Invernizzi, M., Peringer, P., & Pulgarin, C. (2001). Photochemical biological treatment of a real industrial biorecalcitrant wastewater containing 5 amino-6-methyl-2-benzimidazolone. *Water Science and Technology, 44*(5), 93–101.
214. Sekomo, C. B., Rousseau, D. P. L., Saleh, S. A., & Lens, P. N. L. (2012). Heavy metal removal in duckweed and algae ponds as a polishing step for textile wastewater treatment. *Ecological Engineering, 44*, 102–110.
215. Shafqat, M., Khalid, A., Mahmood, T., Siddique, M. T., Han, J. I., & Habteselassie, M. Y. (2017). Evaluation of bacteria isolated from textile wastewater and rhizosphere to simultaneously degrade azo dyes and promote plant growth. *Journal of Chemical Technology and Biotechnology, 92*(10), 2760–2768.
216. Shah, M. P., Patel, K. A., Nair, S. S., & Darji, A. (2013). Optimization of environmental parameters on microbial degradation of reactive black dye. *Journal of Bioremediation and Biodegradation, 4*, 10–15.
217. Shanmugam, B. K., Easwaran, S. N., Lakra, R., et al. (2017). Metabolic pathway and role of individual species in the bacterial consortium for biodegradation of azo dye: A biocalorimetric investigation. *Chemosphere, 188*, 81–89.
218. Shanmugam, B. K., Easwaran, S. N., Mohanakrishnan, A. S., Kalyanaraman, C., & Mahadevan, S. (2019). Biodegradation of tannery dye effluent using Fenton's reagent and bacterial consortium: A biocalorimetric investigation. *Journal of Environmental Management, 242*, 106–113.
219. Shanmugam, B.K., Easwaran, S.N., Mohanakrishnan, A.S., Kalyanaraman, C., Mahadevan, S. (2018). Biodegradation of tannery dye effluent using Fenton's reagent and bacterial consortium: a biocalorimetric investigation,*Journal of Environmental Management 242*:106–113.
220. Sharma, S. C. D., et al. (2016). Decolorization of azo dye methyl red by suspended and co immobilized bacterial cells with mediators anthraquinone-2,6-disulfonate and Fe_3O_4 nanoparticles. *International Biodeterioration and Biodegradation, 112*, 88–97.
221. Sharma, D. K., Saini, H. S., Singh, M., et al. (2004). Biological treatment of textile dye Acid violet-17 by bacterial consortium in an up-flow immobilized cell bioreactor. *Letters in Applied Microbiology, 38*(5), 345–350.
222. Sharma, K. P., Sharma, S., Sharma, S., Singh, P. K., Kumar, S., Grover, R., & Sharma, P. K. (2007). A comparative study on characterization of textile wastewaters (untreated and treated) toxicity by chemical and biological tests. *Chemosphere, 69*, 48–54.
223. Shelley, T. R. (1994). Dye pollution clean-up by synthetic mineral. *International Dyer, 79*, 26–31.
224. Si, J., Cui, B. K., & Dai, Y. C. (2013). Decolorization of chemically different dyes by white-rot fungi in submerged cultures. *Annals of Microbiology, 63*, 1099–1108.
225. Singh, L. (2017). Biodegradation of synthetic dyes: A mycoremediation approach for degradation/decolourization of textile dyes and effluents. *Journal of Applied Biotechnology & Bioengineering, 3*, 430–435.
226. Singh, S., Pakshirajan, K., & Daverey, A. (2010). Enhanced decolourization of Direct Red-80 dye by the white rot fungus Phanerochaete chrysosporium employing sequential design of experiments. *Biodegradation, 21*(4), 501–511.
227. Singh, R. L., Singh, P. K., & Singh, R. P. (2015). Enzymatic decolorization and degradation of azo dyes—A review. *International Biodeterioration & Biodegradation, 104*, 21–31.
228. Singh, S. N. (2014). *Microbial degradation of synthetic dyes in wastewaters* (367 p). Springer. ISBN: 978-3-319-10942-8
229. Sivakumar, D. (2014). Role of *Lemna minor* Lin in treating the textile industry wastewater. *International Journal of Environmental Ecology and Geology Mining of Engineering, 8*, 203–207.

230. Sivakumar, D., Shankar, D., Vijaya Prathima, A. J., & Valarmathi, M. (2013). Constructed wetland treatment of textile industry wastewater using aquatic macrophytes. *International Journal of Environmental Sciences, 3*, 1223–1232.
231. Slama, H. B., Chenari Bouket, A., Pourhassan, Z., Alenezi, F. N., Silini, A., Cherif-Silini, H., & Belbahri, L. (2021). Diversity of synthetic dyes from textile industries, discharge impacts and treatment methods. *Applied Sciences, 11*(14), 6255.
232. Soares Graça, M. B., Costa-Ferreira, M., & De amorim, M. T. (2001). Decolorization of an anthraquinone-type dye using a laccase formulation. *Bioresource Technology, 79*, 171–177.
233. Solanki, K., Subramanian, S., & Basu, S. (2013). Microbial fuel cells for azo dye treatment with electricity generation: A review. *Bioresource Technology, 131*, 564–571.
234. Solis, M., Solis, A., Pérez, H. I., Manjarrez, N., & Flores, M. (2012). Microbial decolouration of azo dyes: A review. *Process Biochemistry, 47*(12), 1723–1748.
235. Srinivasan, S., Sadasivam, S. K., Gunalan, S., Shanmugam, G., & Kothandan, G. (2019). Application of docking and active site analysis for enzyme linked biodegradation of textile dyes. *Environmental Pollution, 248*, 599–608.
236. Sumandono, T., Saragih, H., Migirin, Watanabe, T., et al. (2015). Decolorization of Remazol Brilliant Blue R by new isolated white rot fungus collected from tropical rain forest in East Kalimantan and its ligninolytic enzymes activity. *Procedia Environmental Sciences, 28*, 45–51.
237. Sun, J., Bi, Z., Hou, B., Cao, Y. Q., & Hu, Y. Y. (2011). Further treatment of decolorization liquid of azo dye coupled with increased power production using microbial fuel cell equipped with an aerobic biocathode. *Water Research, 45*(1), 283–291.
238. Sun, J., Li, W., Li, Y., Hu, Y., & Zhang, Y. (2013). Redox mediator enhanced simultaneous decolorization of azo dye and bioelectricity generation in air-cathode microbial fuel cell. *Bioresource Technology, 142*, 407–414.
239. Sun, Y., Zheng, H., Tan, M., Wang, Y., Tang, X., Feng, L., & Xiang, X. (2014). Synthesis and characterization of composite flocculant PAFS–CPAM for the treatment of textile dye wastewater. *Journal of Applied Polymer Science, 131*, 1–8.
240. Syafalni, S., Abustan, I., Dahlan, I., Wah, C. K., & Umar, G. (2012). Treatment of dye wastewater using granular activated carbon and zeolite filter. *Modern Applied Science, 6*, 37–51.
241. Tahir, U., Nawaz, S., Hassan Khan, U., & Yasmin, A. (2021). Assessment of biodecolorization potentials of biofilm forming bacteria from two different genera for Mordant Black 11 dye. *Bioremediation Journal, 25*(3), 252–270.
242. Tan, L., Ning, S., Zhang, X., et al. (2013). Aerobic decolorization and degradation of azo dyes by growing cells of a newly isolated yeast Candida tropicalis TL-F1. *Bioresource Technology, 138*, 307–313.
243. Thanavel, M., Bankole, P. O., Selvam, R., Govindwar, S. P., & Sadasivam, S. K. (2020). Synergistic effect of biological and advanced oxidation process treatment in the biodegradation of Remazol yellow RR dye. *Science and Reports, 10*(1), 1–9.
244. Thapa, I., & Gaur, S. (2021). Decolorization of azo dyes by newly isolated Citrobacter sp. strain EBT-2 and effect of various parameters on decolourization. *Journal of Applied Biology & Biotechnology, 9*(6), 92–99.
245. Thiruppathi, K., Rangasamy, K., Ramasamy, M., & Muthu, D. (2021). Evaluation of textile dye degrading potential of ligninolytic bacterial consortia. *Environmental Challenges, 4*, 100078. ISSN 2667-0100
246. Toh, Y. C., Yen, J. J. L., Obbard, J. P., & Ting, Y. P. (2003). Decolourisation of azo dyes by White-rot fungi (WRF) isolated in Singapore. *Enzyme and Microbial Technology, 33*, 569–575.
247. Tomei, M. C., Pascual, J. S., & Angelucci, D. M. (2016). Analysing performance of real textile wastewater bio-decolourization under different reaction environments. *Journal of Cleaner Production, 129*, 468–477.
248. Un, U. T., & Aytac, E. (2013). Electrocoagulation in a packed bed reactor-complete treatment of color and cod from real textile wastewater. *Journal of Environmental Management, 123*, 113–119.

249. Undp, A. (2012). *Africa human development report 2012 towards a food secure future (No. 267636).* United Nations Development Programme (UNDP).
250. Unnikrishnan, S., Khan, M. H., Ramalingam, K. (2018). Dye-tolerant marine Acinetobacter baumannii-mediated biodegradation of reactive red. *Water Science and Engineering,11*(4), 265-275.
251. Vandevivere, P. C., Bianchi, R., & Verstraete, W. (1998). Review treatment and reuse of wastewater from the textile wet-processing industry: Review of emerging technologies. *Journal of Chemical Technology and Biotechnology, 72*, 289–302.
252. Varjani, S., Rakholiya, P., Ng, H. Y., You, S., & Teixeira, J. A. (2020). Microbial degradation of dyes: an overview. *Bioresource Technology, 314*.
253. Velayutham, K., Madhava, A. K., Pushparaj, M., Thanarasu, A., Devaraj, T., Periyasamy, K., & Subramanian, S. (2018). Biodegradation of Remazol Brilliant Blue R using isolated bacterial culture (*Staphylococcus* sp. K2204). *Environmental Technology, 39*(22), 2900–2907. https://doi.org/10.1080/09593330.2017.1369579
254. Verma, A., Dhiman, K., & Shirkot, P. (2017). Biodegradation of Malachite green by extracellular bacterial laccase and its phytotoxicity studies. *International Journal of Pure & Applied Bioscience, 5*, 913–922.
255. Vitor, V., & Corso, C. R. (2008). Decolorization of textile dye by Candida albicans isolated from industrial effluents. *Journal of Industrial Microbiology and Biotechnology, 35*(11), 1353–1357.
256. Welham, A. (2000). The theory of dyeing (and the secret of life). *Journal of the Society of Dyers and Colourists, 116*, 140–143.
257. World Health Organization and Unicef. (2017). *Progress on drinking water, sanitation and hygiene.*
258. Yadav, D. N., Naz, I., Kishore, K. A., & Saroj, D. (2020). Evaluation of tire derived rubber (TDR) f ixed biofilm reactor (FBR) for remediation of methylene blue dye from wastewater. *Environmental Technology*, 1–27.
259. Yang, R. T. (2013). *Gas separation by adsorption processes.* Butterworth Heinemann.
260. Yanmis, D., Adiguzel, A., Nadaroglu, H., et al. (2016). Purification and characterization of laccase from thermophilic *Anoxybacillus gonensis* P39 and its application of removal textile dyes. *Romanian Biotechnological Letters, 21*(3), 11485–11496.
261. Yaseen, D. A., & Scholz, M. (2016). Shallow pond systems planted with *Lemna minor* treating azo dyes. *Ecological Engineering, 94*, 295–305.
262. Yaseen, D. A., & Scholz, M. (2019). *Textile dye wastewater characteristics and constituents of synthetic effluents: A critical review 16* (p. 2). Springer.
263. Yusuf, M., Shabbir, M., & Mohammad, F. (2017). Natural colorants: Historical, processing and sustainable prospects. *Natural Products and Bioprospecting, 7*(1), 123–145.
264. Van Der Zee, F., & Villaverde, S. (2005). Combined anaerobic–aerobic treatment of azo dyes—A short review of bioreactor studies. *Water Research, 39*(1425), 1440.
265. Zhang, Z. B., Liu, C. Y., Wu, Z. Z., Xing, L., & Li, P. F. (2006). Detection of nitric oxide in culture media and studies on nitric oxide formation by marine microalgae. *Medical Science Monitor, 12*, 75–85.
266. Zhuang, M., et al. (2020). Azo dye degrading bacteria tolerant to extreme conditions inhabit nearshore ecosystems: Optimization and degradation pathways. *Journal of Environmental Management, 261*, Article 110222.
267. Zucca, P., Neves, C. M. B., Simões, M. M. Q., et al. (2016). Immobilized lignin peroxidase like metalloporphyrins as reusable catalysts in oxidative bleaching of industrial dyes. *Molecules, 21*, E964.

Application of Biochar for Remediation of Textile Industry Waste

Karan Sathish, Nishtha Naudiyal, Hema Kandpal, Aastha Joshi, Shweta Saraswat, and Madhaiyan Sakthivel

1 Introduction

The textile industry plays a substantial role in environmental harm, currently generating 92 million tons of waste annually, a number anticipated to escalate to 134 million tons by 2030 if current methods continue. This waste crisis is fueled by fast fashion and a linear production-consumption-disposal model, compounded by ineffective waste collection systems, lack of Extended Producer Responsibility (EPR), and inadequate recycling infrastructure. Consequently, most textile waste ends up in landfills or incinerators [48]. Additionally, the textile industry is one of the largest water consumers. Producing 1 kg of textiles through wet processing requires 50–100 L of water, leading to significant wastewater generation [28]. This wastewater contains chemically complex effluents with dyes, mordants, additives, and heavy metals from processes such as dyeing, bleaching, and finishing [20]. These effluents often exceed the treatment capacity of conventional systems, resulting in inadequate wastewater management and persistent environmental problems [12]. Addressing these challenges demands a transition to sustainable production practices, improved waste management, and stringent regulatory frameworks to minimize the environmental footprint of the textile industry.

Textile solid waste can be recycled using mechanical methods to reclaim fibers, chemical methods like hydrolysis and glycolysis to regenerate monomers, and

K. Sathish (✉) · N. Naudiyal · H. Kandpal · A. Joshi · S. Saraswat (✉)
Department of Environmental Sciences, College of Basic Science and Humanities, G. B. Pant University of Agriculture and Technology, Pantnagar, Uttarakhand-263145, India
e-mail: karansathish555@gmail.com

S. Saraswat
e-mail: shwetasaraswat.bbe@gbpuat.ac.in

M. Sakthivel
Department of Plant Pathology, College of Basic Science and Humanities, G. B. Pant, University of Agriculture and Technology, Pantnagar, Uttarakhand-263145, India

R. A. Bhat et al. (eds.), *Biocentric Approaches in Textile Waste Management*, Textile Science and Clothing Technology, https://doi.org/10.1007/978-3-032-04974-2_8

biochemical processes employing enzymes or microorganisms to decompose waste sustainably. Thermal methods, including pyrolysis and gasification, convert waste into energy but may emit greenhouse gases [5, 48]. For wastewater, physical methods like sedimentation and filtration remove solids, while chemical processes such as coagulation and advanced oxidation degrade pollutants. Biological treatments, including activated sludge and wetlands, naturally break down organic matter, and membrane technologies like reverse osmosis offer efficient water purification [1, 47]. Pyrolysis is particularly effective, transforming textile waste into valuable resources while also remediating wastewater pollutants [23].

Pyrolysis is an efficient method for converting organic materials into biochar, a carbon-rich material with excellent properties for treating textile wastewater. Conducted in oxygen-limited environments at temperatures between 300 and 800 °C, pyrolysis produces biochar with high porosity and surface area, ideal for adsorbing pollutants like heavy metalsand dyes [40, 44]. Biochar removes pollutants through mechanisms like physical adsorption, chemical interactions, and ion exchange, outperforming traditional materials in constructed wetlands [4]. Its regeneration capability ensures sustainability, while its production sequesters carbon, making it a carbon-negative technology. This dual benefit of waste management and effective pollutant removal underscores the promise of pyrolysis-derived biochar as a sustainable solution for wastewater remediation [40]. Optimizing pyrolysis conditions can further enhance its potential in addressing the textile industry's environmental challenges.

This chapter explores textile waste types, biochar production from textile solid waste, and its mechanisms for removing wastewater pollutants. It highlights biochar's efficacy in addressing dyes, organic pollutants, and heavy metals through case studies, underscoring its potential as a sustainable solution for managing textile waste and mitigating environmental impacts.

2 Textile Waste and Types of Textile Wastes

A textile is a material made up of any type of fabric. The production process of textiles includes various processes such as spinning and weaving, etc. Textile is not limited to the clothing sector, it also includes furnishing materials, industry materials, etc. The textile industry is one of the main industries with respect to economy and employment. In India, the textile sector accounts for more than 2% of the GDP and is the second largest employment provider after the agriculture sector [34]. However, the textile industry is also a most polluting industry. According to Sillanpää and Ncibi [41] five percent of worldwide waste comes from textile industries. It produces solid waste and wastewater. Solid waste is generated at different steps, from raw material acquisition to manufacturing, distribution, and packaging. Textile consumption has been continuously increasing due to fast fashion. The waste is produced due to overconsumption of cloth that is finally dumped as waste due to changes in fashion. Seasonal changes in trends influence people to buy more than what they need.

This behavior causes more consumption and increases waste production. Almost 92 million tonnes of waste were produced in 2015 due to global fashion and estimated it to increase up to 148 million tonnes by 2030 [15]. The textile industry also generates large amounts of liquid waste or wastewater that contains many pollutants like heavy metals and chemicals used in the manufacturing process. Other than this textile generation also contributes to air pollution. It is responsible for 10% of global CO_2 emissions. The type of waste produced varies from factory to factory depending on fabric type and chemicals used in the process. The synthetic fiber consumption is more than natural fiber. The textile is produced from natural materials like cotton, silk, jute, etc., or synthetic materials like polyester, acrylic and nylon etc. Polyester fiber is mostly consumed for textile and second largest share come from cotton [19]. The non-biodegradable synthetic fibers are even more harmful to the environment as they accumulate in dumping grounds. The industry also uses various resources like fossil fuels, water, and energy in textile production. Therefore, it is necessary to understand textile waste to invent new ways to recirculate the waste into a valuable product that will reduce environmental concerns and increase sustainability.

Textile waste is generated in the form of liquid and solid waste (Fig. 1). Solid waste is classified into three categories: production waste, pre-consumer waste, and post-consumer waste. The manufacturing waste consists of fabric scrap, fiber cuttings, yarns, etc. The pre-consumer waste is composed of faulty fabric that has some manufacturing defects. The unsold and damaged fabric is also included in this category. The post-consumer waste is the textile waste generated by households, such as garments that are not needed or discarded. The consumer discards the clothes and household furnishing material that is damaged, worn out, and become out of fashion. The liquid waste includes chemicals and heavy metals produced from dying, bleaching, and printing. Chemicals like dyes, acids, surfactants, alkalis, and hydrogen peroxide are present in effluent wastewater. The aromatic dyes are complex in structure and can't be degraded easily [7]. Chemicals like polyvinyl alcohol (PVA), carboxymethyl cellulose (CMC), polyacetate, and polycyclic acids are added to the fiber to provide strength. To remove non-cellulosic material and impurities such as oil, fats, and various chemicals like soaps, detergent, sodium hydroxide, ethers, and glycerol are used. Chemicals like H_2O_2 and peracetic acid are used as bleaching agents, which improve whiteness in fiber and remove unwanted color. Cold or hot caustic soda (NaOH) is used to improve appearance and dye affinity with fiber. Dyes are employed to impart color to fibers, with azo dyes being the most widely utilized in the textile industry [22]. Consumption of these chemicals in the textile generation process causes the presence of many pollutants such as surfactants, salts, bleaching agents, detergents, and phenols in wastewater effluents.

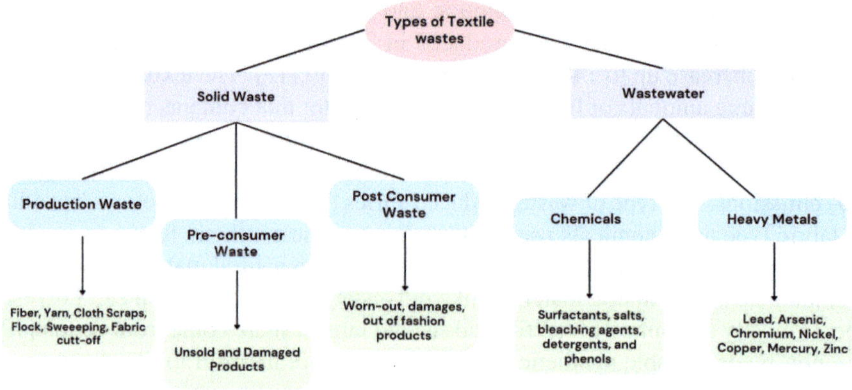

Fig. 1 Types of waste generated from the textile industry

3 Biochar

Biochar, a carbon-rich material produced through the pyrolysis of various organic waste feedstocks, has garnered considerable attention for its diverse applications in pollutant removal and environmental restoration [8]. It is produced through the thermal decomposition of organic materials such as forestry residues, animal manure, and crop waste under oxygen-limited conditions [11]. The use of porous materials for adsorption is among the most promising and cost-effective methods for removing dissolved pollutants, offering an alternative to energy-intensive technologies [29]. With its microporous structure and remarkable sorption capacity, biochar functions as an efficient adsorbent for eliminating a wide range of organic and inorganic contaminants [46]. As a biomass-derived material, biochar is gaining attention due to its renewable nature, diverse feedstock options, large surface area, high porosity, functional surface groups, and high efficiency in treating a range of pollutants [8]. Vyavahare et al. [46] highlight that converting agro-industrial waste biomass into biochar provides a cost-effective and efficient method for treating fabric dyes and wastewater effluents.

Production of Biochar

Various types of agro-industrial waste biomasses, such as plant residues, industrial waste, animal manure, sewage sludge, and forest biomass, have been utilized for biochar production [46]. Thermochemical conversion methods, including hydrothermal carbonization, pyrolysis, gasification, and torrefaction, are commonly used for biochar preparation [21]. Traditional pyrolysis systems, characterized by slower heating rates, require extended material retention times and yield high production outputs (e.g., 94% at 300 °C and 23% at 750 °C for hardwood). In contrast, fast pyrolysis significantly reduces residency time and produces a variety of byproducts, including bio-liquids and syngas, alongside biochar, depending on parameters like temperature, reaction time, heating rate, and vapor residence time [11]. Biomass with high moisture content is directly converted into a carbon-rich solid material known as hydrochar [39]. Gabhane et al. [18] explain that gasification is an effective thermochemical process for converting biomass into energy fuels, while simultaneously producing biochar as a byproduct. Torrefaction, a mild form of pyrolysis, generates biochar from biomass at lower temperatures compared to conventional pyrolysis [9].

Mechanisms Involved in Textile Wastewater Remediation by Biochar

Water pollution from dye manufacturing is one of the most severe environmental issues caused by the textile industry [42]. Textile wastewater is often discharged into the environment due to the absence of cost-effective and sustainable methods to remove dyes from water. Biochar has been shown to be highly effective for dye adsorption in textile wastewater effluent [29]. Yu et al. [55] outline various dye removal methods, with adsorption being the most favored for water decontamination [42]. Biochar, recognized globally for its effectiveness and cost-efficiency, plays a key role in adsorption [36]. Its adsorption capacity is influenced by factors like pH, porosity, surface functional groups, mineral composition, and surface charge, which must be optimized for specific contaminants [8] (Fig. 2).

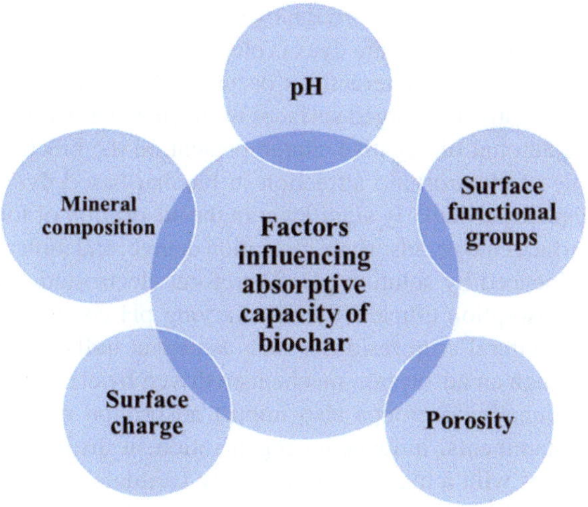

Fig. 2 Factors influencing absorptive capacity of biochar

According to insights from various research, the dyes are adsorbed through a variety of processes, such as π–π interaction, pore filling, hydrophobicity, electrostatic attraction, and hydrogen affinity [35]. The pH of biochar varies based on the pyrolysis temperature and feedstock used [24]. Praveen et al. [35] reported that the pH of the solution significantly affects the biosorptive removal of methylene blue dye by biochar derived from agricultural wastes, with dye removal efficiency increasing from 40 to 90% as the pH changed from 2 to 9. Studies on the degradation and adsorption of acid red dye revealed that decolorization rates increased with solution temperature. pH is a significant element that has consistently been observed to influence the behavior of biochar as an adsorbent [42]. Studies have shown a direct link between biochar's surface area and its adsorption capacity for organic dyes, as a larger surface area offers more active sites for dye molecules [9]. Biochar dosage significantly impacts removal efficiency, with increased dosage enhancing the removal of heavy metals. A study on wet torrefied microalgal biochar found that a slight increase in dosage raised removal efficiency by 80%, stabilizing at a maximum dosage [42]. The negative surface charge of biochar, with a point of zero charge at 8.5, facilitates electrostatic interactions during dye removal [35].

Additionally, biochar porosity significantly impacts metal sorption [50]. The mesoporous and microporous structures in high-surface-area biochar enhance dye adsorption by offering additional adsorption sites and diffusion pathways [9]. Biochar porosity and surface area are highly dependent on pyrolysis temperature, with higher temperatures typically resulting in larger pore sizes and increased surface area [24]. Surface functional groups, including carboxyl (–COOH), hydroxyl (–OH), and phenolic (–Ph) groups, play a crucial role in facilitating the adsorption of dye molecules onto the biochar surface. Numerous studies have demonstrated that these functional groups greatly impact biochar's adsorption capacity and selectivity for organic dyes [9]. Biochar produced at lower temperatures contains a higher density of functional groups, making it more effective at adsorbing polar organic compounds through hydrogen bonding. The primary mechanism driving the adsorption of cationic dyes is electrostatic interaction [35].

Electrostatic interactions occur predominantly between cationic pollutants and the negatively charged surfaces of biochar. The extent of adsorption is influenced by the amount of negative charge present on the biochar, which highlights the critical role of electrostatic attraction in biochar-based dye adsorption [45]. This mechanism is particularly significant in the adsorption of ionizable species, as the biochar surface charge, adsorbate ionization degree, and molecular conformation are strongly influenced by solution pH. Therefore, electrostatic interactions play a pivotal role in adsorption efficiency under varying pH conditions [6]. Biochar produced from agricultural crop residues, such as peanut hulls, effectively removed cationic dyes through an adsorption mechanism driven by electrostatic interactions [35].

Functional groups also impact adsorption mechanisms differently for various contaminants, influencing regeneration in distinct ways. For example, functional groups with a high electron density facilitate π-interactions and hydrogen bonding during dye adsorption, resulting in improved regeneration efficiency [9]. π-π stacking refers to noncovalent interactions between compounds with unsaturated π-electrons.

Although weaker than hydrogen bonding, this interaction plays a role in adsorption processes involving aromatic compounds [6, 46].

Biochar and other adsorbents containing exchangeable ions, such as Na^+, K^+, Ca^{2+}, and Mg^{2+}, facilitate ion exchange with target pollutants during adsorption [6]. The mineral content of biochar significantly influences its cation exchange capacity [46]. An alternative adsorption mechanism is ion exchange, as shown by the increased concentration of metal cations released by biochar into the equilibrium solution after methylene blue dye adsorption. The rise in Na^+ and K^+ ions suggest their role in ion exchange. Furthermore, biochar derived from calcium-rich crab shells effectively adsorbed malachite green and Congo red dyes, indicating that calcium ion interactions enhanced dye adsorption [42]. Adsorbent materials such as biochar, with large volumes of micro- and mesopores, provide a high specific surface area for adsorption [6]. Biochar with higher pore volumes adsorbs more pollutants through internal diffusion, while low-porosity biochar is less effective [46]. A study on malachite green dye using activated biochar from *Opuntia ficus-indica* showed enhanced adsorption due to the pore-filling effect. The removal of malachite green, methylene blue, and Rhodamine B dyes was limited by intramolecular diffusion rates [42]. Biochar primarily removes toxic heavy metals and contaminants via adsorption [50]. Conversely, greater aromaticity and hydrophobicity, characterized by a lower O/C ratio, are associated with enhanced adsorption capacity for organic molecules. However, this trend does not apply to the adsorption of ionic dyes or situations where non-carbonized matter fractions are preferred [42]. The mechanisms of textile waste removal by biochar are summarized in Table 1.

4 Removal of Pollutants from Textile Wastewater Using Biochar

Textile industries contribute to environmental degradation worldwide by discharging the textile wastewater in the ecosystem. Important procedures in textile industries including desizing, bleaching, dyeing and finishing produce wastewater with significant levels of suspended particles, heavy metals, synthetic dyes and chemical and biological oxygen demand (COD and BOD); many of which are hazardous and persistent in nature. Synthetic dyes are difficult to break down because of their diverse and durable nature and thus frequently contaminate soil and water. Furthermore, in conventional wastewater treatment, azo dyes may degrade into hazardous amines, increasing the dangers to the environment. Secondary treatment is challenging for untreated textile wastewater because they disturb aquatic habitats and interfere with microbial activity [13, 52].

To lessen these effects, 'adsorption' is a dependable and economical technique for removing contaminants of textile wastewater. Although activated carbon as an adsorbent works well, its expensive cost of production makes biochar more convenient adsorbent. Biochar technology turns waste into useful resources and provides a

Table 1 Biomass-based mechanisms for pollutant removal

S. No.	Biomass used	Pollutant	Mechanism	Reference
1.	Poplar	Bisphenol A	Adsorptive/catalytic synergy	[56]
2..	Sludge	Phenolic compounds	Coupling mechanism	[51]
3.	Water hyacinth	Phenol	Degradation via an electron-transfer pathway	[27]
4.	Parthenium hysterophorus	Hexavalent chromium	Adsorption mechanism	[32]
5.	Peanut shell	Acetaminophen	Electron-transfer mechanism	[49]
6.	Grass	Benzene	Adsorption mechanism	[31]
7.	Water hyacinth	Antibiotics	Degradation	[14]
8.	Dewatered sewage sludge	2,4-Dichlorophenol	Electron transfer degradation	[10]
9.	Typha	Antibiotic	Adsorption mechanism	[26]
10.	Barley bagasse	Ciprofloxacin	Adsorption mechanism	[30]
11.	Spinach leaf	Microplastics and methylene blue dye	Degradation	[17]
12.	Giant mud crab shell	Methyl violet	Adsorption mechanism	[16]
13.	Prussian blue/ wood	Persistent organic pollutant	Radical, non-radical and electron transfer degradation	[61]

sustainable textile wastewater treatment option. Plants, animals, marine trash, etc. are among the natural waste components that make up biochar. Effective dye adsorption and degradation are made possible by its large surface area, porosity and modifiable functionality. In contrast to costly techniques like ozonation and reverse osmosis, biochar-based adsorption is scalable, affordable, renewable, efficient and environmentally benign due to which biochar has the potential to remove contaminants and is considered as a potential platform for environmentally friendly wastewater treatment technologies [43, 44].

Removal of Dyes and Organic Pollutants

Biochar has demonstrated efficacy in removing dyes and organic pollutants (OPs) from textile wastewater using adsorption process through electrostatic interactions, pore filling, hydrogen bonding, ion exchange, hydrophobic interactions, coulombic attraction and π-π donor–acceptor mechanism. OPs even at low concentrations can be effectively removed by sophisticated methods like photocatalysis and electrocatalysis including Fenton reactions. Kinetic parameters viz. solution pH, particle size, temperature, and coexisting ions, starting pollutant concentration, dosage and reaction time are some of the variables that affect biochar's adsorption effectiveness.

Table 2 Type of biochar used for removal of dyes from Textile wastewater

Type of biochar	Targeted dye	References
Peanut straw biochar	Methyl Violet	[2]
Kenaf fibre biochar	Methylene Blue	[2]
Rice straw-derived biochar	Malachite Green	[2]
Sugarcane bagasse biochar	Malachite Green	[2]
Bamboo hydrochars	Congo Red	[2]
Groundnut shell biochar	Basic Red 09	[36]
Switch grass biochar	Congo Red	[36]
Korean cabbage biochar	Congo Red	[29]
Korean cabbage biochar	Crystal Violet	[29]
Pecan nutshell biochar	Reactive Red 141	[29]
Palm Kernel shell biochar	Crystal Violet	[29]
Sugarcane bagasse biochar	Malachite Green	[29]

Functionalized biochar that contains functional groups comprising of oxygen, hybridized carbon and heteroatoms like phosphorus, nitrogen, and Sulphur, have remarkable capacity to adsorb a variety of OPs. This functionalized biochar engages with Ops through redox mechanisms, chemical reactions and physical adsorption. Furthermore, OPs are effectively degraded by sulphate radicals and graphitic nitrogen improves photo-degradation effectiveness, whereas functional groups like –OH and –COOH on the surface of biochar encourage hydrophobic interactions with OPs. The addition of chemical agents like magnetic minerals, nano-metal oxides, etc. to biochar enhances its catalytic and adsorption capabilities. Because of their doped metals, oxygen-containing functional groups, and enduring free radicals, these materials exhibit strong catalytic activity. For instance, magnetic biochar through processes including surface precipitation, ion exchange, and Fenton-type reactions makes it easier to remove OPs. Higher pyrolysis temperatures greatly improve catalytic performance and the efficiency of OPs breakdown.

To sum up, biochar is a promising technology for the removal of dyes and OPs, providing sustainable and affordable wastewater treatment options. Additional investigation into functionalized biochar, including magnetic and photocatalytic varieties, may improve their effectiveness and expand their uses [37, 57]. The types of biochar used for dye removal from textile wastewater are summarized in Table 2.

Removal of Heavy Metal

In wastewater treatment, biochar has become a popular and affordable adsorbent, especially for the removal of heavy metals from aqueous solutions. Many research shows how effective it is; some even show that biochar is more effective than activated carbon. Heavy metals can be eliminated by processes such physical sorption,

ion exchange, complexation, precipitation, and electrostatic interactions. Adsorption mechanisms are complicated and occasionally metal-specific due to these considerations as well as the variety of heavy metal characteristics. The specific surface areaand availability of functional groups of biochar are the main factors which affects adsorption of heavy metals; that is in turn highly dependent on the feedstock material. The effectiveness of biochar is further clarified and its use in wastewater treatment is encouraged by mathematical sorption models. Heavy metal ions may be effectively adsorbed and physically held inside the pores of biochar due to its porous nature and large surface area through electrostatic interactions. Negatively charged biochar surfaces engage with these positively charged heavy metal ions. The functional groups such as carboxylate (–COOH) and hydroxyl (–OH), further increases the biochar's ability to form complexes with a variety of heavy metals. Research emphasizes that in order tomaximize the adsorption efficiency of biochar, its manufacturing and application conditions must be optimized as the sorption processes are intimately related to modifications like temperature changes and chemical activation. For instance, the formation of graphene-like structures is favored at temperatures higher than 400 °C, which increases the surface area of biochar and strengthens electrostatic forces. The pH of the solution influences the surface charge on the biochar, degree of ionization and heavy metal speciation, have an impact on sorption efficiency. Higher pH values encourage heavy metal adsorption by ion exchange and surface complexation, whereas lower pH levels may protonate functional groups on biochar, decreasing their interaction with heavy metal ions. Furthermore, chemical treatment, such as impregnation with other chemicals, boosts the adsorption potential of biochar and further improves its capacity to sorb metals. Another significant process is precipitation, in which heavy metal compounds develop on the surface of the biochar or in the solution and are efficiently immobilized by this procedure. In conclusion, biochar is a viable option for heavy metal removal in textile wastewater and significantly contribute to the solution of water pollution issues by using its natural qualities and streamlining production procedures [25, 33].

Removal of Other Pollutants

The excessive release of nitrate and phosphate from organic contaminants, dyes and textile industrial processes has led to severe water pollution, including eutrophication, posing risks to aquatic ecosystems and human health. Ammonium-nitrogen (NH_4–N) pollution exacerbates eutrophication and jeopardizes water quality through leaching. Solutions for NH_4–N recovery are provided by sustainable methods such as struvite crystallization and biochar adsorption, which allow for their repurposing as slow-release fertilizers in agriculture. Modified biochar has greater adsorption capacity in comparison to unmodified biochar and greatly improve NH_4–N adsorption because of enhanced ion exchange, electrostatic interactions and surface complexation. Researchers have observed that biochar made from sewage sludge and brewers' leftover grain have a high ammonia removal effectiveness (above 60%) which is because of their advantageous surface area and chemical characteristics. Therefore, modified biochar are useful instruments for sustainable textile wastewater treatment since they not only reduce ammonium pollution but also aid in nitrogen recovery and carbon dioxide emission reduction [3, 59].

Modified biochar offers an effective solution for nitrate (NO_3^-) and phosphate (PO_4^{3-}) removal from eutrophic water due to enhanced surface area, functional groups, and charge properties. According to studies, chemical adsorption processes in aluminum-modified biochar greatly increase the adsorption capabilities for NO_3^- and PO_4^{3-} while magnesium–aluminum modified biochar is efficient in co-adsorbing phosphate, nitrate and ammonium. This is because metal doping causes structural changes which in turn improve adsorption effectiveness. With potential uses as slow-release fertilizers to increase soil fertility and absorb carbon, modified biochars provide a sustainable method of eutrophic water restoration. In order to address complicated nutrient contamination in real-world situations, more research on co-adsorption processes and practical deployment is required [53, 54, 58].

5 Conclusion and Future Perspectives

Biochar-based solutions have shown great promise to solve textile wastewater related environmental concerns, especially when it comes to the removal of heavy metals, synthetic dyes and other persistent contaminants. Dye, heavy metal and other pollutants can be effectively adsorbed and catalytically degraded by its porous structure, wide surface area and changeable functional groups. Pore filling, π-π interactions, electrostatic interactions, hydrogen bonding, etc. are some of the adsorption mechanisms of biochar. Even modified biochar through complexation, ion exchange and chemical precipitation, has demonstrated better effectiveness in removing heavy metals, organic and other contaminants. Notwithstanding these encouraging outcomes, a number of obstacles still exist. Important topics requiring further study include: (1) optimization of biochar production and modification in order to improve the performance of biochar, (2) to make sure biochar applications don't harm the environment, (3) to get a deeper understanding of interactions at the molecular level, advanced characterization methods such as FTIR and X-ray spectroscopy should be employed, (4) to assess biochar's effectiveness, potential for regeneration and environmental effects under various circumstances, long-term field research is required. In summary, biochar provides a flexible, environment friendly and effective platform for treating wastewater from textiles and cleaning up the environment. This technology can be extremely important in reducing water pollution and fostering environmental sustainability on a worldwide scale by resolving existing constraints and developing functionalized biochar designs.

References

1. Adane, T., Adugna, A. T., & Alemayehu, E. (2021). Textile industry effluent treatment techniques. *Journal of Chemistry, 2021*(1), 5314404.
2. Ambaye, T. G., Vaccari, M., van Hullebusch, E. D., Amrane, A., & Rtimi, S. J. I. J. O. E. S. (2021). Mechanisms and adsorption capacities of biochar for the removal of organic and inorganic pollutants from industrial wastewater. *International Journal of Environmental Science and Technology, 18*(10), 3273–3294.
3. An, Q., Li, Z., Zhou, Y., Meng, F., Zhao, B., Miao, Y., & Deng, S. (2021). Ammonium removal from groundwater using peanut shell based modified biochar: Mechanism analysis and column experiments. *Journal of Water Process Engineering, 43*, Article 102219.
4. Ayadi, M., Passaseo, D., Bonaccorso, G., Fichera, M., Renai, L., Venturini, L., Colzi, I., Fibbi, D., & Del Bubba, M. (2024). Biochar from co-pyrolysis of biological sludge and sawdust in comparison with the conventional filling media of vertical-flow constructed wetlands for the treatment of domestic-textile wastewater. *Water Science & Technology, 89*(5), 1252–1263.
5. Baloyi, R. B., Gbadeyan, O. J., Sithole, B., & Chunilall, V. (2024). Recent advances in recycling technologies for waste textile fabrics: A review. *Textile Research Journal, 94*(3–4), 508–529.
6. Barquilha, C. E., & Braga, M. C. (2021). Adsorption of organic and inorganic pollutants onto biochars: Challenges, operating conditions, and mechanisms. *Bioresource Technology Reports, 15*, Article 100728.
7. Behera, M., Nayak, J., Banerjee, S., Chakrabortty, S., & Tripathy, S. K. (2021). A review on the treatment of textile industry waste effluents towards the development of efficient mitigation strategy: An integrated system design approach. *Journal of Environmental Chemical Engineering, 9*(4), Article 105277.
8. Chandola, D., & Rana, S. (2022). Biochar for environmental remediation. In *Biochar-productive technologies, properties and applications*. IntechOpen.
9. Chavan, R., Patil, R., &Chougale, A. (2024). Efficient dye removal strategies: Exploring the role of biochar. *ChemSci Advances, 1*(1), 4–13.
10. Cui, J., Wei, L., Ning, C., Zhang, F., Cui, J., & Xiangli, P. (2024). Highly-efficient persulfate activation by nitrogen-doped biochar derived from dewatered sewage sludge for 2, 4-dichlorophenol removal: Process and mechanism. *Journal of Environmental Chemical Engineering, 12*(3), Article 112561.
11. Dar, A. A., Mohd, Y. R., Javid, M., Waseem, Y., Khursheed, A. W., & Dheeraj, V. (2019). Biochar: Preparation, properties and applications in sustainable agriculture. *International Journal of Theoretical & Applied Sciences, 11*(2), 29–40.
12. Deshmukh, S. P., Kale, D. P., Kar, S., et al. (2020). Ultrasound assisted preparation of rGO/TiO_2 nanocomposite for effective photocatalytic degradation of methylene blue under sunlight. *Nano-Structures & Nano-Objects, 21*, Article 100407.
13. Dey, S., & Islam, A. (2015). A review on textile wastewater characterization in Bangladesh. *Resources and Environment, 5*(1), 15–44.
14. Duan, W., Zhang, M., & Zhou, R. (2024). Efficient degradation of antibiotic wastewater by biochar derived from water hyacinth stems via periodate activation: Pyridinic N and carbon structures improved the electron transfer process. *Water Science & Technology, 89*(1), 212–224.
15. Echeverria, C. A., Handoko, W., Pahlevani, F., & Sahajwalla, V. (2019). Cascading use of textile waste for the advancement of fibre reinforced composites for building applications. *Journal of Cleaner Production, 208*, 1524–1536.
16. Faizal, A. N. M., Putra, N. R., Aziz, A. H. A., Agi, A., & Zaini, M. A. A. (2024). Giant mud crab shell biochar: A promising adsorbent for methyl violet removal in wastewater treatment. *Journal of Cleaner Production, 447*, Article 141637.
17. Gaber, M. M., Samy, M., & Shokry, H. (2024). Effective degradation of synthetic micropollutants and real textile wastewater via a visible light-activated persulfate system using novel spinach leaf-derived biochar. *Environmental Science and Pollution Research, 31*(17), 25163–25181.

18. Gabhane, J. W., Bhange, V. P., Patil, P. D., Bankar, S. T., & Kumar, S. (2020). Recent trends in biochar production methods and its application as a soil health conditioner: A review. *SN Applied Sciences, 2*, 1–21.
19. Hanoğlu, A., Çay, A., & Yanık, J. (2019). Production of biochars from textile fibres through torrefaction and their characterisation. *Energy, 166*, 664–673.
20. Kallawar, G. A., & Bhanvase, B. A. (2024). A review on existing and emerging approaches for textile wastewater treatments: Challenges and future perspectives. *Environmental Science and Pollution Research, 31*(2), 1748–1789.
21. Khan, A. A., Gul, J., Naqvi, S. R., Ali, I., Farooq, W., Liaqat, R., AlMohamadi, H., Štěpanec, L., &Juchelková, D. (2022). Recent progress in microalgae-derived biochar for the treatment of textile industry wastewater. *Chemosphere, 306*, 135565.
22. Kishor, R., Purchase, D., Saratale, G. D., Saratale, R. G., Ferreira, L. F. R., Bilal, M., Chandra, R., & Bharagava, R. N. (2021). Ecotoxicological and health concerns of persistent coloring pollutants of textile industry wastewater and treatment approaches for environmental safety. *Journal of Environmental Chemical Engineering, 9*(2), Article 105012.
23. Lee, H. S., Jung, S., Lin, K. A., Kwon, E. E., & Lee, J. (2023). Upcycling textile waste using pyrolysis process. *The Science of the Total Environment, 859*(Pt 2), Article 160393.
24. Li, H., Dong, X., da Silva, E. B., de Oliveira, L. M., Chen, Y., & Ma, L. Q. (2017). Mechanisms of metal sorption by biochars: Biochar characteristics and modifications. *Chemosphere, 178*, 466–478.
25. Liu, Z., Xu, Z., Xu, L., Buyong, F., Chay, T. C., Li, Z., et al. (2022). Modified biochar: Synthesis and mechanism for removal of environmental heavy metals. *Carbon Research, 1*(1), 8.
26. Liu, J., Wen, J., Hu, J., Ma, Y., Wang, X., & Li, H. (2024). Functionalized typha biochar for antibiotic removal via low-carbon integrated method: Performance and mechanism analyses. *CLEAN–Soil, Air, Water, 52*(4), 2300179.
27. Lu, L., Tang, D., Luo, Z., Mo, H., Sun, Y., Hu, J., & Sun, J. (2024). Water hyacinth derived hierarchical porous biochar absorbent: Ideal peroxydisulfate activator for efficient phenol degradation via an electron-transfer pathway. *Environmental Research, 242*, Article 117773.
28. Madhav, S., et al. (2018). A review of textile industry: Wet processing, environmental impacts, and effluent treatment methods. *Environmental Quality Management, 27*(3), 31–41.
29. Mamane, H., Altshuler, S., Sterenzon, E., & Vadivel, V. K. (2020). Decolorization of dyes from textile wastewater using biochar: A review. *Acta Innovations, 37*, 36–46.
30. Meseguer, V. F., Ortuño, J. F., Aguilar, M. I., Lloréns, M., Pérez-Marín, A. B., & Fuentes, E. (2024). Ciprofloxacin uptake from an aqueous solution via adsorption with K_2CO_3-activated biochar derived from Brewing Industry Bagasse. *Processes, 12*(1), 199.
31. Na, Y., Weon, S. H., Lee, G. W., Kim, H. J., Lee, S. H., Kim, Y. H., et al. (2024). Evaluation of benzene adsorption onto grass-derived biochar and comparison of adsorption capacity via RSM (Response Surface Methodology). *Journal of Composites Science, 8*(4), 132.
32. Naudiyal, N., & Srivastava, R. K. (2024). Low-cost biochar from *Parthenium hysterophorus* for efficient Cr^{6+} removal: A sustainable wastewater treatment solution. *International Journal of Environment and Climate Change, 14*(11), 605–618.
33. Patra, J. M., Panda, S. S., & Dhal, N. K. (2017). Biochar as a low-cost adsorbent for heavy metal removal: A review. *International Journal of Research in BioSciences, 6*(1), 1–7.
34. Prakash, Y., Charwak, B., & Kumar, P. V. (2020). Textile industry in new India: Challenges and opportunities. *International Journal of Indian Culture and Business Management, 21*(4), 435–458.
35. Praveen, S., Jegan, J., Bhagavathi Pushpa, T., Gokulan, R., & Bulgariu, L. (2022). Biochar for removal of dyes in contaminated water: An overview. *Biochar, 4*(1), 10.
36. Praveen, S., Gokulan, R., Pushpa, T. B., and Jegan, J. (2021). Techno-economic feasibility of biochar as biosorbent for basic dye sequestration. *Journal of the Indian Chemical Society, 98*, 100107.https://doi.org/10.1016/j.jics.2021.100107
37. Qiu, B., Shao, Q., Shi, J., Yang, C., & Chu, H. (2022). Application of biochar for the adsorption of organic pollutants from wastewater: Modification strategies, mechanisms and challenges. *Separation and Purification Technology, 300*, Article 121925.

38. Rago, Y. P., Surroop, D., & Mohee, R. (2018). Torrefaction of textile waste for production of energy-dense biochar using mass loss as a synthetic indicator. *Journal of Environmental Chemical Engineering, 6*(1), 811–822.
39. Sathish, K., & Saraswat, S. (2024). Characterization and use of hydrochars from wheat straw, fruit peels, and sewage sludge: A potential biofuel source. *Pantnagar Journal of Research, 22*(2), 470–478.
40. Shah, V., Gajbhiye, P., Mehta, J., Tudu, B., Sillanpaa, M., Siddiqui, M. I. H., Kumar, A., & Shah, M. A. (2024). Effectiveness of *Canna indica* leaves and stalk biochar in the treatment of textile effluent. *AIP Advances, 14*(3).
41. Sillanpää, M., &Ncibi, C. (2019). The circular economy: Case studies about the transition from the linear economy. Academic Press.
42. Srivatsav, P., Bhargav, B. S., Shanmugasundaram, V., Arun, J., Gopinath, K. P., & Bhatnagar, A. (2020). Biochar as an eco-friendly and economical adsorbent for the removal of colorants (dyes) from aqueous environment: A review. *Water, 12*(12), 3561.
43. Suleman, M., Zafar, M., Ahmed, A., Rashid, M. U., Hussain, S., Razzaq, A., et al. (2021). Castor leaves-based biochar for adsorption of safranin from textile wastewater. *Sustainability, 13*(12), 6926.
44. Sutar, S., Patil, P., & Jadhav, J. (2022). Recent advances in biochar technology for textile dyes wastewater remediation: A review. *Environmental Research, 209*, Article 112841.
45. Vyavahare, G. D., Gurav, R. G., Jadhav, P. P., Patil, R. R., Aware, C. B., & Jadhav, J. P. (2018). Response surface methodology optimization for sorption of malachite green dye on sugarcane bagasse biochar and evaluating the residual dye for phyto and cytogenotoxicity. *Chemosphere, 194*, 306–315.
46. Vyavahare, G., Sutar, S., Gurav, R., Patil, R., Patil, D., & Jadhav, J. (2021). Application of biochar for the treatment of textile dyes and wastewater. In: *Biochar and its application in bioremediation* (pp. 169–191).
47. Wang, X., Jiang, J., & Gao, W. (2022). Reviewing textile wastewater produced by industries: Characteristics, environmental impacts, and treatment strategies. *Water Science and Technology, 85*(7), 2076–2096.
48. Wojnowska-Baryła, I., Bernat, K., Zaborowska, M., & Kulikowska, D. (2024). The growing problem of textile waste generation—The current state of textile waste management. *Energies, 17*(7).
49. Wu, F., Yue, X., Xu, Y., Gao, J., Zhang, C., Li, Z., & Zhang, H. (2025). Peanut-shell-derived biochar mediated peracetic acid activation for the elimination of acetaminophen via an electron-transfer regime. *Separation and Purification Technology, 353*, Article 128150.
50. Yaashikaa, P. R., Kumar, P. S., Varjani, S., & Saravanan, A. J. B. R. (2020). A critical review on the biochar production techniques, characterization, stability and applications for circular bioeconomy. *Biotechnology Reports, 28*, Article e00570.
51. Yang, Y., Song, Z., Ren, W., Vongsvivut, J., Wang, Z., Ren, N., et al. (2024). Sludge biochar accelerates transformative phenolic compounds removal from wastewater via the coupling mechanism. *Applied Catalysis B: Environment and Energy, 359*, Article 124470.
52. Yaseen, D. A., & Scholz, M. (2019). Textile dye wastewater characteristics and constituents of synthetic effluents: A critical review. *International Journal of Environmental Science and Technology, 16*, 1193–1226.
53. Yin, Q., Ren, H., Wang, R., & Zhao, Z. (2018). Evaluation of nitrate and phosphate adsorption on Al-modified biochar: Influence of Al content. *Science of the Total Environment, 631*, 895–903.
54. Yin, Q., Wang, R., & Zhao, Z. (2018). Application of Mg–Al-modified biochar for simultaneous removal of ammonium, nitrate, and phosphate from eutrophic water. *Journal of Cleaner Production, 176*, 230–240.
55. Yu, K. L., Lee, X. J., Ong, H. C., Chen, W. H., Chang, J. S., Lin, C. S., Show, P. L., & Ling, T. C. (2021). Adsorptive removal of cationic methylene blue and anionic Congo red dyes using wet-torrefied microalgal biochar: Equilibrium, kinetic and mechanism modeling. *Environmental Pollution, 272*, Article 115986.

56. Yu, X., Wang, L., Shen, X., Wu, Y., Xu, L., Zhang, Y., et al. (2025). New insight into the S and N co-doped poplar biochar for efficient BPA removal via peroxymono sulfate activation: S for adsorptive removal and N for catalytic removal. *Separation and Purification Technology, 354*, Article 128809.
57. Zeghioud, H., Fryda, L., Djelal, H., Assadi, A., & Kane, A. (2022). A comprehensive review of biochar in removal of organic pollutants from wastewater: Characterization, toxicity, activation/functionalization and influencing treatment factors. *Journal of Water Process Engineering, 47*, Article 102801.
58. Zhang, Z., Huang, G., Zhang, P., Shen, J., Wang, S., & Li, Y. (2023). Development of iron-based biochar for enhancing nitrate adsorption: Effects of specific surface area, electrostatic force, and functional groups. *Science of The Total Environment, 856*, Article 159037.
59. Zhang, J., & Wang, Q. (2016). Sustainable mechanisms of biochar derived from brewers' spent grain and sewage sludge for ammonia–nitrogen capture. *Journal of Cleaner Production, 112*, 3927–3934.
60. Zhao, Y., Chen, W., Liu, F., & Zhao, P. (2022). Hydrothermal pretreatment of cotton textile wastes: Biofuel characteristics and biochar electrocatalytic performance. *Fuel, 316*, Article 123327.
61. Zheng, D., Wang, Y., Jia, X., Yao, W., Wang, S., Li, Z., et al. (2024). Developing Prussian blue/wood-derived biochar catalyst for persistent organic pollutant degradation: Preparation, characterization, and mechanism. *Chemosphere, 351*, Article 141150.

Insights of Conventional Approaches and Microbial-Biocatalysts for Management of Textile Wastes

Harsh Sable, Aashi Sable, Vandana Singh, Sahil Kohli, and Muhammad Shafiq

1 Introduction

In recent decades, environmental contamination has emerged as a significant issue in the contemporary world. Owing to the relentless rise of industry and urbanization, a substantial quantity of harmful and dangerous substances is generated, which cannot be disposed of in the environment under duress. Heightened human activity and elevated pollution levels significantly impact our ecology. Numerous chemicals generated by various industries, including aromatics, dyes, and pigments, other are very poisonous and pose a significant environmental danger. The textile manufacturing is the second biggest and one of the most developing sectors in India, generating substantial waste throughout the printing, dyeing, and finishing methods. Dyes are synthetic pollutants with complicated structures that render Since they are

Harsh Sable and Aashi Sable authors are contributed equally to this work.

H. Sable
Department of Forensic Science, SSAHS, Sharda University, UP, Greater Noida, India

A. Sable
University of Kassel in Witzenhausen, Kassel, Germany

Hochschule Fulda, University of Applied Sciences, Fulda, Germany

V. Singh (✉)
Department of Microbiology, SSAHS, Sharda University, UP, Greater Noida, India
e-mail: vandana.singh@sharda.ac.in

S. Kohli
Research and Development Cell, Department of Chemistry, Manav Rachna International Institute of Research and Studies, Faridabad, Haryana, India

M. Shafiq
Department of Pharmacology, Research Institute of Clinical Pharmacy, Shantou University Medical College, Shantou, China

R. A. Bhat et al. (eds.), *Biocentric Approaches in Textile Waste Management*, Textile Science and Clothing Technology, https://doi.org/10.1007/978-3-032-04974-2_9

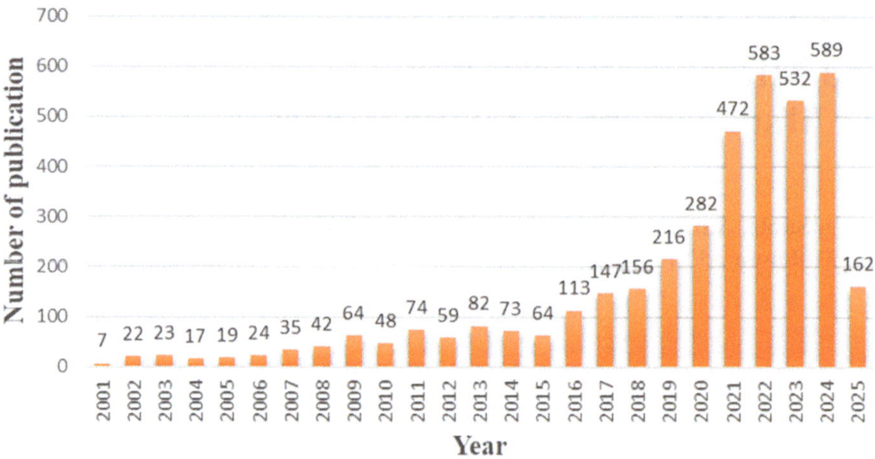

Fig. 1 Publication trends from 2001 to 2025 the present in the textile wastewater bioremediation utilizing microbial and enzymatic treatments. *Source* ScienceDirect, keywords: bioremediation, textile dye, enzymatic, microbial

insoluble in water, they are not biodegradable. Over 100,000 dyes that are commercially accessible, with roughly 700,000 tones created yearly globally. It is believed that during synthesis and dyeing, roughly 10–15% of all dyes used in the textile and paint industries is discharged into global water bodies [58, 74]. Figure 1 illustrates a notable increase in publication trends regarding the biodegradation of textile dye wastewater, indicating a growing interest in this effective wastewater treatment method.

Consumption by the conventional textile industry is around 100 ltr of water to produce almost textile materials of 1 kg. During this procedure, the colours fail to adhere fully to the cloths, with losses in wastewater ranging from 2% for simple dyes to Reactive dyes at 50%, which results in considerable contamination of ground surface water in dyeing industry regions. The many processing stages utilized in the textile industry result in the production of textile effluent. This picture emphasizes the significance of economical and environmentally sustainable biological treatments using various microbes to decolorize and breakdown synthetic dyes present in textile effluents [4, 31].

The chemical makeup of the dye influences its pigmentation, determining the lightness or darkness of each hue. The coloration results from the absorption of light at certain wavelengths within the visible spectrum; the dye is a molecule that absorbs particular radiations of light, after which the complementary hues are reflected. Table 1 consolidates the primary categories of the textile industry dyes, the corresponding fibre or substrate types for each dye class, the nature of interactions between the dye and the fibreor substrate, as well as the dyeing or application techniques employed [14, 62, 80].

Table 1 Lists the primary dye classes used in the textile industry, along with the kinds of fibres to which each class's dyes are applied, the ways in which dyes interact with fibres, and the techniques for applying or dying them

Classes	Fiber type	Interaction between dye and fiber	Method of application
Direct dye	Cotton, rayon, leather, nylon	Intermolecular forces	Neutral or slightly alkaline baths containing additional electrolytes
Acid dye	Nylon, wool, silk	Electrostatics; hydrogen bond	Neutral to acid dye baths
Basic dye	Modified nylon, polyester	Electrostatics	Acid baths
Sulfur dye	Cotton, rayon	Covalent bond	Aromatic substrate covered with sodium sulfide and reoxidized to sulfur-containing products, insoluble in fiber
Reactive dye	Cotton, nylon, silk, wool	Covalent bond	Under the influence of heat and pH of the medium, which must be alkaline, the dye reacts with the fiber functional group, with which it covalently bonds
Vat dye	Cotton, rayon	Impregnation and oxidation	Water-insoluble dyes are solubilized by reduction with sodium hydrosulfite and then exhausted into the fiber and reoxidized
Dispersed dye	Polyester, polyamide, acetate, plastic, acrylic	Hydrophobic—solid-state mechanism	High or low-temperature pressure transport methods

The synthetic dyes removal from textile effluent in industries is often achieved by physical and chemical approaches. Industries utilize a variety of physio-chemical techniques, including ion exchange, ozonation, oxidation, adsorption, membrane filtration, and coagulation processes. These approaches are constrained by elevated costs, less adaptability, restricted efficiency, interaction with other wastewater constituents, and challenges in management of waste. The use of activated carbon as an adsorbent is an effective method for dye removal; nevertheless, it is also a costly procedure. As a result, there is an increasing and advancing interest in biological therapies. Biological treatments are eco-friendly processes using different microorganisms, including bacteria, yeasts, fungus, and algae, for the breakdown and decolorization of pollutants, and may totally mineralize some azo dyes under environmental circumstances. Bioremediation may be the most effective method for removing industrial dyes from wastewater treatment. Bioremediation is an ecologically acceptable, non-hazardous, cost-effective, and much more dependable alternative to traditional textile disposal waste methods [18, 20, 33, 34, 65].

The increasing volume of articles about the bioremediation of effluents derived from textile dyes reflects a worldwide apprehension about this matter. This comprehensive study aims to tackle environmental concerns arising from the wastewater of the textile industry, which contains synthetic Dyes. The text examines contemporary methodologies and research outcomes regarding the remediating wastewater for synthetic dyes by several possible microorganisms [21, 37, 40, 44]. The process of dye biodegradation and the primary parameters influencing the efficacy of biological therapy via microbial activity are elucidated comprehensively. Numerous breakthroughs in biological treatment strategies for the elimination synthetic dyes derived from wastewater are discussed. This chapter consolidates the primary and latest findings stated in the literature regarding dye bioremediation using biocatalysts from textile industry wastewater. The detrimental effects of dyes utilized in this industrial sector on the health of humans and animals are examined in this context, along with conventional techniques for treating dye-containing industrial effluents, the fundamentals of the biocatalysts involved in enzymatic bioremediation employed in this process, and their byproducts [3, 8, 14, 24, 45, 74].

Detrimental Effects of Textile Dyes on Animal and Human Health

Effluents from the textile sector are regarded as the most harmful substances due to their substantial volume and toxicity. The wastewater generated by the textile industry is anticipated to contain dyes concentrations ranging from 10 to 200 mg L^{-1}, in addition to various inorganic compounds, additives, and organic chemicals. Despite these effluents being treated, approximately 90% of the dyes are still released into waterbodies without any chemical modification. The breakdown of these dyes is hindered by their foreign characteristics, aromatic structure, high heat resistance, and stability under light exposure [28, 38, 45].

The primary worry of dye discharge is the detection of teratogenic, mutagenic, genotoxic, and carcinogenic consequences, as shown by animal research. Carcinogenicity pertains to the generation in which ions bind to RNA and DNA, causing mutations, and the emergence of cancers. 2-naphthylamine and benzidine dyes are linked to a significant prevalence cancer of bladder. Azure-B dye may integrate into the helix structure of DNA and may exhibit cytotoxic effects as it inhibits monoamine oxidase A (MAO-A), a specific enzyme that plays a significant role in the (central nervous system) CNS and is essential for human behaviour. Sudan 1 dye, frequently utilized in the textile sector, is prohibited in most European nations and the United States, although it is also employed in food products, like paprika. This dye is altered by biocatalysts in carcinogenic aromatic amines when it is found in animal and human bodies. In addition, adverse reactions to dyes may still include congenital malformations, nausea, headaches, and irritation of the lungs and skin [40, 55, 59, 62, 77].

Triphenylmethane dyes exhibit phytotoxicity in agricultural crops, cytotoxicity in humans, and induce tumour formation in many fish species. The violet crystal dye is a potent carcinogen, capable of producing malignancies in fish, including reticular cell sarcoma and hepatocellular carcinoma in several organs. The primary environmental issues associated with the disposal of synthetic dyes include: (1) surface water contamination, resulting in diminished light penetration, which adversely affects photosynthesis and leads to oxygen deficiency; (2) nonbiodegradable organic dyes accumulation within the food chain; (3) contamination of water and soil; and (4) inhibition of development and growth in various agriculturally significant crops [5, 38, 72].

The literature describes a relationship between the development of microalgae and the concentration of dyes, which causes the growth to be completely suppressed. The research used varying doses of three dyes to assess the *Chlorella vulgaris*, a kind of green algae, rate of specific growth when exposed to these colours. In aquatic ecosystems, the trophic transfer of nutrients and energy is disrupted when microalgae development is inhibited, which makes such studies significant. Of the thousands of dyes examined in effluents, over 100 possess the potential to generate carcinogenic amines. Nonetheless, these potentially hazardous dyes continue to be marketed and utilized, particularly in small textile manufacturing facilities. In various regions globally, the pressures of export and low-cost labour support the operation of small-scale factories that illegally discharge toxic dyes into bodies of water [62, 68, 69, 79].

Textile Industry Effluents: Conventional Methods of Treatment

Physical methods, including sorption techniques, membrane filtration (reverse osmosis, nanofiltration, electrodialysis), and Chemical methods like flocculation or coagulation, when combined with flotation and filtration, as well as flocculation through precipitation, electroflotation, and electrokinetic coagulation, are utilized for removal of numerous dyes; however, they do not lead to their degradation. These procedures just facilitate the diminution of colour concentration, transforming them from one chemical form to another, so generating secondary pollution. The removal of wastewater dyes can be accomplished through a variety of methods, including photocatalytic degradation, adsorption, electrochemical treatment, biodegradation, and chemical oxidation (Fig. 2). When sunlight is used in the process, photocatalysts exhibit low cost, high efficiency, speed, and improved environmental performance [24, 45, 50, 61, 78].

Chitosan is one of the natural minerals used in physical dye adsorption procedures. The most common polymer on Earth, chitosan, is a modified natural biopolymer that is made by deacetylating chitin, which is obtained from significant biomass produced by both plants and animals, including crustacean shells, arthropods, lobsters, crabs, squid, and shrimp. Although adsorption is one of the most effective ways to remove

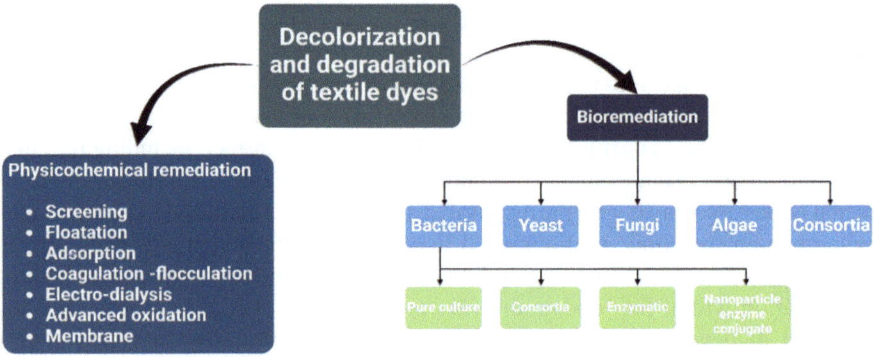

Fig. 2 Textile dye decolorization and biodegradation have been accomplished using a variety of accessible treatment techniques

dyes, the residue left behind after the procedure has to be treated further. Biological processes significantly contribute to the removal of dyes from wastewater, alongside chemical and physical methods. Phytoremediation is a biological method for the removal of dyes from wastewater of industries, offering advantages over chemical and physical removal techniques. Plants remove textile dyes through accumulation, subsequent degradation, and adsorption, facilitated by biocatalysts [5–7, 10, 61].

In scenarios requiring continuous application of chemical products, the utilization of microorganisms presents a simpler and cost-effective approach, as they can be introduced once into the effluent for treatment, given their capacity for multiplication. In this context, activated sludge is frequently employed in bioreactors for effluent treatment, representing one of the most utilized processes in the textile industry. A potential biological approach for effluent treatment involves the application of bacterial cultures. The isolation of pure cultures from textile wastewater is typically not conducted due to the slow and labour-intensive nature of the process. Mixed bacterial cultures are frequently employed, as their cooperative interactions yield enhanced outcomes in the mineralization and discoloration of toxic aromatic amines [13, 35, 63, 71]. Table 2 provides details on the major outcomes of the treatment procedures that have been investigated for the removal of textile dyes from industrial effluents, as documented in the literature.

Principles of Enzymatic Bioremediation

Bioremediation strategies have gained significant importance globally because to their high public acceptability, cost-effectiveness relative to traditional treatment approaches, abundant enzyme supply, and minimum environmental impact. The investigation of biocatalysts for bioremediation is highly regarded because of their capacity to operate throughout extensive temperature and pH ranges, even in the

Table 2 Examples of treatment processes used to remove textile dyes

Name of dyes	Treatment method	Main results	Reference
Malachite Green, Reactive Red 198, and Direct Yellow 31	Chitosan adsorption	The amount of dye adsorbed depends on the mass of the adsorbent and decreased with its increase	[61]
Reactive Yellow 138, Reactive Red 231, and Navy HEXL® Procion	Electrolysis, carried out in a filter-press cell, under galvanostatic conditions	Complete discoloration (99%) was observed in all cases	[47]
Methylene Blue (MB)	Photocatalytic degradation of organic dyes with nanocomposites	Under sun light, synthesized nanocompounds demonstrated strong catalytic activity for the reduction of methylene blue, with an efficiency of up to 90.1% and a straightforward, inexpensive process	[12]
Acid Blue 25	Adsorption	It was found that the absorbent material reduced absorption with alkaline solutions and reached an equilibrium constant after 270 min. Comparing 30 mg of quaternary chitosan granules to the control, the mortality rate of *Daphnia similis* decreased from 50 to 10%	[42]
Procion Red HE-3B (RR120)	Photoelectrocatalysis	With up to 100% decolorization in 30 min at a concentration of 10 mg L^{-1} of the dye RR120, the treatment demonstrated efficacy. Only at low concentrations is the efficiency efficient; as concentration rises, some decolorization takes place before stabilizing	[36]

presence of pollutants and salty concentrations. Bioremediation via biocatalysts is an ecological, affordable, and revolutionary technology with significant potential. The procedure involves investigating the standard traits of microorganisms or GMOs (genetically modified organisms) that may generate specialized biocatalysts to metabolize or catalyze pollutants, converting the hazardous variant into a harmless version and sometimes into novel compounds [22, 26, 38, 46, 81].

Biocatalysts participating in bioremediation processes include hydrolases, dehalogenases, and laccases. Laccases are biocatalysts that catalyze the oxidation of phenolic substances, aromatic amines and their derivatives. Dehalogenases decompose various halogenated compounds by severing C–X bonds (X = halogen atom, e.g., Cl). Hydrolases cleave chemical bonds using water, transforming bigger molecules into smaller ones and reducing their toxicity. These biocatalysts catalyze the cleavage of S–N, C–O, S–P, C–P, C–N, C–C, and S–S bonds [1, 16, 19, 70].

Biocatalysts may be used in either immobilized or free forms, with the latter offering benefits such as prolonged operating stability, facile recovery, and reusability in industrial applications, hence enhancing process efficiency and reducing total costs. Immobilization involves attaching the enzyme to an insoluble support matrix to preserve optimal shape, hence ensuring enhanced stability of the enzyme. The process of bioremediation with microbial biocatalysts may be protracted, and to yet, only a limited number of bacterial species have shown the ability to manufacture biocatalysts with significant biodegradation potential. Consequently, the use of GMOs is more prevalent owing to their capacity to generate substantial quantities of biocatalysts under optimum settings [5, 17, 23, 25].

Biocatalysts derived from aerobic bacteria, including *Alcaligenes*, *Pseudomonas*, *Mycobacterium*, *Rhodococcus*, and *Sphingomonas*, are frequently employed in the bioremediation of hydrocarbons and pesticides, whereas biocatalysts produced by anaerobic bacteria are predominantly utilized for the bioremediation of trichloroethylene (TCE) decolorization, chloroform, and polychlorinated biphenyls (PCBs) (Fig. 2). The primary biocatalysts used in bioremediation methods are cytochrome P450 family biocatalysts, dehydrogenases, hydrolases, proteases, laccases, dehalogenases, and lipases. Fungi may biodegrade, often facilitated by biocatalysts like manganese peroxidases, laccases, azoreductases, and lignin peroxidases. White rot fungus may degrade textile dyes via laccases and peroxidases [2, 48, 49, 51, 81].

In the remediation of effluents from textile industry, biocatalysts interact with dyes, producing precipitates that may be readily extracted or chemically converted into manageable chemicals. The rate of dye breakdown by biocatalysts is contingent upon the dye's chemical structure, pH, metal ion concentration, salt concentration, and temperature of wastewater. The enzymatic degradation of contaminants in textile effluents has several benefits, including substrate selectivity and specificity, as well as being an accessible and effective approach that adheres to the principles of green chemistry. The primary challenges associated with Enzymatic breakdown for the treatment of wastewater include the necessity for substantial enzyme quantities, elevated costs, thermal instability, inhibition of proteolytic degradation of specific biocatalysts, enzymatic activity, and the generation of not desired by-products [30, 39, 52, 64, 76].

Certain enumerated issues may be addressed, at least in part, by immobilizing efficient biocatalysts inside economical matrices, facilitating their separation and reutilization, as well as employment in continuous bioreactors. Utilizing biocatalysts is often more beneficial than employing cells for regulating reactions in the process of biodegradation. The elevated expense of biocatalysts, attributed to the pursuit of highly purified enzymatic solutions, is expected to diminish as technological advancements and methodologies improve, alongside the exploration of more economical growth substrates for microorganism cultivation [30, 39, 52, 54].

Key Biocatalysts Involved in Textile Dye Bioremediation

Because of its adaptability and effectiveness in breaking down persistent organic pollutants, enzyme-mediated bioremediation has become well-known and is used in environmental, biotechnological, and industrial processes. The intracellular and extracellular metabolites of certain bacterial, fungal, algal, and plant species may be extracted to produce these biocatalysts. Numerous biocatalysts, such as riboflavin reductase, azoreductase, peroxidases, laccase, NADH-DCIP reductase, reductase, tyrosinase, veratryl alcohol oxidase, lignin peroxidase, and aminopyrine *N*-demethylase, are produced by dye-decolorizing microorganisms, according to published reports. Among such biocatalysts, peroxidases, azoreductases, and laccases are the primary ones that cause azo dyes to discolour [15, 43]. Table 3 presents studies concerning the degradation of dyes by biocatalysts generated by microorganisms. Many of the tested substances can decolorize dyes and reduce their toxicity. For instance, peroxidase from horseradish facilitates a reduction in the toxic effect of methyl orange dye.

Azoreductases are regarded as the primary degradation biocatalysts generated by bacteria. These biocatalysts may be categorized into two types: (1) membrane-bound or (2) intracellular, which are hydrophilic, thermostable, and generally more effective in degrading azo dyes. Laccases are copper-dependent biocatalysts that use oxygen to decompose lignin and other aromatic chemicals, such as textile dyes. Due to their elevated redox potential, fungal laccases are used to remediate numerous xenobiotics, including textile dyes, in water and soil, in contrast to plant or bacterial laccases. Peroxidases contribute to the breakdown of azo dyes and are classified as oxidoreductases that include heme. Peroxidases exist in plants, microbes, and mammals. The activity mechanism of these biocatalysts resembles that of laccases, facilitating color degradation without generating harmful by-products. Peroxidases specifically target synthetic dyes, decomposing their components by the oxidative polymerization of phenolic chemicals to produce insoluble polymers. An interaction among oxide-reducing biocatalysts may markedly diminish the toxic effects of dyes [56, 57, 67, 75]. Biocatalysts are proteins that are significantly influenced by changes in pH, minor fluctuations in the medium's pH may lead to modifications in the ionization state of the site of activity and the charge distribution within the protein structure, potentially impacting its substrate affinity. Consequently, a primary issue of enzymatic therapy is the deactivation of the biocatalyst, mostly resulting from enzyme denaturation induced by the medium's pH or severe temperatures, which may modify the conformation of the active site of the enzyme. Notwithstanding the many advancements in enzymatic engineering, biocatalysts remain costly and/or unstable; hence, their industrial use often necessitates immobilization inside a matrix [29, 53, 60, 66].

Kassas and Mohammed [27] examine how textile waste effluent affects *Chlorella vulgaris*. The *C. vulgaris* culture reached C_{max} and l_{max} at wastewater volumes of 5.0% and 17.5%, respectively. COD and Colour reductions were highest with 17.5%

Table 3 Key findings from research on degradation byproducts and enzyme-mediated dye bioremediation

Study objective(s)	Results and by-products of degradation	Reference
Utilizing the *Oudemansiella canarii* isolate to generate laccase and assessing its capacity to break down Congo Red (CR)	The red dye was effectively decolored by the *O. canarii* laccase. Accumulation of different intermediates, such as derivatives of naphthalene, during degradation. Compared to CR, these products are less hazardous	[29]
Drimaren Red CL-5B (Reactive Red 195) is a revolutionary bioinformatics amalgamation and bacterial remediation method that uses non-native strains to decolorize and degrade azo dyes	The breakdown of the dye was confirmed by the gas chromatography–mass spectrometry (GC–MS) study of the degradation products, which showed the creation of low molecular weight metabolites. Prior to large-scale bioremediation, testing for microbiological toxicity, cytotoxicity, and phytotoxicity must be conducted	[73]
Copper alginate laccase is used to break down dyes such as Indigo Carmine (IC), Remazol Brilliant Blue R (RBBR), Bromophenol Blue (BB), Crystal Violet (CV), Malachite Green (MG), Congo Red (CR), Direct Blue 15 (DB), and Direct Red 23 (DR) in an airlift bioreactor	Rapid, 100% IC and RBBR decolorization. MG, BB, and CV had discoloration percentages of 82, 64.4%, and 48.5%, respectively. CR, BD, and DR azo dye discoloration percentages are 64, 54, and 22%, respectively. The primary breakdown product was determined to be isatin sulfonic acid	[76]
Examine *Aspergillus niger's* capacity to detoxify and discolor Congo Red (CR) dye	85% elimination of CR is high. Adsorption and enzymatic biodegradation work together to produce 97% of the discolouration. Degradation of amines in solution is shown by*A. niger* detoxification The metabolites produced via CR biodegradation are less harmful than the crude dye, per the findings of phytotoxicity and microtoxicity analyses	[11]
Analyze how well an immobilized enzyme removes Methylene Blue (MB) dye	In comparison to the pure nanocarrier and the free enzyme (81 and 36% removal, respectively), the immobilized enzyme exhibited the greatest removal efficiency (99%). There was no presentation of the by-products' toxicity analysis results	[9]

textile waste effluent. *C. vulgaris* in textile waste effluent showed promise for colour, and COD removal, and biomass production [27].

The research conducted by Martins et al. [41] aims to accomplish two main objectives: (i) to evaluate the pollutant removal efficacy of *Chlorella vulgaris* from two separate undiluted textile wastewater samples and (ii) to analyze the composition of microalgal biomass for potential valorization within a circular economy framework. Microalgae exhibited successful growth rates between 0.234 ± 0.005 and

$0.290 \pm 0.003\ d^{-1}$, with average productivities ranging from 78 ± 3 to 112.39 ± 0.07 mg DW $L^{-1}\ d^{-1}$ [41].

Using *Bacillus cereus* EBT1 under SSF, Red M5B dye, a dye combination, and actual textile effluent are adsorbed onto (DIW-YB) distillery industry waste-yeast biomass and then bioremediated was the focus of the study conducted by Kadam et al. [32]. Adsorption of textile dyestuffs on DIW-YB was permitted. *B. cereus* was used to decolorize DIW-YB adsorbed dyestuffs under SSF. To guarantee that Red M5B was decolorized, enzyme analysis was performed. They used GC–MS, UV–Vis spectroscopy, HPLC, and FTIR to examine the metabolites that remained after the breakdown of the dye. DIW-YB demonstrated an adsorption capacity of 87% for Red M5B, 70% for a dye's combination, and 81% for a textile wastewater sample. *B. cereus* decolorized up to 98% of the DIW-YB adsorbed Red M5B in 36 h. In contrast, *B. cereus* was able to successfully decrease the value of the DIW-YB adsorbed combination of textile dyes by 70% and textile wastewater by 100%, according to the American Dye Manufacture Institute. It is possible that extracellular biocatalysts like laccase and azoreductase are involved in the breakdown of dyes since their induction occurs. A degradation pathway for Red M5B by B. cereus was proposed based on GC–MS analysis (Fig. 3) [32].

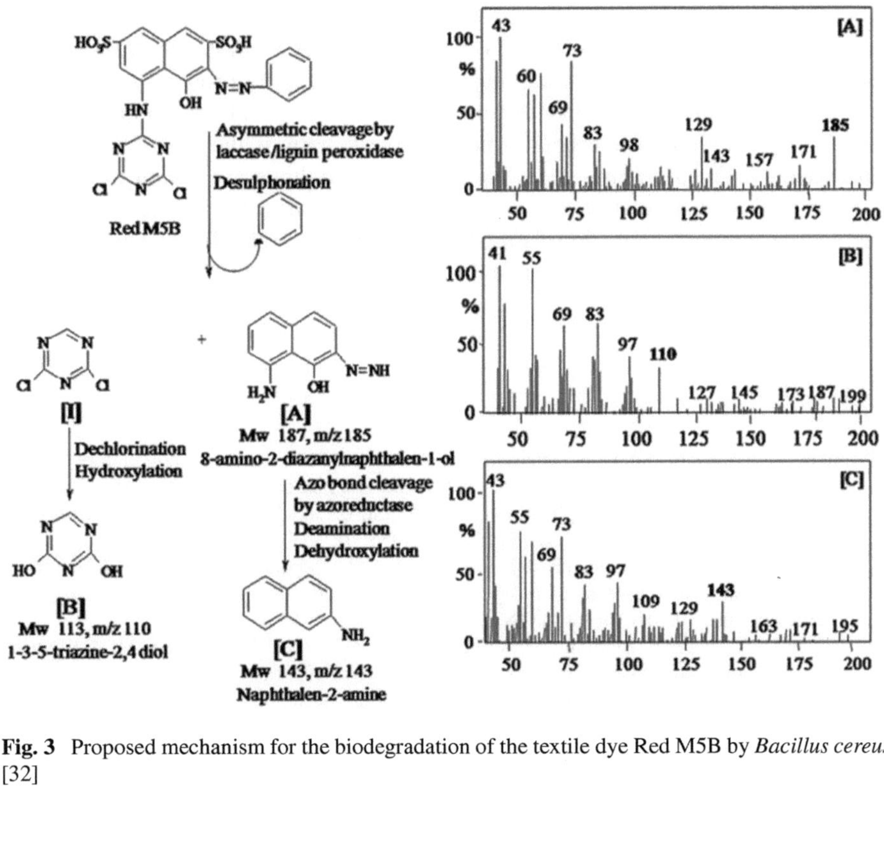

Fig. 3 Proposed mechanism for the biodegradation of the textile dye Red M5B by *Bacillus cereus* [32]

2 Conclusion

A substantial amount of textile dyes continues to be released into rivers without undergoing chemical alteration, despite traditional effluent treatment methods. The pollution caused by dyes from textile industry effluents is detrimental to animal and human health, exhibiting genotoxic, carcinogenic, and mutagenic effects. Furthermore, these dyes adversely affect the survival of aquatic species, as they can bioaccumulate in the food chain, imparting toxicity to soil and water and disrupting the growth of agriculturally significant crops. A thorough examination of dye discharge is essential due to its possible toxic effect, as well as the possibility of manufacturers surreptitiously discharging untreated effluents containing poisonous colors into water bodies. Research must be conducted to enhance effluent treatment procedures that are both environmentally sustainable and efficient, using contemporary scientific advances. Photocatalytic degradation, among the existing approaches, yields favorable outcomes, is cost-effective, and utilizes sunshine as a clean energy source. Alongside this procedure, there exists phytoremediation, regarded as an environmentally sound process, and enzymatic remediation. The biocatalysts used in textile industry wastewater enzymatic bioremediation mostly include laccases, peroxidases, and azoreductases. Enzymatic bioremediation and conventional treatment may produce by-products that are as harmful as the original substances. However, in some instances, less harmful intermediate molecules are produced, as shown in this chapter. Consequently, significant attention must be given to these by-products or secondary products, necessitating their identification, quantification, and appropriate management and processing. The crucial aspect of dye therapy is the need for increased corporate investment to implement scientific study findings. An option would be to conduct testing in simulation stations, akin to an industrial size. Moreover, genetic engineering has profoundly transformed bioremediation by enabling the modification of organisms or their metabolites to enhance their efficacy in pollutant degradation.

References

1. Adak, A., Tiwari, R., Singh, S., Sharma, S., & Nain, L. (2016). Laccase production by a novel white-rot fungus *Pseudolagarobasidium acaciicola* LA 1 through solid-state fermentation of parthenium biomass and its application in dyes decolorization. *Waste Biomass Valorization, 7*, 1427–1435. https://doi.org/10.1007/S12649-016-9550-0
2. Afrin, S., Shuvo, H. R., Sultana, B., Islam, F., Rus'd, A. A., Begum, S., & Hossain, M. N. (2021). The degradation of textile industry dyes using the effective bacterial consortium. *Heliyon, 7*.https://doi.org/10.1016/J.HELIYON.2021.E08102
3. Ahsan, M. A., Satter, F., Siddique, M. A. B., Akbor, M. A., Ahmed, S., Shajahan, M., & Khan, R. (2019). Chemical and physicochemical characterization of effluents from the tanning and textile industries in Bangladesh with multivariate statistical approach. *Environmental Monitoring and Assessment, 191*.https://doi.org/10.1007/S10661-019-7654-2
4. Ajaz, M., Shakeel, S., & Rehman, A. (2020). Microbial use for azo dye degradation—A strategy for dye bioremediation. *International Microbiology, 23*, 149–159. https://doi.org/10.1007/S10123-019-00103-2

5. Al-Tohamy, R., Ali, S. S., Li, F., Okasha, K. M., Mahmoud, Y. A. G., Elsamahy, T., Jiao, H., Fu, Y., & Sun, J. (2022a). A critical review on the treatment of dye-containing wastewater: Ecotoxicological and health concerns of textile dyes and possible remediation approaches for environmental safety. *Ecotoxicology and Environmental Safety, 231*. https://doi.org/10.1016/J.ECOENV.2021.113160
6. Al-Tohamy, R., Kenawy, E. R., Sun, J., & Ali, S. S. (2020). Performance of a newly isolated salt-tolerant yeast strain *Sterigmatomyces halophilus* SSA-1575 for azo dye decolorization and detoxification. *Frontiers in Microbiology, 11*.https://doi.org/10.3389/FMICB.2020.01163
7. Amin, S., Rastogi, R. P., Chaubey, M. G., Jain, K., Divecha, J., Desai, C., & Madamwar, D. (2020). Degradation and toxicity analysis of a reactive textile diazo dye-direct red 81 by newly isolated *Bacillus* sp. DMS2. *Frontiers in Microbiology, 11*. https://doi.org/10.3389/FMICB.2020.576680
8. Aragaw, T. A. (2024). A review of dye biodegradation in textile wastewater, challenges due to wastewater characteristics, and the potential of alkaliphiles. *Journal of Hazardous Materials Advances, 16*, Article 100493. https://doi.org/10.1016/J.HAZADV.2024.100493
9. Ariaeenejad, S., Motamedi, E., & Hosseini Salekdeh, G. (2021). Application of the immobilized enzyme on magnetic graphene oxide nano-carrier as a versatile bi-functional tool for efficient removal of dye from water. *Bioresource Technology, 319*, Article 124228. https://doi.org/10.1016/J.BIORTECH.2020.124228
10. Asgher, M., Noreen, S., & Bilal, M. (2017). Enhancing catalytic functionality of Trametes versicolor IBL-04 laccase by immobilization on chitosan microspheres. *Chemical Engineering Research and Design, 119*, 1–11. https://doi.org/10.1016/J.CHERD.2016.12.011
11. Asses, N., Ayed, L., Hkiri, N., & Hamdi, M. (2018). Congo Red decolorization and detoxification by *Aspergillus niger*: Removal mechanisms and dye degradation pathway. *BioMed Research International, 2018*, 3049686. https://doi.org/10.1155/2018/3049686
12. Ayyob, M., Ahmad, I., Hussain, F., Kashif Bangash, M., Awan, J. A., & Jaubert, J. N. (2020). A new technique for the synthesis of lanthanum substituted nickel cobaltite nanocomposites for the photo catalytic degradation of organic dyes in wastewater. *Arabian Journal of Chemistry, 13*, 6341–6347. https://doi.org/10.1016/J.ARABJC.2020.05.036
13. Batra, V., Kaur, I., Pathania, D., Sonu, Chaudhary, V. (2022). Efficient dye degradation strategies using green synthesized ZnO-based nanoplatforms: A review. *Applied Surface Science Advances, 11*, 100314. https://doi.org/10.1016/J.APSADV.2022.100314
14. Benkhaya, S., M'rabet, S., & El Harfi, A. (2020). A review on classifications, recent synthesis and applications of textile dyes. *Inorganic Chemistry Communications, 115*.https://doi.org/10.1016/J.INOCHE.2020.107891
15. Bento, R. M. F., Almeida, M. R., Bharmoria, P., Freire, M. G., & Tavares, A. P. M. (2020). Improvements in the enzymatic degradation of textile dyes using ionic-liquid-based surfactants. *Separation and Purification Technology, 235*, Article 116191. https://doi.org/10.1016/J.SEPPUR.2019.116191
16. Bhandari, S., Poudel, D. K., Marahatha, R., Dawadi, S., Khadayat, K., Phuyal, S., Shrestha, S., Gaire, S., Basnet, K., Khadka, U., & Parajuli, N. (2021). Microbial biocatalysts used in bioremediation. *Journal of Chemistry, 2021*, 8849512. https://doi.org/10.1155/2021/8849512
17. Bilal, M., Asgher, M., Parra-Saldivar, R., Hu, H., Wang, W., Zhang, X., & Iqbal, H. M. N. (2017). Immobilized ligninolytic biocatalysts: An innovative and environmental responsive technology to tackle dye-based industrial pollutants—A review. *Science of The Total Environment, 576*, 646–659. https://doi.org/10.1016/J.SCITOTENV.2016.10.137
18. Chang, J. S., Chou, C., Lin, Y. C., Lin, P. J., Ho, J. Y., & Lee Hu, T. (2001). Kinetic characteristics of bacterial azo-dye decolorization by *Pseudomonas luteola*. *Water Research, 35*, 2841–2850. https://doi.org/10.1016/S0043-1354(00)00581-9
19. Chen, H., Hopper, S. L., & Cerniglia, C. E. (2005). Biochemical and molecular characterization of an azoreductase from *Staphylococcus aureus*, a tetrameric NADPH-dependent flavoprotein. *Microbiology Society, 151*, 1433–1441. https://doi.org/10.1099/mic.0.27805-0
20. Chen, K. C., Huang, W. T., Wu, J. Y., & Houng, J. Y. (1999). Microbial decolorization of azo dyes by Proteus mirabilis. *Journal of Industrial Microbiology and Biotechnology, 23*, 686–690. https://doi.org/10.1038/SJ.JIM.2900689

21. Chen, X., Ning, X. an, Lai, X., Wang, Y., Zhang, Y., He, Y., 2021. Chlorophenols in textile dyeing sludge: pollution characteristics and environmental risk control. *Journal of Hazardous Materials, 416*, 125721. https://doi.org/10.1016/j.jhazmat.2021.125721
22. Chowdhary, P., More, N., Yadav, A., & Bharagava, R. N. (2018). Ligninolytic biocatalysts: An introduction and applications in the food industry. In *Biocatalysts in Food Biotechnology: Production, Applications, and Future Prospects* (pp. 181–195).https://doi.org/10.1016/B978-0-12-813280-7.00012-8
23. Darwesh, O. M., Matter, I. A., & Eida, M. F. (2019). Development of peroxidase enzyme immobilized magnetic nanoparticles for bioremediation of textile wastewater dye. *Journal of Environmental Chemical Engineering, 7*, Article 102805. https://doi.org/10.1016/J.JECE.2018.11.049
24. de Souza, T. N. V., de Carvalho, S. M. L., Vieira, M. G. A., da Silva, M. G. C., & Brasil, D. D. S. B. (2018). Adsorption of basic dyes onto activated carbon: Experimental and theoretical investigation of chemical reactivity of basic dyes using DFT-based descriptors. *Applied Surface Science, 448*, 662–670. https://doi.org/10.1016/J.APSUSC.2018.04.087
25. Dutta, S., Adhikary, S., Bhattacharya, S., Roy, D., Chatterjee, S., Chakraborty, A., Banerjee, D., Ganguly, A., Nanda, S., & Rajak, P. (2024). Contamination of textile dyes in aquatic environment: Adverse impacts on aquatic ecosystem and human health, and its management using bioremediation. *Journal of Environmental Management, 353*.https://doi.org/10.1016/j.jenvman.2024.120103
26. El Enshasy, H. A., Hanapi, S. Z., Abdelgalil, S. A., Malek, R. A., & Pareek, A. (2017). Mycoremediation: Decolourization potential of fungal ligninolytic biocatalysts. (pp. 69–104). https://doi.org/10.1007/978-3-319-68957-9_5
27. El-Kassas, H. Y., & Mohamed, L. A. (2014). Bioremediation of the textile waste effluent by *Chlorella vulgaris*. *Egyptian Journal of Aquatic Research, 40*, 301–308. https://doi.org/10.1016/J.EJAR.2014.08.003
28. Gita, S., Shukla, S. P., Saharan, N., Prakash, C., & Deshmukhe, G. (2019). Toxic effects of selected textile dyes on elemental composition, photosynthetic pigments, protein content and growth of a freshwater chlorophycean alga *Chlorella vulgaris*. *Bulletin of Environmental Contamination and Toxicology, 102*, 795–801. https://doi.org/10.1007/S00128-019-02599-W
29. Iark, D., Buzzo, A. J. R., Garcia, J. A. A., Côrrea, V. G., Helm, C. V., Corrêa, R. C. G., Peralta, R. A., Peralta Muniz Moreira, R. F., Bracht, A., & Peralta, R. M. (2019). Enzymatic degradation and detoxification of azo dye Congo red by a new laccase from *Oudemansiella canarii*. *Bioresource Technology, 289*, 121655. https://doi.org/10.1016/J.BIORTECH.2019.121655
30. Ihsanullah, I., Jamal, A., Ilyas, M., Zubair, M., Khan, G., & Atieh, M. A. (2020). Bioremediation of dyes: Current status and prospects. *Journal of Water Process Engineering, 38*, Article 101680. https://doi.org/10.1016/J.JWPE.2020.101680
31. Jamee, R., & Siddique, R. (2019). Biodegradation of synthetic dyes of textile effluent by microorganisms: An environmentally and economically sustainable approach. *European Journal of Microbiology and Immunology (Bp), 9*, 114. https://doi.org/10.1556/1886.2019.00018
32. Kadam, A. A., Kamatkar, J. D., Khandare, R. V., Jadhav, J. P., & Govindwar, S. P. (2013). Solid-state fermentation: Tool for bioremediation of adsorbed textile dyestuff on distillery industry waste-yeast biomass using isolated *Bacillus cereus* strain EBT1. *Environmental Science and Pollution Research, 20*, 1009–1020. https://doi.org/10.1007/S11356-012-0929-6/TABLES/5
33. Kalyani, D. C., Telke, A. A., Dhanve, R. S., & Jadhav, J. P. (2009). Ecofriendly biodegradation and detoxification of Reactive Red 2 textile dye by newly isolated Pseudomonas sp. SUK1. *Journal of Hazardous Materials, 163*, 735–742. https://doi.org/10.1016/J.JHAZMAT.2008.07.020
34. Karim, M. E., Dhar, K., & Hossain, M. T. (2018). Decolorization of textile reactive dyes by bacterial monoculture and consortium screened from textile dyeing effluent. *Journal of Genetic Engineering and Biotechnology, 16*, 375–380. https://doi.org/10.1016/J.JGEB.2018.02.005
35. Khan, M. D., Singh, A., Khan, M. Z., Tabraiz, S., & Sheikh, J. (2023). Current perspectives, recent advancements, and efficiencies of various dye-containing wastewater treatment

technologies. *Journal of Water Process Engineering, 53*.https://doi.org/10.1016/J.JWPE.2023.103579

36. Khan, S. U., Perini, J. A. L., Hussain, S., Khan, H., Khan, S., & Boldrin Zanoni, M. V. (2020). Electrochemical preparation of Nb_2O_5 nanochannel photoelectrodes for enhanced photoelectrocatalytic performance in removal of RR120 dye. *Chemosphere, 257*, Article 127164. https://doi.org/10.1016/J.CHEMOSPHERE.2020.127164
37. Kumar, D., Patel, Z., Pandit, P., Pandit, R., Patel, A., Joshi, M., & Joshi, C. (2021). Textile industry wastewaters from Jetpur, Gujarat, India, are dominated by Shewanellaceae, Bacteroidaceae, and Pseudomonadaceae harboring genes encoding catalytic biocatalysts for textile dye degradation. *Frontiers in Environmental Science, 9*. https://doi.org/10.3389/FENVS.2021.720707
38. Lellis, B., Fávaro-Polonio, C. Z., Pamphile, J. A., & Polonio, J. C. (2019). Effects of textile dyes on health and the environment and bioremediation potential of living organisms. *Biotechnology Research and Innovation, 3*, 275–290. https://doi.org/10.1016/J.BIORI.2019.09.001
39. Lin, J., Ye, W., Xie, M., Seo, D. H., Luo, J., Wan, Y., & Van der Bruggen, B. (2023). Environmental impacts and remediation of dye-containing wastewater. *Nature Reviews Earth & Environment, 4*, 785–803. https://doi.org/10.1038/S43017-023-00489-8
40. Madhav, S., Ahamad, A., Singh, P., & Mishra, P. K. (2018). A review of textile industry: Wet processing, environmental impacts, and effluent treatment methods. *Environmental Quality Management, 27*, 31–41. https://doi.org/10.1002/tqem.21538
41. Martins, R. A., Salgado, E. M., Gonçalves, A. L., Esteves, A. F., & Pires, J. C. M. (2024). Microalgae-based remediation of real textile wastewater: Assessing pollutant removal and biomass valorisation. *Bioengineering, 11*.https://doi.org/10.3390/BIOENGINEERING11010044
42. Mendes, C. R., Dilarri, G., Stradioto, M. R., de Faria, A. U., Bidoia, E. D., & Montagnolli, R. N. (2019). The addition of a quaternary group in biopolymeric material increases the adsorptive capacity of Acid Blue 25 textile dye. *Environmental Science and Pollution Research, 26*, 24235–24246. https://doi.org/10.1007/S11356-019-05652-7/TABLES/4
43. Mishra, S., & Maiti, A. (2019). Applicability of biocatalysts produced from different biotic species for biodegradation of textile dyes. *Clean Technologies and Environmental Policy, 21*(4), 763–781. https://doi.org/10.1007/S10098-019-01681-5
44. Nagarajan, D., Lee, D. J., Varjani, S., Lam, S. S., Allakhverdiev, S. I., & Chang, J. S. (2022). Microalgae-based wastewater treatment-microalgae-bacteria consortia, multi-omics approaches and algal stress response. *Science of Total Environment, 845*.https://doi.org/10.1016/j.scitotenv.2022.157110
45. Natarajan, S., Bajaj, H. C., & Tayade, R. J. (2018). Recent advances based on the synergetic effect of adsorption for removal of dyes from waste water using photocatalytic process. *Journal of Environmental Sciences (China), 65*, 201–222. https://doi.org/10.1016/J.JES.2017.03.011
46. Oladoye, P. O., Ajiboye, T. O., Omotola, E. O., & Oyewola, O. J. (2022). Methylene blue dye: Toxicity and potential elimination technology from wastewater. *Results in Engineering, 16*.https://doi.org/10.1016/J.RINENG.2022.100678
47. Orts, F., del Río, A. I., Molina, J., Bonastre, J., & Cases, F. (2018). Electrochemical treatment of real textile wastewater: TrichromyProcion HEXL®. *Journal of Electroanalytical Chemistry, 808*, 387–394. https://doi.org/10.1016/J.JELECHEM.2017.06.051
48. Patel, A., Patel, V., Patel, H., Trivedi, U., & Patel, K. (2020). White rot fungi: Nature's scavenger. In *Microbial bioremediation and biodegradation* (pp. 267–307).https://doi.org/10.1007/978-981-15-1812-6_11
49. Periasamy, D., Mani, S., & Ambikapathi, R. (2019). White rot fungi and their biocatalysts for the treatment of industrial dye effluents. (pp. 73–100). https://doi.org/10.1007/978-3-030-25506-0_4
50. Rápó, E., Aradi, L. E., Szabó, Á., Posta, K., Szép, R., & Tonk, S. (2020). Adsorption of remazol brilliant violet-5R textile dye from aqueous solutions by using eggshell waste biosorbent. *Science and Reports, 10*. https://doi.org/10.1038/S41598-020-65334-0

51. Rathour, R. K., Sharma, D., Ullah, S., Mahmoud, E.-H. M., Sharma, N., Kumar, P., Bhatt, A. K., Ahmad, I., & Bhatia, R. K. (2024). Bacterial–microalgal consortia for bioremediation of textile industry wastewater and resource recovery for circular economy. *Biotechnology for the Environment, 1*(1), 1–21. https://doi.org/10.1186/S44314-024-00005-2
52. Routoula, E., & Patwardhan, S. V. (2020). Degradation of anthraquinone dyes from effluents: A review focusing on enzymatic dye degradation with industrial potential. *Environmental Science and Technology, 54*, 647–664. https://doi.org/10.1021/ACS.EST.9B03737/SUPPL_FILE/ES9B03737_SI_001.PDF
53. Roy, U., Sengupta, S., Das, P., Bhowal, A., & Datta, S. (2018). Integral approach of sorption coupled with biodegradation for treatment of azo dye using Pseudomonas sp.: Batch, toxicity, and artificial neural network. *3 Biotech, 8*. https://doi.org/10.1007/S13205-018-1215-1
54. Sable, H., Kumar, V., Mishra, R., Kumar, S., Pandit, S., Singh, V., Roy, A., Ranjan, N., Sharma, K., & Verma, R. (2024a). Biosorption of hexavalent chromium by *Shewanella putrefaciens* MTCC 8104: Response surface methodology optimization and mechanistic insights. *Indian Journal of Microbiology*, 1–17.https://doi.org/10.1007/S12088-024-01365-9/TABLES/5
55. Sable, H., Kumar, V., & Singh, V. (2024). β2-microglobulin biomarkers' roles in assessing cadmium toxicity. *International Journal of Medical Toxicology & Legal Medicine, 26*, 112–116. https://doi.org/10.5958/0974-4614.2023.00057.8
56. Sable, H., Kumar, V., Singh, V., Rustagi, S., Chahal, S., & Chaudhary, V. (2024). Strategically engineering advanced nanomaterials for heavy-metal remediation from wastewater. *Coordination Chemistry Reviews, 518*, Article 216079. https://doi.org/10.1016/j.ccr.2024.216079
57. Sable, H., Kumar, V., Singh, V., Rustagi, S., Chaudhary, V., & Pandit, S. (2024). Review—nanosystems-enhanced electrochemical biosensors for precision in one health management. *Journal of the Electrochemical Society, 171*, Article 037527. https://doi.org/10.1149/1945-7111/AD32A1
58. Sable, H., Singh, V., Khosla, A., & Chaudhary, V. (2025). Detection and analysis of cadmium prevalence in soil-water matrices of industrial settings of Delhi: Bioavailability and geochemical study. *Journal of the Electrochemical Society*. https://doi.org/10.1149/1945-7111/ADB459
59. Sable, H., Singh, V., Kumar, V., Roy, A., Pandit, S., Kaur, K., Rustagi, S., & Malik, S. (2024). Toxicological and bioremediation profiling of nonessential heavy metals (mercury, chromium, cadmium, aluminium) and their impact on human health: A review. *Toxicologie Analytique et Clinique*. https://doi.org/10.1016/J.TOXAC.2024.03.096
60. Sarkar, S., Banerjee, A., Chakraborty, N., Soren, K., Chakraborty, P., & Bandopadhyay, R. (2020). Structural-functional analyses of textile dye degrading azoreductase, laccase and peroxidase: A comparative in silico study. *Electronic Journal of Biotechnology, 43*, 48–54. https://doi.org/10.1016/J.EJBT.2019.12.004
61. Subramani, S. E., & Thinakaran, N. (2017). Isotherm, kinetic and thermodynamic studies on the adsorption behaviour of textile dyes onto chitosan. *Process Safety and Environmental Protection, 106*, 1–10.https://doi.org/10.1016/J.PSEP.2016.11.024
62. Sharma, J., Sharma, S., & Soni, V. (2021). Classification and impact of synthetic textile dyes on Aquatic Flora: A review. *Regional Studies in Marine Science, 45*.https://doi.org/10.1016/J.RSMA.2021.101802
63. Sharma, K., Sonu, S., Sudhaik, A., Ahamad, T., Kaya, S., Thakur, S., Le, Q. V., Nguyen, V. H., Singh, A., Nguyen, L. H., Singh, P., & Raizada, P. (2024). Mechanistic insights into the reaction pathway for efficient cationic dye photocatalytic degradation and the importance of the enhanced charge isolation over dual Z-scheme $CeO_2/BiOCl/Ag_2WO_4$ photocatalyst. *Journal of Water Process Engineering, 66*, Article 105918. https://doi.org/10.1016/J.JWPE.2024.105918
64. Singh, A., Pal, D. B., Mohammad, A., Alhazmi, A., Haque, S., Yoon, T., Srivastava, N., & Gupta, V. K. (2022). Biological remediation technologies for dyes and heavy metals in wastewater treatment: New insight. *Bioresource Technology, 343*.https://doi.org/10.1016/j.biortech.2021.126154

65. Singh, A. K., Bilal, M., Iqbal, H. M. N., & Raj, A. (2021). Lignin peroxidase in focus for catalytic elimination of contaminants-a critical review on recent progress and perspectives. *International Journal of Biological Macromolecules, 177*, 58–82. https://doi.org/10.1016/j.ijbiomac.2021.02.032
66. Singh, R. L., Singh, P. K., & Singh, R. P. (2015). Enzymatic decolorization and degradation of azo dyes—A review. *International Biodeterioration and Biodegradation, 104*, 21–31. https://doi.org/10.1016/j.ibiod.2015.04.027
67. Singh, V., Kasana, E., Batra, J., Sable, H., & Saxena, R. (2022). Antimicrobial resistance: Mechanisms, screening techniques and biosensors. *Journal of Pharmaceutical Negative Results*, 1724–1735. https://doi.org/10.47750/PNR.2022.13.S06.227
68. Singh, V., Sable, H., & Vaishali (2024). Bioremediation of emerging pollutants: A sustainable remediation approach. *Emerging Contaminants* 335–361.https://doi.org/10.1016/B978-0-443-18985-2.00013-4
69. Singh, V., Sable, H., Ghalley, M., & Tenzin, K. (2024). Environmental toxicology, fate and risk assessment of emerging contaminants. *Occurrence, Distribution and Toxic Effects of Emerging Contaminantsx*, 134–167.https://doi.org/10.1201/9781003335757-6
70. Solayman, H. M., Hossen, M. A., Abd Aziz, A., Yahya, N. Y., Leong, K. H., Sim, L. C., Monir, M. U., & Zoh, K. D. (2023). Performance evaluation of dye wastewater treatment technologies: A review. *Journal of Environmental Chemical Engineering, 11*. https://doi.org/10.1016/J.JECE.2023.109610
71. Soni, V., Sonu, Sudhaik, A., Singh, P., Thakur, S., Ahamad, T., Nguyen, V. H., Thi, L. A. P., Quang, H. H. P., Chaudhary, V., & Raizada, P. (2024). Visible-light-driven photodegradation of methylene blue and doxycycline hydrochloride by waste-based S-scheme heterojunction photocatalyst Bi5O7I/PCN/tea waste biochar. *Chemosphere, 347*, 140694. https://doi.org/10.1016/J.CHEMOSPHERE.2023.140694
72. Souza, A. G. V., Maria, T. C., Saran, L. M., Alves, L. M. C., Souza, A. G. V., Maria, T. C., Saran, L. M., & Alves, L. M. C. (2022). Enzymatic bioremediation of dyes from textile industry effluents. *The Toxicity of Environmental Pollutants*. https://doi.org/10.5772/INTECHOPEN.103064
73. Srinivasan, S., & Sadasivam, S. K. (2018). Exploring docking and aerobic-microaerophilic biodegradation of textile azo dye by bacterial systems. *Journal of Water Process Engineering, 22*, 180–191. https://doi.org/10.1016/J.JWPE.2018.02.004
74. Srivastava, A., Rani, R. M., Patle, D. S., & Kumar, S. (2022). Emerging bioremediation technologies for the treatment of textile wastewater containing synthetic dyes: A comprehensive review. *Journal of Chemical Technology & Biotechnology, 97*, 26–41. https://doi.org/10.1002/JCTB.6891
75. Talreja, R. K., Sable, H., Chaudhary, V., Kadian, S., Singh, M., Kishore, J., Kumar, M., Chaudhary, V., & Khosla, A. (2024). Challenges in lab-to-clinic translation of 5th/6th generation intelligent nanomaterial-enabled biosensors. *ECS Sensors Plus*.https://doi.org/10.1149/2754-2726/AD9F7E
76. Teerapatsakul, C., Parra, R., Keshavarz, T., & Chitradon, L. (2017). Repeated batch for dye degradation in an airlift bioreactor by laccase entrapped in copper alginate. *International Biodeterioration and Biodegradation, 120*, 52–57. https://doi.org/10.1016/J.IBIOD.2017.02.001
77. Thukral, P., Chowdhury, R., Sable, H., Kaushik, A., & Chaudhary, V. (2023). Sustainable green synthesized nanoparticles for neurodegenerative diseases diagnosis and treatment. *Materials Today: Proceedings, 73*, 323–328. https://doi.org/10.1016/J.MATPR.2022.10.315
78. Verma, A. K., Dash, R. R., & Bhunia, P. (2012). A review on chemical coagulation/flocculation technologies for removal of colour from textile wastewaters. *Journal of Environmental Management, 93*, 154–168. https://doi.org/10.1016/J.JENVMAN.2011.09.012
79. Wollmann, F., Dietze, S., Ackermann, J. U., Bley, T., Walther, T., Steingroewer, J., & Krujatz, F. (2019). Microalgae wastewater treatment: Biological and technological approaches. *Engineering in Life Sciences, 19*, 860–871. https://doi.org/10.1002/ELSC.201900071

80. Yaseen, D. A., & Scholz, M. (2019). Textile dye wastewater characteristics and constituents of synthetic effluents: A critical review. *International Journal of Environmental Science and Technology, 16*, 1193–1226. https://doi.org/10.1007/S13762-018-2130-Z
81. Yesilada, O., Birhanli, E., & Geckil, H. (2018). Bioremediation and decolorization of textile dyes by white rot fungi and laccase biocatalysts. (pp. 121–153). https://doi.org/10.1007/978-3-319-77386-5_5

Biocentric Approaches to Textile Waste Management: Sustainable Solutions and Environmental Perspectives

Vandana Singh, Vaishali Kumar, Ankush Verma, Soumya Pandit, and Manmeet Kaur

1 Introduction

Although it is a major worldwide sectors of the economy, the textile business is among the most contaminating one as it dumps enormous amounts of partly or wastewater that is untreated into natural environments. Among the hazardous contaminants carried in these waste products are surfactants, chemical-based colors, phenols, and toxic metals [106]. In the course of manufacture, about 10–15% of the generated dyes and pigments find their way into the surroundings. Because artificial dyes defy biological degradation while exhibiting carcinogenic abilities as well as other negative impacts on marine organisms, they cause enormous worry. This makes it difficult for the community to create economically viable, as well as effective plans for textile disposal and restoration [43, 110].

Conventional physicochemical methods, such as adsorption, coagulation, advanced oxidation processes and flocculation, are employed for textile effluent

V. Singh (✉)
Department of Microbiology, SSAHS, Sharda University, Greater Noida, India
e-mail: vandana.singh@sharda.ac.in

V. Kumar
Department of Forensic Science, SSAHS, Sharda University, Greater Noida, India

A. Verma
Department of Allied Health Sciences, SSAHS, Sharda University, Greater Noida, India
e-mail: ankush.verma@sharda.ac.in

S. Pandit
Department of Life Sciences, School of Basic Sciences and Research, Sharda University, Greater Noida, India
e-mail: soumya.pandit@sharda.ac.in

M. Kaur
Department of Microbiology, Punjab Agricultural University, Ludhiana, Punjab, India
e-mail: manmeet1-mb@pau.edu

R. A. Bhat et al. (eds.), *Biocentric Approaches in Textile Waste Management*, Textile Science and Clothing Technology, https://doi.org/10.1007/978-3-032-04974-2_10

treatment [80]. However, these techniques have few limitations, such as incomplete removal of pollutants, the byproducts of secondary waste, high operational costs [51, 79]. As a result, the exploration of biological approaches, which utilize the biosorptive, enzymatic, and metaboliccapabilities of plants, enzymes, microbes, has gained attention [20, 39, 108].

Microbial remediation is a cornerstone of biological approaches, leveraging the biodegradative potential of algae, fungi, bacteria to transform pollutants released by textile industries into non-toxic or less harmful forms [19, 47, 78]. For instance, bacterial species comprising *Pseudomonas aeruginosa*, *Bacillus subtilis*, and *Shewanella oneidensis* have been extensively studied for their ability to degrade azo dyes through azoreductase activity [63, 97]. Fungal species, including *Trametes versicolor* and *Phanerochaete chrysosporium*, exhibit the potential of dye degradation through lignin-modifying enzymes such as manganese peroxidases, lignin peroxidases, laccases, etc. Algae, such as *Chlorella vulgaris* and *Spirulina platensis*, not only exhibit biosorption properties for heavy metals but also facilitate pollutant remediation via photosynthetic bioaccumulation [1, 32, 33].

Isolated enzymes are used in enzymatic remediation to catalyze the breakdown of pollutants, providing a targeted remediation approach. Various enzymes such as laccases from *Pleurotus ostreatus* and peroxidases such as those from Horseradish peroxidase have been shown to be capable of breaking down the recalcitrant dye molecules efficiently and converting them into simpler and less toxic intermediates. Development in enzyme encapsulation techniques has raised biocatalysts' versatility and reliability. OMics technology in association with synthetic biology has produced microbial strains able to break down toxins. Effectively propagated in recombinant *Escherichia coli* strains, azoreductase and laccases overproduced to effectively clean wastewater from textiles [19, 51, 110].

Phycoremediation, an emerging biological approach, utilizes microalgae to remove both organic and inorganic contaminants. Algal species such as *Scenedesmus obliquus* and *Chlamydomonas reinhardtii* have biosorptive and bioaccumulative properties for heavy metals and hazardous chemicals, making them attractive options for textile wastewater treatment. Additionally, they improve the effectiveness of bioremediation systems by eliminating nutrients and giving oxygen [8, 49].

The biological methods used for textile waste elimination and remedial action are thoroughly covered in this part (Fig. 1). It emphasizes phycoremediation, enzymatic catalysis, and microbial degradation—among other processes causing pollution to deteriorate. It also addresses the integration of ecological techniques with physicochemical ones to generate hybrid systems, thereby overcoming limitations such as extended treatment durations and inadequate pollutant elimination [39, 68, 110]. This chapter aims to help find innovative and long-term approaches for reducing the environmental effect of textile sector pollution by offering insights on present research and prospective paths.

Fig. 1 A summarised form of biocentric mechanisms for textile waste management, covering target pollutants, biological groups (microbes, plants, earthworms, insects), and degradation mechanisms. It includes enzymatic processes, biosorption, biodegradation, and photosynthesis. Advanced strategies such as genetic engineering, synergistic approaches, and nanotechnology are also highlighted for sustainable waste remediation

2 Composition and Characteristics of Textile Waste

The textile sector produces many waste streams including complicated contaminants that seriously affect human health and the environment. Three main types may be distinguished among these waste streams: dyes, fibers, and chemical compounds. They come from different phases of textile manufacture: fiber the production process, using dyes, printing purposes and finishing up. The complex chemical makeup and stubborn character of these toxins make their control a major scientific and ecological difficulty [21, 33, 110].

Types of Textile Wastes

1. **Dyes**: Particularly well-known for their high dye concentration—which may account for up to 20% of the effluent volume—textile wastewater Because of their vivid colors, durability, and simplicity of application, synthetic dyes—such as azo dyes, reactive dyes, alongside anthraquinone dyes—are used somewhat widely. Their resilience under adverse conditions, however, makes them difficult

for natural decomposition mechanisms to work through. For instance, azo dyes—which make around 60–70% of all dye produced—have azo (–N=N–) connections resistant to degradation under aerobic circumstances. Through azoreductase function, examines have demonstrated that Pseudomonas aeruginosa may breakdown azo dyes into aromatic amines, which are then additionally metabolized under an oxygen-deprived environment. Apart from synthetic colors, dyes employed for textile printing add to the landfill as well. In contrast to dyes, the pigments are unable to dissolve colorants that linger in wastewater as well as are complicated to extract using traditional treatment techniques. These colorants cause great turbidity along with toxicity in aquatic bodies in addition to giving fabrics striking colors [37, 54, 107].

2. **Fibers**: Manufacturing artificial fibers in combination with organic ones, the textile industry contributes to solid waste and microplastic pollution. Although they are biodegradable, if discarded in enough organic materials involving cotton and wool might increase solid waste issues. Unlike natural disintegration materials, polymers such as acrylic, nylon, and polyester wind up as little bits of plastic all about. During cleaning processes, synthetic textiles eliminate microfibers in waterways that eventually accumulate in rivers and lakes. A polyester clothing may release up to 1900 microfibers, for example, throughout one wash period. These fibers are consumed by aquatic life, and ultimately finds passage inside the dietary chain together with ecosystems and damage human well-being [70, 81, 109].
3. **Chemicals**: In addition to auxiliary chemicals, textile handling uses surfactants, chemicals, and even bleaching chemicals very extensively. In recognition of mordants used to bind colors to fabrics, for such as chromium is a basic component. Standard in textile effluent, hexavalent chromium is highly carcinogenic despite low levels and very toxic. Studies reveal that algae such as *Chlorella vulgaris* might biosorb and simultaneously bioaccumulate chromium ions as well, therefore reducing their longevity in the environment and toxic impacts. The organic pollutant presence and impact on the availability of oxygen in aquatic habitats of textile-related waste products characterize their primary chemical oxygen demand (COD) as well as biological oxygen demand (BOD). Furthermore, numerous microbes determine challenging to survive within these circumstances, hence the considerable salt while alkaline satisfied of these contaminants impede biological mitigation processes [14, 74, 108].

Toxicity and Environmental Challenges

The combined acts of chemical-based intricacy, resistance, especially the combined effects of the elements in textile waste generate contaminants that are harmful. By absorbing light in the visible range, dyes in textile waste products lower the diffusion of light into aquatic bodies. The reason behind this interferes with the process of photosynthesis in algae along with aquatic plants, therefore producing less oxygen and consequent pollution. Research findings on dye-contaminated rivers and lakes in India, for instance, found a substantial decrease within dissolved oxygen that led to fish mortality and the breakdown of aquatic food chains [4, 7, 51]. Furthermore poisonous, mutagenic, or carcinogenic are certain synthetic colors and their breakdown products, aromatic amines. Various health problems, which comprises tumors, allergies, and reproductive dysfunction have been related to prolonged proximity to these substances by contact with the skin or drinking water. Moreover, adding to toxicity in textile waste are heavy metals. Commonly employed in dye fixing, chromium builds up in sediments and moves via bioaccumulation in aquatic life into the food chain. Researchers studying rivers contaminated with chromium found higher chromium concentrations in fish tissues, therefore underscoring the hazards to human consumers. Another big problem is soil pollution, especially in places where irrigation uses treated textile effluent. These effluents' high salt, alkalinity, and toxic components destroy soil structure, lower activity of microbes, and stop plant development [75, 77, 89, 104].

Sources and Volume of Waste Generated

Knowing the composition, harmful consequences and sources of textile waste helps one design targeted remedial approaches. The complex and varied nature of textile waste requires a multidisciplinary approach integrating chemical, physical, as well as biological procedures to adequately lower its negative environmental consequences. From post-consumer trash disposal to fiber management, textile waste comes from many stages of manufacture [63, 85].

The textile industry generates significant amounts of waste, including pre-consumer waste, which includes fabrics, scraps, and rejected goods, and post-consumer waste, which includes textiles and abandoned clothes. The industry is expected to generate 148 million tons of solid trash by 2030, resulting in 92 million textile discards. The durability of synthetic textiles in post-consumer waste contributes to environmental issues. The textile industry also generates a significant volume of discarded water during dyeing, with a daily production of 1.5 million liters of harmful metals, chemicals, and colors. These wastes contribute to environmental issues and the need for sustainable practices [4, 6, 67].

3 Conventional Textile Waste Management Techniques

Although the usual traditional management of textile waste disposal has been crucially important in reducing environmental effects, they are seen as insufficient for the complex character of textile contaminants. Potential substitutes that may either enhance or substitute conventional techniques include emerging biological

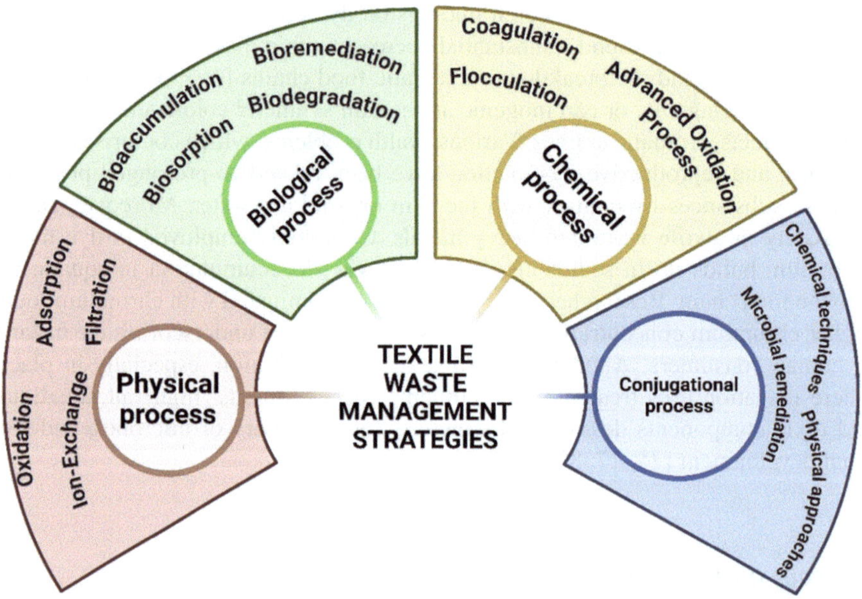

Fig. 2 Illustrating the classification of textile waste management strategies, including specific examples within each category (physical, biological, chemical, and conjugational)

tools consisting of bacterial digestion, enzymatic interventions, and phytoremediation, therefore opening the path for ecologically conscious textile waste management (Fig. 2) [94, 110].

Physical Methods

- **Sedimentation and Filtration**: One often utilized main treatment method for suspended particles from textile wastewater is sedimentation. Under gravity, particles like trash and fibers get settled during the sedimentation tanks; the filtered water is then released or further treated. To eliminate finer particles, sand filters and the membrane technologies—among other filtration systems—are used. For certain dissolved contaminants and particles that are suspended, membrane filtration, including ultrafiltration (UF) plus reverse osmosis (RO)—offers improved effectiveness in eliminating them. These techniques, however, are not very effective in handling compounds that dissolve or colors [44, 108].
- **Adsorption**: Adsorption is the deposition of dissolved contaminants where the larger surface area along with the adsorption capacity of activated carbon makes it quite popular. The elimination of synthetic dyes such as reactive and dispersion

dyes from textile effluent requires this approach. However recuperation of activated carbon requires energy, hence incorrect disposal of used adsorbents might cause secondary pollution [15, 66].
- **Flotation**: Microbubbles introduced into sewage employing dissolved air flotation (DAF) devices adhere to hydrophobic particles such as oil and grease, floating to the top for treatment. Suitable for certain contaminants, flotation ignores water-soluble dyes as well as heavy metals [36, 39, 65].

Chemical Methods

- **Coagulation and Flocculation**: For destabilization of particles of colloidal nature in wastewater, coagulation calls for adding chemicals include alum, ferric chloride, or polyaluminum chloride. The fragments are subsequently gathered into bigger flocs from flocculants such as polyacrylamide, which may be filtered or sedimentated out. This approach produces a lot of chemical sludge that has to be properly disposed of even if it is efficient at lowering color and turbidity [3, 24].
- **Advanced Oxidation Processes (AOPs)**: AOPs generate extremely reactive hydroxyl radicals ($^{\cdot}OH$) to break down tough organic contaminants including synthetic colors. These are generally used in the textile industries where approaches comprising ozonation, photocatalysis, and Fenton's reagent. Since they demand high energy and particular operating conditions, these techniques efficiently break down complicated dye molecules into smaller, less hazardous chemicals [68, 111].
- **Neutralization and pH Adjustment**: This stage is necessary to satisfy discharge criteria, it does not deal with eliminating additional contaminants such as heavy metals, surfactants, or dyes. The solvents consisting of lime or sulfuric acid are used to neutralize extremely alkaline and acidic pH solvents [11, 27].

4 Biological Methods for Textile Waste Management

Utilizing biological agents to detoxify or degrade pollutants into less harmful compounds is known as bioremediation. Often challenging because of synthetic colors and chemicals, textile effluent may be easily managed by microbial metabolism, enzymatic reactions, and phytoremediation techniques. Employing bacteria, enzymes, algae, and plants, the following techniques treat the complex composition of textile effluent—often comprising synthetic colors, toxic chemicals, and heavy metals. Highlighting the breakdown of pollutants by the metabolic activities of living beings, biological techniques for textile waste disposal provide an environmentally good and creative replacement for traditional methods as illustrated in Fig. 2 [34, 47, 51].

Microbial Remediation

Microbial methods of textile waste treatment have evolved into a necessary component of the answer to the challenges that the textile sector generates. Microbial species provide a natural and effective means of bioremediation via their enzymatic and also metabolic activities of bacteria, fungi, and parasitic organisms, given the complicated structure of textile effluent, which comprises synthetic colors, chemicals, and fibers [42, 97].

1. **Bacteria**: A major class of dyes is azo dyes with one or more azo groups (–N=N–). Complex textile effluents are broken down by bacteria through various different metabolic routes and processes and these dyes are resistant to standard techniques and stable. Azoreductases, which break the azo link into aromatic amines which are capable of converting toxins into non-toxic byproducts, are among the enzymes in charge of azo dye breakdown [63, 74]. *P. aeruginosa* generated azoreductase enzyme breaks the azo bond to release aromatic amines, which are then further broken down. Often rather highly persistent and poisonous, reactive and azo dyes have shown efficient degradation by *Pseudomonas putida* and *Pseudomonas fluorescens*. Suitable for cleaning a range of textile waste contaminants, *Bacillus subtilis* and *Bacillus licheniformis* have been shown to exhibit great tolerance to textile effluents and can breakdown both colors and chemical surfactants. Through aerobic and anaerobic processes, *Acinetobacter spp.* is very successful in breaking down other organic contaminants and in destroying reactive colors. Furthermore, very important for the biodegradation of textile contaminants are bacterial enzymes. Among the main enzymes used in textile waste disposal are azoreductases, which break azo bonds—the fundamental structure of many synthetic colors; These enzymes, laccases and peroxidases, catalyze oxidation events that may degrade dyes and other aromatic molecules; Manganese peroxidases (MnPs) oxidize the lignin-like compounds included in dyes, therefore fragmenting them apart into less hazardous forms [19, 54].

2. **Fungi**

Microbial methods of textile waste treatment have evolved into a necessary component of the answer to the challenges that the textile sector generates. Microbial species provide a natural and effective means of bioremediation via their enzymatic and also metabolic activities of bacteria, fungi, and parasitic organisms, given the complicated structure of textile effluent, which comprises synthetic colors, chemicals, and fibers [1, 33].

White-rot fungi especially are very effective in breaking down a variety of persistent toxins found in textile effluent. These species may generate a variety of extracellular enzymes, that allow them disintegrate complicated chemical molecules ranging from aromatic compounds like dyes, which are usually tolerant to microbial assault [52]. The laccase, lignin peroxidase, and manganese peroxidase enzymes are some of the most significant fungi-produced enzymes. by breaking their chemical bonds, these enzymes oxidize dye molecules therefore making them either non-toxic or simpler for subsequent breakdown. Typically found in rotting wood, white-rot fungus well-known for their lignin-breaking power. Recognized for its capacity to breakdown azo colors, Trametes versicolor is a well-researched white-rot fungus. The process of decolorization was ascribed to laccase's oxidizing action on the phenolic and amine groups found in the dye molecules [69]. Well-known for their capacity to break down a range of textile colors by means of lignin peroxidases and laccases, *Pleurotus ostreatus*—oyster mushroom—is likewise, *Phanerochaete chrysosporium*, often known as white-rot fungus, generates manganese peroxidase, which is very powerful in degrading aromatic dyes and other difficult chemicals. MnP, oxidizes lignin forms in dyes and is especially efficient in fragmenting azo along with anthraquinone dyes; *Pleurotus ostreatus* has been especially known for manufacturing laccase enzymes that have the capacity decolorizing a broad spectrum of colors [31, 91].

3. **Parasites**

Though less often researched in textile waste remedies than bacteria and fungus, parasitic organisms may still be involved in the breakdown and elimination of textile contaminants. Many parasite species create symbiotic interactions with bacteria, fungi, or other microbes, therefore indirectly helping to break down contaminants. Furthermore, certain parasite species may adsorb or bioaccumulate harmful metals, therefore assisting in the elimination of heavy metals from textile wastewater. Heavy metals including lead, cadmium, and chromium are prevalent pollutants in textile effluent, and protozoa, especially species of the genus Tetrahymena, have been shown to bioaccumulate these minerals. These species remove toxic metals via adsorption or bioaccumulation, therefore helping to contribute to the general remedial process even if they do not directly destroy dyes [74]. *Tetrahymena pyriformis* showed capacity for accumulating up to 70% of chromium contained in textile waste. It is possible to purify wastewater contaminated with organic contaminants and heavy metals using this bioaccumulation technique [25]. Through symbiotic interactions with other microbes, parasitic species may also help with bioremediation. Parasitic nematodes, for example, may help break down complex organic contaminants, such as those in textile effluent, by harboring bacteria inside their bodies. Although parasitic species such as protozoa and nematodes have little direct bearing on dye degradation, their function in eliminating heavy metals and pollution transformation helps more general bioremediation initiatives [47, 55] (Table 1).

Phytoremediation of Textile Waste

Another sustainable approach to tackle the waste management of textiles is phytoremediation where the plants are utilized to eliminate the pollutants. Contaminants such as toxic metals, dyes, organic pollutants, etc. persist in textile wastewater and cause major environmental hazards as well as health hazards. The most efficient substitute for conventional techniques is phytoremediation which makes use of plants' inherent

Table 1 Microbial species and their mechanisms for degradation of textile pollutants

Microbial group	Mechanism of degradation	Example of species	Targeted textile waste	References
Bacteria	Enzymatic degradation of dyes, reduction of pollutants, biosorption	*Bacillus subtilis*	Azo dyes, reactive dyes, heavy metals	[54]
	Biotransformation of aromatic compounds into less toxic forms	*Pseudomonas putida*	Azo dyes, PAHs, organic pollutants	[47]
	Oxidation/reduction of textile dyes and pollutants	*Escherichia coli*	Reactive dyes, azo dyes	[41]
	Hydrolysis and enzymatic breakdown of synthetic textile fibers	*Bacillus thuringiensis*	Polyester, nylon, synthetic fibers	[103]
	Biosorption of heavy metals and dyes	*Pseudomonas putida*	Heavy metals (e.g., chromium, lead), dyes	[35]
Fungi	Extracellular enzyme production (laccases, peroxidases)	*Trametes versicolor*	Azo dyes, anthraquinone dyes	[93]
	Degradation of complex organic molecules	*Pleurotus ostreatus*	Lignin-based dyes, cellulose-based textile waste	[31]
	Biosorption and enzymatic breakdown of textile dyes	*Aspergillus niger*	Basic dyes, reactive dyes, and heavy metals	[41]
	Degradation of aromatic compounds and complex dyes	*Phanerochaete chrysosporium*	Aromatic dyes, PAHs, complex synthetic dyes	[91]
	Breakdown of polyester and nylon fibers	*Fusarium oxysporum*	Synthetic fibers (polyester, nylon)	[18]
Parasites	Degradation of cellulose and hemicellulose	*Entomophthora* spp.	Cotton and other cellulose-based textile waste	[48]
	Breakdown of complex organic molecules	*Mermis* spp.	Plant-based fibers (cotton, hemp, flax)	[70, 92]
	Digesting organic material in textile waste	*Trichoderma viride* (parasitic fungus)	Cotton, wool, and natural textile fibers	[17]

(continued)

Table 1 (continued)

Microbial group	Mechanism of degradation	Example of species	Targeted textile waste	References
	Digestion of cellulose and cellulose derivatives	*Steinernema* spp.	Cotton, linen, and other plant-based fibers	[88]
	Degradation of lignocellulosic material	*Nemotoda* spp.	Cellulose-based textile waste (e.g., cotton, linen)	[16]

capacity to absorb, assimilate, and break down, by converting the toxins into less toxic forms. Essential for pollution prevention, plants may address textile wastewater through Rhizofiltration, Phytoextraction, and Phytodegradation [4, 41, 57].

Mechanisms of Phytoremediation

1. **Rhizofiltration**

Plants' utilization of their root systems to collect and filter pollutants from water is known as rhizofiltration. This method is very successful for eliminating synthetic colors and heavy metals from water. Plants' generated root exudates frequently bind contaminants in the nearby soil or water, therefore helping the plant roots to absorb them. Sometimes the pollutants immobilize in the roots, therefore lowering their rate of absorption and toxic substances. Rhizofiltration of textile effluent with chromium has made use of zea mays, or maize. Through its root system, the plant absorbs chromium; the metal is caught in the root zone. Because maize can immobilize chromium, it is a useful technique for eliminating this heavy element from polluted water. Through rhizofiltration, an aquatic plant called *Typha latifolia*—cattail—has demonstrated capacity to eliminate heavy metals consisting of copper and zinc from textile effluent. *Typha latifolia* roots serve as filters, absorbing these metals alongside therefore lessening the water's toxicity [12, 99].

2. **Phytoextraction**

Phytoextraction is the capacity of plants to translocate pollutants from the soil or water to their aerial portions (leaves, stems, etc.), from where they may be collected and eliminated. For the elimination of heavy metals, this procedure is especially successful as the metals build up in the plant tissues and simple extraction and disposal are made possible. For instance, another plant displaying great phytoextraction capacity is *Helianthus annuus*, the sunflower. From polluted water, sunflowers have been proven to absorb copper, cadmium, and heavy metals like lead. Sunflowers may scavenge harmful metals from textile waste by means of the roots they develop and build up in their stems and leaves, where they can be extracted. Similar phytoextraction of heavy metals like mercury, arsenic, and cadmium from contaminated areas uses *Populus* spp. (poplar trees). Poplar trees' great biomass enables notable metal buildup and their capacity to translocate metals to their leaves making them a useful instrument for long-term phytoremediation operations in textile waste treatment [12, 57, 99].

3. **Phytodegradation**

Additionally in the stems, root systems, or leaves, phytodegradation is the direct transformation or breakdown of contaminants within the tissues of a plant. Commonly utilized enzymatic procedures in the breakdown of organic contaminants, including dyes, follow this mechanism. Certain plants create laccases and peroxidases, which may break down complicated organic toxins such azo dyes, into non-toxic byproducts. For instance, phytodegradation is well-known for *Pennisetum purpureum*, elephant grass, to break down azo dyes such Remazol Red [53, 72]. The plant generates laccase enzymes to catalyze the oxidation of the dye, therefore converting it into simpler, less harmful compounds. This procedure allows the dye to be detoxified and its detrimental impact on the environment to be lowered. Likewise, research on *Eucalyptus globulus* (blue gum eucalyptus) has looked at its capacity to break down dyes and other contaminants in textile effluent [5, 98]. Commonly employed in the textile industry, azo and anthraquinone dyes may be broken down by the enzymes secreted by the plant's roots and stems into less toxic molecules.

Phycoremediation of Textile Waste

Using algae to eradicate and break down contaminants, phycoremediation offers a sustainable, viable method for handling textile waste. Algae may efficiently cleanse textile effluent comprising colors, heavy metals, and organic contaminants via many processes including adsorption, bioaccumulation, biological degradation, alongside flocculation [8, 59, 101]. Phycoremediation is likely to be very important in tackling the problems presented by textile-related contamination because of its affordability, sustainable development, and capacity to manage a broad range of pollutants. One of the biggest users of water, the textile sector generates enormous volumes of polluted effluent including heavy metal, chemical, and dye pollutants. Conventional techniques for treating textile wastewater—chemical coagulation, adsorption, and membrane filtration—often are expensive, ineffective, or produce contaminants that are secondary. But phycoremediation offers a sustainable, reasonably priced, biological substitute. By eliminating or breaking down harmful chemicals, algae naturally occurring biochemical activities help to generatereusable water [64, 86].

Mechanisms of Phycoremediation

1. **Adsorption and Biosorption**: Adsorption is the most basic process algae eliminate toxins by. This entails the surface of the algal cell wall binding of contaminants like heavy metals and dyes. Rich in functional groups like carboxyl, hydroxyl, and amino groups, which may interact with and absorb pollutants, algae' cell walls also Similar processes, known as biosorption, involve either passive or active transfer of toxins into the algal cell. For instance, *Chlorella vulgaris* has been demonstrated to efficiently eliminate an azo dye called Reactive Blue 19 from textile effluent via biosorption [34]. The functional groups of the algae help to enable the contact when the dye molecules are absorbed onto their surface. The biosorption process moves quickly and may be tuned by changing pH and temperature [14, 64, 86].
2. **Bioaccumulation and Bioconcentration**: The approach by which algae gather and concentrate contaminants—especially heavy metals—from their surroundings is known as bioaccumulation. Absorbed and assimilated by the cell membrane, heavy metals find their storage in intracellular compartments. Some algal species have developed specific methods to store metals in vacuoles or cell wall complexes, therefore lowering the bioavailability and toxicity of certain metals. An instance of this is the green microorganism *Chlamydomonas reinhardtii* may bioaccumulate copper and nickel among other heavy metals [105]. Absolved from polluted water, these metals build up within the algal cells where they find storage in vacuoles. After that, the metal-loaded algal biomass may be collected either for disposal or metal recovery [38, 86].
3. **Biodegradation**: By means of enzymatic activity, algae also significantly contribute to biodegradation—that is, they break down organic contaminants including textile colors. Many of the enzymes found in algae—peroxidases, laccases, ligninases—can break down complicated chemical molecules into simpler, less harmful compounds. Sometimes algae use these contaminants as a source of carbon, nitrogen, or other necessary elements for development. An instance of this is: Diatom *Phaeodactylum tricornutum* has been shown to breakdown azo dyes like Direct Red 81 [74, 96]. The azo link is broken by the algae using their enzymes, therefore causing the dye to lose color. This technique lessens the dye's toxicity to aquatic life in addition to lessening its visual effect from pollution [8].
4. **Flocculation and Coagulation**: Polymers produced by some macro- and microalgae allow for coagulation and flocculation, which in turn aid in the aggregation of color molecules, heavy metals, and other particle contaminants. Large-scale wastewater treatment finds great efficiency in the procedure as these algae-bound contaminants create flocs that are readily remove from the water [86]. For instance, the green macroalga *Ulva lactuca* generates polysaccharides that aid in the flocculation of heavy metals and textile colors. Larger aggregates formed by the polysaccharides' binding to the contaminants may be filtered or sedimentated apart. Treating effluents with high quantities of suspended particles benefits especially from this technique [22, 28].

5. **Photosynthesis and Oxygen Production**: By means of photosynthesis, algae help to enhance water quality. Algae create oxygen during this process, which helps improve water's aerobic conditions. This encourages the activity of aerobic bacteria, which help to break down organic toxins in turn [95]. Moreover, photosynthesis helps lower carbon dioxide levels in polluted water, therefore enhancing the general water quality. *Scenedesmus obliquus*, for instance, generates oxygen during photosynthesis, therefore improving the water quality. The oxygen preserves aerobic conditions, which encourage the microbial breakdown of contaminants like naturally produced surfactants along with solvents [86, 107] (Table 2).

Enzymatic Approaches

Particularly with regard to textile dyes and chemicals, enzymatic techniques for textile waste treatment are a quite popular and efficient way for control of contaminants. Unlike traditional physical or chemical treatments, enzymatic approaches are selective, ecologically safe, able to break a wide range of contaminants without creating negative side effects. Biological catalysts, enzymes help to break down difficult organic compounds into simpler, less poisonous ones. Because they may decolorize dyes, break down organic pollutants, and increase the biodegradability of textile wastewater, enzymes such as laccases, peroxidases, and cellulases are extensively utilized in textile waste management [26, 69].

Types of Enzymes

1. **Laccases**: The oxidation of a wide range of substrates, such as synthetic dyes, phenolic compounds, and aromatic amines, is facilitated by laccases, which are oxidase enzymes that contain copper. Laccases play a very important part in the process. These enzymes accelerate the electron elimination from substrates to generate radicals that later break down the contaminants. One of the most often used and long-lasting dyes in the textiles business, azo dyes are particularly susceptible to laccases' degradation. Among the many substrates laccases oxidize include lignin, phenols, and azo dyes [50, 84]. The oxidation process is the movement of electrons from the substrate to oxygen that generates radicals attacking the dye molecules, therefore causing their disintegration. For instance, one well-known source of laccases is *Trametes versicolor*, white rot fungus. By breaking the azo linkages, which are usually resistant to standard biodegradation techniques, studies have shown that laccase from this fungus may breakdown several textile dyes, which include Reactive Red and Methylene Blue [93].
2. **Peroxidases:** By hydrogen peroxide (H_2O_2), peroxidases—enzymes—catalyze the oxidation of organic molecules. Including dyes, phenols, and organic toxins prevalent in textile effluent, they are very efficient in breaking down a broad

Table 2 Algal species and their mechanisms for phyco-degradation of textile pollutants

Microalgal group	Mechanism of degradation	Algae	Targeted textile waste	References
Green Algae	Adsorption, bioaccumulation, and biodegradation of dyes	*Chlorella vulgaris*	Azo dyes, reactive dyes, heavy metals	[34]
	Dye decolorization and removal of pollutants	*Scenedesmus obliquus*	Azo dyes, water-soluble dyes, textile effluents	[107]
	Dye decolorization and biosorption	*Chlorella pyrenoidosa*	Reactive dyes, textile dyes	[62]
Blue-Green Algae	Nitrogen fixation and dye degradation	*Spirulina platensis*	Reactive dyes, oil-based textile wastewater	[85]
Red Algae	Bioaccumulation and reduction of heavy metals	*Gracilaria* spp.	Heavy metals, textile dyes, wastewater	[30]
	Removal of textile dyes and metal ions	*Eucheuma* spp.	Heavy metals, azo dyes, synthetic dyes	[102]
Brown Algae	Biodegradation of complex organic pollutants	*Phaeodactylum tricornutum*	Azo dyes, complex synthetic dyes, toxic textile chemicals	[83]
	Absorption and biotransformation of contaminants	*Fucus vesiculosus*	Dyes, heavy metals	[29]
Diatoms	Biosorption of metal ions and dye removal	*Navicula* spp.	Heavy metals (e.g., chromium, lead), dye pollutants	[71]
	Absorption of pollutants and removal of dyes	*Thalassiosira* spp.	Azo dyes, water pollutants, heavy metals	[90]
Chlorophytes	Direct degradation of textile dyes and absorption of contaminants	*Ulva lactuca*	Reactive dyes, heavy metals, organic textile effluents	[28]
	Adsorption and biodegradation of dyes	*Chlamydomonas reinhardtii*	Textile dyes, heavy metals	[49]
Cyanobacteria	Dye degradation via metabolic processes and bioadsorption	*Anabaena sp.*	Azo dyes, anthraquinone dyes, synthetic dyes	[59]
	Absorption and reduction of pollutants	*Nostoc sp.*	Heavy metals, azo dyes	[10]

spectrum of pollutants. By oxidative reactions creating reactive oxygen species (ROS), peroxidases—such as lignin peroxidase and manganese peroxidase—can break both manufactured and natural dyes. To oxidize substrates, peroxidases don electrons from hydrogen peroxide. Attacking the dye molecules, the resultant ROS—hydroxyl radicals—attack them and break them down into smaller, non-toxic molecules. Recalcitrant chemicals like azo and anthraquinone dyes degrade especially well by this process [45]. For instance, the white-rot fungus *Phanerochaete chrysosporium* generates lignin peroxidase, which has been much researched for its capacity to break down textile colors. Laboratory settings have shown that the enzyme may degrade azo dyes like Orange II and Reactive Black 5 [69, 91].

3. **Cellulases:** Enzymes known as cellulases break down intricate cellulose strands into simpler sugars by catalyizing their hydrolysis. Within the framework of textile waste, cellulases handle cotton and other fibers derived from cellulose, therefore breaking down textile fibers and encouraging their biodegradation. Because they enable the release of adsorbed colors from fibers whilst the washing process, these enzymes are also helpful for removing dye from textiles. Between glucose units in cellulose fibers, cellulases operate on the β-1,4-glycosidic linkages [13, 50]. Cellulases improve the biodegradation of textile fibers by dissolving cellulose into decreased oligosaccharides, therefore helping to decolorize textile colors attached to these fibers. For instance, the filamentous fungus *Trichoderma reesei* generates cellulases utilized in textile manufacture for bio-scouring operations. By breaking down the cellulose in cotton and other natural fibers, these enzymes enable the removing of dyes and hence increase the sustainable development of the textile sector [76, 87].
4. **Ligninases:** Enzymes known as ligninases, which include manganese peroxidases and lignin peroxidases, are responsible for the degradation of lignin, which is a complex polymer that is present in the cell walls of plants. Though lignin itself is not usually a textile pollution, ligninase enzymes are very efficient in breaking down aromatic molecules like the phenolic structures in textile colors. Often utilized in tandem with other enzymes, like laccases and peroxidases, these ones help break down organic contaminants and dyes [56, 100]. Ligninases disrupt the carbon–carbon bonds in aromatic rings, hence degrading lignin. These bonds' oxidative breakage produces reactive intermediates capable of further breaking down intricate organic contaminants like textile colors. For instance, the well-known fungus *Ganoderma lucidum* generates ligninases able to breakdown several azo and anthraquinone pigments in textile effluent. Enzymes break down these dyes by attacking their molecules with reactive oxygen species [9, 69].

Mechanisms of Enzymatic Action

Enzymatic treatment of contaminants uses laccases, peroxidases, and ligninases among other enzymes. By means of oxidation processes, that transmit electrons from pollution molecules to molecular oxygen, these enzymes create reactive oxygen species (ROS). These ROS attacks contaminants or color molecules to split them into smaller, less harmful pieces. Laccases, for instance, cut the azo link in azo dyes, therefore decolorizing the dye and producing non-toxic compounds [60, 73, 86].

Cellulases break down complicated polymers in textiles in simpler sugars or oligosaccharides via hydrolysis processes, therefore helping to degrade fibers and remove dyes linked to them. Cellulases break down cellulose fibers in textile waste treatment to release adsorbed dyes and increase their availability to other biodegradable techniques [13, 76, 100].

In order to accelerate the breakdown of azo dyes that are resistant to degradation, synergistic action often requires the joining of a number of enzymes, including laccases and peroxidases, etc. For instance, it has been shown that the reactive black 5 dye may be successfully degraded by an arrangement of laccase and manganese peroxidase. This is accomplished by breaking azo bonds and other molecular structures [45, 50, 84].

5 Biotechnological Approaches

Biotechnology has evolved as a major instrument for tackling textile waste management difficulties recently. Combining genetic engineering, nanotechnology, and synergistic bio-nano techniques has produced ground-breaking answers for the efficient breakdown of toxins like heavy metals, dyes, and dangerous compounds from textile effluences. The scientific underpinnings of these methods are explored in this part, with particular attention on how coupled systems, nanomaterials, and genetic changes interact to improve the effectiveness and environmental sustainability of textile waste treatment [41, 47, 82].

A key component of environmental biotechnology, genetic engineering lets microorganisms be changed for particular uses in pollution reduction. Microorganisms may be provided with novel metabolic pathways, improved resistance to harmful substances, or optimal capacity to generate enzymes that facilitate the breakdown of complex contaminants, like as dyes and industrial chemicals, by introducing or changing certain genes [38, 69, 106, 111].

Microorganisms must clone and express genes for dye degradation if they are to break down textile dyes. *Escherichia coli*, a normally lacking laccase activity non-native host, may have laccase expression raised by genetic engineering. Through effective oxidation and breakdown of colors, this helps these bacteria cleanse textile effluents [69, 75, 82].

Another field in which genetic engineering might alter microorganisms is metabolic pathway engineering for the breakdown of harmful substances. For instance, the addition of certain metal-binding proteins or stress response genes could improve the tolerance and detoxification capacity of a microorganism towards heavy metals. In bacteria, metal-binding proteins—including metallothioneins—are absolutely essential for the sequestration and detoxification of heavy metals [61, 69].

In contrast, nanotechnology is the study and practice of creating and manipulating materials on a very small scale, usually between one and one hundred nanometers. Nanomaterials, which include nanoparticles, nanotubes, and nanocomposites, possess a variety of distinctive characteristics. These characteristics include enhanced surface area, reactivity, and catalytic potential. As a result, these nanomaterials are very efficient at adsorbing and degrading textile contaminants, such as dyes and hazardous compounds [18, 46].

Because of their great surface area and porosity, nanomaterials are very successful adsorbent for dye removal. Nanomaterials that are based on carbon, such as activated carbon and graphene oxide, have a large surface area and functional groups, which enables them to adsorb organic dye molecules by several mechanisms, including electrostatic attraction and van der Waals forces [2, 40].

Ultimately, in terms of lowering textile pollutants and enhancing wastewater management, genetic engineering and nanotechnology have demonstrated encouraging outcomes.

Under UV light, nanomaterials including titanium dioxide (TiO_2) nanoparticles are extensively employed as photocatalysts to degrade organic contaminants including textile colors. Reactive oxygen species (ROS) produced by these nanoparticles breakdown complicated organic molecules into less dangerous molecules. TiO_2 nanoparticles excite electrons in the conduction band under UV light, which produces ROS on the surface of the nanoparticles. Attacks on the chemical bonds let these reactive ROS start the oxidative breakdown of dye molecules.

Bringing together nanotechnology with biological systems like bacteria, fungus, and algae generates hybrid systems that use the complimentary benefits of both nanomaterials and microorganisms, hence enhancing the pollution degradability. While bacteria help to biodegrade, detoxify, and mineralize toxins, nanoparticles may increase the stability, surface contact, and catalytic activity of microorganisms. Nanomaterials have the potential to serve as catalysts or transporters for microorganisms or enzymes in bio-nano hybrid systems, hence improving their ability to interact with and destroy contaminants [95, 100, 111].

Combining the advantages of biological processes with nanomaterials, synergistic bio-nanotechnologies provide a potent, multifarious method for treating textile waste. Combining nanomaterials with microorganisms lets one more effectively remove pollutants by improving microbial growth, activity, and stability under demanding environmental circumstances. Interacting with microbial cell membranes, functionalized nanoparticles may improve microbial biodegradation and help to absorb toxins [46, 58].

Nanoparticles are combined into bioreactor systems in bio-nanosano bioreactors for textile wastewater treatment to improve microbial breakdown of textile contaminants. These systems promote enzyme activity, assist the immobilization of microorganisms, and raise general treatment efficiency by means of nanomaterials. Activated carbon-immobilized silver nanoparticles in a bio-nano bioreactor system improved dye and heavy metal removal, while the nanoparticles' antibacterial properties bolstered the viability and productivity of the microbial culture [23].

6 Challenges and Limitations

Biological approaches are acknowledged as sustainable and environmentally acceptable options for textile waste treatment; yet, they encounter several problems in large-scale deployment. These include the intricacies of large-scale bioreactors, maintaining a stable microbial population, and tackling cost and time limitations. Scale-up challenges are essential for the effective use of biological techniques in the textile sector. Large-scale bioreactors need an even distribution of microbes, oxygen, and nutrients, resulting in dead zones where pollutants are inadequately destroyed. Ensuring a stable and active microbial population is essential for the efficacy of biological therapies, since bigger systems are more susceptible to contamination and microbial competition, potentially disrupting the treatment process. Financial and temporal considerations provide considerable obstacles to the extensive implementation of biological approaches. Significant initial setup expenses, labor-intensive procedures, and ongoing operating and maintenance costs may restrict their widespread adoption. Large-scale bioreactors need substantial financial expenditure for development, installation, and maintenance, along with a continuous supply of nutrients, microbial cultures, or enzymes for extensive systems.

Biological therapies, especially microbial and enzymatic procedures, may need more time to attain full or adequate breakdown of contaminants than chemical ones. The prolonged treatment duration might significantly hinder enterprises requiring swift wastewater processing to sustain operational efficiency. Maintaining biological systems may be costly, since it requires ongoing monitoring, modifications to nutritional and ambient conditions, and management of microbial inoculums or plant systems, which accumulate extra expenses over time. This is especially applicable to phytoremediation systems, because sustaining plant vitality and optimizing pollution absorption requires meticulous control of the plants' growth circumstances, including water and nutrient provision. The resilience of biological systems is a significant problem regarding their long-term use in textile waste treatment. Biological systems, including microbial communities, enzymatic therapies, and plants, exhibit significant sensitivity to environmental variations and harmful compounds, potentially resulting in diminished efficacy over time. Microbial instability, enzyme instability, and phytoremediation system instability are issues that must be resolved for biological techniques to gain practicality and broader acceptance in the textile sector.

7 Future Prospects

Sustainable textile waste management is the way of the future, and biological approaches have a lot of potential. The textile sector must address increasing demand to minimize its environmental impact, necessitating breakthroughs in biotechnology, the incorporation of biological processes with circular economy concepts, and the implementation of appropriate legislation to attain sustainable solutions. This section examines the future potential of biological approaches for textile waste treatment and provides suggestions for their further implementation and enhancement.

Research is concentrating on improving the efficacy of microbes, enzymes, and plants engaged in the breakdown of textile waste. Genetically modified microbes and innovative bioreactor designs provide possible solutions to existing restrictions. Genetic modification of microbes may improve their capacity to decompose textile contaminants, specifically targeting colors, chemicals, and fibers prevalent in textile waste. Microbial colonies and bioaugmentation, including bacteria, fungus, and algae, are assemblages of microorganisms with synergistic metabolic pathways that enhance the degradation of textile contaminants.

Enzyme engineering may increase enzymatic methods for textile waste remediation by altering enzyme structures to improve their resilience and activity in challenging textile waste conditions. Improvements in enzyme synthesis, stabilization, and immobilization methods may lower operating expenses linked to enzymatic treatments and enhance their commercial relevance.

8 Conclusion

Biological approaches for textile waste management, such as microbial degradation, enzymatic treatments, and phytoremediation, offer significant advantages over conventional physical and chemical methods. These methods are more eco-friendly, energy-efficient, and specific, targeting only the pollutants without the need for toxic reagents or producing harmful by-products, unlike traditional techniques that often generate secondary waste. Microorganisms like bacteria and fungi can effectively degrade complex textile dyes and chemicals under mild conditions, while phytoremediation can recover valuable resources, such as heavy metals and dyes, making the process more sustainable. While biological methods may face challenges such as longer treatment times and scalability issues in industrial settings, their long-term benefits, lower environmental impact, reduced operational costs, and the potential for resource recovery make them a superior, more sustainable alternative to traditional methods. With continued research and innovation, these limitations can be addressed, reinforcing the importance of biological approaches for sustainable textile waste management.

References

1. Abdelghaffar, F. (2021). The performance of yeast, fungi, and algae biomass in dye elimination. In S. S. Muthu, A. Khadir (Eds.), *Advanced removal techniques for dye-containing wastewaters, sustainable textiles: Production, processing, manufacturing and chemistry* (pp. 217–236). Springer Singapore. https://doi.org/10.1007/978-981-16-3164-1_8

2. Ahmed, A., Usman, M., Yu, B., Ding, X., Peng, Q., Shen, Y., & Cong, H. (2020). Efficient photocatalytic degradation of toxic Alizarin yellow R dye from industrial wastewater using biosynthesized Fe nanoparticle and study of factors affecting the degradation rate. *Journal of Photochemistry and Photobiology B: Biology, 202*, Article 111682.
3. Aleem, M., Cao, J., Li, C., Rashid, H., Wu, Y., Nawaz, M. I., Abbas, M., & Akram, M. W. (2020). Coagulation-and adsorption-based environmental impact assessment and textile effluent treatment. *Water, Air, & Soil Pollution, 231*, 1–8.
4. Alegbe, E. O., & Uthman, T. O. (2024). A review of history, properties, classification, applications and challenges of natural and synthetic dyes. *Heliyon.*
5. Ansari, N., Ali, A., Akhtar, M. S., Hasan, S., Khatoon, T., Khan, A. R., Wabaidur, S. M., & Rahman, Q. I. (2024). Green synthesis of zinc oxide marigold shaped clusters using Eucalyptus globulus leaf extract as robust photocatalyst for dyes degradation under sunlight. *Materials Science in Semiconductor Processing, 173*, Article 108087.
6. Aragaw, T. A. (2024). A review on biodegradation of textile dye wastewater: Challenges due to wastewater characteristics and the potential of alkaliphiles. *Journal of Hazardous Materials Advances, 100493.*
7. Ardila-Leal, L. D., Poutou-Piñales, R. A., Pedroza-Rodríguez, A. M., & Quevedo-Hidalgo, B. E. (2021). A brief history of colour, the environmental impact of synthetic dyes and removal by using laccases. *Molecules, 26*, 3813.
8. Arutselvan, C., Narchonai, G., Pugazhendhi, A., Kumar Seenivasan, H., LewisOscar, F., & Thajuddin, N. (2022). Phycoremediation of textile and tannery industrial effluents using microalgae and their consortium for biodiesel production. *Journal of Cleaner Production, 367*, 133100.
9. Asad, S., Gu, P., Peng, C., Huang, H., Jiang, F., Patabedige, N., Karunarathna, S. C., & Hapuarachchi, K. K. (2024). Biotechnological potential of *Ganoderma* species: Current progress and future prospects. *New Zealand Journal of Botany*, 1–60.https://doi.org/10.1080/0028825X.2024.2376924
10. Atoku, D. I., Ojekunle, O. Z., Taiwo, A. M., & Shittu, O. B. (2021). Evaluating the efficiency of Nostoc commune, *Oscillatoria limosa* and *Chlorella vulgaris* in a phycoremediation of heavy metals contaminated industrial wastewater. *Scientific African, 12*, Article e00817.
11. Azanaw, A., Birlie, B., Teshome, B., & Jemberie, M. (2022). Textile effluent treatment methods and eco-friendly resolution of textile wastewater. *Case Studies in Chemical and Environmental Engineering, 6*, Article 100230.
12. Bakshe, P., & Jugade, R. (2023). Phytostabilization and rhizofiltration of toxic heavy metals by heavy metal accumulator plants for sustainable management of contaminated industrial sites: A comprehensive review. *Journal of Hazardous Materials Advances, 10*, Article 100293.
13. Bernal, S. P., Lira, M. M., Jean-Baptiste, J., Garcia, P. E., Batista, E., Ottoni, J. R., & Passarini, M. R. (2021). Biotechnological potential of microorganisms from textile effluent: Isolation, enzymatic activity and dye discoloration. *Anais da Academia Brasileira de Ciencias, 93*, Article e20191581.
14. Brar, A., Kumar, M., Vivekanand, V., & Pareek, N. (2019). Phycoremediation of textile effluent-contaminated water bodies employing microalgae: Nutrient sequestration and biomass production studies. *International Journal of Environmental Science and Technology, 16*, 7757–7768.
15. Chai, W. S., Cheun, J. Y., Kumar, P. S., Mubashir, M., Majeed, Z., Banat, F., Ho, S.-H., & Show, P. L. (2021). A review on conventional and novel materials towards heavy metal adsorption in wastewater treatment application. *Journal of Cleaner Production, 296*, Article 126589.
16. Chauhan, H., Sweta, T., Ali, M., & George, L. (2023). Application of nematode derived enzymes in industries-a review. *Vidya-A Journal of Gujarat University, 2*, 215–223.
17. Dąbrowska, G. B., Garstecka, Z., Olewnik-Kruszkowska, E., Szczepańska, G., Ostrowski, M., & Mierek-Adamska, A. (2021). Comparative study of structural changes of polylactide and poly (ethylene terephthalate) in the presence of Trichoderma viride. *International Journal of Molecular Sciences, 22*, 3491.

18. Darwesh, O. M., Li, H., & Matter, I. A. (2022). Nano-bioremediation of textile industry wastewater using immobilized CuO-NPs myco-synthesized by a novel Cu-resistant Fusarium oxysporum OSF18. *Environmental Science and Pollution Research, 30*, 16694–16706. https://doi.org/10.1007/s11356-022-23360-7
19. Dell'Anno, F., Rastelli, E., Sansone, C., Brunet, C., Ianora, A., & Dell'Anno, A. (2021). Bacteria, fungi and microalgae for the bioremediation of marine sediments contaminated by petroleum hydrocarbons in the omics era. *Microorganisms, 9*, 1695.
20. DeVoy, J. E., Congiusta, E., Lundberg, D. J., Findeisen, S., & Bhattacharya, S. (2021). Post-consumer textile waste and disposal: Differences by socioeconomic, demographic, and retail factors. *Waste Management, 136*, 303–309.
21. Dissanayake, D. G. K., Weerasinghe, D. U., Thebuwanage, L. M., & Bandara, U. (2021). An environmentally friendly sound insulation material from post-industrial textile waste and natural rubber. *Journal of Building Engineering, 33*, Article 101606.
22. Divakar, S., Paul, T., Bollepalli, D., Nachiappan, K., Thiagarajan, K., & Chandrasekaran, R. (2024). Phycoremediation of tannery effluents—An approach to circular economy. In G. A. Ravishankar, A. R. Rao, & S. Kim, (Eds.), *Algae mediated bioremediation* (pp. 423–438). Wiley. https://doi.org/10.1002/9783527843367.ch20
23. Donkadokula, N. Y., Kola, A. K., Naz, I., & Saroj, D. (2020). A review on advanced physico-chemical and biological textile dye wastewater treatment techniques. *Reviews in Environmental Science & Biotechnology, 19*, 543–560. https://doi.org/10.1007/s11157-020-09543-z
24. Dotto, J., Fagundes-Klen, M. R., Veit, M. T., Palácio, S. M., & Bergamasco, R. (2019). Performance of different coagulants in the coagulation/flocculation process of textile wastewater. *Journal of cleaner production, 208*, 656–665.
25. Dutta, S., Adhikary, S., Bhattacharya, S., Roy, D., Chatterjee, S., Chakraborty, A., Banerjee, D., Ganguly, A., Nanda, S., & Rajak, P. (2024). Contamination of textile dyes in aquatic environment: Adverse impacts on aquatic ecosystem and human health, and its management using bioremediation. *Journal of Environmental Management, 353*, Article 120103.
26. Ekanayake, M. S., Udayanga, D., Wijesekara, I., & Manage, P. (2021). Phytoremediation of synthetic textile dyes: Biosorption and enzymatic degradation involved in efficient dye decolorization by *Eichhornia crassipes* (Mart.) Solms and *Pistia stratiotes* L. *Environmental Science and Pollution Research, 28*, 20476–20486.
27. Ekawati, E., Pradipta, J., & Yulia, E. (2023). Automation system architecture of pH neutralization process in batik wastewater treatment plant. *Journal of Physics: Conference Series*, 012020.
28. El-Ahmady El-Naggar, N., Rabei, N. H., & El-Malkey, S. E. (2020). Eco-friendly approach for biosorption of Pb^{2+} and carcinogenic Congo red dye from binary solution onto sustainable Ulva lactuca biomass. *Scientific Reports, 10*, 16021.
29. El-Naggar, N.E.-A., Hamouda, R. A., Saddiq, A. A., & Alkinani, M. H. (2021). Simultaneous bioremediation of cationic copper ions and anionic methyl orange azo dye by brown marine alga Fucus vesiculosus. *Scientific Reports, 11*, 3555.
30. El-Naggar, N.E.-A., & Rabei, N. H. (2020). Bioprocessing optimization for efficient simultaneous removal of methylene blue and nickel by *Gracilaria* seaweed biomass. *Scientific Reports, 10*, 17439.
31. El-Ramady, H., Abdalla, N., Fawzy, Z., Badgar, K., Llanaj, X., Törős, G., Hajdú, P., Eid, Y., & Prokisch, J. (2022). Green biotechnology of oyster mushroom (*Pleurotus ostreatus* L.): A sustainable strategy for myco-remediation and bio-fermentation. *Sustainability, 14*, 3667.
32. El-Sheekh, M. M., El Shafay, S. M., El-Shanshoury, A.E.-R.R., Hamouda, R., Gharieb, D. Y., & Abou-El-Souod, G. W. (2023). Impact of immobilized algae and its consortium in biodegradation of the textile dyes. *International Journal of Phytoremediation, 25*, 687–696. https://doi.org/10.1080/15226514.2022.2103093
33. Fayyad, R. J., Muslim, S. N., & Ali, A. N. M. (2020). Application strategies for using fungi and algae as bioremediators: A review. *Plant Archives, 20*, 788–792.

34. Fazal, T., Rehman, M. S. U., Javed, F., Akhtar, M., Mushtaq, A., Hafeez, A., Din, A. A., Iqbal, J., Rashid, N., & Rehman, F. (2021). Integrating bioremediation of textile wastewater with biodiesel production using microalgae (*Chlorella vulgaris*). *Chemosphere, 281*, Article 130758.
35. Haq, I., Kalamdhad, A. S., & Pandey, A. (2022). Genotoxicity evaluation of paper industry wastewater prior and post-treatment with laccase producing *Pseudomonas putida* MTCC 7525. *Journal of Cleaner Production, 342*, Article 130981.
36. Isawi, H., Sadik, M. A., & Nasr, F. A. (2024). Combined electrocoagulation/flotation technique and membrane desalination for textile wastewater reuse. *Journal of Environmental Chemical Engineering, 12*, Article 113661.
37. Jamee, R., & Siddique, R. (2019). Biodegradation of synthetic dyes of textile effluent by microorganisms: An environmentally and economically sustainable approach. *European Journal of Microbiology and Immunology, 9*, 114–118.
38. Jose, S., & Archanaa, S. (2019). Phycoremediation of textile wastewater: Possibilities and constraints. In S. K. Gupta, & F. Bux (Eds.), *Application of microalgae in wastewater treatment* (pp. 291–319). Springer International Publishing. https://doi.org/10.1007/978-3-030-13913-1_14
39. Juanga-Labayen, J. P., Labayen, I. V., & Yuan, Q. (2022). A review on textile recycling practices and challenges. *Textiles, 2*, 174–188.
40. Kadhim, M. J., Mahdi, M. A., Selman, A. M., Al-Ani, S. K. J., Hassan, J. J., & Ahmed, N. M. (2023). The most important parameters that affect the photocatalytic activity of ZnO nanostructures against organic dyes: A review. *Iranian Journal of Catalysis, 13*.
41. Kannan, S., Palanichamy, J., Sugitha, T., & Mayilsami, C. (2022). Bioremediation of textile dyeing industry effluent from small scale industries using a microbial consortium of *Bacillus* sp., *Escherichia coli*, and *Aspergillus niger*. *Journal of Applied Biology and Biotechnology, 10*, 100–106.
42. Kapoor, R. T., Danish, M., Singh, R. S., Rafatullah, M., HPS, A. K. (2021). Exploiting microbial biomass in treating azo dyes contaminated wastewater: Mechanism of degradation and factors affecting microbial efficiency. *Journal of Water Process Engineering, 43*, 102255.
43. Kasavan, S., Yusoff, S., Guan, N. C., Zaman, N. S. K., & Fakri, M. F. R. (2021). Global trends of textile waste research from 2005 to 2020 using bibliometric analysis. *Environmental Science and Pollution Research, 28*, 44780–44794.
44. Keskin, B., Ersahin, M. E., Ozgun, H., & Koyuncu, I. (2021). Pilot and full-scale applications of membrane processes for textile wastewater treatment: A critical review. *Journal of Water Process Engineering, 42*, Article 102172.
45. Khalid, N., Kalsoom, U., Ahsan, Z., & Bilal, M. (2022). Non-magnetic and magnetically responsive support materials immobilized peroxidases for biocatalytic degradation of emerging dye pollutants—A review. *International Journal of Biological Macromolecules, 207*, 387–401.
46. Khan, A., Roy, A., Bhasin, S., Emran, T. B., Khusro, A., Eftekhari, A., Moradi, O., Rokni, H., & Karimi, F. (2022). Nanomaterials: An alternative source for biodegradation of toxic dyes. *Food and Chemical Toxicology, 164*, Article 112996.
47. Kuanar, A., Kabi, S. K., Rath, M., Dhal, N. K., Bhuyan, R., Das, S., & Kar, D. (2022). A comparative review on bioremediation of chromium by bacterial, fungal, algal and microbial consortia. *Geomicrobiology Journal, 39*, 515–530. https://doi.org/10.1080/01490451.2022.2035019
48. Kumar, A., Yadav, A. N., Mondal, R., Kour, D., Subrahmanyam, G., Shabnam, A. A., Khan, S. A., Yadav, K. K., Sharma, G. K., & Cabral-Pinto, M. (2021). Myco-remediation: A mechanistic understanding of contaminants alleviation from natural environment and future prospect. *Chemosphere, 284*, Article 131325.
49. Kumar, C. P., Sylas, V. P., Mechery, J., Ambily, V., Kabeer, R., & Sunila, C. T. (2023). Phycoremediation of cashew nut processing wastewater and production of biodiesel using *Planktochlorella nurekis* and *Chlamydomonas reinhardtii*. *Algal Research, 69*, Article 102924.

50. Kumar, V., Pallavi, P., Sen, S. K., & Raut, S. (2024). Harnessing the potential of white rot fungi and ligninolytic enzymes for efficient textile dye degradation: A comprehensive review. *Water Environment Research, 96*, Article e10959. https://doi.org/10.1002/wer.10959
51. Kumar, V., Singh, V., & Pandit, S. (2025). Advanced omics approach and sustainable strategies for heavy metal microbial remediation in contaminated environments. *Bioresource Technology Reports,* 102040.
52. Latif, W., Ciniglia, C., Iovinella, M., Shafiq, M., & Papa, S. (2023). Role of white rot fungi in industrial wastewater treatment: A review. *Applied Sciences, 13*, 8318.
53. Lellis, B., Fávaro-Polonio, C. Z., Pamphile, J. A., & Polonio, J. C. (2019). Effects of textile dyes on health and the environment and bioremediation potential of living organisms. *Biotechnology Research and Innovation, 3*, 275–290.
54. Li, H., Wang, Y.-T., Wang, Y., Wang, H., Sun, K., & Lu, Z. (2019). Bacterial degradation of anthraquinone dyes. *Journal of Zhejiang University-Science B, 20*, 528–540.
55. Li, S.-N., Zhang, C., Li, F., Ren, N.-Q., & Ho, S.-H. (2023). Recent advances of algae-bacteria consortia in aquatic remediation. *Critical Reviews in Environmental Science and Technology, 53*, 315–339. https://doi.org/10.1080/10643389.2022.2052704
56. Lopes, L. S., Vieira, N., da Luz, J. M. R., Marliane de Cássia, S. S., Cardoso, W. S., & Kasuya, M. C. M. (2020). Production of fungal enzymes in Macaúba coconut and enzymatic degradation of textile dye. *Biocatalysis and Agricultural Biotechnology, 26*, Article 101651.
57. Mandal, R. R., Bashir, Z., Mandal, J. R., & Raj, D. (2024). Potential strategies for phytoremediation of heavy metals from wastewater with circular bioeconomy approach. *Environmental Monitoring and Assessment, 196*, 502.
58. Marimuthu, S., Antonisamy, A. J., Malayandi, S., Rajendran, K., Tsai, P.-C., Pugazhendhi, A., & Ponnusamy, V. K. (2020). Silver nanoparticles in dye effluent treatment: A review on synthesis, treatment methods, mechanisms, photocatalytic degradation, toxic effects and mitigation of toxicity. *Journal of Photochemistry and Photobiology B: Biology, 205*, Article 111823.
59. Mathimani, T., Alshiekheid, M. A., Sabour, A., Le, T. H. T., & Xia, C. (2024). Appraising the phycoremediation potential of cyanobacterial strains Phormidium and Oscillatoria for nutrient removal from textile wastewater (TWW) and synchronized biodiesel production from TWW-tolerant biomass. *Environmental Research, 241*, Article 117628.
60. Mishra, A., Takkar, S., Joshi, N. C., Shukla, S., Shukla, K., Singh, A., Manikonda, A., & Varma, A. (2022). An integrative approach to study bacterial enzymatic degradation of toxic dyes. *Frontiers in Microbiology, 12*, Article 802544.
61. Mishra, S., & Maiti, A. (2019). Applicability of enzymes produced from different biotic species for biodegradation of textile dyes. *Clean Technologies and Environmental Policy, 21*, 763–781.
62. Mishra, S., Srivastava, P. K., Singh, V., & Sharma, M. (2020). Development of sustainable phycoremediation with the help of *Chlorella pyrenoidosa* for textile wastewater analysis using response surface methodology.
63. Moyo, S., Makhanya, B. P., Zwane, P. E. (2022). Use of bacterial isolates in the treatment of textile dye wastewater: A review. *Heliyon, 8.*
64. Nair, V. K., Selvaraju, K., Samuchiwal, S., Naaz, F., Malik, A., & Ghosh, P. (2023). Phycoremediation of synthetic dyes laden textile wastewater and recovery of bio-based pigments from residual biomass: An approach towards sustainable wastewater management. *Processes, 11*, 1793.
65. Nippatla, N., & Philip, L. (2019). Electrocoagulation-floatation assisted pulsed power plasma technology for the complete mineralization of potentially toxic dyes and real textile wastewater. *Process Safety and Environmental Protection, 125*, 143–156.
66. Palumbo, L., Fiorito, S., Epifano, F., Sharifi-Rad, M., Genovese, S., & Collevecchio, C. (2023). Solid-phase adsorption methodologies of naturally occurring anthraquinones: A review. *Phytochemical Analysis, 34*, 153–162. https://doi.org/10.1002/pca.3203
67. Patel, H., Yadav, V. K., Yadav, K. K., Choudhary, N., Kalasariya, H., Alam, M. M., Gacem, A., Amanullah, M., Ibrahium, H. A., & Park, J.-W. (2022). A recent and systemic approach towards microbial biodegradation of dyes from textile industries. *Water, 14*, 3163.

68. Paździor, K., Bilińska, L., & Ledakowicz, S. (2019). A review of the existing and emerging technologies in the combination of AOPs and biological processes in industrial textile wastewater treatment. *Chemical Engineering Journal, 376*, Article 120597.
69. Periasamy, D., Mani, S., & Ambikapathi, R. (2019). White rot fungi and their enzymes for the treatment of industrial dye effluents. In A. N. Yadav, S. Singh, S. Mishra, & A. Gupta (Eds.), *Recent advancement in white biotechnology through fungi, fungal biology* (pp. 73–100). Springer International Publishing. https://doi.org/10.1007/978-3-030-25506-0_4
70. Rahman, S. S., Siddiqua, S., & Cherian, C. (2022). Sustainable applications of textile waste fiber in the construction and geotechnical industries: A retrospect. *Cleaner Engineering and Technology, 6*, Article 100420.
71. Rahman, T. T., Jiang, T., Zhang, C., Rao, Y., Sims, R. C., & Hou, L. (2024). Diatoms in wastewater treatment: Potentials, applications, and values of biomass. *Critical Reviews in Environmental Science and Technology, 54*, 557–580. https://doi.org/10.1080/10643389.2023.2259278
72. Rajput, A., Shukla, S. K., Kumar, R., Jha, G., Kalia, V., & Mangla, B. (2024). A sustainable catalytic approach for wastewater bodies: An innovation and technological point of view. In S. K. Shukla, C. M. Hussain, S. Patra, & M. Choudhary (Eds.), *Electrocatalytic materials for renewable energy* (pp. 353–376). Wiley. https://doi.org/10.1002/9781119901310.ch13
73. Rápó, E., & Tonk, S. (2021). Factors affecting synthetic dye adsorption; desorption studies: A review of results from the last five years (2017–2021). *Molecules, 26*, 5419.
74. Rathour, R. K., Sharma, D., Ullah, S., Mahmoud, E.-H.M., Sharma, N., Kumar, P., Bhatt, A. K., Ahmad, I., & Bhatia, R. K. (2024). Bacterial–microalgal consortia for bioremediation of textile industry wastewater and resource recovery for circular economy. *Biotechnology for the Environment, 1*, 6. https://doi.org/10.1186/s44314-024-00005-2
75. Raza, N., Rizwan, M., & Mujtaba, G. (2024). Bioremediation of real textile wastewater with a microalgal-bacterial consortium: An eco-friendly strategy. *Biomass Conversion and Biorefinery, 14*, 7359–7371.
76. Rehman, A. U., Masood, H., Abbas, S. H., Nawaz, H., Ahmad, K., Abdullah, A., Khan, S. M. (2025). Practical aspect of handling the use of cellulases in the textile industry. In M. Arshad (Ed.), *Enzymes in textile processing: A climate changes mitigation approach, SDGs and textiles* (pp. 151–170). Springer Nature Singapore. https://doi.org/10.1007/978-981-97-8058-7_6
77. Routoula, E., & Patwardhan, S. V. (2020). Degradation of anthraquinone dyes from effluents: A review focusing on enzymatic dye degradation with industrial potential. *Environmental Science and Technology, 54*, 647–664. https://doi.org/10.1021/acs.est.9b03737
78. Sable, H., Kumar, V., Mishra, R., Kumar, S., pandit, S., Singh, V., Roy, A., Ranjan, N., Sharma, K., & Verma, R. (2024). Biosorption of hexavalent chromium by Shewanella putrefaciens MTCC 8104: Response surface methodology optimization and mechanistic insights. *Indian Journal of Microbiology,*1–17.
79. Sable, H., Kumar, V., Mishra, R., Singh, V., Roy, A., Rai, A. K., Ranjan, N., Rustagi, S., & Pandit, S. (2024). Bamboo stem derived biochar for biosorption of Cadmium (II) ions from contaminated wastewater. *Environmental Nanotechnology, Monitoring & Management, 21*, Article 100936.
80. Sable, H., Singh, V., Kumar, V., Roy, A., Pandit, S., Kaur, K., Rustagi, S., & Malik, S. (2024). Toxicological and bioremediation profiling of nonessential heavy metals (mercury, chromium, cadmium, aluminium) and their impact on human health: A review. *Toxicologie Analytique et Clinique.*
81. Sadrolodabaee, P., Claramunt, J., Ardanuy, M., & de la Fuente, A. (2021). A textile waste fiber-reinforced cement composite: Comparison between short random fiber and textile reinforcement. *Materials, 14*, 3742.
82. Saeed, M. U., Hussain, N., Sumrin, A., Shahbaz, A., Noor, S., Bilal, M., Aleya, L., & Iqbal, H. M. (2022). Microbial bioremediation strategies with wastewater treatment potentialities–A review. *Science of the Total Environment, 818*, Article 151754.

83. Santaeufemia, S., Abalde, J., & Torres, E. (2021). Efficient removal of dyes from seawater using as biosorbent the dead and living biomass of the microalga *Phaeodactylum tricornutum*: Equilibrium and kinetics studies. *Journal of Applied Phycology, 33*, 3071–3090.
84. Sarkar, S., Banerjee, A., Chakraborty, N., Soren, K., Chakraborty, P., & Bandopadhyay, R. (2020). Structural-functional analyses of textile dye degrading azoreductase, laccase and peroxidase: A comparative in silico study. *Electronic Journal of Biotechnology, 43*, 48–54.
85. Selvaraj, D., & Arivazhagan, M. (2024). Synergistic effects of Spirulina platensis cultivation in textile wastewater towards nutrient removal and seed germination study. *Environmental Pollution, 357*, Article 124435.
86. Selvaraj, D., Dhayabaran, N. K., & Mahizhnan, A. (2023). An insight on pollutant removal mechanisms in phycoremediation of textile wastewater. *Environmental Science and Pollution Research, 30*, 124714–124734.
87. Sen, A., Kapila, R., Chaudhary, S., & Nigam, A. (2021). Biotechnological applications of microbial enzymes to replace chemicals in the textile industry—A review. *Textile Association, 82*, 68–73.
88. Shahid, M., Shaukat, F., Shahid, A., Sohail, A., & Nadeem, M. (2023). Biopesticides: A potential solution for the management of insect pests.
89. Sharma, J., Sharma, S., & Soni, V. (2021). Classification and impact of synthetic textile dyes on Aquatic Flora: A review. *Regional Studies in Marine Science, 45*, Article 101802.
90. Shrivastava, S., & Marghade, D. (2024). Diatoms: A sustainable tool for phycoremediation of industrial wastewater. In *Marine biomass: Biorefinery, bioproducts and environmental bioremediation* (p. 149).
91. Sierra-Solache, R. E., Muro, C., Maciel, A., Illescas, J., Díaz, M. C., Carbajal-Franco, G., & Hernández, O. A. (2020). Water recovery from textile wastewater treatment by encapsulated cells of *Phanerochaete chrysosporium* and ultrafiltration system. *Biologia, 75*, 1717–1729. https://doi.org/10.2478/s11756-020-00466-2
92. Singh, I. (2021). Basic concept of insect-pests management or integrated insect-pests management. In *Recent advancements and research in biological sciences* (p. 245).
93. Tišma, M., Žnidaršič-Plazl, P., Šelo, G., Tolj, I., Šperanda, M., Bucić-Kojić, A., & Planinić, M. (2021). Trametes versicolor in lignocellulose-based bioeconomy: State of the art, challenges and opportunities. *Bioresource Technology, 330*, Article 124997.
94. Uddin, F. (2021). Environmental hazard in textile dyeing wastewater from local textile industry. *Cellulose, 28*, 10715–10739.
95. Ugya, A. Y. (2021). The efficiency and antioxidant response of microalgae biofilm in the phycoremediation of wastewater resulting from tannery, textile, and dyeing activities. *International Aquatic Research, 13*, 289–300.
96. Upadhyay, M., Hansa, A., Devi, A., Lajayer, B. A., Price, G. W., & Sharma, R. (2024). Status and future scope of algal biomass-based remediation for various environmental contaminants. *Journal of Water Process Engineering, 65*, Article 105809.
97. Varjani, S., Rakholiya, P., Ng, H. Y., You, S., & Teixeira, J. A. (2020). Microbial degradation of dyes: An overview. *Bioresource Technology, 314*, Article 123728.
98. Velmurugan, G., Chohan, J. S., Paramasivam, P., & Maranan, R. (2024). Green synthesis of silver nanoparticles from southern *Eucalyptus globulus*: Potent antioxidants and photocatalysts for rhodamine B dye degradation. *Desalination and Water Treatment, 320*, Article 100687.
99. Viancelli, A., & Michelon, W. (2024). Bioremediation, phytoremediation and mycoremediation of wastewater. In M. P. Shah (Ed.), *Microbial remediation of hazardous chemicals from water & industrial wastewater treatment plant, environmental science and engineering* (pp. 357–371). Springer International Publishing. https://doi.org/10.1007/978-3-031-62898-6_16
100. Wakade, G., Lin, S., Saha, P., Kumari, U., & Daniell, H. (2023). Abatement of microfibre pollution and detoxification of textile dye—Indigo by engineered plant enzymes. *Plant Biotechnology Journal, 21*, 302–316. https://doi.org/10.1111/pbi.13942

101. Waqas, M. R., Nadeem, S. M., Khan, M. Y., Ahmad, Z., Ali, L., Asghar, H. N., & Khalid, A. (2022). Phycoremediation of textile effluents with enhanced efficacy of biodiesel production by algae and potential use of remediated effluent for improving growth of wheat. *Environmental Science and Pollution Research, 29*, 46118–46126.
102. Wibiyan, S., Royani, I., Ahmad, N., & Lesbani, A. (2024). Assessing the efficiency, selectivity, and reusability of ZnAl-layered double hydroxide and *Eucheuma* cottonii composite in removing anionic dyes from wastewater. *Inorganic Chemistry Communications, 170*, Article 113347.
103. Wu, S., Zhong, J., Lei, Q., Song, H., Chen, S.-F., Wahla, A. Q., Bhatt, K., & Chen, S. (2023). New roles for Bacillus thuringiensis in the removal of environmental pollutants. *Environmental Research, 236*, Article 116699.
104. Xu, C., Yu, H., Zhang, S., Shen, C., Ma, C., Wang, J., & Li, F. (2024). Cleaner production evaluation system for textile industry: An empirical study from LCA perspectives. *Science of the Total Environment, 913*, Article 169632.
105. Xu, K., Du, M., Yao, R., Luo, J., Chen, Z., Li, C., Lei, A., & Wang, J. (2024). Microalgae-mediated heavy metal removal in wastewater treatment: Mechanisms, influencing factors, and novel techniques. *Algal Research,* 103645.
106. Yalcin-Enis, I., Kucukali-Ozturk, M., & Sezgin, H. (2019). Risks and management of textile waste. In K. M. Gothandam, S. Ranjan, N. Dasgupta, & E. Lichtfouse (Eds.), *Nanoscience and biotechnology for environmental applications, environmental chemistry for a sustainable world* (pp. 29–53). Springer International Publishing. https://doi.org/10.1007/978-3-319-97922-9_2
107. Yang, Y., Zhao, J., Song, M., Yu, J., Yu, X., Ding, B., & Chen, X. (2023). Analysis of photosynthetic pigments pathway produced by CO_2-toxicity-induced Scenedesmus obliquus. *Science of The Total Environment, 867*, Article 161309.
108. Ye, W., Ye, K., Lin, F., Liu, H., Jiang, M., Wang, J., Liu, R., & Lin, J. (2020). Enhanced fractionation of dye/salt mixtures by tight ultrafiltration membranes via fast bio-inspired co-deposition for sustainable textile wastewater management. *Chemical Engineering Journal, 379*, Article 122321.
109. Yousef, S., Tatariants, M., Tichonovas, M., Kliucininkas, L., Lukošiūtė, S.-I., & Yan, L. (2020). Sustainable green technology for recovery of cotton fibers and polyester from textile waste. *Journal of Cleaner Production, 254*, Article 120078.
110. Yousef, S., Tatariants, M., Tichonovas, M., Sarwar, Z., Jonuškienė, I., & Kliucininkas, L. (2019). A new strategy for using textile waste as a sustainable source of recovered cotton. *Resources, Conservation and Recycling, 145*, 359–369.
111. Zhang, Y., Shaad, K., Vollmer, D., & Ma, C. (2021). Treatment of textile wastewater using advanced oxidation processes—A critical review. *Water, 13*, 3515.

Enzyme-Based Bioremediation of Textile Dyes and Effluents: A Bird's-Eye View on Current Status and Limitations

Prem Rajak, Abhratanu Ganguly, Sayantani Nanda, Moutushi Mandi, Kanchana Das, Gopal Biswas, Pritam Maitra, and Nurunnisa Khatun

1 Introduction

Textiles play a vital role in human life, providing clothing, shelter, and comfort for human well-being. However, the textile industry is also one of the largest polluters globally, generating massive amounts of waste [1]. The production process of textiles involves various stages, including raw material extraction, manufacturing, processing, and finishing, all of which contribute to environmental degradation. The textile industry significantly contribute to the global CO_2 emissions, while also consuming vast amounts of water and energy [2]. Furthermore, the rapid growth of textile industries has resulted in an alarming amount of textile waste production, contaminating various components of the environment [3]. This waste often ends up in landfills or incinerators, contributing to pollution and waste management issues. Textile wastes contain a wide range of contaminants that pose significant environmental issues. These contaminants include heavy metals such as lead, mercury, and arsenic, which are known to be toxic and carcinogenic. Additionally, textile wastes may contain volatile organic compounds (VOCs), polycyclic aromatic hydrocarbons (PAHs), and other hazardous chemicals that can cause respiratory problems, skin irritation, GI issues, and other health problems [1]. Azo dyes found in textile waste are mutagenic in nature and pose notable health threats to humans and other organisms [4]. The release of these contaminants into the environment can occur during various stages of the textile production, including dye synthesis, dyeing, and finishing treatments. Moreover, the washing and disposal of dyed textiles can also lead to the

P. Rajak (✉) · A. Ganguly · S. Nanda · N. Khatun
Toxicology Research Laboratory, Department of Animal Science, Kazi Nazrul University, Asansol, West Bengal, India
e-mail: prem.rajak@knu.ac.in

M. Mandi · K. Das · G. Biswas · P. Maitra
Toxicology Research Unit, Department of Zoology, The University of Burdwan, Purba Bardhaman, West Bengal, India

R. A. Bhat et al. (eds.), *Biocentric Approaches in Textile Waste Management*, Textile Science and Clothing Technology, https://doi.org/10.1007/978-3-032-04974-2_11

release of these contaminants into waterways and soil. To mitigate these risks, there is a growing need for the development and use of sustainable and eco-friendly textile dyes that are free from hazardous contaminants, as well as the implementation of stricter regulations and standards for textile dyeing and production.

Textile dye effluent, released into waterways during the textile manufacturing process, has devastating effects on aquatic organisms and ecosystems [5]. The effluent contains a cocktail of toxic chemicals, which can alter the physical and chemical properties of water, leading to reduced oxygen levels, increased turbidity, and altered pH. This can result in the disruption of aquatic food chains, with many organisms, including fish, crustaceans, and microorganisms, experiencing reduced growth rates, increased mortality, and altered behavior. The bioaccumulation of toxic chemicals from textile dye effluent can also have long-term effects on aquatic ecosystems, leading to the development of resistant bacterial strains and the contamination of human food sources [1]. Moreover, the exposure to textile dye effluent has also been linked to various human health problems, including skin irritation, respiratory issues, and increased cancer risk, highlighting the need for effective treatment and management of textile dye effluent to protect both environmental and human health [6]. The textile industry's environmental impact is further exacerbated by the use of synthetic materials, such as polyester and nylon, which are non-biodegradable and can take hundreds of years to decompose. To mitigate these issues, there is a growing need for sustainable textile production practices, including the use of eco-friendly materials, reducing waste, and promoting recycling and upcycling of textile waste.

2 Textile Waste Contaminants

The contamination of textile waste is an escalating issue, since the textile industry produces substantial waste that poses a threat to the environment and animals. The methods of textile manufacture utilize dyes, finishes, and other chemicals that may pollute land and groundwater. Synthetic fibers from textiles can decompose into microplastics, which may be consumed by soil organisms and pollute the food chain [7]. Textile waste, including fabric remnants and microplastics, poses a threat to animals by entangling and suffocating them. Animals may consume microplastics, leading to physical injury, obstructions, and toxic effects. Textile waste can accumulate in ecosystems, disrupting habitats and altering ecological dynamics. The textile industry emits several prevalent air pollutants, including particulate matter, encompassing PM10 and PM2.5, which are emitted throughout textile manufacturing, especially during the spinning, weaving, and knitting stages [8]. Textile finishing processes such as dyeing, printing, and bleaching release Volatile Organic Compounds (VOCs), including formaldehyde, benzene, and toluene. Nitrogen oxides (NO_x) and sulfur dioxide (SO_2) are emitted during the burning of fossil fuels for energy production in textile mills. During textile manufacture and processing, airborne fibers such as cotton, wool, and synthetic materials may be emitted [9].

The textile industry is among the most significant consumers of water and producer of wastewater, containing pollutants such as heavy metals, dyes, finishes, surfactants, bleaching agents, and other organic compounds. These pollutants can severely impact aquatic ecosystems, polluting water bodies,and harming aquatic organisms [7]. Moreover, pollutants in textile wastewater can adversely affect human health, leading to skin irritation, allergic responses, and potential carcinogenic consequences. The untreated effluents discharged by the textile sector encompass a variety of organic contaminants, with textile dyes being the most predominant [10]. More than 60% of textile dyes are azo dyes, which are identified by having one or more azo groups. They are also the most commonly used dyes in the textile industry [11]. Textile dyeing techniques may result in the release of 15–50% of unbound azo dyes into wastewater [12]. Certain textile industries process their wastewater to eliminate free azo dyes, whereas others release untreated industrial effluents directly into aquatic systems, resulting in significant ecotoxicological risks and detrimental effects on living organisms. In developing nations, farmers traditionally irrigate their fields with wastewater containing untreated industrial effluents, adversely affecting soil quality and crop germination rates [6] (Fig. 1).

Azo dyes released in water restrict light penetration, thus affecting the growth of algae and aquatic plants. Moreover, dyes consumed by fish and other organisms can be converted into hazardous intermediates within their body, adversely affecting the health [13]. Humans and other mammals can be exposed to azo dyes of industrial effluents via oral intake or direct dermal contact. Several organs of the body may be injured by azo dyes and intermediate products [14]. Some, bacteria can break down azo dyes and generate carcinogenic amines. Due to the presence of substantial volumes of dye-containing wastewater, proper management strategies are essential to prevent pollution and enhance sustainability [15]. It is important to get rid of

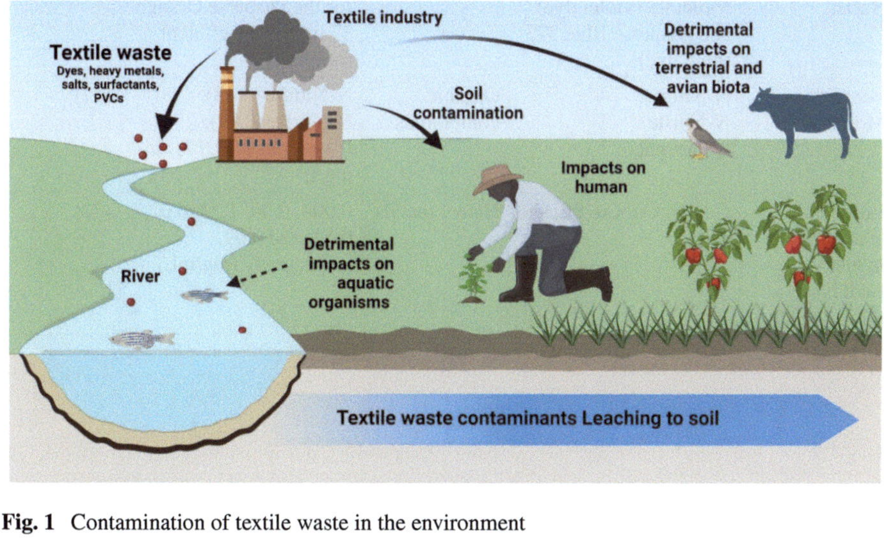

Fig. 1 Contamination of textile waste in the environment

these pollutants from textile wastewater, and several remedial technologies have been designed to remove pollutants from textile wastewater. These include physical, chemical, and biological methods. The efficacy of various treatments may differ based on the type and quantity of contaminants in textile wastewater, underscoring the necessity for continuous study and the advancement in the development of innovative technologies to alleviate the environmental and health effects of textile wastewater pollutants [16]. Efficient management and regulation of textile wastewater pollutants are crucial for minimizing environmental contamination and safeguarding both environmental and human health (Fig. 1) (Table 1).

Table 1 Contamination of water with toxic chemicals linked to the textile wastes

Study site	Contaminants	Category of contaminants	Concentrations	Reference
Fez City, Morocco	Cu, Ni, Pb, Cd, and Cr (III)	Heavy metals	Cu (0.0126), Ni (0.01), Pb (0.031), Cd (0.024), and Cr (III) (0.025) in mg/L	[17]
Taihu Lake Basin, China	Sb	Heavy metal	0.48–21.4 μg/L	[18]
Chinese southeastern coastal area	Sb, Hg, Fe, Mn, Zn, Cr, and As	Heavy metals	0.289, 0.009, 0.579, 0.277, 0.035, 0.016, and 0.013 mg/L respectively	[19]
Cristais River, São Paulo, Brazil	Disperse azo dyes C.I. Disperse Blue 373, C.I. Disperse Orange 37 and Disperse Violet 93	Textile dyes	1.65 ng/L to 316 μg/L	[20]
Cristais River, Sao Paulo, Brazil	Disperse Orange 37 (orange dye), Disperse Violet 93 (violet dye) and Disperse Blue 373 (blue dye)	Textile dyes	57.9 μg/L of Disperse Blue 373 and 316 μg/L for the Disperse Orange 37 in the raw effluent sample	[21]
Fez City, Morocco	N-Nitrate, N-Nitrite, P-Phosphate	Organic compounds	N-Nitrate (4.18), N-Nitrite (2.99), P-Phosphate (0.798) in mg/L	[17]
Effluents textile industry, Ludhiana, Punjab (India)	Cd, Cr, Co, Cu, Pb, and Zn	Heavy metals	0.004, 0.06, 1.72, 0.02, 0.13, and 0.09, respectively (mg/mL)	[22]

3 Bioremediation of Textile Dyes

Bioremediation Using Microalgae

Textile production is among the oldest and most vital manufacturing sectors globally, yielding thousands of tons of textile products each year [23]. The release of untreated textile effluents containing elevated levels of nitrogen and phosphorus can cause significant eutrophication, leading to diminished oxygen levels in aquatic environments and adversely affecting marine biodiversity and ecosystems [24]. The discharge of textile dye effluents into public water bodies constitutes a significant environmental and health issue. Over the last twenty years, numerous physicochemical decolorization methods have been developed. Nevertheless, only a limited number have gained acceptance within the textile industry. Their failure to adopt has primarily resulted from high costs, low efficiency, and ineffectiveness with a diverse range of dyes. The ability of microorganisms to carry out dye decolorization has garnered significant attention. Microalgae are regarded as a significant resource for the decolorization of textile dye effluent. Microalgae are economically feasible to cultivate, exhibit significant resilience to unfavorable environmental circumstances, and demonstrate rapid growth rates [25]. Throughout the treatment process, microalgae absorb CO_2, hence reducing greenhouse gas emissions into the environment. Post-microalgal treatment, the biomass can be employed for multiple applications, including fertilizers, biofuels (such as bioethanol, biogas, or biodiesel), energy generation, and the extraction of important pigments like carotenoids, chlorophyll-a, and chlorophyll-b. Various research has demonstrated the efficacy, environmental sustainability, and economic viability of microalgae treatment for textile dyes. Oyebamiji et al. [26] investigated the efficacy of six microalgal strains in diminishing heavy metal concentrations and colour intensity in various dilutions of textile wastewater (0.25–16% wastewater). *Micractinium sp.* attained the peak production of 1.35 gDW L^{-1} at a concentration of 1.0% wastewater. Tested microalgal species reduced colour by 47.10–70.03% and removed Selenium (Se) and Lead (Pb). Furthermore, Hernández-Zamora et al. [27] examined the efficacy of *C. vulgaris* in eliminating Congo Red (CR) dye. Microalgae were subjected to different doses of CR (5–25 mg L^{-1}) for a duration of 4 days under regulated circumstances. After 96 h, dye removal efficiency ranged from 58 to 83%, with greater values attained at lower concentration ratios. Another study by Lim et al. [28] examined the efficacy of *Chlorella vulgaris* as a decolorizing agent for textile effluent. High-rate algal ponds were employed to evaluate microalgal proliferation in diverse concentrations of textile wastewater (20–100% WW) over a period of 10 days. Specific growth rates varied from 0.12 to 0.21 d^{-1}, with peak rates seen at 80% (0.21 d^{-1}) and 100% (0.17 d^{-1}) concentrations. The most substantial color reduction occurred at 20% (34.69 $\pm$ 1.85%) and 40% (36.30 $\pm$ 3.56%) concentrations. A study by Ghazal et al. [29] assessed five local microalgae strains, *Nostoc elepsosporum*, *Nostoc linkia*, *Anabaena flos-aquae*, *Chlorella vulgaris*, and *Anabaena variabilis*, for their effectiveness in growth and the removal of red colour and heavy metals, specifically chromium, lead, iron, copper, molybdenum, and arsenic, from textile wastewater effluent released by a textile factory located in the Sixth of October City, Industrial Region, Giza Governorate, Egypt. The results indicated that all examined microalgae strains successfully eliminated the color of the textile wastewater effluent over a 4-week incubation period under constant light. Microalgae-based therapy methods are effective, economical, and environmentally sustainable, offering a feasible alternative to traditional treatment approaches. Additional research and development in this domain are essential to fully exploit the potential of microalgae in bioremediation of textile dyes, mitigate the industry's environmental impact, and generate value from by-products.

Bioremediation Using Fungi

Textile waste contains natural fibers, synthetic polyesters, textile dyes, and microplastics. Synthetic fibers can readily enter into the soil and water bodies and leads to eutrophication, degradation and pollution. Now days diverse groups of fungi are used to manage these textile wastes by the process of bio-transformation, immobilization, biosorption and reducing the heavy metals [30]. Textile wastes also contains hazardous synthetic dyes like azo dyes and anthraquinone dyes. However it also contains several harmful contaminants like perchloroethylene, toluene, xylene, trichloroethylene, and perchloroethylene [31]. Fungi produces oxidative enzymes which potentially digest complex compounds viz., lignin peroxidase, heme peroxidase, dye-decolorizing peroxidase, etc. Moreover, fungul laccase can degrade phenolic and non-phenolic compounds. Fungi also produce enzymes, like aryl-alcohol oxidase and glyoxal oxidase, generating hydrogen peroxide which effectively degrade dyes. These compounds play important role in different biotechnological applications and textile wastewater treatment [32].

Many studies revealed that ligninolytic fungi has the ability to assimilate lignin and different types of aromatic compounds with non-specific extracellular ligninolytic enzyme system. These enzymes can easily oxidize persistent organic pollutants including synthetic dyes. Beside the enzyme-dependent degradation, textile pollutants can also be converted into intermediate compounds by active culture of white rot fungi [33]. It is well established that Pleurotus flabellatus, a species of fungi collected from Tanzania, efficiently eliminated effluent produced by textile dyes. On other hand it is also documented that other different white-rot fungi strains *Phanerochaete chrysosporium, Pleurotus ostreatus,* and *Pleurotus sajor-caju* were effective in the treatment of textile waste in artificial medium [34].

Wood decay fungi are members of filamentous fungi group which produces hyphae-network and explore external environment. They can utilize the polyphenols produced from solid waste of textile industries and convert them into eco-friendly carbon and maintain carbon cycle. This fungi group is also capable to reduce the toxic potential of heterogenous mixture of solid waste of textiles like highly processed natural fibres such as viscose, synthetic petroleum-derivatives, polyester, elastane, cotton and azo-dyes by breaking their bonds. This potential of filamentous fungi can be considered a tool for bioremediation [35].

White rot fungi are also competent for complete mineralization of lignocellulose by the help of cellulases, low-specificity peroxidases, laccases, and other CAZymes (carbohydrate-active enzymes). Another group of Wood decay fungi are Brown rot fungi. They are capable of altering lignin with the help of oxidative Fenton reaction to facilitate the capture of the cellulose component of woody resources through CAZymes. Both the mechanisms utilized by two groups of fungi are non-specific and extracellular. These two mechanisms can be employed for textile waste bioremediation [36].

Poly-ethylene terephthalate, commonly known as PET is a derivative of linen. Cotton, rayon and PET are mixed in all type of textile products. The byproducts are highly susceptible to biological degradation. Different genera of fungi, such as *Fusarium* sp., *Verticillium* sp., *Chaetomium* sp., *Aspergillus* sp *Alternaria* sp., *Myrothecium* sp., *Penicillium* sp., *Stachybotrys* sp., *Memnoniella* sp., and *Trichoderma* sp., etc., are used to treat textile waste [37].

A previous study by Singh [38] demonstrated that common dyes used in textile industries are Rhodamine B and Methyl violet 2B. It is important to decolorize the synthetic effluents before it get mixed with environment. *Schizophyllum commune* and *Pezizomycotina sp.* have potential to decolourize both the dyes. Both strains can degrade dye in a wide pH and temperature range.

Bioremediation Using Bacteria

Bacteria are essential for cleaning up oil spills and other hydrocarbon pollutants because they break down hydrocarbons in the environment, efficiency for Soil bacteria varied from 0.13–50%, whereas marine bacteria have an efficiency of 0.003–100% [39]. Several bacterial-based processes have been developed for heavy metal remediation in wastewater treatment, including *ex-situ* techniques such as microbial bioreactors, fuel cells, and desalination units, as well as in-situ approaches such as built floating wetlands and bio-augmentation systems [40]. In contrast to *in-situ* remediation, which treats the pollution on-site, *ex-situ* remediation involves transferring contaminated soil from the pollution site to a different area for microbial treatment to immobilize metal ions [41]. Microbes like *E. asburiae* and *B. cereus* can immobilize heavy metals, reducing pollution in the environment [42]. Bacteria are classified as autotrophs (using inorganic resources) or heterotrophs (using organic substances) such as *Pseudomonas*, *Flavobacterium*, *Archromobactor,* and *Alkaligenes.* Aerobic bacteria biodegrade organic waste into CO_2, H_2O, and energy, whereas anaerobic bacteria breakdown pollutants without oxygen, producing CO_2, N_2, and H_2S [43]. Bacterial consortia have been found in studies to be effective metal removal systems. For example, Acinetobacter and Arthrobacter decreased chromium by 78%, and *Micrococcus luteus* removed lead at a rate of 1965 mg/g [44].

Microorganisms such as Bacillus subtilis, *B. megaterium*, *Aspergillus niger*, and *Penicillium* sp. have been reported to absorb lead, chromium, and cadmium [45]. Notably, zeolite-immobilized *Desulfovibrio desulfuricans* had substantial removal rates for Cr^{6+} (99.8%), Cu (98.2%), and Ni (90.1%). These findings demonstrate the potential of microbial consortia in bioremediation, providing a possible solution to metal contamination [44]. *Pseudomonas* sp. (CY63) isolated from lead-contaminated soils, showed remarkable lead accumulation ability. This strain demonstrated a remarkable 158.9% lead accumulation capacity, emphasizing its potential for bioremediation applications [46]. Bacterial decolorization of azo dyes usually occurs through anaerobic cleavage of azo linkages, resulting in colorless aromatic amines that are subsequently broken down into harmless compounds in aerobic systems. Depending on the pH, temperature, dye structure, concentration, and contact time, different bacterial strains like *Bacillus megaterium, Micrococcus luteus, Aeromonas hydrophila and Bacillus cereus* may reach 100% degradation of dyes [47]. Gene transfer allows *Escherichia coli* to biodegrade azo textile dyes. The gene for azoreductase enzyme, which breaks down azo dyes, can be transplanted from *Pseudomonas luteola.* Additionally, the CotA laccase enzyme gene from *Bacillus subtilis* can be introduced, allowing *E. coli* to decolorize azo and anthraquinone dyes. This genetic alteration enables *E. coli* to breakdown textile dyes, demonstrating the potential for bioremediation applications. Gene transfer improves *E. coli's* ability to degrade contaminants [6]. Bacteria have natural capabilities that allow them to chelate heavy metals and so remove them from contaminated settings. This process is aided by the release of exopolysaccharides, as seen in the marine bacteria *Enterobacter cloacae.* Bacteria's capacity to digest toxins and heavy metals demonstrates it's potential as a useful tool for bioremediation and environmental cleanup. This natural mechanism provides a viable option to reduce environmental pollution [48] (Table 2).

Bioremediation Using Enzymes

Recently numerous biochemical approaches have been employed to treat textile wastes yielding diverse levels of effectiveness, one of the major way is achieved by treating textile wastes with the help of enzymes [57]. The key enzymes implicated in bioremediation are cytochrome P450s, laccases hydrolases, and azoreductases, which exhibit considerable promise in degrading various type of textile wastes [58]. One of the main waste compound of textile industries are Azo dyes, characterized by their electron-deficient (–N=N–) chromophoric group, which is often accompanied by multiple electron-withdrawing substituents [59]. Azoreductase facilitates the reduction of azo dyes to colorless amines via a reductive cleavage process, requiring low molecular weight reducing agents like FADH or NADH to drive the redox reaction [59]. As cytoplasmic azoreductases are non-diffusable thus membrane-bound azoreductases play crucial role in bioremediation [59]. Laccases facilitate the oxidation of diverse substrates by transferring electrons to molecular oxygen, ultimately leading to the reduction of oxygen to water [47]. Laccases play

Table 2 Bioremediation of textile dyes

Microorganisms	Dye	Degradation efficiency (%)	Mechanism involved	Reference
Aspergillus bombycis	Reactive Red 31	99.02	Ligninolytic enzyme	[49]
Coelastrella sp.	Xanthene Rhodamine B	80	Biodegradation	[50]
Bacillus sp. YZU1	Reactive black 5	95	Azoreductase	[51]
Pseudomonas sp.	Reactive red BL1	99.28	Amino pyrine N-demethylase and NADH-DCIP reductase	[52]
Micrococcusglutamicus	Reactive green 19A	100	Oxidoreductive enzymes	[53]
Phormidiumvalderianum	Acid red 119	90	Bioadsorption	[54]
Coriolopsis sp	Cotton blue	79.6	Ligninolytic enzyme	[55]
Micrococcus sp.	Malachite green	100	Laccase and NADH-DCIP reductase	[56]

a pivotal role in bioremediation of industrial effluents by oxidizing a wide range of pollutants, including phenols, aromatic amines, polycyclic aromatic hydrocarbons, plastics, synthetic dyes, and pharmaceutically active compounds, thereby facilitating their degradation and detoxification [60].

Peroxidases, a class of hemiproteins, contain a heme group comprising an iron (Fe^{3+}) cation and protoporphyrin IX, enabling them to facilitate electron transfer and redox catalysis [6]. Unlike laccases, peroxidases utilize hydrogen peroxide (H_2O_2) as an electron acceptor, which is subsequently reduced, allowing the enzymes to oxidize a diverse array of organic and inorganic substrates [6]. Lignin peroxidase and Mn peroxidase are the most extensively investigated peroxidases for textile dye decoloration [6]. Notably, lignin peroxidase possesses a high redox potential, enabling it to catalyze the H_2O_2-mediated oxidation of non-phenolic aromatic lignin compounds and various phenolic substances, resulting in the formation of radical cations [6]. Mn peroxidases can degrade lignin and cellulose compounds along with recalcitrant aromatic contaminants [47].

CYP450 catalyze a wide range of reactions including hydroxylation, dealkylation, sulfurization, epoxidation, deamination, dehalogenation, peroxidation, and N-oxide reduction, facilitating the transport, metabolism and breakdown of various organic wastes of textile industry (Natural fibers, Fabric scraps, Yarn and thread wastes, chemicals) by enabling the oxidation of inert substrates by cleaving robust C–H bonds and forming stronger O–H bonds, thereby rendering the substrates more reactive and susceptible to further metabolic transformations [61].

A multitude of enzymes derived from bacteria, fungi, and plants facilitate the biodegradation of hazardous organic pollutants into less toxic or harmless compounds such as CO_2, CH_4, H_2O, and biomass, thereby maintaining environmental equilibrium [62]. This bioremediation process empowered by microbial hydrolytic enzymes (Hydrolases/Lipases) is an eco-friendly biotechnological approach that leverages microbial breakdown and environmental reactions, including hydrolysis, to transform toxic compounds into less harmful substances, providing a safe, cost-effective, and environmentally sustainable solution for pollutant elimination [6].

A major drawback of enzymatic bioremediation is its vulnerability to enzyme inactivation by inhibitors present in the harsh conditions of polluted environments. Nevertheless, this approach remains a promising and attractive alternative to traditional physicochemical methods, offering effective removal of recalcitrant pollutants, generation of highly reactive free radicals, and treatment of highly diluted contaminants, thereby providing a valuable solution for environmental remediation [6, 63] (Fig. 2).

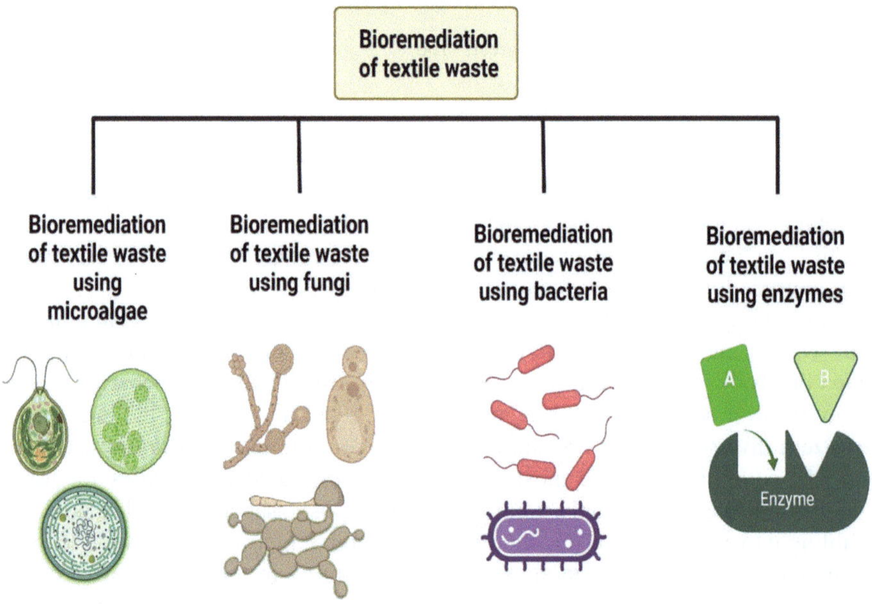

Fig. 2 Bioremediation of textile waste contaminants

4 Mechanism of Bioremediation

Colourful trendy apparel is one of the essential amenity for contemporary human society. These varying colours are generated by textile industries through application of heterogeneous dyes. However, the major concern is that, somewhat 10–15% of the used dyes are reaching the environment and causing contamination to nearby water bodies [64]. Among various groups of dyes mostly used groups are azo, nitro, phenylmethane, phthaleins, xanthene, and various metal containing dyes [7]. Upon reaching into the water bodies through industrial effluents, it can pose severe risk to organisms as well as human beings due to the presence of noxious heavy metals, phenolic derivatives, and other carcinogenic substances [65]. Thus, removal of these toxic substances from environment is very much essential to maintain a healthy niche for aquatic and other biological organisms including human.

To treat the harmful industrial effluents, various physical (membrane technology) and chemical (oxidation, photochemical oxidation, coagulation, ion-pair extraction) approaches are employed [7]. Besides their effectiveness in treating wastewater, higher cost margins of these processes and risk of generating secondary contaminants cannot be avoided. In this perspective, bioremediation, a cost effective and eco-friendly process is widely adopted to remove toxicants from polluted wastewater using the high metabolic competence of microorganisms [66]. The principle of bioremediation stands upon the ability of various microorganisms such as, bacteria, fungi and algae to use components as energy and nutrient source from hazardous effluents [44]. Bioremediation process using microorganisms starts with proper identification and selection of suitable microbial species those are proficient in breaking down any particular pollutant. For example, several bacteria species of both aerobic and anaerobic category such as, *Pseudomonus sp.*, *Bacillus subtilis*, *Lactobacillus sp.*, *Clostridium sp.*, *Rhabdobacter sp.*, *Micrococcus sp.* and also *E. coli* have the ability to degrade azo dyes [67]. For decolourization and mineralization of azo dyes, *Psedomonas sp* and *Bascidiomycetes* are widely used [68]. Besides, several species of fungi such as white rot fungus (*Phanerochaete chrysosporium*) and turkey tail fungus (*Trametes versicolor*) synthesizes ligninolytic enzymes laccase, lignin peroxidase and manganese peroxidase, which can effectively degrade complex dye molecules and pesticides [69].

Conventional modes of degradation by microorganisms encompasses some key events, which are: adsorption or bio absorption, breakdown by enzymatic action and mineralization. In bioadsorption, the dyes are get adsorbed on the surface of specific microorganisms. These adsorbents are from diverse groups of microorganisms such as fungi, microalgae, or bacteria but in adsorption, degradation of dyes into fragments are not occurred rather it causes decolouration of textile dyes [70]. Next, more common and efficient mode of biodegradation accomplishes through enzymatic action. Here specific enzymes breakdown complex contaminant and dyes into simple and less harmful molecules. For example, electron deficient azo dyes are used in reducing environment form aromatic amines by azoreductase enzymes which are further transformed into innocuous end products in oxidizing environment [67].

The end products can now be utilized as a carbon or nitrogen source during their metabolism [71, 72]. Another report by Ajaz et al. [71] revealed the generation of three carbon compound pyruvate by *Alcaligenes aquatilis* 3c bacterium from enzymatic degradation of synazol red 6HBN through metabolic actions desulfonation, carboxylation, and oxidative deamination. Another approach of degradation includes mineralization. White rot fungi are reported as a biodegrading organism of synthetic dyes by mineralization. This fungi can synthesize many secretory oxidoreductase enzymes which are able to break down lignin and other closely related aromatic composites. One of this bacterial enzyme laccase is widely used to remove phenol and pigments from contaminated water [70].

Besides microorganisms, algae and plants play a crucial role in the bioremediation of textile dye tinted waste water. Most algal species use the mode of adsorption for bioremediation. However, *Oscillatoria* and *Chlorella* are reported as degrading organisms of azo dyes initially in aromatic amines and then into simpler organic compounds and carbon dioxide. Many plants like *Brassica juncea, Phaseolus mungo, and Sorghum vulgare* have been documented for their decolorizing ability of textile effluents up to 79%, 53% and 57%, respectively [53]. Nowadays, in bioremediation procedure of textile waste, an advanced approach is in use; where, a group of symbiotically associated microorganisms were employed and this is named as microbial consortia [64]. In this microbial consortia, dye degradation is accomplished is several mechanisms including enzyme degradation, intermediate formation, making biofilm, and they also have the ability to uptake nutrient from degraded substrates [73]. So, the mechanism of bioremediation is a multifaceted process that employs diverse metabolic pathways of plants, microorganisms, and algae to degrade and detoxify harmful substances from textile effluents and utilize them as nutrient source. However, microbial resistance to certain pollutants and arrested growth and metabolic activity of microorganism due to high level of toxicants are the major challenges in bioremediation process.

5 Challenges with Bioremediation

Microalgae have demonstrated significant potential in addressing environmental pollution by detoxifying both organic and inorganic pollutants. This capability has been effectively utilized in several large-scale wastewater treatment technologies, including those targeting textile contaminants [74]. Despite the advancements in conventional wastewater treatment systems, these methods often fall short in removing emerging contaminants (ECs), such as textile pollutants, pharmaceuticals, personal care products, and industrial chemicals, which increasingly threaten both environmental and human health [74]. There is an urgent need for innovative and efficient solutions to overcome this limitation—particularly cost-effective, eco-friendly treatment systems that can be applied across various types of wastewater. Microalgae-based treatment systems have shown promise due to their dual functionality. In addition to aiding in the degradation or removal of wastewater, microalgae can simultaneously remove nutrients like nitrogen and phosphorus, which are commonly present in wastewater. This integrated approach offers a sustainable, economically viable solution for reducing contaminant pollution in waterways, while also mitigate eutrophication and contribute to broader environmental conservation efforts [74].

Bioremediation using fungi, commonly referred to as mycoremediation, is an innovative and promising approach that utilizes the diverse metabolic and enzymatic abilities of fungi to address environmental pollution. Fungi are mainly suited for this purpose due to their ability to degrade and transform a wide range of pollutants, including hydrocarbons, pesticides, heavy metals, dyes, plastics, and other synthetic compounds. This capacity stems from their production of extracellular enzymes, such as lignin peroxidases, manganese peroxidases, and laccases, which can break down complex organic molecules and detoxify contaminants [75]. Moreover, fungi's extensive mycelial networks allow them to access and metabolize pollutants in soil, water, and even air, making them versatile agents for environmental cleanup [76, 77]. Despite its potential, mycoremediation faces significant challenges that must be addressed to ensure its broader application in real-world scenarios. These challenges span scientific, technical, environmental, and regulatory domains, underscoring the complexity of translating laboratory success into large-scale field applications [78, 79].

Bioremediation using bacteria utilizes the metabolic and enzymatic capabilities of microorganisms to degrade, detoxify, or transform environmental pollutants into less harmful substances. This approach is highly adaptable, targeting contaminants such as hydrocarbons, heavy metals, pesticides, and industrial chemicals in soil, water, and air [80, 81]. Despite its promise, large-scale implementation presents significant challenges. Bacteria require specific environmental conditions, such as optimal temperature, pH, oxygen, moisture, and nutrient levels, to function efficiently [82]. Some pollutants, including heavy metals and complex hydrocarbons, exhibit high toxicity or resistance to bacterial degradation, which limits treatment efficiency [83]. Field applications are further hindered by site heterogeneity, uneven pollutant distribution, and competition with native microbial communities. Scaling up from laboratory settings to real-world scenarios introduces logistical challenges, including the effective delivery of bacterial inoculants and monitoring their performance across contaminated areas [84]. Additionally, the degradation process may produce toxic by-products that require further treatment [85]. Economic and regulatory barriers, such as the high costs of bacterial cultivation, nutrient supplementation, and public concerns—particularly regarding genetically modified strains—further complicate adoption [86]. Overcoming these challenges will require coordinated efforts in interdisciplinary research, bacterial engineering, integrated remediation strategies, and the development of supportive regulatory frameworks to realize the full potential of bacterial bioremediation as a sustainable pollution management tool [87] (Fig. 3).

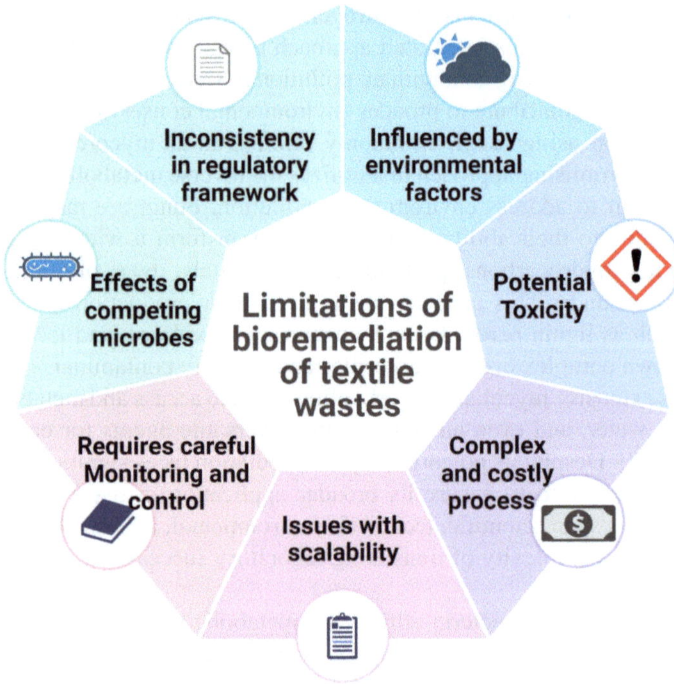

Fig. 3 Several limitations associated with the bioremediation of textile wastes

6 Conclusion

Textile waste poses a significant environmental and health risk due to its large volume, non-biodegradable nature, and toxic chemical composition. However, recent advancements in microbial and enzymatic remediation technologies offer a promising solution to this problem. Microorganisms such as bacteria, fungi, and yeast, as well as enzymes like laccases, peroxidases, and cellulases, have shown great potential in breaking down and degrading textile waste. These biological agents can effectively degrade synthetic dyes, polymers, and other pollutants, reducing the environmental impact of textile waste. Furthermore, microbial and enzymatic remediation methods are eco-friendly, cost-effective, and can be easily scaled up for industrial applications. Therefore, the use of microbes and enzymes for textile waste remediation is a sustainable and innovative approach that can help mitigate the environmental and health risks associated with textile waste, and promote a more circular and environmentally friendly textile industry.

Acknowledgements Authors are thankful to BioRender.com, which was used to prepare illustrations for this work.

References

1. Dutta, S., Adhikary, S., Bhattacharya, S., Roy, D., Chatterjee, S., Chakraborty, A., Banerjee, D., Ganguly, A., Nanda, S., & Rajak, P. (2024). Contamination of textile dyes in aquatic environment: Adverse impacts on aquatic ecosystem and human health, and its management using bioremediation. *Journal of Environmental Management., 27*(353), Article 120103.
2. Leal Filho, W., Perry, P., Heim, H., Dinis, M. A., Moda, H., Ebhuoma, E., & Paço, A. (2022). An overview of the contribution of the textiles sector to climate change. *Frontiers in Environmental Science, 5*(10), Article 973102.
3. Zhou, Q., Le, Q. V., Meng, L., Yang, H., Gu, H., Yang, Y., Chen, X., Lam, S. S., Sonne, C., & Peng, W. (2022). Environmental perspectives of textile waste, environmental pollution and recycling. *Environmental Technology Reviews, 11*(1), 62–71.
4. de Lima, R. O., Bazo, A. P., Salvadori, D. M., Rech, C. M., de Palma, O. D., & de Aragão, U. G. (2007). Mutagenic and carcinogenic potential of a textile azo dye processing plant effluent that impacts a drinking water source. *Mutation Research/Genetic Toxicology and Environmental Mutagenesis, 626*(1–2), 53–60.
5. Athira, N., & Jaya, D. S. (2018). The use of fish biomarkers for assessing textile effluent contamination of aquatic ecosystems: A review. *Nature Environment & Pollution Technology,17*(1).
6. Lellis, B., Fávaro-Polonio, C. Z., Pamphile, J. A., & Polonio, J. C. (2019). Effects of textile dyes on health and the environment and bioremediation potential of living organisms. *Biotechnology Research and Innovation, 3*(2), 275–290. https://doi.org/10.1016/j.biori.2019.09.001
7. Periyasamy, A. P. (2024). Recent advances in the remediation of textile-dye-containing wastewater: Prioritizing human health and sustainable wastewater treatment. *Sustainability, 16*(2), 495. https://doi.org/10.3390/su16020495
8. Issac, M. N., & Kandasubramanian, B. (2021). Effect of microplastics in water and aquatic systems. *Environmental Science and Pollution Research, 28*, 19544–19562.
9. Le Marechal, A. M., Križanec, B., Vajnhandl, S., & Valh, J. V. (2012). Textile finishing industry as an important source of organic pollutants. In *POPs* (pp. 29–54).
10. Azanaw, A., Birlie, B., Teshome, B., & Jemberie, M. (2022). Textile effluent treatment methods and eco-friendly resolution of textile wastewater. *CSCEE, 6*, Article 100230.
11. Al-Tohamy, R., Ali, S. S., Li, F., Okasha, K. M., Mahmoud, Y. A. G., Elsamahy, T., Jiao, H., Fu, Y., & Sun, J. (2022). A critical review on the treatment of dye-containing wastewater: Ecotoxicological and health concerns of textile dyes and possible remediation approaches for environmental safety. *Ecotoxicology and Environmental Safety, 231*, Article 113160.
12. Kumaravel, R., & Shanmugam, V. K. (2024). Biomimetic and ecological perspective towards decolorization of industrial important Azo Dyes using bacterial cultures–A review. *Sustainable Chemistry for the Environment,* 100130.
13. Goswami, D., Mukherjee, J., Mondal, C., & Bhunia, B. (2024). Bioremediation of azo dye: A review on strategies, toxicity assessment, mechanisms, bottlenecks and prospects. *Science of The Total Environment*, 176426.
14. Ngo, A. C. R., & Tischler, D. (2022). Microbial degradation of azo dyes: Approaches and prospects for a hazard-free conversion by microorganisms. *International Journal of Environmental Research and Public Health, 19*(8), 4740.
15. Singh, G. B., Vinayak, A., Mudgal, G., & Kesari, K. K. (2024). Azo dye bioremediation: An interdisciplinary path to sustainable fashion. *Environmental & Technology Innovation*, 103832.
16. Singh, B. J., Chakraborty, A., & Sehgal, R. (2023). A systematic review of industrial wastewater management: Evaluating challenges and enablers. *Journal of Environmental Management, 348*, Article 119230.
17. Talouizte, H., Merzouki, M., Benlemlih, M., & Bendriss, A. M. (2020). Chemical characterization of specific micropollutants from textile industry effluents in Fez City, Morocco. *Journal of Chemistry, 2020*(1), 3268241.
18. Li, F., Guo, Z., Mao, L., Feng, J., Huang, J., & Tao, H. (2023). Impact of textile industries on surface water contamination by Sb and other potential toxic elements: A case study in Taihu

Lake Basin, China. *International Journal of Environmental Research and Public Health, 20*(4), 3600.
19. Li, F., Zhong, Z., Gu, C., Shen, C., Ma, C., Liu, Y., Yin, S., & Xu, C. (2021). Metals pollution from textile production wastewater in Chinese southeastern coastal area: Occurrence, source identification, and associated risk assessment. *Environmental Science and Pollution Research, 28*, 38689–38697.
20. Carneiro, P. A., Umbuzeiro, G. A., Oliveira, D. P., & Zanoni, M. V. (2010). Assessment of water contamination caused by a mutagenic textile effluent/dyehouse effluent bearing disperse dyes. *Journal of Hazardous Materials, 174*(1–3), 694–699.
21. Oliveira, D. P., Carneiro, P. A., Sakagami, M. K., Zanoni, M. V., & Umbuzeiro, G. A. (2007). Chemical characterization of a dye processing plant effluent—Identification of the mutagenic components. *Mutation Research/Genetic Toxicology and Environmental Mutagenesis, 626*(1–2), 135–142.
22. Kaur, J., Bhatti, S. S., Bhat, S. A., Nagpal, A. K., Kaur, V., & Katnoria, J. K. (2021). Evaluating potential ecological risks of heavy metals of textile effluents and soil samples in vicinity of textile industries. *Soil Systems, 5*(4), 63.
23. Thakker, A. M., & Sun, D. (2023). Sustainable development goals for textiles and fashion. *Environmental Science and Pollution Research International, 30*, 101989–102009.
24. Marazzi, F., Fornaroli, R., Clagnan, E., Brusetti, L., Ficara, E., Bellucci, M., & Mezzanotte, V. (2023). Wastewater from textile digital printing as a substrate for microalgal growth and valorization. *Bioresource Technology, 375*, Article 128828.
25. Martins, R. A., Salgado, E. M., Gonçalves, A. L., Esteves, A. F., & Pires, J. C. M. (2024). Microalgae-based remediation of real textile wastewater: Assessing pollutant removal and biomass valorisation. *Bioengineering, 11*, 44.
26. Oyebamiji, O. O., Boeing, W. J., Holguin, F. O., Ilori, O., & Amund, O. (2019). Green microalgae cultured in textile wastewater for biomass generation and biodetoxification of heavy metals and chromogenic substances. *Bioresource Technology Reports, 7*, Article 100247.
27. Hernández-Zamora, M., Cristiani-Urbina, E., Martínez-Jerónimo, F., Perales-Vela, H. V., Ponce-Noyola, T., Montes-Horcasitas, M. D., & Cañizares-Villanueva, R. O. (2015). Bioremoval of the azo dye Congo Red by the microalga *Chlorella vulgaris*. *Environmental Science and Pollution Research, 22*, 10811–10823.
28. Lim, S. L., Chu, W. L., & Phang, S. M. (2010). Use of *Chlorella vulgaris* for bioremediation of textile wastewater. *Bioresource Technology Reports, 101*, 7314–7322.
29. Ghazal, F. M., Mahdy, E. S., Abdelfattah, M. S., El-Sadany, A. E., & Doha, N. (2018). The use of microalgae in bioremediation of the textile wastewater effluent. *Natural Science, 16*, 98–104.
30. Pundir, A., Thakur, M. S., Prakash, S., Kumari, N., Sharma, N., Parameswari, E., He, Z., Nam, S., Thakur, M., Puri, S., Puranik, S., Kumar, S., Madhu, & Kumar, M. (2024). Fungi as versatile biocatalytic tool for treatment of textile wastewater effluents. *Environmental Sciences Europe, 36*(1), 185. https://doi.org/10.1186/s12302-024-01007-3
31. Pous, N., Balaguer, M. D., Colprim, J., & Puig, S. (2018). Opportunities for groundwater microbial electro-remediation. *Microbial Biotechnology, 11*(1), 119–135. https://doi.org/10.1111/1751-7915.12866
32. Palanivelan Ramachandran, P. R., Rajakumar Sundharam, R. S., Jayanthi Palaniyappan, J. P., & Munusamy, A. P. (2013). *Potential process implicated in bioremediation of textile effluents: A review.*
33. Blánquez, P., Sarrà, M., & Vicent, T. (2008). Development of a continuous process to adapt the textile wastewater treatment by fungi to industrial conditions. *Process Biochemistry, 43*(1), 1–7. https://doi.org/10.1016/j.procbio.2007.10.002
34. Nilsson, I., Möller, A., Mattiasson, B., Rubindamayugi, M. S. T., & Welander, U. (2006). Decolorization of synthetic and real textile wastewater by the use of white-rot fungi. *Enzyme and Microbial Technology, 38*(1–2), 94–100. https://doi.org/10.1016/j.enzmictec.2005.04.020
35. Hazelgrove, L., & Moody, S. C. (2024). Successful cultivation of edible fungi on textile waste offers a new avenue for bioremediation and potential food production. *Scientific Reports, 14*(1), 11510. https://doi.org/10.1038/s41598-024-61680-5

36. Harper, R., & Moody, S. C. (2023). Filamentous Fungi are potential bioremediation agents of semi-synthetic textile waste. *Journal of Fungi, 9*(6), 661. https://doi.org/10.3390/jof9060661
37. Wojnowska-Baryła, I., Bernat, K., & Zaborowska, M. (2022). Strategies of recovery and organic recycling used in textile waste management. *International Journal of Environmental Research and Public Health, 19*(10), 5859. https://doi.org/10.3390/ijerph19105859
38. Rajender Singh, O. P. A. A. R. (2017). *Decolourization of textile dyes by ligninolytic fungi isolated from spent mushroom substrate.*
39. Ganesan, M., Mani, R., Sai, S., Kasivelu, G., Awasthi, M. K., Rajagopal, R., Azelee, N. I., Selvi, P. K., Chang, S. W., & Ravindran, B. (2022). Bioremediation by oil degrading marine bacteria: An overview of supplements and pathways in key processes. *Chemosphere, 303*, 134956
40. Sreedevi, P. R., Suresh, K., & Jiang, G. (2022). Bacterial bioremediation of heavy metals in wastewater: A review of processes and applications. *Journal of Water Process Engineering, 48*, Article 102884.
41. Ayilara, M. S., & Babalola, O. O. (2023). Bioremediation of environmental wastes: The role of microorganisms. *Frontiers in Agronomy, 5*, 1183691.
42. Abo-Alkasem, M. I., Hassan, N. M., & Abo Elsoud, M. M. (2023). Microbial bioremediation as a tool for the removal of heavy metals. *Bulletin of the National Research Centre, 47*(1), 31.
43. Thirumalaivasan, N., Gnanasekaran, L., Kumar, S., Durvasulu, R., Sundaram, T., Rajendran, S., Nangan, S., & Kanagaraj, K. (2024). Utilization of fungal and bacterial bioremediation techniques for the treatment of toxic waste and biowaste. *Frontiers in Materials, 11*, 1416445.
44. Tarekegn, M. M., Salilih, F. Z., & Ishetu, A. I. (2020). Microbes used a tool for bioremediation of heavy metals from the environment. *Cogent Food & Agriculture, 6*, 1783174. https://doi.org/10.1080/23311932.2020.1783174
45. Rocco, D. H., Freire, B. M., Oliveira, T. J., Alves, P. L., de Oliveira Júnior, J. M., Batista, B. L., Grotto, D., & Jozala, A. F. (2024). *Bacillus subtilis* as an effective tool for bioremediation of lead, copper and cadmium in water. *Discover Applied Sciences, 6*(8), 430.
46. Ashkan, M. F. (2023). Lead: Natural occurrence, toxicity to organisms and bioremediation by lead-degrading bacteria: A comprehensive review. *Journal of Pure & Applied Microbiology, 17*(3).
47. Ihsanullah, I., Jamal, A., Ilyas, M., et al. (2020). Bioremediation of dyes: Current status and prospects. *Journal of Water Process Engineering, 38*, Article 101680. https://doi.org/10.1016/j.jwpe.2020.101680
48. Tegene, B. G., & Tenkegna, T. A. (2020). Mode of action, mechanism and role of microbes in bioremediation service for environmental pollution management. *Journal of Biotechnology & Bioinformatics Research, 2*, 1–8.
49. Khan, R., & Fulekar, M. H. (2017). Mineralization of a sulfonated textile dye Reactive Red 31 from simulated wastewater using pellets of *Aspergillus bombycis. Bioresource and Bioprocessing, 4*, 1–11.
50. Baldev, E., MubarakAli, D., Ilavarasi, A., Pandiaraj, D., Ishack, K. A. S. S., & Thajuddin, N. (2013). Degradation of synthetic dye, rhodamine B to environmentally non-toxic products using microalgae. *Colloids and Surface, B, 105*, 207–214.
51. Wang, H.-C., Cheng, H.-Y., Wang, S.-S., Cui, D., Han, J.-L., Hu, Y.-P., Su, S.-G., & Wang, A.-J. (2016). Effcient treatment of azo dye containing wastewater in a hybrid acidogenic bioreactor stimulated by biocatalyzed electrolysis. *Journal of Environmental Sciences (China), 39*, 198–207.
52. Kalyani, D. C., Patil, P. S., Jadhav, J. P., & Govindwar, S. P. (2008). Biodegradation of reactive textile dye Red BLI by an isolated bacterium Pseudomonas sp. SUK1. *Bioresource Technology, 99*, 4635–4641.
53. Saratale, R. G., Saratale, G. D., Chang, J. S., & Govindwar, S. P. (2011). Bacterial decolorization and degradation of azo dyes: A review. *Journal of the Taiwan Institute of Chemical Engineers, 42*, 138–157.
54. Sharma, B., Dangi, A. K., & Shukla, P. (2018). Contemporary enzyme based technologies for bioremediation: A review. *Journal of Environment Management, 210*, 10–22.

55. Munck, C., Thierry, E., Graßle, S., Chen, S. H., & Ting, A. S. Y. (2018). Bioflm formation of flamentous fungi *Coriolopsis* sp. on simple muslin cloth to enhance removal of triphenylmethane dyes. *Journal of Environment Management, 214*, 261–266.
56. Du, L. N., Zhao, M., Li, G., Xu, F. C., Chen, W. H., & Zhao, Y. H. (2013). Biodegradation of malachite green by Micrococcus sp. strain BD15: Biodegradation pathway and enzyme analysis. *International Biodeterioration and Biodegradation, 78*, 108–116.
57. Deng, D., Lamssali, M., Aryal, N., et al. (2020). Textiles wastewater treatment technology: A review. *Water Environment Research, 92*, 1805–1810. https://doi.org/10.1002/wer.1437
58. Bhandari, S., Poudel, D. K., Marahatha, R., et al. (2021). Microbial enzymes used in bioremediation. *Journal of Chemistry, 2021*, 1–17. https://doi.org/10.1155/2021/8849512
59. Sarkar, S., Banerjee, A., Halder, U., et al. (2017). Degradation of synthetic azo dyes of textile industry: A sustainable approach using microbial enzymes. *Water Conservation Science and Engineering, 2*, 121–131. https://doi.org/10.1007/s41101-017-0031-5
60. Arregui, L., Ayala, M., Gómez-Gil, X., et al. (2019). Laccases: Structure, function, and potential application in water bioremediation. *Microbial Cell Factories, 18*, 200. https://doi.org/10.1186/s12934-019-1248-0
61. Lin, S., Wei, J., Yang, B., et al. (2022). Bioremediation of organic pollutants by white rot fungal cytochrome P450: The role and mechanism of CYP450 in biodegradation. *Chemosphere, 301*, Article 134776. https://doi.org/10.1016/j.chemosphere.2022.134776
62. Sharma, A., Sharma, T., Sharma, T., et al. (2019). Role of microbial hydrolases in bioremediation. (pp. 149–164).
63. Kumar, L., & Bharadvaja, N. (2019). Enzymatic bioremediation: A smart tool to fight environmental pollutants. In *Smart bioremediation technologies* (pp. 99–118). Elsevier.
64. Rathour, R. K., Sharma, D., Ullah, S., Mahmoud, E. H., Sharma, N., Kumar, P., Bhatt, A. K., Ahmad, I., & Bhatia, R. K. (2024). Bacterial–microalgal consortia for bioremediation of textile industry wastewater and resource recovery for circular economy. *Biotechnology for the Environment, 1*(1), 6. https://doi.org/10.1186/s44314-024-00005-2
65. Chen, X., Ning, X., Lai, X., Wang, Y., Zhang, Y., & He, Y. (2021). Chlorophenols in textile dyeing sludge: Pollution characteristics and environmental risk control. *Journal of Hazardous Materials, 416*, Article 125721. https://doi.org/10.1016/j.jhazmat.2021.125721
66. Bala, S., Garg, D., Thirumalesh, B. V., Sharma, M., Sridhar, K., Inbaraj, B. S., & Tripathi, M. (2022). Recent strategies for bioremediation of emerging pollutants; a review for a green and sustainable environment. *Toxics, 10*, 484. https://doi.org/10.3390/toxics10080484
67. Zafar, S., Bukhari, D. A., & Rehman, A. (2022). Azo dyes degradation by microorganisms—An efficient and sustainable approach. *Saudi Journal of Biological Sciences, 29*(12), Article 103437.
68. Hsueh, C. C., & Chen, B. Y. (2007). Comparative study on reaction selectivity of azo dye decolorization by *Pseudomonas luteola. HAZMAT, 141*(3), 842–849.
69. Chowdhary, P., Shukla, G., Raj, G., Ferreira, L. F., & Bharagava, R. N. (2019). Microbial manganese peroxidase: A ligninolytic enzyme and its ample opportunities in research. *SN Applied Sciences, 1*(1), 45. https://doi.org/10.1007/s42452-018-0046-3
70. Jamee, R., & Siddique, R. (2019). Biodegradation of synthetic dyes of textile effluent by microorganisms: An environmentally and economically sustainable approach. *European Journal of Microbiology and Immunology (Bp), 9*(4), 114–118. https://doi.org/10.1556/1886.2019.00018
71. Ajaz, M., Rehman, A., Khan, Z., Nisar, M. A., & Hussain, S. (2019). Degradation of azo dyes by *Alcaligenes aquatilis* 3c and its potential use in the wastewater treatment. *Amb Express*, 1–12
72. Sponza, D. T., & Işik, M. (2002). Decolorization and azo dye degradation by anaerobic/aerobic sequential process. *Enzyme and Microbial Technology, 31*(1–2), 102–110.
73. Saha, P., & Rao, K. V. (2024). Biodegradation of commercial textile reactive dye mixtures by industrial effluent adapted bacterial consortium VITPBC6: A potential technique for treating textile effluents. *Biodegradation, 35*(2), 173–193. https://doi.org/10.1007/s10532-023-10047-0

74. Sutherland, D., & Ralph, P. J. (2019). Microalgae in wastewater treatment: A review of emerging technologies for the treatment of emerging contaminants. *Science of the Total Environment, 681*, 1–12.
75. Singh, H. (2006). *Mycoremediation: Fungal bioremediation.* Wiley-Interscience.
76. Anastasi, A., Varese, G. C., & Marchisio, V. F. (2010). Fungal biodiversity and mycoremediation of polycyclic aromatic hydrocarbons. *Mycological Progress, 9*(3), 457–465.
77. Harms, H., Schlosser, D., & Wick, L. Y. (2011). Untapped potential: Exploiting fungi in bioremediation of hazardous chemicals. *Nature Reviews Microbiology, 9*(3), 177–192.
78. Deshmukh, R., Khardenavis, A. A., & Purohit, H. J. (2016). Diverse metabolic capacities of fungi for bioremediation. *Indian Journal of Microbiology, 56*(3), 247–264. https://doi.org/10.1007/s12088-016-0584-6
79. Huang, D. L., Zeng, G. M., Jiang, X. Y., & Feng, C. L. (2016). Mycoremediation of organic pollutants: A review. *Journal of Hazardous Materials, 301*, 1–13.
80. Das, N., & Chandran, P. (2011). Microbial degradation of petroleum hydrocarbon contaminants: An overview. *Biotechnology Research International,* 941810. https://doi.org/10.4061/2011/941810
81. Gentry, T. J., Rensing, C., & Pepper, I. L. (2004). New approaches for bioaugmentation as a remediation technology. *Critical Reviews in Environmental Science and Technology.* https://doi.org/10.1080/10643380490452362
82. Abatenh, E., Gizaw, B., Tsegaye, Z., & Wassie, M. (2017). The role of microorganisms in bioremediation: A review. *Open Journal of Environmental Biology, 2*(1), 38–46.
83. Chandra, R., Bharagava, R. N., Yadav, S., & Mohan, D. (2018). Accumulation and detoxification of heavy metals through microorganisms: Mechanisms and approaches. *Environmental Engineering & Management Journal.*
84. Azubuike, C. C., Chikere, C. B., & Okpokwasili, G. C. (2016). Bioremediation techniques–classification based on site of application: Principles, advantages, limitations, and prospects. *World Journal of Microbiology and Biotechnology, 32*, 1–18.
85. Megharaj, M., Naidu, R., & Sethunathan, N. (2011). Bioavailability and bioremediation of organic contaminants in soil. *Advances in Agronomy.*
86. Ghosal, D., Ghosh, S., Dutta, T. K., & Ahn, Y. (2016). Current state of knowledge in microbial degradation of polycyclic aromatic hydrocarbons (PAHs): A review. *Frontiers in Microbiology.* https://doi.org/10.3389/fmicb.2016.01369
87. Singh, A., Kuhad, R. C., & Ward, O. P. (2011). Advances in applied bioremediation. *Soil Biology Series.* https://doi.org/10.1007/978-3-540-89621-0
88. Pande, V., Pandey, S. C., Sati, D., Bhatt, P., & Samant, M. (2022). Microbial interventions in bioremediation of heavy metal contaminants in agroecosystem. *Frontiers in Microbiology, 13*, Article 824084. https://doi.org/10.3389/fmicb.2022.824084
89. Velusamy, S., Roy, A., Sundaram, S., & Kumar, M. T. (2021). A review on heavy metal ions and containing dyes removal through graphene oxide-based adsorption strategies for textile wastewater treatment. *The Chemical Record, 21*(7), 1570–1610.

Biodecontamination of Textile Waste Through Microbial Enzymes

J. Immanuel Suresh and S. Divyeswari

1 Introduction

The textile industry is one of the oldest and largest industries globally, which contributes to economic development and employment. However, it is also a major source of environmental pollution throughout the lifecycle of garments, from synthetic fiber production and dye usage to waste disposal. The industry generates complex wastewater containing dyes, chemicals, and other pollutants, making it one of the most polluting industrial sectors. Textile effluents disrupt ecological balance, affecting aquatic life, and poses health risks due to their toxic and carcinogenic properties [1]. Fast fashion trends exacerbate this issue by promoting mass production, rapid turnover, and excessive waste generation. Discarded textiles, often containing synthetic fibers and chemical treatments, present a significant environmental challenge due to their resistance to degradation. Although natural fibers like cotton and wool biodegrade more readily, the processing and dyeing of these fabrics can hinder this process [2].

Need for Novel Strategies in Textile Waste Treatment

Conventional effluent treatment methods include physical, chemical, and biological approaches. However, conventional methods often face limitations such as high costs, toxic by-products, and incomplete pollutant removal. Traditional wastewater treatment methods often fall short in effectively managing the solid and liquid waste generated by textile production, especially in the wet processing stage, which is a major source of pollutant discharge [3]. Its operations generate substantial amounts

J. Immanuel Suresh (✉) · S. Divyeswari
Department of Microbiology, The American College, Madurai, Tamil Nadu, India
e-mail: immanuelsuresh@americancollege.edu.in

R. A. Bhat et al. (eds.), *Biocentric Approaches in Textile Waste Management*, Textile Science and Clothing Technology, https://doi.org/10.1007/978-3-032-04974-2_12

of wastewater containing toxic chemicals and dyes, particularly azo dyes, constituting 60–70% of all dyes used. These dyes are notorious for their complex chemical structures, making them resistant to conventional degradation methods and posing severe risks to ecosystems and human health. Skin disorders, neurological complications, and aquatic toxicity are just some of the adverse effects associated with untreated textile effluents [4]. In recent years "bioremediation", a biological method leveraging microorganisms and their enzymes, has emerged as a sustainable and cost-effective solution. Microbial enzymes, such as laccases and peroxidases, can efficiently degrade dyes and other pollutants into non-toxic end products [5]. Microbial bioremediation techniques utilize microorganisms in various forms—such as dried powders, nanosized particles, whole cells or purified enzymes are used to perform functions like biosorption, where dyes are absorbed, or through enzymatic degradation, which breaks pollutants into non-toxic or simpler mineralized forms [6]. For example, key enzymes from bacteria and fungi such as azoreductase, laccase, and peroxidase have shown exceptional effectiveness in degrading complex dye molecules [7], while *Bacillus sp.*, and fungi like *Fusarium sp.*, specialize in breaking down cellulose in cotton, other enzymes such as keratinases for wool and protease degrade synthetic fibers like nylon [2]. Microbial enzymes, whether intracellular or extracellular, play a central role in this process. Known for their precision and efficiency, these enzymes target specific pollutants in textile wastewater without disrupting other components. Their versatility positions them as indispensable tools in bioremediation [6]. This chapter explores the potential of microorganisms involved and their enzymes used for bioremediation of textile waste. It examines the role of microbial systems, their enzymatic mechanisms, and the advancements driving eco-friendly solutions for textile waste treatment. By integrating biological innovations, the textile industry can move towards more sustainable practices, mitigating its environmental impact while preserving valuable resources (Fig. 1).

2 Textile Waste and Its Biodegradability

Biodegradable textile fibers, whether natural or synthetically modified, can be broken down by microbial enzymes, with degradation influenced by factors such as crystallinity, polymerization, and environmental conditions [8]. Natural fibers like cotton, linen, and wool vary in biodegradability, with linen degrading the fastest and wool being more resistant due to its complex molecular structure. Blended materials, such as linen/PET, show slower degradation, as PET resists enzymatic breakdown. Conditions such as soil temperature, moisture, and microbial activity impact the biodegradation of textiles. Cotton and rayon degrade rapidly under warm, moist conditions, with half-lives ranging from 22 to 40 days [9, 10]. In contrast, synthetic fibers like polyester and PLA resist degradation, with PLA remaining largely unchanged even after long burial. Microbial degradation of cotton, which contains cellulose, is primarily driven by fungi such as *Trichoderma sp*, *Fusarium sp*, and *Aspergillus spp.* However, the presence of azo dyes can hinder cotton degradation, as these

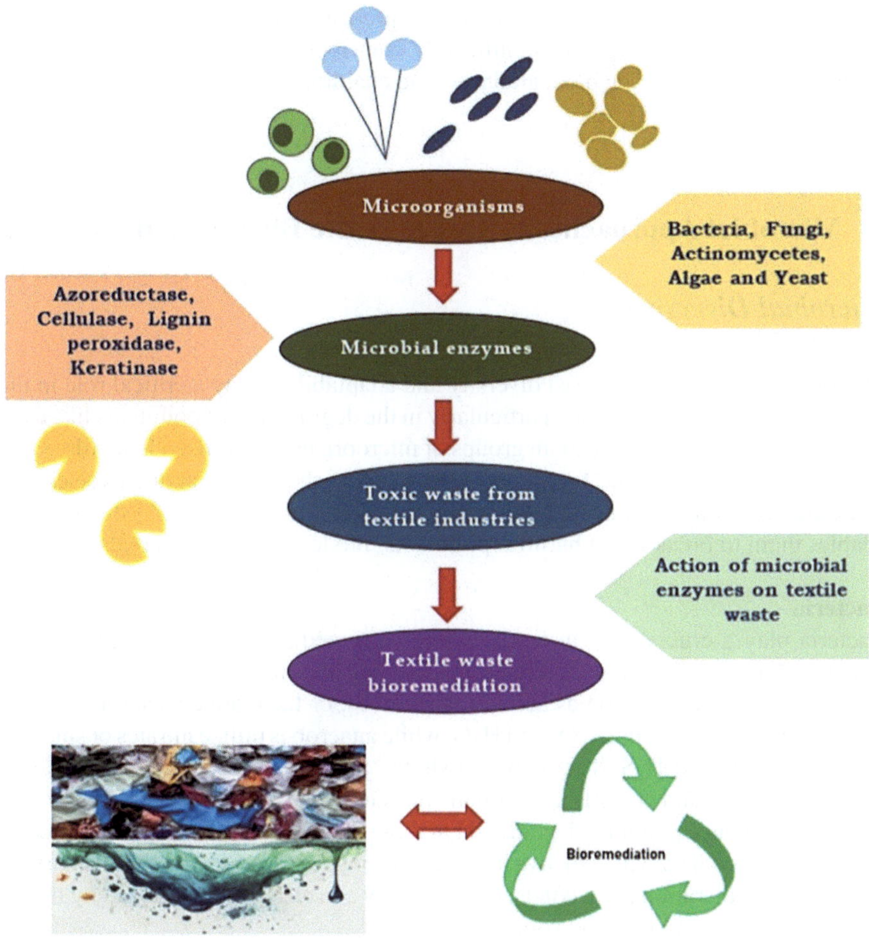

Fig. 1 Schematic presentation of microbial enzymes in textile waste remediation

synthetic dyes are more resistant to microbial reduction [11]. Fiber modification techniques, such as adding resins or finishes, can further slow biodegradation by making the fibers less hydrophilic and more resistant to microbial attack. Disposal and recycling of these textile waste, particularly in nonwoven products, benefit from understanding the biodegradation rates of various fibers. Fungi and bacterial species capable of hydrolyzing ester bonds in synthetic fibers like PET and nylon provide potential for developing enzymatic recycling strategies for textile waste [12]. Wool, which contains keratin, requires specialized enzymes for its degradation. Bacteria and fungi, such as *Bacillus sp* and *Chrysosporium sp*, are known to break down keratin, offering potential for bioremediation in wool textile waste [13]. Similarly, polyamide fibers like nylon can be degraded by enzymes from microbial species such as *Flavobacterium sp.* [14]. Bioremediation approaches, leveraging natural and engineered microbial enzymes, can significantly contribute to sustainable textile waste management by enhancing biodegradation and enabling the recycling of both natural and synthetic fibers [12].

3 Microbial Approaches to Textile Waste Bioremediation

Microbial Diversity

Microorganisms, with their vast diversity and adaptability, play a critical role in the bioremediation of textile waste, particularly in the degradation of pollutants like dyes and chemicals [4]. The three main groups of microorganisms utilized in textile waste treatment include bacteria, fungi, and yeasts, while the role of actinomycetes and algae are also being evaluated. Each group offers unique enzymatic systems which enables them to break down harmful pollutants into less toxic compounds.

Bacteria
Bacteria play a crucial role in textile waste bioremediation, converting pollutants into less harmful compounds. Heterotrophic bacteria like *Pseudomonas sp.* and *Alcaligenes sp.* can function as aerobes, anaerobes, or facultative bacteria. Aerobes degrade organic matter into CO_2 and H_2O, while anaerobes utilize nitrates or sulfates, producing CO_2 and H_2S. The activated sludge process relies on aerobic bacteria, offering efficient and cost-effective treatment. Genetically modified strains are also utilized to enhance pollutant degradation making them more efficient in bioremediation [15]. Bacterial enzymes such as laccases and oxygenases are effective in breaking down dyes by disrupting azo bonds, [16] and they can also assist in degrading textile fibers like cotton and nylon in solid textile waste.

Fungi
Fungi, belonging to a diverse kingdom of eukaryotes, are effective tools for bioremediation, particularly in treating textile industrial waste [15]. White-rot fungi, such as *Phanerochaete chrysosporium* and *Trametes versicolor*, produce extracellular enzymes like laccases, peroxidases, and cellulases to degrade dyes, phenolic compounds, and textile fibers. These fungi adapt to various conditions, making them efficient in breaking down recalcitrant dyes and complex organic materials like lignin and cellulose [3]. Filamentous fungi excel in solid textile waste remediation due to their extended hyphal growth, which aids in nutrient access. Wood-decay fungi, including white-rot and brown-rot fungi, degrade natural and semi-synthetic textiles like cotton, viscose, and polyester. White-rot fungi achieve lignocellulose mineralization, while brown-rot fungi utilize oxidative Fenton reactions to access cellulose. These versatile mechanisms make fungi as a key agents in textile bioremediation [17].

Yeast

Yeasts, though less studied than bacteria and fungi, have shown promise in dye decolorization. Enzymes like azoreductase in *Galactomyces geotrichum* [18] enable up to 92% degradation of acid scarlet GR [4], while *Candida albicans* transforms carcinogenic dyes like direct violet 51 into non-toxic compounds [19]. Yeast consortia also decolorize lignin-based dyes, demonstrating their utility in treating textile effluents [4].

Algae

Algae including both microalgae and macroalgae, offers a promising approach for textile waste bioremediation, especially in aquatic environments. Algae utilize photosynthesis for growth, and their ability to degrade textile pollutants comes through enzymatic degradation, biosorption, and bioassimilation processes. Adane et al. [3] discussed that species such as *Shewanella* algae degrade Acid Red 27, and *Enteromorpha sp.* removes Basic Red 46. While, Cladophora and Vaucheria adsorb Malachite Green and Acid Orange II, respectively. *Chlorella sp.*, and *Oscillatoria sp.*, possess enzymes like azo reductase [20] which utilize azo dyes as carbon and energy sources and degrades them into aromatic amines, these are further converted into simpler organic and inorganic compounds [6]. Algae's photoautotrophic nature minimizes external carbon requirements, offering cost-effective and sustainable wastewater treatment. Algal biomass is also explored as a alternative to activated carbon for biosorption, enhancing its value in bioremediation [3].

Actinomycetes

Actinomycetes' ability to grow in diverse environments and their robustness make them an attractive option for textile waste bioremediation. Species from the *Streptomyces* genus, are known for their ability to degrade various textile dyes through a combination of biosorption and biodegradation mechanisms. *Streptomyces sp.* strains, such as *Streptomyces sp.* S27, has demonstrated high efficiency in decolorizing azo and triphenylmethane dyes [21, 22]. These microorganisms also produce extracellular enzymes capable of breaking down dye molecules, contributing to the detoxification of textile effluents [4].

Mechanisms of Textile Pollutant Remediation

Textile waste remediation involves biosorption and biodegradation processes. Biosorption is a rapid, non-metabolic process where dyes bind to microbial cell walls. Here, dead cells are usually preferred due to their resilience and reusability. Factors like pH, biomass type, and dye chemistry can also impact the adsorption efficiency. Biosorbents from algae, bacteria, and fungi are one of the cost-effective solutions for bioremediation [23]. Biodegradation is also a critical process in bioremediation, begins with microorganisms acting as "biodestructors" by adhering to

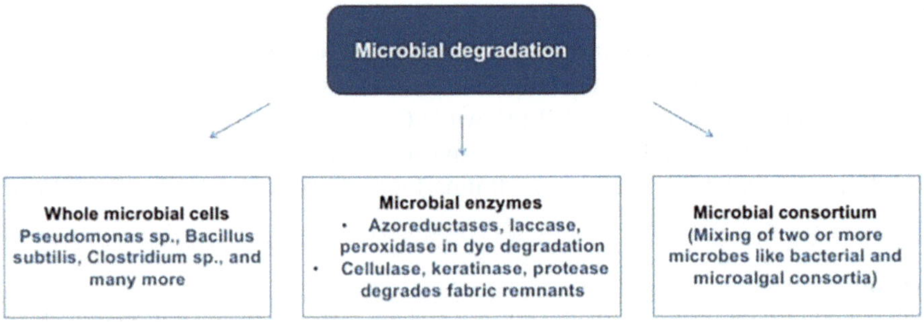

Fig. 2 Various methods involved in microbial degradation of textile waste

material surfaces through adhesion or aggregation. This triggers microbial proliferation, enzyme production, and polymer breakdown into simpler, degradable products. In soil, particularly its upper layers (5–15 cm), is rich in microorganisms, with up to 10^8 per gram, providing an ideal environment for biodegradation processes. Cellulose, which is the most abundant biopolymer, found in natural fibers like cotton, linen, and jute, is susceptible to microbial attack, especially in its amorphous regions. Their degradation rate depends on factors like fiber crystallinity, orientation, substitution, and other non-cellulosic substances. Bacteria and fungi are key agents, with bacteria degrading fibers from the surface inward, while fungi penetrating deeper layers [9]. Microorganisms use several mechanisms to degrade textile pollutants, each contributing to the efficient breakdown of harmful substances in textile waste. Microbial degradation can be facilitated by whole microbial cells, their enzymes, or microbial consortium. The enzymatic approach has gained widespread recognition over the past decade for its ability to yield non-toxic end products, reduced sludge generation, and enhanced reaction rates, making it a faster and more sustainable alternative to conventional methods. In textile effluents, dye degradation begins with the cleavage of the azo bond (–N=N–) by anaerobic azo reductase, converting the dye into corresponding amines. This process involves co-enzymes like FMN-independent reductases and NADPH-dependent reductases. The initial reductive cleavage results in a colorless solution, which is further broken down into amines and other by-products through oxidative cleavage [7] (Fig. 2).

4 Overview of Specific Microbial Enzymes in Textile Waste Treatment

Industrial enzymes are primarily obtained from animal (e.g., chymosin, pepsin, trypsin, pancreatin), plant (e.g., ficin, papain, bromelain), and microbial sources (e.g., peroxidase, laccase, cutinase). Microbial enzymes have emerged as indispensable tool in industries like pharmaceuticals, food, detergents, textiles, leather, paper, biofuels, and beverages due to their superior stability, activity, and large-scale production capabilities through fermentation. This has driven a significant focus on identifying novel microbial strains to develop enzymes with specific and enhanced properties [24]. Several notable microbial enzymes involved in textile waste management are,

Cellulase

Cellulose is an insoluble polysaccharide of D-glucose linked by β-1,4-glycosidic bonds, is hydrolyzed by cellulase enzymes found in diverse organisms. These enzymes are classified into endocellulase-cleaves internal bonds, exocellulase-processes chain ends into smaller sugars, and β-glucosidases [24]. Commercial cellulase blends these enzyme types for broader utility. Whilecellulases, derived from microbes like *Trichoderma reesei* (Celluclast) and *Humicola sp.* (Denimax Ultra), are becoming increasingly important in handling cotton textile waste. These enzymes help in processes like biostoning, biopolishing, and removing excess dye from fabrics, minimizing the environmental footprint of the textile industry [25]. Other cellulase-producing organisms, such as *Cellulomonas sp*, *Pseudomonas sp*, *E. coli*, *Bacillus sp*, and *Serratia marcescens*, are often isolated from alkaline soils and play a significant role in waste management [26]. These enzymes helps in degrading cotton-based textiles, de-ink fabrics, and improve fabric softness, all while reducing the impact of textile waste [27].

Peroxidase

Peroxidase are enzymes that reduce peroxides, such as hydrogen peroxide (H_2O_2), as part of antioxidative systems [25]. Found in bacteria, fungi, yeast, and actinomycetes, these enzymes are crucial for detoxifying heavy metals, degrading toxicants, regulating hormones, and catalyzing biochemical reactions in clinical and industrial applications. Their stability under extreme conditions of temperature, pH, and salinity enhances their use in pollutant decomposition, dye decolorization, lignin degradation, and wastewater treatment [5]. In dye decolorization, it provides a sustainable alternative to chemical bleaching agents like H_2O_2 and benzoyl peroxide. Lignin peroxidase (LiP) and manganese peroxidase (MnP), obtained from white rot fungi (*Phanerochaete chrysosporium*) and bacteria (*Serratia marcescens*), decolorize various synthetic dyes, including Reactive Orange-16 (92.52% in 10 days), Reactive-19, and ProcionBrilliant Blue-H-GR. *S. marcescens* also decolorizes up to 82% of real textile wastewater dyes [25]. Versatile peroxidase (VP) is a ligninolytic enzyme which has a combining activities of lignin peroxidase and manganese peroxidase. They are found in species like *Phanerochaete chrysosporium*, *Pleurotus pulmonarius*, and *Bjerkandera* sp., VP efficiently decolorizes recalcitrant dyes such as RBBR when paired with redox mediators. This hybrid enzyme system effectively targets dyes with diverse structures [20].

Laccase

Laccase belong to the oxidase enzyme family and features multicopper sites in their structure. They degrade a variety of textile dyes, including azo, anthraquinone, and triphenylmethane dyes. Through non-specific catalysis involving free radical mechanisms, and by acting on the phenolic groups of azo dyes, laccases breaks down N=N bonds, producing phenolic compounds and aromatic amines. Their optimal activity is generally observed within a pH range of 4–7, with some laccases functioning effectively at alkaline pH levels up to 9. Laccases remain active across a temperature range of 25–60 °C, with certain variants tolerating higher temperatures [28]. Known for their efficiency in breaking down lignin-based and persistent pollutants, laccases have become integral to several industries in treating waste effluents, especially in sectors like textiles, paper, and petrochemicals, they are also applied in polymer synthesis, soil remediation, and beverage processing. Their capacity to degrade synthetic dyes makes them a sustainable choice for decolorizing and treating textile wastewater [29].

Protease

Protease enzymes hydrolyze peptide bonds in proteinaceous by-products from industries like fisheries, leather, and poultry, as well as in dead animals. They are classified into endo- and exopeptidases based on their catalytic mechanism and active site position. Found in fungi (*Aspergillus sp*) and bacteria (*Amycolatopsis sp, Bacillus sp*), proteases are widely used in pharmaceuticals, food production, detergents, leather processing, and wastewater treatment due to their efficiency, cost-effectiveness, and high production. They also exhibit strong bioremediation potential by degrading polymers like PHB and breaking α-ester, γ–ω, and β-ester bonds [5]. Studies have shown that keratinase, a protease enzyme, effectively degrades keratin by disrupting disulfide hydrogen bonds in keratin proteins, making it valuable in applications like detergents, food, and leather processing. These enzymes are produced by microbes like *Bacillus subtilis*, which consists of two domains: an I9 inhibitor (19–77 amino acids) and a peptidase S8 (103–345 amino acids), with calcium ions serving as cofactors to enhance its activity. The enzyme efficiently degrades keratin-rich waste, such as green and animal waste, converting insoluble keratins into biodegradable substances, thus addressing environmental pollution caused by untreated keratin waste [30]. This multifunctionality has made them indispensable in food processing, leather production, detergent formulation, and in pharmaceuticals as well [27].

Azoreductase

Azoreductases are bacterial enzymes that degrade azo dyes by cleaving azo bonds (–N=N–), producing aromatic amines, which are further mineralized into CO_2 and H_2O under aerobic conditions. This reduction involves either direct interaction with dyes or indirect electron transfer via redox mediators like NADH, NADPH, FAD, and FMN. These flavoproteins are diverse in structure and function, either by acting in the cytoplasm, on membranes, or through extracellular ways. Some strains, such as *Pseudomonas putida* WLY, use extracellular azoreductases for dye decolorization, while *Sphingomonas sp.* BN6 shows membrane-bound activity. Though typically less effective in oxygen-rich environments, strains like *Citrobacter sp.* KCTC 18061P demonstrate efficient decolorization under both aerobic and anaerobic conditions [31].

Cutinase

Cutinases are extracellular esterases secreted by phytopathogenic fungi and pollen, which are responsible for hydrolyzing ester bonds in cutin, the structural polyester in plant cuticles. In addition, these enzymes can efficiently catalyze the breakdown of synthetic esters and triacylglycerols, exhibiting lipase-like activity without the need for interfacial activation. Their broad substrate specificity makes them valuable for applications in detergent formulations, triglyceride synthesis, polymerization processes, agrochemical production, and in biodegradation of plasticsas well [32]. Cutinases are also utilized for cotton fiber bioscouring and wax removal at low temperatures. Enzymes from *Fusarium solani* efficiently degrade cotton waxes within a short time, offering performance similar to solvent-based methods. They remain effective even in environments with high levels of detergents and also boost pectinase activity, aiding in the removal of pectin from textile fibers [25].

Other Enzymes

Catalase

Catalase are enzymes that breaks down residual hydrogen peroxide in textile bleaching processes, converting it into water and oxygen. Catalase produced by *Aspergillus fumigatus* is used to clean remnant biobleaching solutions containing peroxides [33]. This helps to reduce the chemical load in effluents, ensuring cleaner and safer discharge, while minimizing environmental impact [25].

Glucose Oxidase

These enzymes oxidize chromophores in textile effluents, transforming them into colorless, toxic substances. This process helps in reducing the environmental impact of textile effluents by making them safer and easier to treat, with improved sustainability in textile production. Glucose oxidases produced by *Bacillus subtilis* are used in biobleaching, as they act on chromophores and transform them into colorless products [25, 34].

5 Recombinant Microbial Systems for Efficient Enzyme Production

Enzymes for industrial applications are sourced from animals, plants, and microbes, but their natural abundance is often insufficient, leading to the adoption of recombinant techniques for large-scale production. Various hosts, including *Escherichia coli*, *Bacillus sp.*, *Saccharomyces cerevisiae*, *Pichia pastoris*, filamentous fungi, and mammalian cells, are utilized for heterologous protein expression. While *E. coli* remains a popular system due to its rapid growth, cost-efficiency, scalability, and genetic tools, but it lacks pathways for complex post-translational modifications, limiting its suitability for certain proteins. In such cases, alternative prokaryotic hosts like *Bacillus sp.*, and lactic acid bacteria (*Streptococcus sp.*, *Lactococcus sp.*, *Lactobacillus sp.*) are employed. For eukaryotic proteins, yeast systems like *Pichia pastoris* are preferred, offering high expression levels under the aox1 promoter, with efficient secretion, minimal endogenous contamination, scalability, and their ability to perform post-translational modifications such as glycosylation and proteolytic processing. Moreover, integrating multiple expression cassettes into the *P. pastoris* genome eliminates plasmid-associated challenges, making it a preferred system for biotechnological applications [32].

6 Applications of Microbial Enzymes in Textile Waste Remediation

Liquid Textile Waste Remediation

The management of textile waste, both liquid and solid, can be significantly improved using microbial enzymes for bioremediation. Liquid waste, generated during textile production, contains high levels of dyes, heavy metals, grease, and other recalcitrant chemicals which could pose a significant health hazard for both humans and aquatic environments.

Dyes

Textile dyes which primarily consists of azo groups are degraded by several key enzymes like azoreductases, laccases, lignin peroxidases, manganese peroxidases, tyrosinase, DCIP–NADH reductases, aminopyrine N-demethylase, and riboflavin reductases [31]. Bacterial azo reductases break azo bonds under anaerobic conditions but generate toxic aromatic amines, necessitating further mineralization. Advanced enzymes like dye-decolorizing peroxidase (DyP) from Trametes cucumeris, capable of degrading over 30 synthetic dyes, and versatile peroxidase (VP), which combines lignin and manganese peroxidase activities, offer broader degradation potential. Fungal laccases from *Trametes versicolor* and *Phlebia radiata* achieve over 80% dye decolorization without producing harmful by-products. Additionally, MnP from *Phanerochaete sordida* effectively degrades azo and anthraquinone dyes, while LiP from *P. chrysosporium* targets methylene blue. The incorporation of enzyme-redox mediator systems further enhances the sustainable treatment of textile effluents [20].

Heavy Metals

Textile, wool, and leather tanning industries generate effluents rich in toxic metals and metal complex dyes (e.g., Cr, Co, Cu complexes) from processes using mordants like dichromates. Bioremediation using microorganisms offers effective solutions for pollutant removal. *Bacillus amyloliquefaciens* and *Bacillus methylotrophicus* reduce Cr(VI) to less toxic Cr(III) via chromate reductase enzymes, with optimal activity at pH 7, while *Bacillus firmus* TE7 also oxidizes As(III) to As(V) [35]. *Shewanella sp.* J18 143 enzymatically degrades Cr and Co complex azo dyes, with maximum efficiency at pH 8–9 under controlled dye concentrations. For mercury, *Aspergillus flavus* accumulates Hg intracellularly, aided by oxidative enzymes and organic acids, with immobilized cells showing superior performance. These strategies demonstrate the potential of microbial and enzymatic approaches in mitigating heavy metal and dye pollution [36].

Microplastics

During washing, synthetic fibers release microplastics, which persist in aquatic environments due to plastics' long degradation times (50–600 years).Microbial enzymes offer promising solutions for plastic degradation in the textile industry, where synthetic fibers like polyethylene terephthalate (PET) dominate. Enzymes such as hydroxylases, laccases, peroxidases, and reductases from actinomycetes, bacteria, and fungi degrade polyethylene (PE), while cutinases, esterases, lipases, and ureases target polyurethane (PU). PET degradation involves hydrolases from *Thermobifida fusca* and cutinases and esterases from *Streptomyces scabies*. Polystyrene (PS) breakdown is facilitated by esterases from *Lentinus tigrinus* and from*Bacillus sp.*, and *Pseudomonas sp.*, as well. Nylon degradation relies on manganese peroxidase from white-rot fungi and hydrolases from *Flavobacterium sp.*, and *Pseudomonassp.*, These enzymes often extracellular ones, depolymerize plastics into monomers, while intracellular enzymes facilitates further mineralization of these pollutants from textiles [37].

Phenolic Compounds

Phenols and phenolic compounds, widely used in industries such as textiles, paper, plastic, and pharmaceuticals, which are toxic pollutants that pose environmental and health hazards, even at low concentrations. The enzymatic degradation of phenolic compounds has emerged as an effective bioremediation strategy. Immobilized laccase has demonstrated complete removal of 0.4 mM phenol in 40 min at 35 °C and pH 4.5 [38]. Additionally, peroxidases, with their haem cofactor and redox-activated cysteine/selenocysteine residues, are highly effective in transforming phenolic compounds due to their easily accessible active sites. Immobilized peroxidases have also shown the ability to degrade 99.9% of 1 mM phenol in 5 h under ambient temperature with pH 7.0 conditions [39, 40].

Solid Textile Waste Remediation

Solid waste from textile industries, including pre-consumer (fiber, yarn, and fabric residues) and post-consumer waste (natural and synthetic fibers, zippers, and buttons), presents a significant environmental challenge due to its slow degradation rate. Conventional methods like landfilling and incineration are widely used but have severe drawbacks, including leachate contamination, methane emissions, and the release of toxic compounds like dioxins during incineration. Recycling and microbial bioremediation provides sustainable alternatives. Pre-consumer waste, which is largely biodegradable, can be decomposed by microbial enzymes such as cellulases, proteases, and laccases. These enzymes break down fibers into reusable raw materials or convert them into value-added products like composites and adsorbents. For post-consumer waste, microbial treatments focus on degrading synthetic fibers like polyester, which are resistant to natural decomposition [41].

Cotton

Cotton waste, primarily composed of 95% cellulose, can be effectively treated using cellulases and oxidative enzymes. Cellulases, produced by bacteria like *Cellulomonas fimi Thermomonospora fusca* and fungi such as *Trichoderma spp.* and *Aspergillus spp.*, hydrolyze cellulose into fermentable sugars like glucose. Factors such as enzyme activity, substrate composition, and pre-treatment methods can influence hydrolysis efficiency. Further, Lytic polysaccharide monooxygenases (LPMOs) can enhance cellulose degradation by oxidizing glycosidic linkages, making fibers more susceptible to enzymatic hydrolysis [16].

Nylon

Polyamide 6.6 (nylon) degradation has been studied using laccase-mediator systems and proteases, with nylon oligomers biodegraded by microorganisms like *Flavobacterium sp.* and *Pseudomonas aeruginosa.* Lignolytic fungi oxidases also contribute to polyamide depolymerization [12]. Past studies have also shown that three lignin-degrading basidiomycetes (*Phanerochaete chrysosporium, Trametes versicolor*, and IZU-154) can degrade nylon-66 polymer under nutrient-limited conditions, similar to lignin degradation [42]. A nylon-degrading enzyme was purified from nitrogen-starved fungalIZU-154 cultures through chromatography techniques. Interestingly, this enzyme shares properties with manganese-dependent peroxidase (Mn peroxidase), a key lignin-degrading enzyme, but its nylon-degradation mechanism differs significantly. Unlike the Mn peroxidase reaction, which requires manganese, hydrogen peroxide, and an α-hydroxy acid (e.g., lactate), nylon degradation does not need exogenous H_2O_2 or lactate and is inhibited by lactate and superoxide dismutase. Nylon degradation involves oxidative cleavage of the CN and CC bonds near nitrogen atoms, producing end groups such as CHO, CH_3, and CONH. The process likely begins with enzymatic attack on methylene groups near nitrogen, followed by auto-oxidative reactions contributing to nylon degradation [43].

PET Fiber Waste

Being a synthetic polymer, Polyethylene terephthalate (PET) is widely used in textiles such as polyester fibers, blended fabrics, non-woven materials, and technical textiles like seat belts and geotextiles, poses environmental challenges due to its non-biodegradable nature and resistance to hydrolysis, especially in highly crystalline forms. Recent advancements in enzymatic treatments have highlighted hydrolases as effective tools for PET fiber waste management. Cutinases, belonging to carboxylic ester hydrolases, can hydrolyze PET, enhancing surface hydrophilicity and facilitating degradation. *Thermobifida fusca* [44] cutinase was the first enzyme identified to depolymerize amorphous PET, followed by thermostable actinomycetes cutinases like LC-cutinase. A breakthrough occurred with the discovery of *Ideonella sakaiensis* [45], a mesophilic bacterium producing PETase and MHETase, which synergistically hydrolyze PET into its monomers, terephthalic acid (TPA) and ethylene glycol (EG). Structural insights into PETase, including its unique disulfide bonds and substrate-binding cleft, have enabled engineering of variants with improved catalytic efficiency, including activity against biobased PET alternatives like polyethylene furanoate (PEF). Additionally LC-cutinase variants were capable of depolymerizing 90% of post-consumer PET at 72 °C, yielding TPA for recycling into high-quality PET products. These advancements, coupled with PET micronization prior to enzymatic treatment, are promising steps toward sustainable PET fiber waste valorization in the textile industry [16].

Leather

The leather industry generates substantial pollution due to the use of toxic chemicals during processes like dehairing, tanning, and pickling. A promising alternative is enzymatic dehairing using keratinase—a protease enzyme, which is capable of breaking down keratin—a highly stable, insoluble structural protein found in feathers, hair, and nails. Key microbes involved in keratin degradation include Bacillus species, such as *Bacillus licheniformis* and *Bacillus subtilis*, as well as gram-negative bacteria like *Pseudomonassp.* These keratinolytic microbes produce keratinase that catalyze the hydrolysis of the keratin proteins. The enzyme cleaves the strong disulfide bonds and peptide linkages that give keratin its structural integrity, making it resistant to most proteolytic enzymes [46]. These microbial enzymes hydrolyze keratin, providing an eco-friendly solution for processing feather waste and improving the sustainability of leather production.

Wool

Wool is a natural fiber harvested from sheep, goats, rabbits, and other animals as well. It goes through a series of processing steps to be converted into fabric suitable for making clothing [47]. The biodegradation resistance of wool keratin stems from its dense sulfur cross-links and a water-repellent membrane that prevents microbial and enzymatic penetration [2, 12]. Riessen and Antranikian identified *Thermoanaerobacter keratinophilus*, a bacterium that thrives at 70 °C under anaerobic, thermophilic conditions, utilizing wool as a carbon source [48]. This bacterium produces an extracellular protease which is capable of breaking down keratin in wool, with potential for use in the textile industry [49].

Silk

Silk is a natural proteinaceous fiber produced by *Bombyx mori*, primarily consists of fibroin, a filamentous protein, bound by sericin, a gum-like protein. Before weaving, sericin is removed through degumming. Fibroin is composed of heavy and light chains linked by a single disulfide bond. Silk biodegrades similarly to wool but is more susceptible to microbial enzymes due to fewer disulfide bonds. Proteases such as proteinase K and alkaline proteases from *Bacillus sp.*, efficiently hydrolyze its peptide bonds without requiring reducing agents. Silk also contains small peptides that inhibit bacterial and fungal proteinases, enhancing its durability making its degradation quite difficult. Due to the structural similarity between silk's amino acids (glycine and L-alanine) and L-lactic acid in poly(L-lactide), enzymes like serine proteases from

Amycolatopsis sp., can degrade both silk fibroin and poly(L-lactide). This process releases lactic acid, which can be easily metabolized by microorganisms [49, 50].

Textile Waste Valorization

Textile waste valorization plays a vital role in the circular economy by transforming discarded textiles into valuable products. Cotton, due to its high cellulose content, is a promising feedstock for biorefineries to produce biofuels, biochemicals, and bioenergy. However, cotton-polyester blends require innovative methods to recover polyester and sugars effectively. Fermentation processes, such as solid-state fermentation (SSF) and submerged fermentation (SmF), have proven effective in producing cellulases for cellulose degradation. SSF is cost-efficient but requires enzyme extraction, while SmF achieves higher cellulase yields at the cost of greater energy inputs. Studies highlight cotton-based textile waste as a sustainable substrate, with *Aspergillus niger* yielding 0.4 FPU/g cellulase activity in SSF and *T. reesei* producing 18.75 FPU/g in SmF using cotton/polyester blends, marking a significant improvement [16]. These methods reduce enzyme costs and contribute to sustainable waste management (Table 1).

7 Recent Developments in Microbial Enzymatic Degradation of Textile Waste

In recent years, microbial enzymes has gained prominence as a sustainable approach for addressing textile waste pollution. With growing demand for eco-friendly and efficient wastewater treatment methods, significant advancements have been made. Several key innovations include the use of extremozymes, genetically modified microorganisms, enzyme immobilization, nanomaterials and phytoremediation to enhance degradation efficiency and system stability. Integration of microbial processes with physicochemical treatments has further improved the effectiveness of dye removal, with promising long-term solutions.

Enzyme Immobilization Techniques

Enzyme immobilization overcomes the limitations of free enzymes, such as instability and difficulty in recovery, by attaching them to solid supports. This enhances enzyme stability, activity, and reuse potential while protecting them from degradation. Immobilization methods depend on factors like enzyme type and operational needs [51]. Immobilized enzymes like laccase and bilirubin oxidase show promise

Table 1 Applications of microbial enzymes in textile waste remediation

S. No.	Enzyme	Microbial source	Target pollutant	Function	References
1.	Azoreductase	*Galactomyces geotrichum*	Acid scarlet GR	Cleavage of azo bonds	[18]
2.	Laccase	*Trametes versicolor* and *Phlebia radiata*	Dyes and phenolic compounds	Oxidizes organic substrates	[20, 38]
3.	Peroxidase (lignin peroxidase, manganese peroxidase)	*Phanerochaete chrysosporium* and *Serratia marcescens,*	Reactive-107, Reactive-19, Ranocid fast blue, Procion brilliant blue-H-GR	Converts peroxides into harmless forms	[25]
4.	Chromate reductase	*Bacillus amyloliquefaciens*	Heavy metals	Reduces Cr(VI) to less toxic Cr(III)	[35]
5.	Cutinase	*Thermobifida fusca*and*Ideonella sakaiensis*	PET fiber	Hydrolyzes PET into monomers	[44, 45]
6.	Cellulase	*Cellulomonas fimi*and*Trichoderma spp.,*	Cotton	Breaks down cellulose into glucose	[16]
7.	Protease (keratinase)	*Bacillus sp., Thermoanaerobacter keratinophilus* and *Amycolatopsis sp.*	Leather, wool and silk	Degrades keratin by breaking disulfide bonds	[46, 48, 50]
8.	Catalase	*Aspergillus fumigatus*	Remnant biobleaching solution	Breaks down residual hydrogen peroxide	[33]

in dye decolorization for wastewater treatment. Laccase immobilized on magnetic chitosan and Fe_3O_4 nanoparticles effectively decolorized dyes like Direct Blue 15, Evans Blue, and Congo Red, with retained activity over multiple cycles. Bilirubin oxidase immobilized on silica material exhibited excellent stability and efficacy under varying pH conditions, emphasizing their role in sustainable dye remediation [5].

Extremozymes

Extremophiles are microorganisms that thrive in extreme conditions such as high temperature, salinity, pH, and nutrient scarcity, producing specialized enzymes called extremozymes which have significant industrial and biotechnological applications. For instance, cellulase, widely used in bioethanol production, textile biopolishing, and denim biostoning, is produced by extremophiles like *Pseudoalteromonas sp.* DY3 and *Thermus thermophilus* HB8. Similarly, thermostable laccases from *Bacillus subtilis, Aquifex aeolicus* and *Thermus thermophilus* HB strain 27., is utilized in dye decolorization and biomass delignification, with thermostable variants optimized through heterologous expression in *E. coli* for cost-effective production [52]. Another thermostable amidase (TvgC) from *Thermovenabulum gondwanense*, degrades nylon-6 and nylon-6,6 with a 1.2 wt% conversion rate for nylon-6, matching engineered nylonases. Its high stability (93 °C melting point) makes it ideal for nylon recycling [53].

Genetic Engineering

Microbial engineering tackles synthetic fiber degradation by focusing on hydrolysable bonds for enzyme-catalyzed depolymerization. For instance, engineered hydrolases such as PETase and MHETase from *Ideonella sakaiensis* degrade polyethylene terephthalate (PET), while site-directed mutagenesis enhances enzyme efficiency [2]. Through recombinant DNA technology (RDT), laccase enzymes have been engineered for higher yield and stability. A recombinant laccase, expressed in *Pichia pastoris* GS115 from *E. coli* K12 (pHBM905BDM), achieved 98.1% decolorization of Congo red and 98.5% of Malachite green with mediators, and 90.03% of Remazol BBR without mediators. This enzyme also effectively treated textile wastewater [54]. Additionally, laccase from *Bacillus vallismortis* fmb-103, expressed in *E. coli* BL21 DE3, demonstrated improved enzymatic activity, stability across pH ranges, and thermal resilience, aiding in Malachite green decolorization with mediators, highlighting its potential for bioremediation [5].

8 Conclusion

Microbial enzymes provide a sustainable solution for managing textile industry waste, offering advantages like adaptability, cost-effectiveness, and continuous availability. They efficiently degrade pollutants, transform recalcitrant dyes, and enable eco-friendly wastewater treatment while reducing chemical, water, and energy use. Advances in genetic engineering, synthetic biology, and enzyme immobilization are driving innovation, enhancing industrial applications and promoting a bio-based economy. Addressing research gaps in degradation pathways, process inhibitors, and by-product safety through interdisciplinary collaboration will pave the way for sustainable waste management and a greener future.

References

1. Gupta, N., Tripathi, A. K., & Harsh, N. S. K. (2011). Bioremediation of cotton-textile effluent using fungi. *Bulletin of Environment, Pharmacology and Life Sciences, 1*(1), 15–19.
2. Tripathi, M., Sharma, M., Bala, S., Thakur, V. K., Singh, A., Dashora, K., Hart, P., & Gupta, V. K. (2024).Recent technologies for transforming textile waste into value-added products: A review. In: *Current Research in Biotechnology, 7*, Article 100225. Advance online publication. https://doi.org/https://doi.org/10.1016/j.crbiot.2024.100225.
3. Adane, T., Adugna, A. T., & Alemayehu, E. (2021). Textile industry effluent treatment techniques. *Journal of Chemistry, 1–14.* https://doi.org/10.1155/2021/5314404
4. Tripathi, M., Singh, S., Pathak, S., Kasaudhan, J., Mishra, A., Bala, S., Garg, D., Singh, R., Singh, P., Singh, P. K., Shukla, A. K., & Pathak, N. (2023). Recent strategies for the remediation of textile dyes from wastewater: A systematic review. *Toxics, 11*(11), 940. https://doi.org/10.3390/toxics11110940
5. Rathore, S., Varshney, A., Mohan, S., & Dahiya, P. (2022). An innovative approach of bioremediation in enzymatic degradation of xenobiotics. *Biotechnology & Genetic Engineering Reviews, 38*(1), 1–32. https://doi.org/10.1080/02648725.2022.2027628
6. Patel, H., Yadav, V. K., Yadav, K. K., Choudhary, N., Kalasariya, H., Alam, M. M., Gacem, A., Amanullah, M., Ibrahium, H. A., Park, J.-W., et al. (2022). A recent and systemic approach towards microbial biodegradation of dyes from textile industries. *Water, 14*, 3163. https://doi.org/10.3390/w14193163
7. Kathing, C., & Saini, G. (2022). A review of various treatment methods for the removal of dyes from textile effluent. *Recent Progress in Materials, 4*(4), 21. https://doi.org/10.21926/rpm.2204028
8. Satyanarayana, K. G., Arizaga, G. G. C., & Wypych, F. (2009). Biodegradable composites based on lignocellulosic fibers-an overview. *Progress in Polymer Science, 34*, 982–1021.
9. Arshad, K., Skrifvars, M., Vivod, V., Volmajer, J., & Voncina, B. (2014). Biodegradation of natural textile materials in soil. *Tekstilec., 57*, 118–132. https://doi.org/10.14502/Tekstilec2014.57.118-132
10. Warnock, M., Davis, K., Wolf, D., & Gbur, E. (2009). Biodegradation of three cellulosic fabrics in soil. *Summ. Ark. Cotton Res, 582*, 208–211.
11. Szostak-Kotowa, J. (2004). Biodeterioration of textiles. *International Biodeterioration and Biodegradation, 53*, 165–170.
12. Wojnowska-Baryła, I., Bernat, K., & Zaborowska, M. (2022). Strategies of recovery and organic recycling used in textile waste management. *International Journal of Environmental Research and Public Health, 19*(10), 5859. https://doi.org/10.3390/ijerph19105859
13. Lange, L., Huang, Y., & Busk, P. K. (2016). Microbial decomposition of keratin innature-a new hypothesis of industrial relevance. *Applied Microbiology and Biotechnology, 100*(5), 2083–2096. https://doi.org/10.1007/s00253-015-7262-1
14. Negoro, S. (2000). Biodegradation of nylon oligomers. *Applied Microbiology and Biotechnology, 54*, 461–466.
15. Thirumalaivasan, N., Gnanasekaran, L., Kumar, S., Durvasulu, R., Sundaram, T., Rajendran, S., Nangan, S., & Kanagaraj, K. (2024). Utilization of fungal and bacterial bioremediation techniques for the treatment of toxic waste and biowaste. *Frontiers in Materials, 11*, 1416445. https://doi.org/10.3389/fmats.2024.1416445
16. Subramanian, K., Sarkar, M. K., Wang, H., Qin, Z.-H., Chopra, S. S., Jin, M., Kumar, V., Chen, C., Tsang, C.-W., & Lin, C. S. K. (2022). An overview of cotton and polyester, and their

blended waste textile valorisation to value-added products: A circular economy approach—Research trends, opportunities and challenges. *Critical Reviews in Environmental Science and Technology, 52*(21), 3921–3942. https://doi.org/10.1080/10643389.2021.1966254
17. Harper, R., & Moody, S. C. (2023). Filamentous fungi are potential bioremediation agents of semi-synthetic textile waste. *Journal of Fungi (Basel, Switzerland), 9*(6), 661. https://doi.org/10.3390/jof9060661
18. Guo, G., Tian, F., Zhao, Y., Tang, M., Liu, W., Liu, C., Xue, S., Kong, W., Sun, Y., & Wang, S. (2019). Aerobic decolorization and detoxification of acid scarlet GR by a newly isolated salt-tolerant yeast strain *Galactomyces geotrichum* GG. *International Biodeterioration and Biodegradation, 145*, 104818.
19. Vitor, V., & Corso, C. R. (2008). Decolorization of textile dye by *Candida albicans* isolated from industrial effluents. *Journal of Industrial Microbiology & Biotechnology, 35*(11), 1353–1357. https://doi.org/10.1007/s10295-008-0435-5
20. Palanivelan, R., Rajakumar, S., Jayanthi, P., & Ayyasamy, P. M. (2013). Potential process implicated in bioremediation of textile effluents: A review. *Advances in Applied Science Research., 4*, 131–145.
21. Adenan, N. H., Lim, Y. Y., & Ting, A. S. Y. (2022). Removal of triphenylmethane dyes by *Streptomyces bacillaris:* A study on decolorization, enzymatic reactions and toxicity of treated dye solutions. *Journal of Environmental Management, 318*, 115520. https://doi.org/10.1016/j.jenvman.2022.115520
22. Dong, H., Guo, T., Zhang, W., Ying, H., Wang, P., Wang, Y., & Chen, Y. (2019). Biochemical characterization of a novel azo-reductase from *Streptomyces sp.* application in eco-friendly decolorization of azo-dyes wastewater. *International Journal of Biological Macromolecules, 140*, 1037–1046.
23. Sudarshan, S., Harikrishnan, S., RathiBhuvaneswari, G., Alamelu, V., Aanand, S., Rajasekar, A., & Govarthanan, M. (2023). Impact of textile dyes on human health and bioremediation of textile industry effluent using microorganisms: Current status and future prospects. *Journal of Applied Microbiology, 134*(2), lxac064. https://doi.org/10.1093/jambio/lxac064
24. Tatta, E. R., Imchen, M., Moopantakath, J., & Kumavath, R. (2022). Bioprospecting of microbial enzymes: Current trends in industry and healthcare. *Applied Microbiology and Biotechnology, 106*(5–6), 1813–1835. https://doi.org/10.1007/s00253-022-11859-5
25. Sen, A., Kapila, R., Chaudhary, S., & Nigam, A. (2021). Biotechnological applications of microbial enzymes to replace chemicals in the textile industry-a review. *Journal of the Textile Association, 82/2*, 68–73.
26. Sethi, S., Datta, A., Gupta, B. L., & Gupta, S. (2013). Optimization of cellulase production from bacteria isolated from soil. *ISRN Biotechnology, 2013*, 1–7. https://doi.org/10.5402/2013/985685
27. Pandey, K., Singh, B., Pandey, A. K., Badruddin, I. J., Pandey, S., Mishra, V. K., & Jain, P. A. (2017). Application of microbial enzymes in industrial waste water treatment. *International Journal of Current Microbiology and Applied Sciences, 6*, 1243–1254. https://doi.org/10.20546/ijcmas.2017.608.151
28. Das, S., Cherwoo, L., & Singh, R. (2023). Decoding dye degradation: Microbial remediation of textile industry effluents. *Biotechnology notes (Amsterdam, Netherlands), 4*, 64–76. https://doi.org/10.1016/j.biotno.2023.10.001
29. Hasan, M., Nabi, F., & Mahmud, R. (2015). Benefits of enzymatic process in textile wet processing. *International Journal of Fiber Textile Research, 5*, 16–19.
30. Razzaq, A., Shamsi, S., Ali, A., Ali, Q., Sajjad, M., Malik, A., & Ashraf, M. (2019). Microbial proteases applications. *Frontiers in Bioengineering and Biotechnology, 7*, 110. https://doi.org/10.3389/fbioe.2019.00110
31. Muhammad, I., David, C., Azeem, K., Sabir, H., Muhammad, M., & Muhammad, A. (2014). Microbial biotechnology for decolorization of textile wastewaters. *Reviews in Environmental Science and Bio/Technology., 14*. https://doi.org/10.1007/s11157-014-9344-4
32. Araujo, R., Casal, M., & Cavaco-Paulo, A. (2008). Application of enzymes for textile fibres processing. *Biocatalysis and Biotransformation, 26*, 332–349.

33. Mehra, S., & Chadha, P. (2020). Bioaccumulation and toxicity of 2-naphthalene sulfonate: An intermediate compound used in textile industry. *Toxicology Research, 9*(2), 127–136.
34. Kahoush, M., Behary, N., Cayla, A., Mutel, B., Guan, J., & Nierstrasz, V. (2020). Influence of remote plasma on PEDOT: PSS-coated carbon felt for improved activity of glucose oxidase. *Journal of Applied Polymer Science, 137*(14), 48521. https://doi.org/10.1002/app.48521
35. Rath, B. P., Das, S., Das, M. P., & Thatoi, H. (2014). Optimization of extracellular chromate reductase production by *Bacillus amyloliquefaciens* (CSB 9) isolated from chromite mine environment. *Biocatalysis and Agricultural Biotechnology, 3*(3), 35–41.
36. Ghosh, A., Dastidar, M. G., & Sreekrishnan, T. R. (2016). Recent advances in bioremediation of heavy metals and metal complex dyes: Review. *Journal of Environmental Engineering., 142.* https://doi.org/10.1061/(ASCE)EE.1943-7870.0000965
37. Tang, K. H. D., Lock, S. S. M., Yap, P. S., Cheah, K. W., Chan, Y. H., Yiin, C. L., Ku, A. Z. E., Loy, A. C. M., Chin, B. L. F., & Chai, Y. H. (2022). Immobilized enzyme/microorganism complexes for degradation of microplastics: A review of recent advances, feasibility and future prospects. *The Science of the Total Environment, 832*, 154868. https://doi.org/10.1016/j.scitotenv.2022.154868
38. Fathali, Z., Rezaei, S., Faramarzi, M. A., & Habibi-Rezaei, M. (2019). Catalytic phenol removal using entrapped cross-linked laccase aggregates. *International Journal of Biological Macromolecules, 122*, 359–366. https://doi.org/10.1016/j.ijbiomac.2018.10.147
39. Wang, S., Fang, H., Yi, X., Xu, Z., Xie, X., Tang, Q., Ou, M., & Xu, X. (2016). Oxidative removal of phenol by HRP-immobilized beads and its environmental toxicology assessment. *Ecotoxicology and Environmental Safety, 130*, 234–239. https://doi.org/10.1016/j.ecoenv.2016.04.022
40. Feng, S., Hao Ngo, H., Guo, W., Woong Chang, S., Duc Nguyen, D., Cheng, D., Varjani, S., Lei, Z., & Liu, Y. (2021). Roles and applications of enzymes for resistant pollutants removal in wastewater treatment. *Bioresource Technology, 335*, 125278. https://doi.org/10.1016/j.biortech.2021.125278
41. Pensupa, N., Leu, S. Y., Hu, Y., Du, C., Liu, H., Jing, H., Wang, H., & Lin, C. S. K. (2017). Recent trends in sustainable textile waste recycling methods: Current situation and future prospects. *Topics in Current Chemistry (Cham), 375*(5), 76. https://doi.org/10.1007/s41061-017-0165-0
42. Deguchi, T., Kakezawa, M., & Nishida, T. (1997). Nylon biodegradation by lignin-degrading fungi. *Applied and Environmental Microbiology, 63*(1), 329–331. https://doi.org/10.1128/aem.63.1.329-331.1997
43. Nomura, N., Deguchi, T., Shigeno-Akutsu, Y., Nakajima-Kambe, T., & Nakahara, T. (2001). Gene structures and catalytic mechanisms of microbial enzymes able to biodegrade the synthetic solid polymers nylon and polyester polyurethane. *Biotechnology and Genetic Engineering Reviews, 18*(1), 125–147.
44. Muller, R. J., Schrader, H., Profe, J., Dresler, K., & Deckwer, W. D. (2005). Enzymatic degradation of poly (ethylene terephthalate): Rapid hydrolyse using a hydrolase from *T. Fusca. Macromolecular Rapid Communications, 26*(17), 1400–1405.
45. Yoshida, S., Hiraga, K., Takehana, T., Taniguchi, I., Yamaji, H., Maeda, Y., Toyohara, K., Miyamoto, K., Kimura, Y., & Oda, K. (2016). A bacterium that degrades and assimilates poly(ethylene terephthalate). *Science (NewYork, N.Y.), 351*(6278), 1196–1199. https://doi.org/10.1126/science.aad6359
46. Akhter, M., Wal Marzan, L., Akter, Y., & Shimizu, K. (2020). Microbial bioremediation of feather waste for keratinase production: An outstanding solution for leather Dehairing in tanneries. *Microbiology Insights, 13*, 1178636120913280. https://doi.org/10.1177/1178636120913280
47. Tesfaw, A., & Assefa, F. (2014). Applications of transglutaminase in textile, wool, and leather processing. *International Journal of Textile Science, 3*, 64–69. https://doi.org/10.5923/j.textile.20140304.02
48. Riessen, S., & Antranikian, G. (2001). Isolation of *Thermoanaerobacter keratinophilus* sp. nov., a novel thermophilic, anaerobic bacterium with keratinolytic activity. *Extremophiles: Life Under Extreme Conditions, 5*(6), 399–408. https://doi.org/10.1007/s007920100209

49. Fedorak, P. M. (2005). Microbial processes in the degradation of fibers. In R. S. Blackburn (Ed.), *Biodegradable and Sustainable Fibres* (pp. 1–35). Woodhead Publishing Limited. https://doi.org/10.1533/9781845690991.1
50. Tokiwa, Y., & Jarerat, A. (2004). Biodegradation of poly(L-lactide). *Biotechnology Letters, 26*(10), 771–777.
51. Abdelhamid, M. A. A., Khalifa, H. O., Yoon, H. J., Ki, M. R., & Pack, S. P. (2024). Microbial immobilized enzyme biocatalysts for multipollutant mitigation: Harnessing nature's toolkit for environmental sustainability. *International Journal of Molecular Sciences, 25*(16), 8616. https://doi.org/10.3390/ijms25168616
52. Priyanka, K., & Neeraj, W. (2022). Extremozymes used in textile industry. *The Journal of the Textile Institute, 113*(9), 2007–2015. https://doi.org/10.1080/00405000.2021.1948251
53. Hoffman, E. R., Rangaswamy, A. M. M., Cleveland, M. E., Keillor, J. W., & Howe, G. W. (2025). Targeted genome mining facilitates the discovery of a promiscuous, Hyperthermostable amidase from *Thermovenabulum Gondwanense* with notable nylon-degrading capacity. *Angewandte Chemie (International Ed. in English), 64*(2), e202414842. https://doi.org/10.1002/anie.202414842
54. Ma, X., Liu, L., Li, Q., Liu, Y., Yi, L., Ma, L., Zhai, C., et al. (2017). High-level expression of a bacterial laccase, CueO from *Escherichia coli* K12 in *Pichia pastoris* GS115 and its application on the decolorization of synthetic dyes. *Enzyme and Microbial Technology, 103*, 34–41. https://doi.org/10.1016/j.enzmictec.2017.04.004

Biochar for Remediation of Textile Industrial Waste

Vaishali Kumar and Vandana Singh

1 Introduction

The textile industry is a large business all over the world that is deeply involved in the life of humankind. Today, textiles are an essential component in our ways of such as clothing, design, household interior, and apparel. The global development process impacted many aspects of life by offering speed, efficiency, and convenience. Nevertheless, it has also produced adverse impacts associated with biosphere contamination resulting from unregulated pollutant emissions from many companies, particularly those handling hazardous and persistent substances. Dyes are used for the pigmentation of many materials, including textile fibers, paper, cosmetics, leather, food, and medicinal items (Fig. 1). Declining industries have relied on synthetic dyes since then and have expanded internationally, producing over 800,000 tons of synthetic dyes annually. The textile sector constitutes around 75% of the worldwide dyes economy and utilizes roughly ten thousand distinct dyes for the printing and/or coloring of various materials. Textile industries mostly reside in emerging nations such as India, Bangladesh, Sri Lanka, and Vietnam, where they augment job opportunities, bolster the economy, and generate foreign currency gains [6, 66]. The textile, paper, and leather industries contribute significantly to the discharge of colored effluent into aquatic environments. Dyes are vital substances used across several sectors, such as textiles, printing, cosmetics, and food, to give color to substrates as illustrated in Fig. 1. This is a type of business that involves intricate manufacturing, consumption processes, and logistics and which has a considerable impact on the worldwide economies provides employment opportunities, and facilitates international

V. Kumar
Department of Forensic Science, Sharda School of Allied Health Sciences, Sharda University, Greater Noida, India

V. Singh (✉)
Department of Microbiology, Sharda School of Allied Health Sciences, Sharda University, Greater Noida, India
e-mail: vandana.singh@sharda.ac.in

R. A. Bhat et al. (eds.), *Biocentric Approaches in Textile Waste Management*, Textile Science and Clothing Technology, https://doi.org/10.1007/978-3-032-04974-2_13

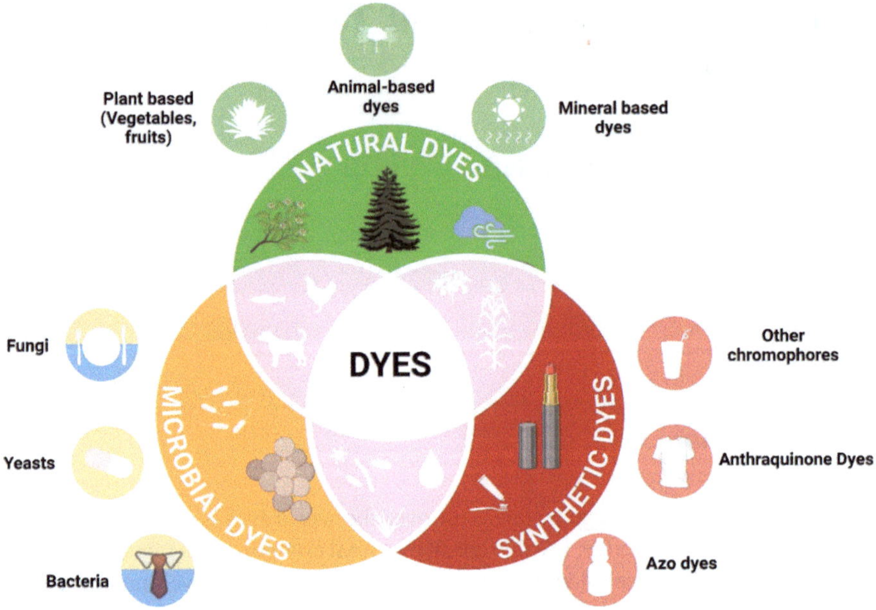

Fig. 1 Illustration of types of dyes; natural dyes, synthetic dyes, and microbial dyes that are used in different sectors such as cosmetics, textile, dietary supplements, etc.

commerce [4, 6]. One of the prominent sectors is concerned with considerable environmental issues related particularly to the processes involved in dyeing of textiles. The large scale of textile manufacturing points to the importance of textile manufacturing globally as an image of an integral part of the international economy and a main source of requisite goods for clothing and shelter. It has considerable influence over all the fields from what people taste the most up to material science or people [5, 26].

Synthetic dyes, mainly from petrochemicals, are preferred in most of the traditional textile dyeing (Fig. 1). While being such an exquisite dye in terms of durability, variety of colors, and economy, the manufacture and use of these dyes is a major cause of serious environmental hazards. Lots of water is used in the dyeing process and a considerable amount of dyes are not adhered to the cloth, resulting in highly pigmented wastewater. Such effluents are usually highly complicated in composition of chemicals that are difficult to degrade and therefore require difficult and expensive treatment [9, 23]. Dyes are chromophoric compounds that may chemically or physically interact with substrates, resulting in the selective absorption of certain light wavelengths and the manifestation of color. These compounds often have conjugated structures that contain π-electrons, which enhance their light-absorbing characteristics and dictate the spectrum of visible colors they emit. The absorption of light transpires via electron transitions within the conjugated circuit when the dye is subjected to energy, usually in the form of light or heat. A variety of synthetic dyes are known as mutagenic, carcinogenic, or hazardous to human health, and also a hazard to aquatic environments [4]. Polluted wastewater may be released into lakes, seas, and rivers and disrupt aquatic ecosystems, jeopardize biodiversity, and cause irreversible ecological harm. Heavy metals intensify the environmental problem that results from many manufactured colors. Brightly colored dye effluent has an aesthetic consequence that lessens the visual quality of natural lakes and streams and may affect tourism and leisure. There has been recent research on innovative dyeing techniques aimed at reducing environmental impact. Biodegradable dyes are used in low temperatures to decrease energy consumption in dyeing. The natural substances used in eco-printing give rise to unique patterns, while enzyme-assisted dyeing fabricates the color uptake of textile fibers using enzymes. Supplementary innovative techniques such as sono dyeing, plasma treatment, and bio-mordants [10, 49].

Given the concern for environmental sustainability and the cumulative effects of traditional textile dyeing methods, there is a need to focus on solving this problem. Biodegradable dye application has come up as a potential choice from all these options. Further pressure on the textile sector to switch from crude to more sustainable processing methods has been put on by the growing awareness of environmental concerns from customers and regulatory entities [23, 61, 66]. As a result, there has been a vast increase in research and development in trying to find a sustainable alternative to traditional dying methods. The allegedly less aggressive method of textile coloring with biodegradable dyes where the source lies in plants, microbes, insects, minerals, etc. This allows the natural decomposition of the dye wastewater and thus mitigates the lasting effects of the dye wastewater. However, the information suggests a leap forward towards bio-based colorants, which, depending on the development of this technology, allow to mitigate the environmental impacts resulting from the clothing sector and contribute to the development of a circular economy [25, 31, 49].

In this chapter excerpts of work in progress on the discovery and implementation of biodegradable textile dyes are provided, and consequently, an in-depth review of the progress made is given. Several sources of these colors will be investigated in this research: the extensive range of natural dyes made from both plants and animals, the chemical constituents of the various soluble dyes, and their properties and dyeing methods. The assessment of the environmental favorable of biodegradable dye demonstrates its potential to eliminate pollution and promote environmental sustainability. The problems and possibilities of deployment of biodegradable dyes in the textile dyeing industry were discussed in this chapter namely, customer acceptability, cost-effectiveness, and availability [9, 49, 66].

2 Natural Dyes

Natural dyes are derived from natural resources like plants, animals, and minerals. This can include indigo (extracted from the indigo plant), turmeric (cochineal, which is produced from insects), and many more. These dyes have been used for millennia, and are intrinsically biodegradable. However, they are easily limited in terms of color variety, consistency, and color preservation. The interest in natural dyes has once more come to life due to increased environmental consciousness and rising consumer demand for sustainable products. There have been various important components for research in this area [20, 39].

Extraction Methods and Analytical Techniques

Many natural sources, such as novel plant species, underutilized plant components (e.g., agricultural wastes), insects, have been recently examined for studying the source. For example, studies have placed a heavy emphasis on optimization of extraction method so as to produce higher yield and purity of dye using eco-friendly solvents and solvents such as ultrasound assisted extraction and supercritical fluid extraction (Table 1) [49]. A variety of studies have been done on the using of mordants (both conventional and biomordant) to improve dye fiber interaction and improve fastness properties. The two critical stages of the transformation of basic plant material to a useful natural dye are extraction and characterisation. The specific dye and its chemical properties influence the choice of the process for extraction strategies, with an attempt to obtain the desired colorant from the material it comes from [20, 39].

One of the most common and environmentally green techniques—aqueous extraction entails submerging or boiling plant matter unto water to extract the pigment. Solvent extraction uses organic solvents such as ethanol or methanol on the other hand if the colour has diminished solubility in water; the organic solvents are chosen carefully based on the dye's polarity. Depending on the pH adjustment extracted colors are enhanced or not, some need enzymatic extraction to penetrate and release bound colorants inside plant cell walls while others don't. The advanced technique of microwave assisted extraction (MAE) as well as supercritical fluid as extraction (SFE) offer the advantage of higher effectiveness and use of smaller amount of solvent [80]. Carbon dioxide is used in supercritical fluid extraction (SFE) and microwave radiation in microwave assisted extraction (MAE) to improve the extraction rate. The dye mixture is then extracted and the components are identified on the basis of analysis. Measurements and dyes are detected spectrophotometry such as UV vis spectroscopy by the measurement of light absorption. Isolated and identified individual color components of a mixture are done using chromatographic techniques like high performance liquid chromatography (HPLC), gas chromatography mass spectrometry (GC–MS), and thin layer chromatography (TLC). Extensive structural investigation is done on the dye molecules with mass spectrometry, infrared (IR) spectroscopy, and nuclear magnetic resonance (NMR). As the chromatic properties of the dye are measured with colorimetry, the physical shape of the dye is measured with microscopy [22, 56]. Biodegradation studies are done to find out the effect of the dye on the environment and elemental analysis is done to understand the elemental composition of the dye. The particular complexity of the natural dyes generally entails a much more thorough characterization requiring a synthesis of these analytical techniques. This method allows researchers to observe the chemical composition

Table 1 A comparison of natural, mineral, biodegradable synthetic, and conventional synthetic dyes, including examples, biodegradability, extraction methods, characterization techniques, and applications in the textile industry

Dye type	Subtype and example	Biodegradability	Method of extraction	Characterization	Application	References
Natural dyes	Plant-Based: Indigo (from Indigofera tinctoria)	Biodegradable (variable, depends on mordant)	Extraction from leaves via fermentation or chemical reduction	UV–vis spectroscopy, HPLC, TLC to identify indigotin and related compounds	Dyeing cotton, historically significant for denim	[2, 58]
	Plant-Based: turmeric (from curcuma longa)	Biodegradable	Extraction from rhizomes using solvents (e.g., ethanol)	HPLC, TLC to quantify curcuminoids	Food coloring, dyeing textiles (traditionally)	[19, 28, 60]
	Plant-based: henna (from lawsonia inermis)	biodegradable	Extraction from leaves by drying and grinding	Microscopic analysis (SEM), chemical tests for lawsone content	Hair dye, mehndi on the skin, polyester fabric dyeing	[50, 63, 70]
	Animal-based: cochineal (from *Dactylopius coccus*)	Biodegradable (chitin may degrade slower)	Extraction of carminic acid from insects	HPLC, spectrophotometry to quantify carminic acid	Food coloring (carmine), textile dyeing	[14, 16, 78]
	Mineral-based: ochre (natural clay)	Not biodegradable (mineral)	Mining and processing of clay	X-ray diffraction, and elemental analysis by AAS, ICP-MS, to determine mineral composition	Pigments for paints are historically used for some textiles	[12, 33, 52]
Mineral dyes/ pigments	Iron oxide (various forms)	Not biodegradable (mineral)	Mining, chemical synthesis	X-ray diffraction, colorimetry	Pigments for paints, coatings, specialized textile applications (canvas)	[4, 49]

(continued)

Table 1 (continued)

Dye type	Subtype and example	Biodegradability	Method of extraction	Characterization	Application	References
	Ultramarine (lapis lazuli/ synthetic)	Natural: not biodegradable; synthetic: mineral, not biodegradable	Natural: mining; synthetic: chemical synthesis	Spectroscopy, microscopy	Pigment for paints, historically used as a dye	[44, 49]
Biodegradable synthetic dyes	Azo Dyes (modified with ester linkages)—example: commercially available dye with specific CAS number	Biodegradable (designed to degrade)	Chemical synthesis	NMR, Mass Spectrometry, UV–Vis to confirm structure and purity. Biodegradation studies to assess breakdown products	Dyeing various textile fibers (cotton, polyester blends)	[32, 69]
	Anthraquinone Dyes (modified with amide linkages)—example: commercially available dye with specific CI number	Biodegradable (designed to degrade)	Chemical synthesis	NMR, Mass Spectrometry, UV–Vis to confirm structure and purity. Biodegradation studies	Dyeing textiles requiring bright shades and good fastness	[8, 13, 38]
Conventional synthetic dyes	Azo dyes—example: reactive blue 19	Not biodegradable	Chemical synthesis	NMR, Mass Spectrometry, UV–Vis to confirm structure and purity	Dyeing cotton and other cellulosic fibers	[17, 32]
	Anthraquinone dyes—Example: disperse blue 3	Not biodegradable	Chemical synthesis	NMR, Mass Spectrometry, UV–Vis	Dyeing polyester and other synthetic fibers	[15, 38, 47]
	Triphenylmethane dyes—example: crystal violet	Not biodegradable	Chemical synthesis	NMR, Mass Spectrometry, UV–Vis	Dyeing silk, wool, acrylics (limited use in textiles due to toxicity concerns in some applications)	[38, 72]

of the dye, improve the adaptability and guarantee the use of this dye ecologically and orthogonously. Advanced analytical techniques including HPLC–MS and NMR have been used by researchers to describe the chemical make-up of the extracted pigments. By focusing on the alteration of dyeing parameters (temperature, pH, duration) for various textile fibers (cotton, wool, silk) and specific colors and durability, research has been done. Giving interest towards utilization of several natural dyes on blended fabric and pre cleaning techniques to improve the dye absorption [21, 49, 66].

Potential Applications

Since natural dyes come with their own set of challenges, however, as they are a renewable and potentially less dangerous source of dye. The reality of research is that the range of colors available is limited, especially in the case of coloring those colors in which there is a vibrant and saturated. However, we are still not aware as much about the characteristics of fastness (especially wash, light) that are being done. Currently, the idea has been tested to examine the use of nanomaterials and biomordants to improve the fastness of natural dyes from a scientific aspect. Currently, via the use of life cycle assessment, an investigation of the environmental effects of natural dye extraction and application is being performed including water and energy. Some study on the scalability and cost of producing natural dyes has taken place Table 1 [4, 9, 36].

3 Synthetic Biodegradable Dyes

Due to chemical engineering, synthetic biodegradable they are genetically modified to match the qualities of traditional dyes whilst being kind to the environment. After these have been through they are intended to decompose back into substances that are harmless. These dyes offer a wider range of hues and better performance compared to natural dyes and thus they become an alternative choice to use in the manufacturing processes [26, 61, 66].

Based on Chromophore

Azo Dyes

The most extensive group of synthetic dyes is those known as azo dyes, defined as having one or more azo (–N = N–) linkages. The general changes to the dyes in order to achieve biodegradable variants involve alteration so those dyes include linkages which are cleavable and are positioned next to the azo group or among the substituent groups of the aromatic rings. That helps in enzymatic or chemical degradation of dye molecule. Preserving the favorable color characteristic of azo dyes and allowing effective biodegradation is a difficult task [55, 62, 68].

Anthraquinone Dyes

A category of considerable importance is anthraquinone dyes with vivid colour and excellent fastness characteristics. Strategic incorporation of biodegradable linkages between the anthraquinone and side chains produces biodegradable anthraquinones, through use of biodegradable ester or amide bond. Cleavage can occur by the foregoing enzymes or given environmental circumstances of the aforementioned linkages [8, 60, 76].

Other Chromophores

Besides azo and anthraquinone dyes, there are growing trends of creating biodegradable alternatives in other classes of chromophores. Dyes based on trypengymethane, indigoids, and phthalocyanines are included. When discussing the basic principle, in every situation, the basic part is the mixing of biodegradable elements in to the dye architecture, whilst conserving the colouristic attributes [45, 74].

Based on Biodegradable Linkage

Ester Linkages

The usual place for ester linkages (–COO–) to be utilized in biodegradable dyes is due to their susceptibility to hydrolysis. Due to common environmental embryos with esterases capable of cleaving the aforementioned linkages, the aforementioned linkages are prone to cleavage of said linkages by enzymes. Esters dyes which contain dyes that could decompose to form smaller, more pale or colorless products can have less ecological footprint [15, 41].

Amide Linkages

Additional means of biodegradability include amide linkages (–CONH–) present. Although they are typically more stable than ester linkages, amides are subject to hydrolysis through action of the enzymes such as amidases in the most severe cases. Specific dye structures that possess amide linkages are integrated within, leading to an equilibrium of durability and biological degradation [41, 53].

Other Labile Bonds

The cleavable connections are still being investigated by scholars and they try to find alternative varieties of cleavable connections for their use in biodegradable dyes. In addition to ethers, acetals, and hemiacetals, it also includes disulfide bonds. The decision of whether or not to choose linkage depends on the particular structure of the dye, the route of intended deterioration, and the task being addressed [47, 79].

Based on Application

The growing environmental issues require a new generation of colorant that can respond to the challenge, and synthetic, biodegradable dyes are a promising candidate. Textile dyeing is an important line of study because it is a good application for replacing environmental impact of the fashion industry with color added to the fibers, cotton, wool, silk and synthetic respectively. Moreover, textiles and also certain biodegradable dyes could function as food colourants, but only if extensively tested and in line with regulatory standards. This can be done in the form of cosmetics, to enable biodegradable colors which will reduce fears of skin irritation and reduce the cosmetic's environmental affect during the production and disposal. Besides, combining these sustainable colorants into packaging inks and coatings ensures that packaging materials considerably reduce the degree of long term pollution [9, 37]. Agri based applications of the biodegradable dyes are included such as use in labeling of the crop or seed, as well as major biomedical applications wherein the dyes can be used for colouring the cells or tissues. Nevertheless, there continue to be challenges, such as meeting the color fastness standards of conventional dyes regarding performance characteristics; increasing costs of manufacture; lack of viable market access; and the changing times of regulatorily relevant regulations and safety standards. However, these challenges have come together to support the growing biodegradable dye market due to the rising awareness of the environment and the advancement in the research of making these biodegradable dyes at reasonably affordable rates [21, 36, 49].

4 Microbial Dyes

There is a strong potential for sustainable dye production using microbial pigments because they have a large color palette and robust methods for production through fermentation.

Microbial Sources and Production

In the investigation, the focus has been on finding innovative microbial strains, bacteria, fungi and algae, that exhibit the capacity to synthesize wide range of pigments. Significant progress has been made in the domains of genetic engineering and metabolic engineering for the improvement in pigment production by microorganisms. Batch and fed batch processes have been studied along with the optimization of growth conditions like carbon source, temperature and pH which have been seen to impact yield of the pigment. Interest in using industrial waste as substrate for microbial dye production is on the rise due to a possible enhancement in its sustainability dimension [4, 37].

Analytical Techniques

In the course of this investigation, researchers have revealed the chemical structure and the properties of microbial pigments in terms of the color, stability and a possibility of application as dyes. The utilization of microbial dyes is researched upon a number of textile fibers with the emphasis placed on the improvement in dyeing conditions, and pre-/post-treatment processes. There have been a number of studies that test the introduction of microbial dyes with some other natural dyes in order to develop new colourations. Microbial dyes offer many advantages, such as an extensive range of hues, high possible yields, and a capability to create dyes in a controlled environment. However, there remain stiles with regard to the affordability, flexibility and development of viable extraction and purification processes. The fastness characteristics of microbial dyes require additional investigation to improve fastness performances towards other textile substrates and their applications [8, 30, 34, 64, 78].

5 Factors Affecting Biodegradability of Dyes

Synthetic dyes' environmental impact is regulated by a complex balancing act between their chemical structure and a mix of environmental conditions and the prevalence of microbe species that are all involved in determining their biodegradability as represented in Fig. 2. Above all, the chemical composition of the dye is important. On the other hand, it is important that there be cleavable linkages, such as ester, amide, or any other labile bond, in order to initiate biodegradation. The linkage nature has a marked effect on the rate and the mechanism of degradation [15, 59]. As a specific example, amide linkages are usually more susceptible to enzymatic hydrolysis than ester linkages. Steric hindrance surrounding these cleavable bonds may play a prominent role in imparting biodegradability, since the presence of larger

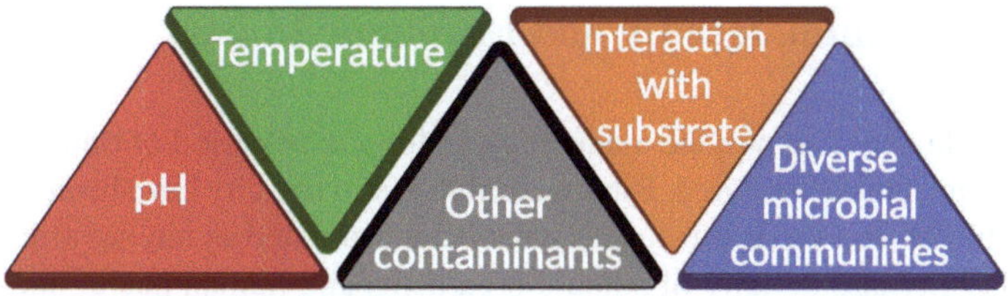

Fig. 2 Factors that affect the process of biodegradation of textile dyes

substituent groups could prevent enzyme access. Additionally, the total dimension and complexity of the dye molecule may influence the possibility of metabolism by microorganisms [41–43].

Factors of the environment are of great influence in determining the biodegradation process. Temperature is an important factor in determining the functionality of degrading enzymes as well as the rate of chemical hydrolysis. The biodegradation requires specific temperature ranges as extreme temperatures can prevent the process. As each enzyme works best at certain pH levels, pH is very important to enzyme function (Fig. 2). Oxygen supply acts as a constraining factor since some biodegradation pathways are requiring oxygen while others demanding anaerobic conditions. Essential nutrients such as nitrogen and phosphorus, specifically, have an important role in determining the development of the bacteria and their movement, thereby influencing the rate at which the dye is degraded [25, 30, 46]. There is significant importance in the existence and structure of microbial communities. Enzymatic competencies vary among different microorganisms and a variety of these microorganisms frequently have better overall degradation of complex dye structures than a single species. It is well established that the existence of particular enzymes, including esterases, amidases, or azoreductases, has an important role in the preliminary degradation of the dye molecule. Because the process of biodegradation is highly dependent on environment, the existence of more pollutants or substances within the environment can significantly interfere with the process of biodegradation by preventing microorganism to use right enzymes available in the environment (Fig. 2). For example, such presence of heavy metals can damage the microorganisms to the extent that the process of dye degradation is hindered. In fact, last the dye to which the substance is applied may also have an impact, such as textile fiber. Availability of the dye to breaking down enzymes may depend on the relationship of the dye and the fiber matrix. Lastly, the biodegradability of synthetic dyes involves a complicated mechanism that depends on such specific factors as chemical structure of the dye, the presence of the environmental conditions, microbial composition and interactions with other environmental medium. This complexity points to the necessity of taking a whole approach to evaluate and increase the impact they have on the environment [3, 30, 46, 54].

6 Environmental, Health, and Economic Impact

Biodegradable dyes can only be made feasible for a long period of time if their environmental impact is considered. Biodegradability and toxicity have been investigated in recent studies and these substances have been evaluated as having impacts on the biodegradability and toxicity of aquatic organisms and human health under various environmental conditions [24, 40, 77]. Bioremediation of wastewater containing biodegradable dyes has been explored with referring to microbial consortia. Life Cycle Assessment (LCA) is a method for assessing the impacts on environment caused in all the phases of the product life cycle during the use of biodegradable dyes. This analysis takes into consideration the water and energy consumption, greenhouse gas emissions along ecotoxicity. LCA separates the dye manufacturing and dye application processes into distinct domains in order to advise on sustainability improvement [18, 57, 73].

The textile industry has a great opportunity, however, to reduce one of its biggest impacts on the environment, moving to biodegradable textile dyes. It offers numerous benefits with respect to human health as well as the economy's sustainability. Additionally, most biodegradable dyes, be it natural derived or synthesized, tend to observe lower toxicity relative to the conventional synthetic colors. As a result, during the production process, the exposure of textile workers to hazardous chemicals is reduced, the probability of allergic reactions and skin irritations among the wearers of dyed textiles decreased, and the air inside houses or buildings becomes cleaner due to the reduced emission of volatile organic compounds (VOC) generally correlated to the use of conventional dyes [7, 27, 78]. Additionally, the application of biodegradable dyes offers a unique opportunity to reduce substances that may be detrimental to infants and children, a particular vulnerability population, when consuming products. Biodegradable dyes can be applied to wastewater treatment processes to give a substantial drop in cost in such processes as these dyes decompose more quickly reducing the burden on treatment facilities. Additionally, companies who follow this environmentally aware approach can gain through improving their reputation and building customers trust about its sustainable products according to the increase in the demand of the sustainable products. It could help in opening specialized markets and giving opportunities for premium pricing [18, 29, 40]. Moving towards biodegradable dyes might aid an increase in and development of local economies as they require the manufacturing of renewable resources needed to make them. As the environmental regulations become stringent, companies using such biodegradable dyes come into a good position regarding compliance, thereby reducing the risk of getting penalty. At a time when regulation is urging for eco-friendly textile production, the search for biodegradable dye technology has a power to drive textile innovation and the development of new dyeing methodologies, new eco-friendly materials, and new thinking in the textile results as a whole. This involves the evidence of environmentally conscious choice and strategic move towards a more productive and economically powerful industry [18, 48, 55].

7 Limitations and Future Directions

Synthetic biodegradable dyes have several avenues of promise but, at the same time, there are many research gaps, limitation and opportunities that call for further efforts. There is also the lack of enough comprehensive data on the biodegradability of specific dye compounds in different environmental situations. Some natural and synthetic dyes have cleavable links but pathways of degradation, rates, and products in soils and water systems are often poorly understood. A determination of their actual environmental impact is hindered by the absence of developed testing methodologies and application of publicly accessible data. There is a trade-off between effects, and biological degradation as discussed in Table 2. However, there continues to be a strong challenge to the utilization of biodegradable dyes in the development for enhanced colorfastness, wash durability, and light rapidity. There is potential to detrimentally affect these integral performance characteristics that are essential to their usefulness in demanding textile environments by a variety of biodegradable modifications. Research for the development of dye structures that can meet critical performance standard while maintaining adequate biodegradability is needed.

The economic efficiency of the production process is another factor that must be also considered. However, the synthesis of some of these biodegradable dyes is less viable than some traditional colors at this time. Economic viability of these dyes requires increasing the productivity and adaptability of synthesis pathways for them. A limitation is the range of biodegradable colors available. When compared with the wide spectrum of colors available from the broader spectrum of conventional synthetic dyes, the availability of biodegradable alternatives is still relatively limited. It is very important to expand the range of colors which can be delivered with biodegradable dyes to satisfy the different requirements of the textile industry. The evaluation of the biodegradability and environmental toxicity of the new dyes has so far been from a regulatory perspective a matter of necessity, given more explicit guidelines and standard testing methodology. Aiming to drive the development and industrialization of the best sustainable colorants, this initiative is working in concert. Several potential research directions seem very promising to find out. The use of computational modeling is anticipated to assist the development of dye compounds that will have improved biodegradability and better effectiveness. In order to increase the ecological responsibility of these colorants, the exploration of novel bio-based feedstocks for the synthesis of dyes can be done. Enzyme mediated degradation of dye chemicals, could develop better and tailored deterioration. A comprehensive environmental impact study of the biodegradable dyes from synthesis to disposal is important to assess their true sustainability and it can be done by using life cycle assessment studies.

Table 2 Biodegradable dyes describing the degradation mechanism along with the advantages and disadvantages

Biodegradable dye	Mechanism of degradation	Advantages	Disadvantages	References
Indigo (from plants)	Microbial degradation through enzymes from bacteria or fungi that break down the indigo molecule	• Natural origin • Significantly non-toxic • Generates unique blue shades	• Requires mordants (sometimes with environmental concerns) • Colorfastness can be an issue • Extraction can be labor-intensive	[1, 11, 35]
Carotenoids (from plants like carrots, etc.)	It may be decomposed by light, enzymes, or oxidative processes	• Vibrant colors (yellows, oranges, reds) • Antioxidant properties in some cases • Generally considered safe	• Can be unstable (sensitive to light, heat) • May require large amounts of plant material • Limited shade range compared to synthetics	[4. 35, 65]
Anthocyanins (from berries, flowers, etc.)	It can be broken down by enzymes or changes in acidity by hydrolysis	• Wide range of colors (reds, purples, blues) • Water-soluble • Potential health benefits (antioxidants)	• Stability can be pH-dependent • Color intensity may vary • Can be challenging to extract in large quantities	[65, 67, 71]
Curcumin (from turmeric)	Broken down by bacteria or fungi degradation in the environment	• Natural yellow color • Anti-inflammatory properties • Relatively safe	• Poor water solubility • Colorfastness can be limited • May not be suitable for all fabrics	[51, 75]

8 Discussion

Biodegradables dyes could be a viable replacement for synthetic dyes with regards to environmental, economic and social benefits. In this way they naturally decompose to eliminate water pollution, ease strain on wastewater treatment facilities and use less energy and make less carbon emissions than synthetic dyes. Moreover, they are shown to be less toxic resulting in cutting down risk to aquatic organisms and human health. All things considered, economic benefits of biodegradable dyes are quite significant considering potential technological advancements in production and realization of economies of scale. It reduces reliance on expensive synthetic chemicals and gives local businesses and communities an opportunity. Shift towards biodegradable dyes offers the possibility of favourable social outcomes by generating local markets via support of the local natural dye assets and fostering sustainable lifestyle. Applications of gene editing and metabolic processes engineering techniques can enhance pigment synthesis of microorganisms, while the inclusion of nanomaterials will tremendously increase dye absorption and fastness properties. Therefore, for responsible manufacturing sources of biodegradable dye and its utilization, there is an urgent need to establish strong raw material supply chains as well as to launch the methods for evaluating the entire life cycle. Therefore, it is essential to collaborate among researchers, industry stakeholders, government entities, and users to help this transition to biodegradable dyes. To enable the research to be put to effective use, academia and industry need to work together on collaborative efforts. Legislative measures supportive for such promotion can be realized. One obvious advantage of biodegradable dyes is related to the level of consumer awareness, as these types of dyes can have a limiting impact on the level of global pollution. The biodegradable dyes present play a major role in setting a sustainable future by having lesser environmental impact and possibilities of creating social and economic benefits. Biodegradable dyes may be adopted as a result of the rising needs of the customers who want to use the environmentally sustainable products, which would increase the economic and social benefits. However, further investigation and progress are needed in order to address issues related to broader implementation.

9 Conclusion

Hence, biodegradable textile dyes must be utilized in order to increase the sustainable nature of the textile industry as a whole. Results from this analysis show that the synthetic colors being used today allow for water pollution and pose a considerable hazard, which indicates that environmental friendly alternatives are necessary. Biodegradable dyes obtained from plants, microbes and minerals have the possibility to reduce these environmental impacts. Despite some issues regarding color extent, fastness, and price, significant advances have been made in recent years. Bidding for novel dye materials, advanced extraction and application procedures, and new dyeing technologies have extended the choice of biodegradable dye leads. With its integration into nanotechnology and biotechnology, bio-based colorants can be beneficial with both improvements in effectiveness and adaptability. Such potential of biodegradable dyes can be fully leveraged through strategic investment in research and development, fostering collaboration between dye producers and end users and through enhancing consumer awareness regarding this type of dyes. The fashion and textile industries, without compromising on color diversity, can adopt biodegradable dyes, thus minimizing environmental impact.

References

1. Affat, S. S. (2021). Classifications, advantages, disadvantages, toxicity effects of natural and synthetic dyes: A review. *University of Thi-Qar Journal of Science, 8*, 130–135.
2. Aggarwal, S. (2025). Advancement in extraction and characterization techniques of natural dyes from dye yielding plants sources: A review. *Pigment & Resin Technology, 54*, 293–301.
3. Ahmed, A., Usman, M., Yu, B., Ding, X., Peng, Q., Shen, Y., & Cong, H. (2020). Efficient photocatalytic degradation of toxic Alizarin yellow R dye from industrial wastewater using biosynthesized Fe nanoparticle and study of factors affecting the degradation rate. *Journal of Photochemistry and Photobiology B: Biology, 202*, Article 111682.
4. Alegbe, E. O., Uthman, T. O. (2024a). A review of history, properties, classification, applications and challenges of natural and synthetic dyes. Heliyon.
5. Aragaw, T. A., (2024). A review on biodegradation of textile dye wastewater: Challenges due to wastewater characteristics and the potential of alkaliphiles. *Journal of Hazardous Materials Advances* 100493.
6. Ardila-Leal, L. D., Poutou-Piñales, R. A., Pedroza-Rodríguez, A. M., & Quevedo-Hidalgo, B. E. (2021). A brief history of colour, the environmental impact of synthetic dyes and removal by using laccases. *Molecules, 26*, 3813.
7. Azam, K., Shezad, N., Shafiq, I., Akhter, P., Akhtar, F., Jamil, F., Shafique, S., Park, Y.-K., & Hussain, M. (2022). A review on activated carbon modifications for the treatment of wastewater containing anionic dyes. *Chemosphere, 306*, Article 135566.
8. Baqi, Y. (2022). Anthraquinone dyes: A synthetic and chemical characterization protocol for an industrial chemistry laboratory course. *Journal of Chemical Education, 99*, 1441–1447. https://doi.org/10.1021/acs.jchemed.1c00865
9. Benkhaya, S., M'rabet, S., & El Harfi, A. (2020). A review on classifications, recent synthesis and applications of textile dyes. *Inorganic Chemistry Communications, 115*, Article 107891.
10. Chattopadhyay, S. N., & Pan, N. C. (2019). Ecofriendly printing of jute fabric with natural dyes and thickener. *Journal of Natural Fibers, 16*, 1077–1088. https://doi.org/10.1080/15440478.2018.1449161
11. Chowdhury, M. F., Khandaker, S., Sarker, F., Islam, A., Rahman, M. T., & Awual, M. R. (2020). Current treatment technologies and mechanisms for removal of indigo carmine dyes from wastewater: A review. *Journal of Molecular Liquids, 318*, Article 114061.
12. Dayet, L. (2021). Invasive and non-invasive analyses of ochre and iron-based pigment raw materials: A methodological perspective. *Minerals, 11*, 210.
13. Deniz, N. G., Sayil, C., Oner, E., Atak, O., Stasevych, M., Zvarych, V., Shyyan, H., & Novikov, V. (2022). Dyeing of polyester fibers with sulfur-and nitrogen-containing anthraquinone derivatives. *Chemical Industry & Chemical Engineering Quarterly, 28*, 47–55.
14. Deveoglu, O. (2020). A review on cochineal (Dactylopius Coccus Costa) dye. *Research Journal of Recent Sciences, 2277*, 2502.
15. Dulo, B., Phan, K., Githaiga, J., Raes, K., & De Meester, S. (2021). Natural quinone dyes: A review on structure, extraction techniques, analysis and application potential. *Waste Biomass Valor, 12*, 6339–6374. https://doi.org/10.1007/s12649-021-01443-9
16. Ferreyra-Suarez, D., Paredes-Vargas, L., Jafari, S. M., García-Depraect, O., & Castro-Muñoz, R. (2024). Extraction pathways and purification strategies towards carminic acid as natural-based food colorant: A comprehensive review. *Advances in Colloid and Interface Science, 323*, Article 103052.
17. Gaffer, H. E., & Althagafi, I. I. (2020). Synthesis of new azobenzene dyes clubbed with thiazolidinone moiety and their applications. *Pigment & Resin Technology, 49*, 207–214.
18. Gupta, R., Kushwaha, A., Dave, D., Mahanta, N.R. (2022). Waste management in fashion and textile industry: Recent advances and trends, life-cycle assessment, and circular economy. *Emerging Trends to Approaching Zero Waste* 215–242.
19. Gupta, V. K. (2019). Fundamentals of natural dyes and its application on textile substrates. *Chemistry and Technology of Natural and Synthetic Dyes and Pigments, 22*, 2019.

20. Haji, A., & Naebe, M. (2020). Cleaner dyeing of textiles using plasma treatment and natural dyes: A review. *Journal of Cleaner Production, 265*, Article 121866.
21. Hosseinnezhad, M., Gharanjig, K., Imani, H., Rouhani, S., & Adeel, S. (2023). Environmentally dyeing of wool yarns using of combination of myrobalan and walnut husk as bio-mordants. *Progress in Color, Colorants and Coatings, 16*, 197–205.
22. Hussain, C. M., & Kecili, R. (2019). *Modern environmental analysis techniques for pollutants*. Elsevier.
23. Ikram, M., Zahoor, M., & El-Saber Batiha, G. (2021). Biodegradation and decolorization of textile dyes by bacterial strains: A biological approach for wastewater treatment. *Zeitschrift für Physikalische Chemie, 235*, 1381–1393. https://doi.org/10.1515/zpch-2020-1708
24. Islam, T., Repon, M. R., Islam, T., Sarwar, Z., & Rahman, M. M. (2023). Impact of textile dyes on health and ecosystem: A review of structure, causes, and potential solutions. *Environmental Science and Pollution Research, 30*, 9207–9242.
25. Ismail, G. A., & Sakai, H. (2022). Review on effect of different type of dyes on advanced oxidation processes (AOPs) for textile color removal. *Chemosphere, 291*, Article 132906.
26. Jamee, R., & Siddique, R. (2019). Biodegradation of synthetic dyes of textile effluent by microorganisms: An environmentally and economically sustainable approach. *European Journal of Microbiology and Immunology, 9*, 114–118.
27. Javaid, R., & Qazi, U. Y. (2019). Catalytic oxidation process for the degradation of synthetic dyes: An overview. *International Journal of Environmental Research and Public Health, 16*, 2066.
28. Kabir, F. N. A., Palleb, S. R., Halim, M. A., Nayan, S. B., Uddin, M. B., & Aziz, M. G. (2024). Extraction of natural food color from tomato, carrot and turmeric. *European Journal of Nutrition & Food Safety, 16*, 175–188.
29. Kadhim, M. J., Mahdi, M. A., Selman, A. M., Al-Ani, S. K. J., Hassan, J. J., Ahmed, N. M. (2023). The most important parameters that affect the photocatalytic activity of ZnO nanostructures against organic dyes: A review. *Iranian Journal of Catalysis* 13.
30. Kapoor, R. T., Danish, M., Singh, R. S., Rafatullah, M., & Hps, A. K. (2021). Exploiting microbial biomass in treating azo dyes contaminated wastewater: Mechanism of degradation and factors affecting microbial efficiency. *Journal of Water Process Engineering, 43*, Article 102255.
31. Khan, A., Roy, A., Bhasin, S., Emran, T. B., Khusro, A., Eftekhari, A., Moradi, O., Rokni, H., & Karimi, F. (2022). Nanomaterials: An alternative source for biodegradation of toxic dyes. *Food and Chemical Toxicology, 164*, Article 112996.
32. Khanum, R., Ali, R. S., Rangaswamy, H. R., Kumar, S. S., Prashantha, A. G., & Jagadisha, A. S. (2023). Recent review on synthesis, spectral Studies, versatile applications of azo dyes and its metal complexes. *Results in Chemistry, 5*, Article 100890.
33. Kora, A. J. (2024). Cosmetic and religious applications of natural nanoclay: an Indian scenario. *Nanoclay-Based Sustainable Materials* 429–447.
34. Kumar, V., Singh, V., & Pandit, S. (2025). Advanced omics approach and sustainable strategies for heavy metal microbial remediation in contaminated environments. *Bioresource Technology Reports* 102040.
35. Li, N., Wang, Q., Zhou, J., Li, S., Liu, J., & Chen, H. (2022). Insight into the progress on natural dyes: Sources, structural features, health effects, challenges, and potential. *Molecules, 27*, 3291.
36. Martins, F. C., Sentanin, M. A., & De Souza, D. (2019). Analytical methods in food additives determination: Compounds with functional applications. *Food Chemistry, 272*, 732–750.
37. Mishra, S., & Maiti, A. (2019). Applicability of enzymes produced from different biotic species for biodegradation of textile dyes. *Clean Technologies and Environmental Policy, 21*, 763–781.
38. Nambela, L., Haule, L. V., & Mgani, Q. (2020). A review on source, chemistry, green synthesis and application of textile colorants. *Journal of Cleaner Production, 246*, Article 119036.
39. Naqvi, S. A. R., Adeel, S., Mia, R., Hosseinnezhad, M., Rather, L. J., & Imran, M. (2024). Modern ecofriendly approach for extraction of luteolin natural dye from weld for silk fabric and wool yarn dyeing. *Sustainable Chemistry and Pharmacy, 39*, Article 101554.

40. Naqvi, S. L. H., Nadeem, M., Ayub, F., Yasar, A., Naqvi, S. H. Z., Tanveer, R. (2024). Social and environmental impacts in textile production. In P. Singh, (Ed.), *Dye pollution from textile industry, SDGs and textiles*. Springer Nature Singapore, pp. 423–453. https://doi.org/10.1007/978-981-97-5341-3_19
41. Natarajan, J., Chatterjee, K., & Madras, G. (2017). Tailored degradation and dye release from poly(ester amides). *Polymer-Plastics Technology and Engineering, 56*, 635–646. https://doi.org/10.1080/03602559.2016.1233263
42. Natarajan, J., Madras, G., & Chatterjee, K. (2017). Poly(ester amide)s from poly(ethylene terephthalate) waste for enhancing bone regeneration and controlled release. *ACS Applied Materials & Interfaces, 9*, 28281–28297. https://doi.org/10.1021/acsami.7b09299
43. Nutan, B., Chandel, A. K. S., Bhalani, D. V., & Jewrajka, S. K. (2017). Synthesis and tailoring the degradation of multi-responsive amphiphilic conetwork gels and hydrogels of poly (β-amino ester) and poly (amido amine). *Polymer, 111*, 265–274.
44. Palamara, E., Kesidis, S., Cifuentes, L. T., Das, P. P., Nicolopoulos, S., & Zacharias, N. (2024). The application of cathodoluminescence (CL) for the characterization of blue pigments. *Heritage, 7*, 3048–3061.
45. Palumbo, L., Fiorito, S., Epifano, F., Sharifi-Rad, M., Genovese, S., & Collevecchio, C. (2023). Solid-phase adsorption methodologies of naturally occurring anthraquinones: A review. *Phytochemical Analysis, 34*, 153–162. https://doi.org/10.1002/pca.3203
46. Paul, D. R., Sharma, R., Nehra, S. P., & Sharma, A. (2019). Effect of calcination temperature, pH and catalyst loading on photodegradation efficiency of urea derived graphitic carbon nitride towards methylene blue dye solution. *RSC advances, 9*, 15381–15391.
47. Penthala, R., Heo, G., Kim, H., Lee, I. Y., Ko, E. H., Son, Y. A. (2020). Synthesis of azo and anthraquinone dyes and dyeing of nylon-6, 6 in supercritical carbon dioxide. *Journal of CO_2 Utilization 38*, 49–58.
48. Periyasamy, A. P. (2024). Recent advances in the remediation of textile-dye-containing wastewater: Prioritizing human health and sustainable wastewater treatment. *Sustainability, 16*, 495.
49. Pranta, A. D., & Rahaman, M. T. (2024). Extraction of eco-friendly natural dyes and biomordants for textile coloration: A critical review. *Nano-Structures & Nano-Objects, 39*, Article 101243.
50. Rabia, S. A., Samad, B. A., Mazhar, H. P., & Alvira, A. A. (2019). An efficient ultrasonic and microwave assisted extraction of organic Henna dye for dyeing of synthetic polyester fabric for superior color strength properties. *Ind Text, 70*, 303–308.
51. Raduly, F. M., Raditoiu, V., Raditoiu, A., & Purcar, V. (2021). Curcumin: Modern applications for a versatile additive. *Coatings, 11*, 519.
52. Rajkhowa, S., Sarma, J., Dutta, R., Ahmed, S., Singh, R. P. (2024). Application of natural dyes in textile industry (a sustainable approach). In P. Singh, (Ed.), *Dye pollution from textile industry, SDGs and textiles*. Springer Nature Singapore, pp. 221–238. https://doi.org/10.1007/978-981-97-5341-3_11
53. Ranganathan, P., Chen, C.-W., Rwei, S.-P., Lee, Y.-H., & Ramaraj, S. K. (2020). Methods of synthesis, characterization and biomedical applications of biodegradable poly (ester amide) s—A review. *Polymer Degradation and Stability, 181*, Article 109323.
54. Rápó, E., & Tonk, S. (2021). Factors affecting synthetic dye adsorption; desorption studies: A review of results from the last five years (2017–2021). *Molecules, 26*, 5419.
55. Rathi, B. S., & Kumar, P. S. (2022). Sustainable approach on the biodegradation of azo dyes: A short review. *Current Opinion in Green and Sustainable Chemistry, 33*, Article 100578.
56. Ribeiro, J. S., & Veloso, C. M. (2021). Microencapsulation of natural dyes with biopolymers for application in food: A review. *Food Hydrocolloids, 112*, Article 106374.
57. Sabina, R., Hussain, N. (2024). Mapping of emerging biotechnological advancement in textile and apparel: A sustainable approach towards circular economy and life cycle assessment. In P. Singh, (Ed.), *Dye pollution from textile industry, SDGs and textiles*. Springer Nature Singapore, pp. 351–404. https://doi.org/10.1007/978-981-97-5341-3_17

58. Sankari, M., Anjum, M. R., Sravya, B., Sowmika, S., Ilakkia, S., Narasimha, G. (2024). Ecofriendly dyes: Extraction, characterization and potential applications. *Journal of Advanced Zoology 45.*
59. Saxena, S., & Jayakannan, M. (2020). Development of L-Amino-acid-based hydroxyl functionalized biodegradable amphiphilic polyesters and their drug delivery capabilities to cancer cells. *Biomacromolecules, 21*, 171–187. https://doi.org/10.1021/acs.biomac.9b01124
60. Shahid, M., Wertz, J., Degano, I., Aceto, M., Khan, M. I., & Quye, A. (2019). Analytical methods for determination of anthraquinone dyes in historical textiles: A review. *Analytica Chimica Acta, 1083*, 58–87.
61. Sharma, J., Sharma, S., & Soni, V. (2021). Classification and impact of synthetic textile dyes on Aquatic Flora: A review. *Regional Studies in Marine Science, 45*, Article 101802.
62. Shi, Y., Yang, Z., Xing, L., Zhang, X., Li, X., & Zhang, D. (2021). Recent advances in the biodegradation of azo dyes. *World Journal of Microbiology and Biotechnology, 37*, 1–18.
63. Sienkiewicz-Szłapka, E., Janysz, A., Wróbel, M. H. (2025). Quantification of lawsone and p-phenylenediamine in natural henna products for hair colouring available in poland. Contact Dermatitis cod. 14739. https://doi.org/10.1111/cod.14739
64. Singh, S., Nimse, S. B., Mathew, D. E., Dhimmar, A., Sahastrabudhe, H., Gajjar, A., Ghadge, V. A., Kumar, P., & Shinde, P. B. (2021). Microbial melanin: Recent advances in biosynthesis, extraction, characterization, and applications. *Biotechnology Advances, 53*, Article 107773.
65. Singh, V., Pandit, C., Roy, A., Pandit, S., Rai, A. K., Rani, A., Ranjan, N., Rustagi, S., Malik, S. (2024). Degradation of food dyes via biological methods: A state-of-the-art review. *Bioresource Technology Reports* 101780.
66. Slama, H. B., Chenari Bouket, A., Pourhassan, Z., Alenezi, F. N., Silini, A., Cherif-Silini, H., Oszako, T., Luptakova, L., Golińska, P., Belbahri, L. (2021). Diversity of synthetic dyes from textile industries, discharge impacts and treatment methods. *Applied Sciences, 11*, 6255.
67. Smerea, O., Bulgaru, V. (2024). Anthocyanins–methods of extraction, stabilization and applications. *Journal of Engineering Sciences* 179–195.
68. Sreedharan, V., Bhaskara Rao, K. V. (2019). Biodegradation of textile azo dyes. In K. M. Gothandam, S. Ranjan, N. Dasgupta, E. Lichtfouse, (Eds.), *Nanoscience and biotechnology for environmental applications, environmental chemistry for a sustainable world*. Springer International Publishing, pp. 115–139. https://doi.org/10.1007/978-3-319-97922-9_5
69. Szadkowski, B., Marzec, A., Rogowski, J., Maniukiewicz, W., & Zaborski, M. (2019). Insight into the formation mechanism of azo dye-based hybrid colorant: Physico-chemical properties and potential applications. *Dyes and Pigments, 167*, 236–244.
70. Tang, Y., He, W., Yang, S., & Liu, L. (2019). Stabilisation and detoxification of henna (*Lawsonia inermis* L.) extract for hair dye cosmetics by spray-drying encapsulation. *Coloration Technology, 135*, 439–450. https://doi.org/10.1111/cote.12433
71. Tena, N., & Asuero, A. G. (2022). Up-to-date analysis of the extraction methods for anthocyanins: Principles of the techniques, optimization, technical progress, and industrial application. *Antioxidants, 11*, 286.
72. Tian, Y., Wu, K., Lin, S., Shi, M., Liu, Y., Su, X., Islam, R. (2024). Biodegradation and decolorization of crystal violet dye by cocultivation with fungi and bacteria. *ACS Omega Acsomega* 3c06978. https://doi.org/10.1021/acsomega.3c06978
73. Xu, C., Yu, H., Zhang, S., Shen, C., Ma, C., Wang, J., & Li, F. (2024). Cleaner production evaluation system for textile industry: An empirical study from LCA perspectives. *Science of the Total Environment, 913*, Article 169632.
74. Yaneva, Z., Ivanova, D., Nikolova, N., & Toneva, M. (2022). Organic dyes in contemporary medicinal chemistry and biomedicine. I. From the chromophore to the bioimaging/bioassay agent. *Biotechnology & Biotechnological Equipment, 36*, 1–14. https://doi.org/10.1080/13102818.2022.2039077
75. Yeamin, M. B., Islam, M. M., Chowdhury, A.-N., & Awual, M. R. (2021). Efficient encapsulation of toxic dyes from wastewater using several biodegradable natural polymers and their composites. *Journal of Cleaner Production, 291*, Article 125920.

76. Yeow, P. K., Wong, S. W., & Hadibarata, T. (2021). Removal of azo and anthraquinone dye by plant biomass as adsorbent—A review. *Biointerface Research in Applied Chemistry, 11*, 8218–8232.
77. Yousaf, A. M., & Aqsa, R. (2023). Integrating circular economy, SBTI, digital LCA, and ESG benchmarks for sustainable textile dyeing: A critical review of industrial textile practices. *Global NEST Journal, 25*, 39–51.
78. Zasada-Kłodzińska, D., Basiul, E., Buszewski, B., & Szumski, M. (2020). Analysis of natural dyes from historical objects by high performance liquid chromatography and electromigration techniques. *Critical Reviews in Analytical Chemistry, 51*, 1–34. https://doi.org/10.1080/10408347.2020.1743640
79. Zhang, X., Malhotra, S., Molina, M., & Haag, R. (2015). Micro-and nanogels with labile crosslinks–from synthesis to biomedical applications. *Chemical Society Reviews, 44*, 1948–1973.
80. Zuber, M., Adeel, S., Rehman, F.-U., Anjum, F., Muneer, M., Abdullah, M., & Zia, K. M. (2020). Influence of microwave radiation on dyeing of bio-mordanted silk fabric using neem bark (*Azadirachta indica*)-Based Tannin Natural Dye. *Journal of Natural Fibers, 17*, 1410–1422. https://doi.org/10.1080/15440478.2019.1576569

Phytoremediation of Textile Industrial Effluents for Sustainable Future

Avni Kumari, Shivani Rawat, Afaq Majid Wani, and Syed Maqbool Geelani

1 Introduction

With the increasing industrialization and urbanization, heavily loaded chemicals have been discharged during the dyeing process which raised a serious environmental concern. A large amount of polluted water containing organic wastes, acid, bases, inorganic substances, colorants, etc. has been released into the water bodies leading to serious environmental damage [1]. Fabrics have been dyed using natural colors taken from fruits, vegetables, and flowers since 3500 BC. Dyes are employed in the creation of paints, textiles, paper, printing inks, and plastics, as well as in the addition of color and pattern to the materials. Chemical dyes, which bond with the cloth and create a rich color that lasts through washing and exposure, are now commonly utilized. Today, there are around 100,000 synthetic dyes commercially accessible in trade, with nearly a million tonnes of colorants manufactured yearly, and approximately 10% of dyes are discharged as waste into the natural surroundings [2]. Depending on the type of cloth or material being coloured, there are many different types of dyes employed in the production of diverse chemical compounds. Acid dyes for coloring animal fibers, basic dyes for coloring paper, direct dyes for use on cotton wool or cotton silk, and pigment dyes for use in paint and inks are only a few of the 3600 various types of dyes available.

There has been an increasing demand for dye production to meet the needs of the growing population for variable designs and distinctive colours for garments as well as other fabrics. These variable designs are created by mixing dyes, synthesizing

A. Kumari · S. Rawat · A. M. Wani (✉)
Department of Forest Biology and Tree Improvement, College of Forestry, Sam Higginbottom University of Agriculture, Technology and Sciences, Prayagraj, India
e-mail: afaqtree@gmail.com

S. M. Geelani
Department of Environmental Engineering and Research, GuruJal Abhipsa-Foundation, Gurugram, India

R. A. Bhat et al. (eds.), *Biocentric Approaches in Textile Waste Management*, Textile Science and Clothing Technology, https://doi.org/10.1007/978-3-032-04974-2_14

them in a reactor, filtering out the impurities and then it is dried and blended. To initiate reactions, several chemical components such as sulphuric acid, chromium, copper, and other metallic elements, as well as solvents, are utilized. Because of their carcinogenic nature, even modest quantities of dyes in water can have a negative influence on the ecosystem [3]. These evolving demands also create an increased release of dye effluent which results in highly fluctuating and diverse streams of wastewater. This massive use of water as a coolant, for cleaning dye equipment, rinsing, and processing dye products is the major component that is quite harmful to the environment. The textile industry ranks among the most significant, producing 60 billion kg of fabric a year and consuming up to 9 trillion gallons of water [4].

Commercially, textile dyes are categorized as- anionic dyes which include acid dyes, reactive dyes, azo dyes, direct dyes, cationic dyes including basic dyes, and non-ionic dyes which are diffused dyes that do not ionize in aqueous solution [5]. Another important class of dyes is azo dyes possessing azo linkages (–N=N–), the chromophoric groups that provide color to the substrate, and sulfonic groups (–SO_3^-). These are known for their durability, ease of preparation, inexpensive and readily available raw ingredients, ability to cover the whole shade range, and good mechanical properties. The commercial applications of the azo linkage colorants are more diverse in their number and structure in textiles, food sectors, pharmaceuticals, petroleum products, and tanning industries. Other classes of dyes include dioxoanthracenes, monastral blues, aromatic aldehydes, and polymethines [6].

Dyestuffs, particularly synthetic ones, are highly resilient in natural environments and thus extremely difficult to treat, posing a long-term threat to the environment. Anionic dyes are highly water soluble and challenging to eliminate by the conventional method. Non-ionic dyes do not ionize in water and hence are not easily removable from the contaminated site. As a result, remediation methods are required to avoid dyestuffs from invading terrestrial, air, and aquatic habitats and to ameliorate polluted land. Thus far, numerous remediation procedures have been explored, including physical, chemical, and biological strategies for dye removal in polluted waste-water. Among various methods such as adsorption, electro-coagulation, advanced oxidation method, and solvent extraction, biological methods are highly valuable because of their eco-friendly nature, energy saving, effectiveness, and with minimum chemical usage. Physico-chemical methods are easy to use but can be cost-effective and are not environment-friendly. By aerobic and anaerobic processes, diverse microorganisms such as bacteria, fungus, algae, and plants transform dye waste into simpler and harmless components. The final products obtained in aerobic treatments are carbon dioxide, water, and biomass, while anaerobic treatments produce carbon dioxide, methane, and biomass [7].

Phytoremediation is a form of bioremediation that has evolved in recent years as an energy-efficient and environmental friendly remediation solution for the decontamination of soil, surface and groundwater, air, or other contaminated media. It uses plants, their related enzymes, and associated bioactive components for the treatment of dye-contaminated sites.

2 Phytoremediation

Phytoremediation is a plant-based strategy that utilizes plants to extract and eliminate chemical elements from the surroundings or to minimise their presence in the soil. Plants may absorb toxins from the environment via their root system, which facilitates mobilization, cleanup, or purification of toxins within plants via different processes such as elimination, confinement, degradation, and so on. In order to collect heavy metals and control their bioavailability, plants extend their root systems into the surrounding soil and create a rhizosphere ecosystem, recovering the contaminated soil and regulating soil fertility. Similar plant qualities have been exploited to effectively remove wastes such as metallic and phenolic compounds, azo dyes and coloring agents, and a variety of other organic and inorganic toxins from contaminated environments [8]. Photosynthesis is also one of the phytoremediation methods to remove carbon dioxide from the polluted air. It appears as an eco-friendly in-situ strategy for minimising soil erosion, enhancing soil nutrients by adding the amount of soil organic matter, and employing plants for chemical breakdown, extraction, conversion, or decontamination [9]. Growing aquatic plants like *Phragmites australis* [10] and free-floating plants like L. minor have both been used in phytoremediation [11].

3 Need for Phytoremediation

The environmental release of dyestuffs has resulted in serious contamination of water ways such as rivers, streams, groundwater, and nearby regions, particularly soils where textile industries are prevalent. The dyes include a chromophoric component that, by trapping sunlight and inhibiting light penetration, impairs the plant productivity of seaweeds and aquatic plants. These dye-based effluents have a high concentration of suspended solid materials that pose turbidity problems in water bodies causing a raise in BOD and COD quantities. These dyes destroy the inherent properties of water resources and endanger marine resources such as vegetation and wildlife by disrupting the environmental integrity, causing substantial ecological issues. Therefore, these dyestuffs need to be treated or removed before their disposal into the water bodies [12].

Additionally, dyes undergo various reactions due to fluctuations in the aquatic environment which alters the chemical makeup and form of the colourants. These structurally changed and degraded dye compounds, such as aromatic amines, may produce xenobiotic chemicals with varying levels of toxicity. Dyestuffs generally are recalcitrant pollutants having complex chemical molecules that resist deterioration, endure or remain undamaged in the natural environment for prolonged periods of time, and are transported vast distances across diverse regions of the globe. A pollutant's recalcitrance is attributed to its greater molecular size due to its extremely dense and convoluted aromatic ring structures, the presence of unique bonds and replacements with other chemicals such as bromine (Br-) or chlorine (Cl-), high stability, and hazardous character.

Many of the colours emitted and their degradation products are hazardous, cancerous, or mutagenic to many living forms, particularly owing to carcinogens such as chlorobenzene, anthracene, and other aromatic chemicals. Dyes are toxic to a variety of aquatic and microbial species, as well as inflicting human ailments, particularly when important human organs are harmed or killed. Even though these coloring agents and their metabolites have low organic loads in wastewater, they must be eliminated due to their potential toxic effects, mutagenicity, and carcinogenicity [13, 14].

4 Phytoremediation Strategies

Several phytoremediation procedures are used to clean up polluted areas, including.

(i) **Phytostabilization**: It is the use of plants to minimize heavy metal toxicity in soil, decreasing metal migration into the environment and lowering the chance of metals infiltrating the food chain. To boost phytostabilization performance, both chemical and biological elements can be added to polluted soil.
(ii) **Phytoextraction**: Plants are used to absorb, store, and breakdown toxins in the environment. Phytoextraction has been the most popular phytoremediation approach for extracting toxic metals and metalloids from contaminated soil in recent years.
(iii) **Phytodegradation**: It includes the absorption, storage, and breakdown of pollutants via enzymes released by plant tissue.
(iv) **Phytostimulation or Rhizodegradation**: It removes contaminants by exploiting rhizosphere symbiotic soil microbes.
(v) **Phytovolatilization**: The practise of employing plants to absorb dye impurities from the ground, transform them to a less poisonous vapour form, and then return them into the ambient environment via the plant's leaves or foliage system is known as phytovolatilization. The pollutants are not entirely removed from the environment using this strategy. It transports contaminants from the soil to the atmosphere, polluting the air with hazardous volatile chemicals. Precipitation may re-distribute these into the soil.
(vi) **Phytofiltration**: It is the technique of employing plant roots, shoots, or seedlings to absorb or adsorb ionized heavy metals from underground water and aqueous waste [9, 15].

5 Selection of Plants for Remediation of Textile Dyes

Selection of a plant for the phytoremediation process is the first step required to remediate contaminated sites. Textile dye phytoremediation technology involves the elimination of dyes through dye breakdown or dye adsorption and/or deposition. Chemicals accumulated in the roots of a plant can be transferred to its branches and leaves. Non-edible plants species are often selected to avoid colourants or their derivatives entering the food chain. The chemical makeup and molecular mass of species, along with the occurrence or lack of sulphur groups in dye contaminants detected in wastewater, should all be examined. Only then will we be able to achieve a high-efficiency phytoremediation technique, because the substances present in the contaminated environment heavily influence plant selection for phytoremediation [16]. Other operational factors like as pH, contact duration, dye concentration, biomass intake, and temperature all have an impact on phytoremediation efficiency. Finally, the success of phytoremediation methods for dye contaminants is determined by factors such as plant growth rate in polluted areas, bulky specific areas that come into contact with water, the possibility of enhanced translocation, and attributes such as bioconcentration factor (BCF) and translocation factor (TF).Plants such as *Brassica juncea, Rheum rhabarbarum, Tagetes patula, Thymus vulgaris, Rosmarinus officinalis, Phragmites australis, Rumex acetosa, Typha angustifolia, Hydrilla verticillata, Nasturtium officinale, Petunia grandiflora Juss., Glandulariapulchella, Armoracia rusticana* (horseradish) are useful for the removal of contaminants released from the textile industries [17].

6 Plant Mechanisms for the Removal of Textile Dyes

Mechanisms Involving Adsorption of Textile Dyes

Plant processes for detoxifying textile dyes, a prevalent type of organic pollution, might be numerous. Plants utilise adsorption and accumulation processes to eliminate textile dyes from their surfaces, in addition to phytodegradation and phytotransformation, which are the most widely described mechanisms used by plants to breakdown organic compounds. Adsorption of xenobiotics to roots, followed by incorporation into the plants, has been proven [18]. Furthermore, the scientists revealed variations in adsorptive capabilities as a result of environmental conditions such as temperature and pH. Increased temperature favoured improved adsorptive elimination of the dye, most likely due to more mobility of the adsorbents. Additionally, the optimum dye removal effectiveness was found at pH 5, which may correlate to the rate of disintegration of the investigated dye with the maximum molecular ionisation. Orange peels, banana skins, neem foliage, peanut shell, and crop residues have all been proposed for colour removal [19]. Methylene Blue and iodine adsorptive characteristics were discovered in untreated pulverized plant leaves of *Salsola vermiculata* from aqueous medium. Moreover, activating these plant leaves with zinc chloride improved their adsorptive characteristics. The influence of pH on the adsorption efficiency of the chemically activated plant was also examined in this plant, which revealed a reduction in adsorption capacity at lower pH values, perhaps due to protons competing with dye molecules for accessible adsorption sites [20]. Consequently, such activation procedures can improve the plant material's intrinsic adsorptive potential. This plant's considerable adsorption of Methylene Blue demonstrates its use in the cleanup of textile colours and effluents.

Plant Stress Response for the Degradation of Textile Dyes

Whereas dye adsorption and deposition are significant phytoremediation processes, they only result in the increasing amount of contaminants from industrial effluent onto and/or into plant surfaces and do not result in the total elimination of the pollutant. Consequently, a plant's phyto-transforming skills are more important since they may be used to either destroy the dye or modify it into non-toxic compounds that can be safely reintroduced into the environment. Because plants are sedentary and exist in a competing and sometimes harsh atmosphere, they have evolved ways to defend themselves against extrinsic environmental stresses, including the decontamination of xenobiotic substances. Plants have several processes that can be used to repair pollutants that bacteria cannot digest. Various aromatic compounds, such as sulfonated anthraquinone derivatives, exist naturally in numerous plant species, implying that these plants have enzymes that can accept and process these aromatic compounds as substrates. The activity of dioxygenases in delivering oxygen across the double bond containing the sulfonate group, resulting in its elimination, appears to be an important step in the removal of sulfonated anthraquinones. Plant degradation of textile dyes can be intracellular, including enzyme reactions inside plant cells, or extracellular, using enzymes produced by the plant in rhizosphere areas [21].

7 Advantages of Phytoremediation

There are advantages of using phytoremediation, which include:

(i) Since phytoremediation is an autotrophic system fuelled by solar energy, it is simple to handle and has a cheap set up and operation cost,
(ii) Toxins may be retrieved from plant tissues, providing an economic potential.
(iii) It can lessen pollution exposure in the ecosystem and its surroundings. The plants involve in phytoremediation of contaminants are harvested and removed from the polluted sites,
(iv) It may be placed on a large-scale field and readily disposed of. It is the least harmful method to preserve the surrounding in its natural state.

(v) It inhibits erosion and metal leaching by stabilising heavy metals, lowering the danger of pollutants spreading, thereby preserving the soil fertility [22].

8 Limitation of Phytoremediation

The drawbacks of using phytoremediation include:

(a) Partial pollution control with long-term poor performance.
(b) Phytoremediation has limited application to many waste types, particularly high-level toxicity wastes.
(c) It is mainly applicable to the upper layer of the soil surface especially area around the roots. Sometimes the contaminants also leached down polluting the groundwater.
(d) In phytoremediation, adequate waste disposal and risk assessment are essential. The contaminated land could affect the survival of the plant species.
(e) There is a risk of undesired invasive plant species being introduced and spreading.
(f) Seasonal variations also influence the effectiveness of the phytoremediation process.
(g) Plant fatalities may occur in very hazardous areas, raising the overall cost of the operation.
(h) A better knowledge of plant behaviour and physiological changes in response to various forms of waste is required [17].

9 Conclusion

Phytoremediation, which involves the utilization of green plants to consume or decolorize dyestuffs from diverse sources across the world, has been extensively examined and studied. However, the technology is still in its early phases of development due to a lack of research on the physiological mechanisms during dyestuff digestion. As a result, much more study is required to better understand dye-degrading plant species, their remedial methods, the usage of different chemicals, and the impact of related bacterial metabolism on phytoremediation procedures. Furthermore, adopting phytoremediation methods to lessen the effects of dye-containing wastes has shown to be a long-term environmental protection option.

References

1. Garg, S., & Roy, A. (2022). Phytoremediation: An alternative approach for removal of dyes. In *Phytoremediation* (pp. 369–386). Academic Press. ISBN 9780323898744, https://doi.org/10.1016/B978-0-323-89874-4.00005-4.

2. Vikrant, K., Giri, B. S., Raza, N., Roy, K., Kim, K. H., Rai, B. N., & Singh, R. S. (2018). Recent advancements in bioremediation of dye: current status and challenges. *Bioresource Technology, 253*, 355–367. ISSN 0960-8524, https://doi.org/10.1016/j.biortech.2018.01.029.
3. Ihsanullah, I., Jamal, A., Ilyas, M., Zubair, M., Khan, G., & Atieh, M. A. (2020). Bioremediation of dyes: Current status and prospects. *Journal of Water Process Engineering, 38*, 101680. ISSN 2214-7144, https://doi.org/10.1016/j.jwpe.2020.101680.
4. Khan, R., Patel, V., & Khan, Z. (2020). Bioremediation of dyes from textile and dye manufacturing industry effluent. In *Abatement of Environmental Pollutants* (pp. 107–125). Elsevier. ISBN 9780128180952. https://doi.org/10.1016/B978-0-12-818095-2.00005-9.
5. Sharma, S., & Kaur, A. (2018). Various methods for removal of dyes from industrial effluents—A review. *Indian Journal of Science and Technology, 11*(12). https://doi.org/10.17485/ijst/2018/v11i12/120847.
6. Beenish, S., Madeeha, J., Azeem, K., Irfan, A., & Ann, D. (2015). Effectiveness of rice agricultural waste, microbes and wetland plants in the removal of reactive black-5 azo dye in microcosm constructed wetlands. *International Journal of Phytoremediation, 17*, 1060–1067.
7. Bhatia, D., Sharma, N., Singh, J., & Kanwar, R. (2017). Biological methods for textile dye removal from wastewater. *Critical Reviews in Environmental Science and Technology, 47*, 1836–1876. https://doi.org/10.1080/10643389.2017.1393263.
8. Yan, A., Wang, Y., Tan, S. N., Mohd Yusof, M. L., Ghosh, S., & Chen, Z. (2020). Phytoremediation: A promising approach for revegetation of heavy metal-polluted land. *Frontiers in Plant Science, 11*, 513099. https://doi.org/10.3389/fpls.2020.00359.
9. Bharathiraja, B., Jayamuthunagai, J., Praveenkumar, R., & Iyyappan, J. (2018). Phytoremediation Techniques for the Removal of Dye in Wastewater. https://doi.org/10.1007/978-981-10-7485-1_12.
10. Hussein, A., & Scholz, M. (2017). Dye wastewater treatment by vertical-flow constructed wetlands. *Ecological Engineering, 101*, 28–38.
11. Usharani, K., & Arunkumar, V. (2023). Bioremoval and resource recovery of nutrients by phytoremediation using aquatic free floating plants Lemna minor. *Ukrainian Journal of Ecology, 13*(5), 13–27. https://doi.org/10.15421/2023_449
12. Dwivedi, P., & Tomar, R. S. (2018). Bioremediation of textile effluent for degradation and decolourization of synthetic dyes: A review. International *Journal of Current Research in Life Sciences, 7*(4), 1948–1951. ISSN: 2319–9490.
13. Kaur, N., & Kaushal, J. (2022). Screening the six plant species for phytoremediation of synthetic textile dye waste water. *Materials Today: Proceedings, 71*(2), 232–238. ISSN 2214-7853, https://doi.org/10.1016/j.matpr.2022.08.512.
14. Tahir, U., Yasmin, A., & Khan, U. H. (2016). Phytoremediation: potential flora for synthetic dyestuff metabolism. *Journal of King Saud University-Science, 28*(2), 119–130. ISSN 1018-3647, https://doi.org/10.1016/j.jksus.2015.05.009.
15. Wang, L., Ji, B., Hu, Y., Liu, R., & Sun, W. (2017). A review on in situ phytoremediation of mine tailings. *Chemosphere*. https://doi.org/10.1016/j.chemosphere.2017.06.025.
16. Mishra, S., Kumar, A., & Shukla. (2015). Study of water quality in Hindon River using pollution index and environmetrics. *India Desalination Water Treatment*. https://doi.org/10.1080/19443994.2015.1098570.
17. Dhanwal, P., Kumar, A., Dudeja, S., Chhokar, V., & Beniwal, V. (2017). Recent advances in phytoremediation technology. *Advances in Environmental Biotechnology*, 227–241.
18. Yadav, S., Yadav, A., Bagotia, N., Sharma, A. K., & Kumar, S. (2021). Adsorptive potential of modified plant-based adsorbents for sequestration of dyes and heavy metals from wastewater—A review. *Journal of Water Process Engineering, 42*, 102148. ISSN 2214-7144, https://doi.org/10.1016/j.jwpe.2021.102148.
19. Shi, Y., Chang, Q., Zhang, T., Song, G., Sun, Y., & Ding, G. (2022). A review on selective dye adsorption by different mechanisms. *Journal of Environmental Chemical Engineering, 10*(6), 108639. ISSN 2213-3437, https://doi.org/10.1016/j.jece.2022.108639.
20. Velusamy, S., Roy, A., Sundaram, S., & Kumar Mallick, T. (2021). A review on heavy metal ions and containing dyes removal through graphene oxide-based adsorption strategies for textile wastewater treatment. *The Chemical Record, 21*(7), 1570–1610.

21. Banerjee, P., Barman, S. R., Sikdar, D., Roy, U., Mukhopadhayay, A., & Das, P. (2017). Enhanced degradation of ternary dye effluent by developed bacterial consortium with RSM optimization, ANN modeling and toxicity evaluation. *Desalination and Water Treatment, 72*, 249–265: ISSN 1944-3986, https://doi.org/10.5004/dwt.2017.20422.
22. Jacob, J. M., Karthik, C., Saratale, R. G., Kumar, S. S., Prabakar, D., Kadirvelu, K., et al. (2018). Biological approaches to tackle heavy metal pollution: A survey of literature. *Journal of Environmental Management, 217*, 56–70. https://doi.org/10.1016/j.jenvman.2018.03.077.

Phycoremediation of Toxic Azo Dyes from Contaminated Environment

Ashish Bedi, Khushi Jain, and Akanksha Jaiswar

1 Introduction

Dyes are gaining increased prominence due to their diverse applications across various industries. The challenge lies in efficiently eliminating dyes from wastewater because of their stable "aromatic molecular structure", which makes them resistant to heat, light, and oxidizing agents. Imparting color to various products or materials, such as paper, leather, textiles, rubber, plastic, and cosmetics, is the major application of dyes. The addition of color enhances the visual appeal of a product and adds value for aesthetic and social reasons. This dyes can be derived from natural sources or synthesized by humans. Synthetic dyes, have huge applications in industries like textiles, tannery, paper, printing, dyeing, food, and cosmetics, are aromatic, water-soluble, and dispersible organic colorants. Despite their contribution to enhancing the visual world, synthetic dyes cause great damage to environment.

Globally, approximately 1 million tons of dyes are produced in a year. A substantial amount of synthetic dyes is discharged into the environment through industries like textiles, pharmaceuticals, and chemicals. Among these, the textile industry stands out as one of the largest consumers of synthetic dyes. The popularity of synthetic dyes in the industry textile and dyeing, is due to their flexibility of synthesis, cost-inexpensive, and various merits such as stability in high-light, temperature-resistant, detergents, and microbes, as well as a huge range of colours in comparison to natural dyes. However, discharge of dyes is the primary environmental concern during the dyeing process, with an efficiency of fixation ranging from 60 to 90%. Approximately

A. Bedi (✉)
School of Life and Environmental Sciences, Deakin University, Geelong, Australia
e-mail: ashish.bedi.del@gmail.com

K. Jain
Central University of Punjab, Bathinda, India

A. Jaiswar
Mossakowski Medical Research Centre, Polish Academy of Sciences, Warsaw, Poland

R. A. Bhat et al. (eds.), *Biocentric Approaches in Textile Waste Management*, Textile Science and Clothing Technology, https://doi.org/10.1007/978-3-032-04974-2_15

'10–15%' of dyes are observed to be released into wastewater during the process of dyeing, contributing to environmental pollution.

In the past, a number of chemical and physical methods have been employed to address this issue, including 'adsorption, chemical precipitation, photolysis, chemical oxidation and reduction, electrochemical treatment, filtration, coagulation, and chemical flocculation'. However, the demerits of such methods are high cost, lower efficiency of decolourization, primarily for synthetic dyes, and complex extraction procedures for their organic metabolites. Additionally, these techniques generate secondary sludge, leading to further pollution issues as "secondary-environmental contaminants". Due to these shortcomings, bioremediation has gained considerable attention as a more sustainable alternative to physicochemical treatment. Bioremediation involves the use of microbes such as Algae, yeast, fungi, and bacteria and plants to accelerate the remediation or transformation of contaminants into less harmful forms.

2 Phyco-Remediation, a Subset of Bioremediation, is a Suitable Approach for Synthetic Dye Removal Using Algae

The key benefits of algae-based treatment of wastewater, surpassing other biological procedures, include:

Rapid Growth Rate

Algae exhibit a high growth rate and have a doubling time of less than 24 h, facilitating the rapid attainment of a substantial cell population during cultivation [17, 50].

Bio-Mitigation and CO_2 Sequestration

Algae possess bio-mitigation and CO_2 sequestration abilities, contributing to environmental sustainability [17, 32].

Resilience to Adverse Weather

The doubling rate of algae remains unaffected from adverse weather conditions, ensuring consistent performance even in challenging environments [32].

No Competition with Food Crops

For cultivable and arable land, Algae-based treatment eliminates concerns about competition with food crops, addressing ethical and environmental considerations [50].

Adaptable Growth Conditions

Algae demonstrate adaptability by thriving in sea-water, waste-water, and both fresh-water and saline-water, enhancing their suitability for diverse settings [50].

Effectiveness of Dead Algae

Dead algae can be as efficient as live algal cells for pollutant removal in wastewater treatment, especially through biosorption, with the added benefit of lower operational costs since they do not require nutrients for survival [22].

Enhanced Bioremediation Efficiency

Algae exhibit enhanced bioremediation efficiency compared to other biological methods, and their ability to generate value-added products adds an economic dimension to the treatment process.

3 Overview of Industrialization and Environmental Hazards

Textile mills and industries release vast quantities of hazardous wastes, characterized by vivid colours and containing organic chemicals, into the environment. The discharged wastewater is a complex mixture of 'sulphur, naphthol, vat dyes, acetic acid, soaps, nitrates, chromium compounds', and heavy metals like 'copper, lead, arsenic, cadmium, mercury, nickel, and cobalt'. This combination results in the high toxicity of wastewater, featuring increased pH and temperature levels, causing severe environmental damage. The presence of colours and oils contributes to increased turbidity, giving the water an undesirable appearance and foul odour [39].

Upon discharge into freshwater bodies, this effluent obstructs sunlight penetration crucial for the photosynthesis process of aquatic flora and fauna, disrupting the ecosystem [9–11, 15, 16]. Furthermore, it disrupts the mechanism of oxygen transfer at the interface of air and water, impeding the self-purification process of water. Agricultural fields irrigated with these effluents experience soil pore clogging, leading to a decreased soil productivity and hardened soil texture, hindering penetration of root [14]. In drains, sewage pipes get corroded and tarnished because of presence of waste-water, similarly in rivers, it adversely impacts quality of drinking water from hand pumps, making the water not suitable for consumption by humans.

Wastewater from textile industry emerges as a potential contributor to degeradation of environment and poses risks of adverse health to mankind. The primary challenge in mitigating textile wastewater lies in the presence of organic chemicals that can react with disinfectants. Evaporation of these chemicals into the air poses health risks, as they can be absorbed through the skin, resulting in allergic reactions and potential harm to children, even during prenatal development.

4 Hazardous Effects of Synthetic Dyes

Water Pollution from Dye Contamination

Water is an indispensable element for various vital activities, playing a pivotal role in agricultural growth and industrial development. Unfortunately, potable water sources are becoming scarce due to factors like population increase, urbanization, and industrialization, leading to contemporary environmental concerns related to excessive exploitation and contamination. Water pollution poses health risks, directly or indirectly to living organisms that are present on Earth, affecting the environment greatly. Dyes, extensively used in industries such as "plastic, printing, pharmaceutical, food, cosmetic, photography, and leather," contribute significantly to these challenges.

The commercial availability of over ten thousand dyes and a production of approximately 7 million metric tons per year underscore the magnitude of the issue. The market of dye, was at the value of 33 billion USD in 2019, is anticipated to increase about 49 billion USD by 2027. The industry of textiles, in particular, consume substantial amounts of fresh-water in the textile process of dyeing, releasing contaminated wastewater. For instance, there is requirement of approximately 2 million litres of water for the production of 8000 kg of fabric. There is a different proportion at which different types of dye molecules bind with fabric, resulting in the discharge of 2–50% of pigments as effluent during processing, leading to pollution of nearby soil and water bodies.

Studies have shown that reactive oxygen species are produced in plants due to the presence of toxic pollutants in water or soil. The textile dyeing industry alone contributes 17–20% of contaminated wastewater, according to a World Bank study. The high-water solubility of dyes poses a challenge in their extraction using traditional methods, and "the release of untreated effluents overloaded with colours into water bodies has destructive ecological consequences". Additionally, there is use of organic solvents for the production of dye that chelate with metals, resulting in wastewater containing "acids, alkalis, salts, and nitro compounds". Even a low pigment concentration of 1.0 mg/L can visibly contaminate water. The presence of colour in effluent poses a significant environmental concern.

Release of Pollutants into the Environment

It is widely acknowledged that pollution in environment stands as potential challenges faced by humankind, with the industry of textile emerging as main contributor because of the substantial discharge of effluent pollutants into water bodies. The persistent color and exceptionally high "biological oxygen demand (BOD)" associated with textile industry effluents are not only environmentally unappealing but also aesthetically unattractive.

The textile industry holds the unfortunate distinction of being the largest producer of colored wastewater. Approximately 20% of synthetic dyes used in textile fiber dyeing remain unfixed to the fibers, leading to their disposal in wastewater. This disposal of unbound azo dyes results in substantial pollution, as these dyes goes into wastewater streams. The azo dyes biodegeradation poses a significant challenge due to their complex synthetic structure.

In general, Azo dyes exhibit harmful effects, including the ability to induce mutations, cytotoxicity, and carcinogenicity. The most substantial prolonged human health risk associated with textile dyes is their ability to cause the genotoxicity, causing DNA or chromosomal damage. Unfortunately, several textile processing plants fail to adequately treat their wastewater to break down azo dyes, leading to the direct release of untreated effluents into water bodies. This discharge poses significant ecotoxicological threats and toxic effects on living organisms. Moreover, conventional wastewater systems struggle to eliminate dyes, heavy metals, and electrolytes. The discharge of untreated textile effluents not only presents an aesthetic nuisance but also triggers various environmental issues. It reduces the reoxygenation capacity of water bodies, inhibits sunlight penetration, affecting the photosynthesis ability of aquatic plants, and increases the demand for oxygen and solid particles in water bodies. Under specific conditions, it may lead to the formation of aromatic amines, known for their carcinogenic nature. Health risks stemming from sporadic and excessive exposure to colored effluents include immunological, neurological, and circulatory disorders, lung edema, eye and skin infections, and allergy problems.

Importantly, environmental damage is not solely contingent on the quantity of disposed dye but also on the composition of the dye mixture, with common constituents exhibiting toxic characteristics. Colored wastewater from synthetic dyes, even at concentrations visible to the naked eye (<1 ppm), reduces the concentration of dissolved oxygen, alters physicochemical and biological parameters, and introduces various pollutants such as heavy metals (chromium, arsenic, zinc) into surface water or groundwater bodies. Synthetic azo dyes, characterized by their non-biodegradability, chemical properties, and structure, often prove recalcitrant, carcinogenic, and toxic, leading to enduring adverse effects on the environment.

When azo dyes are disposed of into surface water or wastewater, they have the potential to bind to suspended organic matter through electrostatic interactions. This binding process leads to their adherence to sediments or wastewater sludge, ultimately increasing their persistence in the environment. The presence of colored water and contaminated sludge creates a pathway for aquatic animals to uptake these hazardous compounds. Subsequently, these compounds can be transferred through the food chain to humans, giving rise to various health disorders, including hypertension, cramps, nausea, bleeding, and ulceration of the skin or mucous membranes. Exposure to azo dyes can cause significant damage to vital organs, including the kidneys, reproductive system, liver, brain, and central nervous system. The potential health risks associated with such exposure underscore the importance of addressing the disposal and persistence of azo-ionic dyes in aquatic environments. Biological methods for treatment.

5 Azo Dyes

Azo compounds, represented as R–N=N–R′, are chemical structures where the –N= N– group signifies the azo group, with R or R′ being aryl or alkyl compounds. According to the International Union of Pure and Applied Chemistry (IUPAC), azo compounds are described as "derivatives of diazene (diimide), HN=NH, wherein both hydrogens are substituted by hydrocarbyl groups, e.g., PhN=NPh azobenzene or diphenyldiazene." The term "azo" stems from "azote," the French term for nitrogen, originating from the Greek words "a" (not) and "zoe" (to live). Historically, the advent of azo dyes marked a significant milestone in the chemical industry's development.

Azo dyes are comprised of a diazotized amine coupled to an amine or a phenol, containing one or more azo linkages. Aromatic amines act as crucial precursors for azo dyes. With their vivid colors, azo compounds constitute roughly two-thirds of all synthetic dyes, rendering them the most extensively used and structurally diverse class of organic dyes in commerce. Historically, a minimum of 3000 azo dyes found application across various industries, including pharmaceuticals, paper, printing inks, paints, varnish, lacquer, wood stains, and colorants for synthetic and natural textile fibers, plastics, leather, hair dyes, waxes, and petroleum.

These dyes form the largest and most versatile class of dyes, accounting for over 50% of worldwide dye production. Presently, there are over 2000 distinct azo dyes in use, with global production exceeding 7×10^5 tons. In 1991, the U.S. Food and Drug Administration (FDA) certified over 3000 tons of azo dyes for use in foods, drugs, and various applications. They constitute a significant portion of FDA-certified colorants, representing a major group.

Due to their light stability and resistance to microbial degradation or fading through washing, azo dyes pose challenges in their elimination from wastewater using conventional treatment methods. Approximately 10% of dyestuff used in textile dyeing processes does not bind to fibers and is consequently released into the environment. Certain components of azo dyes, such as benzidine, have been linked to human bladder cancers. Furthermore, dye workers exposed to azo dyes exhibit an increased incidence of bladder cancer. Thus, azo dyes pose severe health risks, including lethality, genotoxicity, mutagenicity, and carcinogenicity, affecting both humans and animals. The indiscriminate disposal of azo dyes into the environment, particularly from the textile industry, poses a significant threat to human health and the environment.

6 Phycoremediation

Phycoremediation, originating from the Greek term 'phykos' meaning algae and the Latin 'remedium' signifying restoring balance, denotes a biological method for treating wastewater. Essentially, it encompasses the removal or degradation of pollutants like nutrients, heavy metals, and xenobiotics from wastewater by employing algae, both micro and macro, as well as cyanobacteria [35, 36]. Despite the recent coinage of the term 'phycoremediation' by John in 2000, the concept of utilizing algae for wastewater remediation was introduced six decades ago by Oswald and coworkers [37]. Since then, extensive research has been conducted on phycoremediation [1, 38, 46], and it is widely acknowledged as a promising approach [36]. Over the years, algae's diverse range and adaptability to extreme environments have prompted scientists to actively screen and identify strains suitable and promising for wastewater treatment [21]. Phycoremediation represents a versatile technology encompassing wastewater treatment, CO_2 mitigation, and simultaneous biomass generation. The produced biomass holds potential applications as an animal feed additive, feedstock for biofuel and biogas, bio-ore for heavy metals [49] and as a source for extracting vitamins, β carotenes, and fine chemicals like antioxidants [32]. Nevertheless, algae grown in wastewater are unsuitable for use as animal feed and for the extraction of fine chemicals. Yet, their viability as an energy source, such as for biofuel or biogas, remains promising.

Various microorganisms may utilize different pathways to degrade different dyes, influenced by factors like the dye's structure, the microbial system's degradation strategy, and other considerations. Minor structural disparities can have a significant impact on the decolorization process [4]. Microbial removal of dyes typically involves adsorption or degradation [52]. Biosorption refers to the movement of dye molecules from the aqueous phase to the solid phase (the bioadsorbent), whereas biodegradation involves the enzymatic breakdown of the chemical bonds within the dyes [57]. Among the microorganisms studied for their role in dye decolorization and degradation, blue-green algae, green algae, and diatoms are notable [47]. Due to its extensive surface area and strong binding capacity, algae biomass demonstrates significant biosorption efficiency and electrostatic attraction towards pollutants in wastewater. Several studies have illustrated that pollutants and their metabolites in wastewater are effectively adsorbed by the surface of algae [12]. Algal decolorization mechanisms may entail enzymatic degradation, adsorption, or a combination of both processes. Algae predominantly eliminate color by utilizing chromophores for biomass production, transforming colored molecules into non-colored ones, and adsorbing chromophores onto algal biomass [12, 27, 33].

Algae have the ability to degrade azo dyes by inducing the production of azoreductase, which facilitates the reductive cleavage of azo bonds. This process releases aromatic amines, which then undergo degradation into CO_2 and H_2O [3]. The significant surface area and strong binding affinity of algae during biosorption contribute to their enhanced biosorption capacity. Electrostatic attraction and complexation are pivotal factors in algal biosorption [40].The presence of functional groups on algae cell surfaces aids in decolorization by attracting dye contaminants. The biosorption capacity of algae is predominantly influenced by the cell wall, which comprises heteropolysaccharides and lipid compounds containing diverse functional groups like amino, carboxyl, hydroxyl, and phosphate groups. These functional groups facilitate robust attractive forces between dyes and the cell wall. Alginate, found in brown algae, plays a vital role in adsorption by binding to pollutant ions. Metal ions such as K^+, Na^+, Ca^{2+}, and Mg^{2+} present on algal biomass, obtained from seawater, bind to acid functional groups [41].

7 Genetic Alteration of Phyco-Remediator

Phycoremediation continues to be a highly efficient method for tackling dye contamination. Nevertheless, despite various techniques, microalgae frequently encounter challenges in fully and rapidly remediating these dyes. Achieving comprehensive bioremediation may require the utilization of more advanced technologies like genetic engineering, metabolic engineering, or protein engineering. Enhancing the efficacy of phycoremediation involves delving into research at the genetic level and modifying the genetic makeup of microalgae to enhance the process's effectiveness. By conducting in-depth genomic analysis, it becomes possible to devise biomarkers customized for particular contaminated sites.

8 Techniques for Genetic Modification

Progress in science allows for the precise insertion, removal, or substitution of DNA molecule fragments containing one or more nucleotides into an organism's genome. This capability facilitates targeted genetic engineering at both the global (genome) and local (gene) levels [48]. Effector nucleases (TALEN) and clustered regularly interspaced short palindromic repeat (CRISPR)-associated nucleases are frequently employed as transcription activators for gene editing [45].

The pivotal advancement in gene editing occurred with the revelation of the CRISPR/Cas system [13] and ultimate application of the CRISPR/Cas [18] system, where a shared endonuclease (typically Cas9, though other proteins like Cpf1, also known as Cas12, function similarly) is guided to a specific genomic DNA target via a short, readily engineered RNA sequence known as a guide sequence. New genome editing techniques such as RNAi and ZNFs have been utilized to genetically modify microalgae. Improving dye degradation or decolorization can be accomplished by employing genetically modified organisms under different environmental conditions. These GMOs can be produced via gene transfer among distinct species or through gene modification methods [26, 29, 44]. Based on our current literature review, no publications have addressed the genetic modification of phycoremediators. However, genetic alterations have been successfully achieved in other organisms such as *E. coli*, suggesting a similar approach could be adopted for genetically modifying microalgae. Various bacterial species, including *Achromobacter*, *Pseudomonas*, *Dehalococcoides*, *Rhodococcus*, *Comamonas*, *Burkholderia*, *Alcaligenes*, *Bacillus subtillus*, *Aspergillus niger*, *Deinococcusradiodurans*, *Acidithiobacillusferrooxidans*, *Mesorhizobiumhuakuii*, *Pseudomonas* K-62, *Ralstonia*, *Rhodopseudomonas palustris*, and *Sphingomonas*, have been employed in bioremediation processes to convert toxic substances into non-toxic ones [54]. In 2008, Sandhya engineered Escherichia coli SS125 to degrade Remazol red dye by transferring the azoreductase gene from Bacillus latrosporus RRK1 to Escherichia coli DH5a using Plasmid pAZR-SS125 [43]. In 2009, Jin and collaborators developed genetically modified *E. coli* JM109 (pGEX-AZR) in the laboratory, showcasing its ability to decolorize direct blue 71 [25]. The integration of the azoreductase gene into the expression vector pGEX4T-1 enabled this achievement. Following this, the vector was expressed and introduced into E. coli JM109 under the regulation of a lac operon. In 2020, Ajaz and colleagues observed the degradation of Remazol red in the presence of 0.8 mg/L dissolved oxygen, facilitated by the azoreductase gene derived from *Bacillus latrosporus* RRK1 and integrated into *Escherichia coli* [2]. The initial phase in genetically modifying microalgae involves characterizing genes that encode putative enzymes engaged in azo dye degradation within the green alga. Subsequently, these genes can be cloned into expression vectors such as pGenD, facilitating their utilization in the genetic manipulation of microalgae like *Chlamydomonas* to generate mutants capable of upregulating these genes. Azoreductase was overexpressed in *E. coli* YB, enabling this strain to effectively decolorize sulfonated azo dyes [30]. The azoreductase gene has the potential to be transferred from bacteria to microalgae, or alternatively, the existing azoreductase gene within the microalgae can be overexpressed.

Another approach to alter the genetic makeup of microorganisms involves UV irradiation. In 2013, Joshi and Thagard developed a UV-irradiated mutant of Pseudomonas sp. 1F for degrading the sulfonated azo dye Green HE4B (GHE4B). Similarly, UV-irradiated mutants can be created in microalgae, which can then be assessed for their effectiveness in reducing azo dyes.

9 Overview of Synthetic Dyes

Derived from organic compounds, synthetic dyes play an increasingly essential role in modern life, serving diverse purposes across industries like paint production and scientific laboratories. In labs, they are utilized as fluorescent tracers and photoredox catalysts. Chemists have devised numerous strategies to simplify the synthesis of complex synthetic organic dyes [56].

10 Categories of Synthetic Dyes

Dyes are typically categorized based on various criteria: (i) the presence of chromophoric groups in their chemical structures, including azo dyes, anthraquinone dyes, and phthalocyanine dyes, (ii) their method of application or use, such as polyester dyes and cotton reactive dyes, and (iii) their behavior upon dissociation into aqueous solutions, classified as acidic, basic (cationic), direct reactive (anionic), and disperse/nonionic [7, 51]. Reactive dyes, which form a covalent bond with the substrate during application, provide a broad spectrum of shades and demonstrate excellent lightfastness and wash fastness on cotton, making them highly sought-after in the textile sector [20]. The 3 predominant classes of reactive dyes include azo, anthraquinone, and phthalocyanine [55], among which reactive azo dyes, characterized by intense color and high cellulose reactivity, hold particular significance [28]. Acid dyes are applied to nylon, wool, silk, leather, paper, food, and cosmetics, and they encompass chemical groups such as azo, anthraquinone, triphenylmethane, azine, xanthene, nitro, and nitroso [6]. These dyes are generally bright with varying resistance to washing due to structural diversity and may include some metal complexes [20]. Azo dyes, which constitute the largest group of dyes, are characterized by a chromophore of N=N in an aromatic system [5, 12]. They contain functional groups such as –N=N–, –C=O, $-NO_2$, and $O-(C_6H_4)-O$ (quinoid assemblies), while auxochromes include $-NH_3$, –COOH, –OH, and $-SO_3H$, and are further classified into monazo, disazo, trisazo, tetrakisazo, and polyazo dyes [8]. Triarylmethane dyes, existing in both acidic and basic forms, are hydrocarbon derivatives widely used in wool and silk dyeing. Acidic triarylmethane dyes contain at least 2 SO_3H groups, making them suitable for wool and silk fibers, whereas basic triarylmethane dyes are employed in stamping inks, writing, and printing [8]. Anthraquinone dyes, soluble in water due to sulfonic acid groups, are utilized in the wool and silk industries because of their compatibility with auxiliary binding agents, and they are also considered a subclass of acid dyes. Additionally, food coloring dyes, belonging to this category, are used to dye protein fibers and certain nylon fibers at elevated temperatures [12]. Direct dyes are extensively utilized in dyeing protein and synthetic fibers such as nylon and rayon, though they do not exhibit fast-drying characteristics post-application. Basic dyes, which form colored cationic salts in water, strongly adhere to anionic substrates and are recognized for their strong coloring capabilities, particularly on acrylic fibers [3]. Disperse dyes, water-insoluble and nonionic, are commonly applied to polyester, nylon, cellulose, and acrylic fibers. They include chemical groups such as azo, anthraquinone, styryl, nitro, and benzodifuranone. Solvent-soluble dyes, primarily belonging to azo and anthraquinone classes, are used for coloring plastics, gasoline, lubricants, oils, and waxes. Sulfur (sulphur) dyes are employed for coloring cotton, rayon, silk, leather, paper, and wood, while vat dyes, insoluble in water, are used for dyeing cellulosic fibers and are mainly composed of anthraquinone and indigoid groups [3].

11 Potential Environmental and Health Impact

Effluent discharged from dyeing industries significantly contributes to water pollution due to its substantial dye content. The high concentration of dyes in water bodies obstructs sunlight and reduces oxygen levels, thereby affecting aquatic biodiversity and photosynthesis. While the public may tolerate blue, green, or brown water, red and purple hues raise concerns [34]. Natural pigments used in textile coloring degrade readily when exposed to sunlight, heat, and water, resulting in the loss of approximately 15% of textile dyes during the dyeing process. This effluent can lead to mutagenic, carcinogenic, and toxic effects [19]. Synthetic dyes inherently pose health risks, either through their chemical composition or the non-degradable by products formed during their production [23]. Effluents from industries containing dyes disrupt microbial growth and the photosynthetic activities of aquatic flora by reducing light penetration in water bodies. Additionally, discharged dyes decrease dissolved oxygen levels by forming a thin layer on the water's surface, leading to an increase in chemical oxygen demand (COD) and indicating elevated pollution levels [31]. Many dyes are carcinogenic, posing significant hazards to aquatic organisms, including liver, gills, intestine, and gonadal damage, and disrupting aquatic ecosystems [24]. The direct application of untreated dyeing effluents for irrigation results in environmental consequences, including diminished soil fertility, reduced seed germination, and decreased levels of chlorophyll and protein content in plants [42]. Azo dyes have been linked to various health concerns in humans, such as skin irritation, chemosis, contact dermatitis, exophthalmos, lacrimation, rhabdomyolysis, permanent blindness, vomiting, gastritis, acute tubular necrosis, hypertension, vertigo, and, if ingested, edema of the neck, face, tongue, pharynx, and larynx, along with respiratory distress [7]. Certain dyes have been linked to bladder cancer, splenic sarcomas, hepatocarcinoma, and nuclear abnormalities in both humans and laboratory animals [53]. Through chemical processes, dyes undergo changes in their chemical structure, resulting in the creation of new xenobiotic compounds that have a substantial impact on the environment. Moreover, these dyes have been associated with skin irritation, allergies, cancer, neurobehavioral disorders, autoimmune diseases, nausea, vomiting, paralysis, and severe damage to organs such as the liver, kidneys, reproductive system, brain, and central nervous system [53]. Some dyes with monoazo or anthraquinone structures have been associated with hypersensitivity reactions, such as skin irritation or dermatitis [7].

12 Conclusion and Future Direction

The use of dyes has seen a notable increase due to their diverse applications across various industries. However, the discharge of dyes into the environment not only results in aesthetically unpleasant effects but also raises serious environmental concerns. In recent years, there has been a growing interest in the remediation of dye-contaminated wastewater through phycoremediation technology. Phycoremediation technology holds particular significance in developing and underdeveloped countries due to its advantages, including ease of availability, high efficiency, cost-effectiveness, large specific surface area, environmental friendliness, and chemical and physical stability. Algae feature various functional groups on their cell walls, such as amino, carboxyl, hydroxyl, and phosphate groups, which play a key role in the dye removal process. Several operational conditions, including solution pH, contact time, initial dye concentration, adsorbent dosage, and temperature, significantly influence dye removal. Consequently, these factors are carefully considered when evaluating the performance of algal capacity in dye abatement. Despite significant advancements in alga-based removal technologies, notable limitations persist, highlighting the need for further research to enable the effective utilization of holistic phycoremediation removal techniques on a large scale.

References

1. Abdel-Raouf, N., Al-Homaidan, A., & Ibraheem, I. (2012). Microalgae and wastewater treatment. *Saudi Journal of Biological Sciences, 19*, 257–275.
2. Ajaz, M., Shakeel, S., & Rehman, A. (2020). Microbial use for azo dye degradation—A strategy for dye bioremediation. *International Microbiology, 23*, 149–159.
3. Alaguprathana, M., & Poonkothai, M. (2017). Bioremediation of textile dyeing effluent using algae-a review. *Journal of Advances in Microbiology, 7*, 1–12.
4. Ali, H. (2010). Biodegradation of synthetic dyes—A review. *Water, Air, & Soil Pollution, 213*, 251–273.
5. Asad, S., Amoozegar, M., Pourbabaee, A. A., Sarbolouki, M., & Dastgheib, S. (2007). Decolorization of textile azo dyes by newly isolated halophilic and halotolerant bacteria. *Bioresource Technology, 98*, 2082–2088.
6. Axelsson, J., Nilsson, U., Terrazas, E., Aliaga, T. A., & Welander, U. (2006). Decolorization of the textile dyes reactive red 2 and reactive blue 4 using Bjerkandera sp. Strain BOL 13 in a continuous rotating biological contactor reactor. *Enzyme and Microbial Technology, 39*, 32–37.
7. Ayele, A., Getachew, D., Kamaraj, M., & Suresh, A. (2021). Phycoremediation of synthetic dyes: An effective and eco-friendly algal technology for the dye abatement. *Journal of Chemistry, 2021*, 1–14.
8. Benkhaya, S., M'rabet, S. & El Harfi, A. (2020). Classifications, properties, recent synthesis and applications of azo dyes. *Heliyon, 6*.

9. Bharagava, R., Chandra, R., & Rai, V. (2008). Phytoextraction of trace elements and physiological changes in Indian mustard plants (Brassica nigra L.) grown in post methanated distillery effluent (PMDE) irrigated soil. *Bioresource Technology, 99*, 8316–8324.
10. Bharagava, R. N., & Chandra, R. (2010). Biodegradation of the major color containing compounds in distillery wastewater by an aerobic bacterial culture and characterization of their metabolites. *Biodegradation, 21*, 703–711.
11. Bharagava, R. N., & Chandra, R. (2010). Effect of bacteria treated and untreated post-methanated distillery effluent (PMDE) on seed germination, seedling growth and amylase activity in Phaseolus mungo L. *Journal of Hazardous Materials, 180*, 730–734.
12. Bhatia, D., Sharma, N. R., Singh, J., & Kanwar, R. S. (2017). Biological methods for textile dye removal from wastewater: A review. *Critical Reviews in Environmental Science and Technology, 47*, 1836–1876.
13. Bolotin, A., Quinquis, B., Sorokin, A., & Ehrlich, S. D. (2005). Clustered regularly interspaced short palindrome repeats (CRISPRs) have spacers of extrachromosomal origin. *Microbiology, 151*, 2551–2561.
14. Chandra, R., Bharagava, R., Yadav, S., & Mohan, D. (2009). Accumulation and distribution of toxic metals in wheat (Triticum aestivum L.) and Indian mustard (Brassica campestris L.) irrigated with distillery and tannery effluents. *Journal of Hazardous Materials, 162*, 1514–1521.
15. Chandra, R., Bharagava, R. N., Kapley, A., & Purohit, H. J. (2011). Bacterial diversity, organic pollutants and their metabolites in two aeration lagoons of common effluent treatment plant (CETP) during the degradation and detoxification of tannery wastewater. *Bioresource Technology, 102*, 2333–2341.
16. Chandra, R., Bharagava, R. N., Kapley, A., & Purohit, H. J. (2012). Characterization of Phragmites cummunis rhizosphere bacterial communities and metabolic products during the two stage sequential treatment of post methanated distillery effluent by bacteria and wetland plants. *Bioresource Technology, 103*, 78–86.
17. Chen, G., Zhao, L., & Qi, Y. (2015). Enhancing the productivity of microalgae cultivated in wastewater toward biofuel production: A critical review. *Applied Energy, 137*, 282–291.
18. Cong, L., Ran, F. A., Cox, D., Lin, S., Barretto, R., Habib, N., Hsu, P. D., Wu, X., Jiang, W., & Marraffini, L. A. (2013). Multiplex genome engineering using CRISPR/Cas systems. *Science, 339*, 819–823.
19. De Lima, R. O. A., Bazo, A. P., Salvadori, D. M. F., Rech, C. M., De Palma Oliveira, D. & De Aragão Umbuzeiro, G. (2007). Mutagenic and carcinogenic potential of a textile azo dye processing plant effluent that impacts a drinking water source. *Mutation Research/Genetic Toxicology and Environmental Mutagenesis, 626*, 53–60.
20. El Harfi, S. & El Harfi, A. (2017). Classifications, properties and applications of textile dyes: A review. *Applied Journal of Environmental Engineering Science,* 3, 00000–3 N° 3 (2017) 311–320.
21. Fouilland, E. (2012). Biodiversity as a tool for waste phycoremediation and biomass production. *Reviews in Environmental Science and Bio/Technology, 11*, 1–4.
22. Fu, Y., & Viraraghavan, T. (2001). Fungal decolorization of dye wastewaters: A review. *Bioresource Technology, 79*, 251–262.
23. Hamza, S. B. (2022). *Assessment of Physicochemical Properties and Microbial Community of Soil Contaminated with Dye from Textile Industry.* Kwara State University.
24. Jan, S., Mishra, A. K., Bhat, M. A., Bhat, M. A., & Jan, A. T. (2023). Pollutants in aquatic system: A frontier perspective of emerging threat and strategies to solve the crisis for safe drinking water. *Environmental Science and Pollution Research, 30*, 113242–113279.
25. Jin, R., Yang, H., Zhang, A., Wang, J., & Liu, G. (2009). Bioaugmentation on decolorization of CI Direct Blue 71 by using genetically engineered strain Escherichia coli JM109 (pGEX-AZR). *Journal of Hazardous Materials, 163*, 1123–1128.
26. Joutey, N. T., Bahafid, W., Sayel, H., & el Ghachtouli, N. (2013). Biodegradation: Involved microorganisms and genetically engineered microorganisms. *Biodegradation-Life of Science, 1*, 289–320.

27. Khan, R., Bhawana, P., & Fulekar, M. (2013). Microbial decolorization and degradation of synthetic dyes: A review. *Reviews in Environmental Science and Bio/Technology, 12*, 75–97.
28. Khataee, A. R., Dehghan, G., Ebadi, E., & Pourhassan, M. (2010). Central composite design optimization of biological dye removal in the presence of macroalgae Chara sp. *Clean-Soil Air, Water, 38*, 750–757.
29. Kumar, A., Kumar, A., Singh, R., Singh, R., Pandey, S., Rai, A., Singh, V. K., & Rahul, B. (2020). *Genetically engineered bacteria for the degradation of dye and other organic compounds.* Elsevier.
30. Liu, G.-F., Zhou, J.-T., Qu, Y.-Y., & Ma, X. (2007). Decolorization of sulfonated azo dyes with two photosynthetic bacterial strains and a genetically engineered Escherichia coli strain. *World Journal of Microbiology and Biotechnology, 23*, 931–937.
31. Mani, S., Chowdhary, P. & Bharagava, R. N. (2019). Textile wastewater dyes: toxicity profile and treatment approaches. *Emerging and Eco-Friendly Approaches for Waste Management*, 219–244.
32. Mata, T. M., Martins, A. A., & Caetano, N. S. (2010). Microalgae for biodiesel production and other applications: A review. *Renewable and Sustainable Energy Reviews, 14*, 217–232.
33. Murugesan, S., Sridharan, M., & Yoganandam, M. (2016). Biological decolorization and removal of metal from dye industry effluent by microalgae. *Biosciences Biotechnology Research Asia, 6*, 111–120.
34. Mussak, R. A., & Bechtold, T. (2009). Natural colorants in textile dyeing. *Handbook of Natural Colorants*, 315–335.
35. Olguin, E., Rodriguez, D., Sanchez, G., Hernandez, E., & Ramirez, M. (2003). Productivity, protein content and nutrient removal from anaerobic effluents of coffee wastewater in Salvinia minima ponds, under subtropical conditions. *Acta Biotechnologica, 23*, 259–270.
36. Olguín, E. J., & Sánchez-Galván, G. (2012). Heavy metal removal in phytofiltration and phycoremediation: The need to differentiate between bioadsorption and bioaccumulation. *New Biotechnology, 30*, 3–8.
37. Oswald, W., Gotaas, H., Golueke, C., Kellen, W., Gloyna, E., & Hermann, E. (1957). Algae in waste treatment [with discussion]. *Sewage and Industrial Wastes, 29*, 437–457.
38. Oswald, W. J. (1988). Micro-algae and waste-water treatment.
39. Parshetti, G., Parshetti, S., Telke, A., Kalyani, D., Doong, R., & Govindwar, S. P. (2011). Biodegradation of crystal violet by Agrobacterium radiobacter. *Journal of Environmental Sciences, 23*, 1384–1393.
40. Ramesh, B., Saravanan, A., Kumar, P. S., Yaashikaa, P., Thamarai, P., Shaji, A. & Rangasamy, G. (2023). A review on algae biosorption for the removal of hazardous pollutants from wastewater: Limiting factors, prospects and recommendations. *Environmental Pollution*, 121572.
41. Roozegar, M., & Behnam, S. (2019). An eco-friendly approach for copper (II) biosorption on alga Cystoseira indica and its characterization. *Environmental Progress & Sustainable Energy, 38*, S323–S330.
42. Sadek, M. B., Hassan, J., Saif, H. B., Biswas, A., & Sultana, S. (2016). Toxic effect of textile dyeing effluents on germination, growth, yield and nutritional quality of Okra (Abelmoschus esculentus). *International Journal of Ecotoxicology Exobiology, 1*, 82–87.
43. Sandhya, S., Sarayu, K., Uma, B., & Swaminathan, K. (2008). Decolorizing kinetics of a recombinant Escherichia coli SS125 strain harboring azoreductase gene from Bacillus latrosporus RRK1. *Bioresource Technology, 99*, 2187–2191.
44. Saxena, G., Kishor, R., Saratale, G. D., & Bharagava, R. N. (2020). Genetically modified organisms (GMOs) and their potential in environmental management: constraints, prospects, and challenges. *Bioremediation of Industrial Waste for Environmental Safety: Volume II: Biological Agents and Methods for Industrial Waste Management*, 1–19.
45. Sharma, B., & Shukla, P. (2022). Futuristic avenues of metabolic engineering techniques in bioremediation. *Biotechnology and Applied Biochemistry, 69*, 51–60.
46. Shelef, L., Naglik, O., & Bogen, D. (1980). Sensitivity of some common food-borne bacteria to the spices sage, rosemary, and allspice. *Journal of Food Science, 45*, 1042–1044.

47. Singh, P. K., & Singh, R. L. (2017). Bio-removal of azo dyes: A review. *International Journal of Applied Sciences and Biotechnology, 5*, 108–126.
48. Soutullo, S., Sánchez, M., Enríquez, R., Olmedo, R., Jiménez, M., & Heras, M. (2016). Comparative thermal study between conventional and bioclimatic office buildings. *Building and Environment, 105*, 95–103.
49. Spolaore, P., Joannis-Cassan, C., Duran, E., & Isambert, A. (2006). Commercial applications of microalgae. *Journal of Bioscience and Bioengineering, 101*, 87–96.
50. Suganya, T., Varman, M., Masjuki, H., & Renganathan, S. (2016). Macroalgae and microalgae as a potential source for commercial applications along with biofuels production: A biorefinery approach. *Renewable and Sustainable Energy Reviews, 55*, 909–941.
51. Tahir, U., Yasmin, A., & Khan, U. H. (2016). Phytoremediation: Potential flora for synthetic dyestuff metabolism. *Journal of King Saud University-Science, 28*, 119–130.
52. Tang, W., Xu, X., Ye, B.-C., Cao, P., & Ali, A. (2019). Decolorization and degradation analysis of Disperse Red 3B by a consortium of the fungus Aspergillus sp. XJ-2 and the microalgae Chlorella sorokiniana XJK. *RSC Advances, 9*, 14558–14566.
53. Uddin, M. S., Zhou, J., Qu, Y., Guo, J., Wang, P., & Zhao, L. H. (2007). Biodecolorization of azo dye acid red B under high salinity condition. *Bulletin of Environmental Contamination and Toxicology, 79*, 440–444.
54. Zhang, W., Lin, Z., Pang, S., Bhatt, P., & Chen, S. (2020). Insights into the biodegradation of lindane (γ-hexachlorocyclohexane) using a microbial system. *Frontiers in Microbiology, 11*, 522.
55. Zhao, X., Hardin, I. R., & Hwang, H.-M. (2006). Biodegradation of a model azo disperse dye by the white rot fungus Pleurotus ostreatus. *International Biodeterioration & Biodegradation, 57*, 1–6.
56. Ziarani, G., Moradi, R., Lashgari, N. & Kruger, H. (2018). Introduction and importance of synthetic organic dyes. *Metal-Free Synthetic Organic Dyes*, 1–7.
57. Zohoorian, H., Ahmadzadeh, H., Molazadeh, M., Shourian, M. & Lyon, S. (2020). Microalgal bioremediation of heavy metals and dyes. In *Handbook of Algal Science, Technology and Medicine*. Elsevier.

Biological Treatment of Textile Bleaching and Printing Effluents

Sonali Sharma and Shweta Singh

1 Introduction

The textile industry is one of the most water-intensive and pollutive sectors globally, with effluents generated from processes such as bleaching, dyeing, and printing containing high levels of chemical contaminants. Textile bleaching and printing effluents are often rich in hazardous chemicals such as synthetic dyes, heavy metals, and detergents, which pose significant environmental risks if not treated properly. The disposal of untreated or inadequately treated effluents from textile mills can lead to water contamination, toxic effects on aquatic life, and long-term ecosystem damage. In recent decades, there has been growing concern over the environmental impact of conventional chemical treatments used for effluent management. These traditional methods, such as coagulation, flocculation, and chemical oxidation, often result in high operational costs, the generation of secondary waste, and incomplete detoxification of the effluent [2]. As a result, biological treatment has emerged as an alternative that offers a more sustainable and cost-effective solution. Biological treatment methods, which rely on the metabolic activities of microorganisms such as bacteria, fungi, and algae, are gaining prominence in the treatment of textile effluents due to their ability to degrade organic pollutants, including dyes, detergents, and other toxic compounds, into less harmful substances. These biological processes are environmentally friendly, energy-efficient, and can be integrated into existing wastewater treatment plants with relative ease [49]. Furthermore, biological treatments often result in the partial or complete mineralization of dyes and chemicals, thereby reducing the toxicity and improving the overall quality of the effluent. This chapter explores the different biological treatment techniques employed for the treatment

S. Sharma (✉)
School of Studies in Microbiology, Jiwaji University, Gwalior, India
e-mail: sonalisharma169@gmail.com

S. Singh
Pollution Ecology Lab, School of Studies in Microbiology, Jiwaji University, Gwalior, India

R. A. Bhat et al. (eds.), *Biocentric Approaches in Textile Waste Management*, Textile Science and Clothing Technology, https://doi.org/10.1007/978-3-032-04974-2_16

of textile bleaching and printing effluents. It covers a range of methods, including aerobic and anaerobic processes, bio-augmentation, and bioreactor designs, with a focus on their effectiveness, limitations, and potential for industrial-scale implementation. The chapter also discusses recent advancements in microbial biotechnology and genetic engineering, which are enhancing the efficiency of these processes. By examining both traditional and cutting-edge approaches, this chapter aims to provide a comprehensive overview of the current state of biological treatment in the textile industry, offering insights into future trends and the potential for sustainable wastewater management.

Overview of Textile Industry and Wastewater

The textile industry is a major contributor to the global economy, providing employment to over 300 million people worldwide. It plays a pivotal role in industrialization and economic growth in countries like China, India, Bangladesh, and Vietnam [36]. The global textile market was valued at approximately $993.6 billion in 2022 and is projected to grow at a CAGR of 4.6% from 2023 to 2030 [29]. Textile and apparel exports constitute a significant share of GDP in developing nations. Advancements in smart textiles, sustainable materials, and digital production methods are reshaping the sector, contributing to sectors like healthcare, automotive, and defense. The textile industry accounts for 10% of global carbon emissions, making it one of the most polluting industries [71]. Fast fashion exacerbates this, with high energy consumption and reliance on fossil fuels during production processes. It is responsible for nearly 20% of global wastewater due to processes like dyeing and finishing [72]. Cotton production, a key raw material, consumes significant water—approximately 2700 liters to produce one T-shirt [74] Annually, the industry generates around 92 million tonnes of textile waste, much of which ends up in landfills or is incinerated [23]. Synthetic dyes and finishing agents lead to the discharge of hazardous chemicals, impacting aquatic ecosystems and human health (Table 1). Synthetic fibers like polyester contribute to 35% of microplastic pollution in oceans, affecting marine biodiversity.

Environmental Concerns

Textile industries are significant contributors to water pollution due to the discharge of untreated or inadequately treated effluents. These effluents pose severe environmental challenges due to their high chemical oxygen demand (COD), persistent coloration, and toxicity. High Chemical Oxygen Demand (COD) Textile wastewater contains a mixture of organic and inorganic pollutants, including dyes, surfactants, and finishing agents, which lead to elevated COD levels. High COD depletes dissolved oxygen in aquatic systems, causing harm to aquatic life [40]. colour presence of synthetic dyes

Table 1 Types of effluents generated from textile bleaching and printing processes

Effluent type	Source in textile process	Key pollutants	Impact on environment	References
Alkaline wastewater	Sourced from scouring and mercerizing, where alkalis like sodium hydroxide are used to clean and prepare fabrics	High pH, residual alkalis, and organic impurities	Alters water pH, harmful to aquatic life	[21]
Chlorinated effluents	Generated during bleaching operations using chlorine-based agents or hypochlorite	Chlorinated organic compounds, free chlorine, AOX (absorbable organic halogens)	Toxicity to aquatic organisms, bioaccumulation in food chains	[56]
Dye wastewater	Resulting from dyeing and printing processes, including washing residues and dye dispersions	Unused dyes, heavy metals (e.g., chromium, copper), aromatic amines, and salts	High color content reduces light penetration in water bodies, toxicity to aquatic organisms	[65]
Organic wastewater	Released during desizing and washing steps, where starch, synthetic polymers, or oils are removed from fabrics	High COD (chemical oxygen demand), BOD (biological oxygen demand), and residual sizing agents	Oxygen depletion in water bodies, promoting eutrophication	[52]
Acidic wastewater	Neutralization stages and dye fixation processes, particularly with acidic dyes or pH-regulated processes	Low pH, sulphuric acid, and residual dyes	Corrosive effects on aquatic environments, harmful to ecosystems	[45]
Salt-laden effluents	Salt addition in dyeing processes for better dye fixation and print adherence	Sodium chloride, sodium sulphate	Increases salinity of receiving water bodies, harmful to freshwater organisms	[30]
Heavy metal effluents	Metal-complex dyes and mordanting agents used in textile printing processes	Chromium, copper, zinc, and nickel	Bioaccumulation and toxicity to aquatic and terrestrial organisms	[61]

(continued)

Table 1 (continued)

Effluent type	Source in textile process	Key pollutants	Impact on environment	References
Foaming effluents	Surfactants, detergents, and dispersing agents used in bleaching, scouring, or dyeing processes	Surfactants, suspended solids, and oils	Affects water aeration, hampers gas exchange in aquatic systems	[15]
Thermal effluents	Released from processes involving high-temperature water discharge, such as fabric washing or cooling systems	Elevated temperature water	Disrupts aquatic thermal balance, reduces dissolved oxygen levels	[12]

in effluents is a major concern due to their vivid and persistent coloration. These dyes reduce sunlight penetration in water bodies, affecting photosynthetic aquatic organisms and disrupting ecosystems [38]. Toxicity, Many dyes and chemicals used in textiles are non-biodegradable and toxic. Heavy metals, azo dyes, and chlorinated compounds can bioaccumulate, posing risks to human health and aquatic organisms [75]. Environmental degradation has become a pressing global issue, driven by factors such as industrialization, urbanization, and unsustainable resource exploitation. Key concerns include water pollution, air contamination, soil degradation, and the accumulation of hazardous waste. These challenges necessitate innovative and effective treatment solutions to mitigate their impact.

Water Pollution: Contaminants from industrial discharge, agricultural runoff, and untreated wastewater degrade aquatic ecosystems and threaten human health. Advanced treatment technologies, including membrane filtration and bioremediation, are essential to address these issues [18].

Air Pollution: Emissions from vehicles, factories, and power plants release harmful substances such as particulate matter (PM2.5), sulfur oxides (SO_x), and nitrogen oxides (NO_x) into the atmosphere. Effective solutions, such as catalytic converters and renewable energy adoption, are critical for improving air quality [73].

Soil Degradation: The overuse of chemical fertilizers and pesticides has degraded soil quality, reducing agricultural productivity and increasing the risk of erosion. Sustainable practices, like organic farming and phyto-remediation, offer promising solutions [66].

Hazardous Waste: The accumulation of industrial and medical waste poses severe environmental and health risks. Technologies such as waste-to-energy conversion and safe disposal systems are needed to address this challenge [71].

Biological Treatment Methods: A Sustainable and Effective Solution for Textile Wastewater

The textile industry is a significant contributor to global water pollution due to the discharge of wastewater containing dyes, chemicals, and heavy metals. Traditional physicochemical methods for wastewater treatment, such as coagulation, flocculation, and chemical oxidation, are often expensive and generate secondary pollution. In contrast, biological treatment methods are increasingly recognized as sustainable and effective alternatives.

Microbial Treatment—Microorganisms such as bacteria, fungi, and algae play a vital role in degrading organic pollutants and breaking down complex dyes into simpler, less harmful compounds. Bacterial species like *Pseudomonas aeruginosa* and *Bacillus subtilis* have been shown to degrade azo dyes effectively [8]. Similarly, white-rot fungi like *Phanerochaete chrysosporium* utilize ligninolytic enzymes (laccase, manganese peroxidase) to degrade recalcitrant dyes [64].

Enzymatic Treatment—Bioremediation using enzymes such as laccases, azoreductases, and peroxidases has gained attention due to its specificity and efficiency. These enzymes catalyze the breakdown of synthetic dyes under mild conditions, reducing environmental impact [62].

Phytoremediation—The use of plants to remove, stabilize, or degrade contaminants in wastewater is another promising approach. Aquatic plants like *Eichhornia crassipes* (water hyacinth) and *Lemna minor* (duckweed) have demonstrated potential in nutrient uptake and dye removal from textile effluents [76].

Algal Treatment Systems—Microalgae such as *Chlorella vulgaris* and *Spirulina maxima* are effective in adsorbing heavy metals and degrading dyes. Algal systems also contribute to biomass production, which can be utilized in bioenergy generation, making the process economically viable [44].

Integrated Systems—Combining the biological methods with other treatment technologies enhances efficiency. For instance, integrating microbial treatment with constructed wetlands or membrane bioreactors (MBRs) ensures comprehensive pollutant removal and water recovery [54].

2 Composition of Textile Bleaching and Printing Effluents

Textile Bleaching Effluents

Textile bleaching effluents contain various chemicals commonly used in the bleaching process. These chemicals are essential for altering the color of fabrics, but they can pose environmental risks when improperly treated. Below are some common chemicals used in textile bleaching:

Hydrogen Peroxide (H_2O_2): Hydrogen peroxide is widely used in the textile industry as a bleaching agent, especially for cotton and other natural fibers. It is considered an environmentally friendly alternative to chlorine-based bleaches due to its relatively low toxicity and high decomposition rate into water and oxygen. However, improper disposal of hydrogen peroxide can lead to the formation of oxygen-rich compounds, which could impact water quality. Studies show that its application in textile industries has gained popularity due to its effectiveness and reduced environmental impact compared to chlorine-based bleaches [55].

Sodium Hypochlorite (NaOCl): Sodium hypochlorite is a common chlorine-based bleaching agent used in the textile industry. While it effectively whitens fabrics, it also has significant environmental drawbacks. The use of sodium hypochlorite produces toxic by-products such as chlorinated compounds and dioxins, which can contaminate water bodies and harm aquatic life. Despite its environmental impact, sodium hypochlorite remains a popular bleaching agent due to its effectiveness in high-volume industrial processes [28].

Sodium Chlorite ($NaClO_2$): Sodium chlorite, when activated, releases chlorine dioxide (ClO_2) which is a powerful bleaching agent used in textiles. It is often preferred for its ability to bleach at lower temperatures and its efficiency in removing stains and coloring agents from fabrics. However, sodium chlorite can generate toxic chlorinated by-products in effluents, making its disposal a concern for environmental sustainability. It is necessary to neutralize these by-products to prevent pollution [6].

Peracetic Acid (CH_3CO_3H): Peracetic acid is an organic peroxide used in textile bleaching for its high effectiveness, especially in bleaching wool, silk, and other delicate fibers. It is more environmentally friendly than chlorine-based bleaches because it produces fewer harmful by-products. Peracetic acid decomposes into acetic acid and oxygen, which are less harmful to aquatic ecosystems. However, when improperly disposed of, it can still pose risks to local water bodies if not adequately neutralized [11].

Characteristics of Effluents

Effluents are liquid waste or by-products discharged from industrial, agricultural, or domestic sources into the environment. Their characteristics can be broadly classified into physical, chemical, and biological parameters. Physical Characteristics, Temperature, Effluents can have elevated temperatures, which can disrupt aquatic ecosystems by reducing dissolved oxygen levels and affecting the metabolic rates of aquatic organisms [17]. Suspended Solids (SS), these include insoluble particles such as soil, organic debris, and industrial residues. High levels of SS can lead to sedimentation and reduced light penetration in water bodies [68].

Chemical Characteristics-pH: Indicates the acidity or alkalinity of the effluent. Extreme pH levels can be harmful to aquatic life and affect the efficiency of biological treatment processes. Metcalf & Eddy, Inc. [48] Chemical Oxygen Demand (COD), Represents the amount of oxygen required to oxidize organic and inorganic matter in the effluent. High COD levels indicate potential pollution [59]. Biochemical Oxygen Demand (BOD), Measures the amount of oxygen consumed by microorganisms while decomposing organic matter. High BOD indicates organic pollution. Toxic Substances, Effluents may contain heavy metals (e.g., lead, mercury), pesticides, or other toxic chemicals that pose risks to health and ecosystems [60].

Biological Characteristics-Pathogens: Effluents, especially from domestic sources, may contain bacteria, viruses, and parasites, posing risks to public health [13]. Nutrients-High levels of nutrients like nitrogen and phosphorus can cause eutrophication, leading to algal blooms and oxygen depletion in water bodies [20].

Aesthetic Characteristic-Color: Effluents from industries like textiles ad dyes often have intense colors, affecting water aesthetics and light penetration [58]. Organic decomposition or industrial chemicals in effluents can produce unpleasant odors, affecting the surrounding environment [48].

Textile Printing Effluents

The textile industry is one of the largest contributors to water pollution worldwide. Among its processes, textile printing generates significant volumes of effluents containing dyes, chemicals, and other pollutants. These effluents pose serious environmental and health risks due to their high chemical oxygen demand (COD), biological oxygen demand (BOD), and toxic components. Dyes and Pigments, A wide variety of dyes, including reactive, acid, basic, and azo dyes, are used. Many are non-biodegradable and toxic [32]. Heavy Metals: Effluents often contain metals such as chromium, lead, and copper, used in dyeing and printing processes [39]. Organic Compounds High concentrations of surfactants, binders, and solvents contribute to organic pollution [51]. pH Imbalance Effluents often exhibit extreme pH values, affecting aquatic ecosystems. Environmental Impacts Water Pollution

Effluents decrease water quality by introducing toxic substances, reducing dissolved oxygen, and altering pH levels [38]. Soil Contamination Disposal of untreated effluents on land leads to soil infertility and contamination. Toxicity to Aquatic Life, Synthetic dyes and heavy metals inhibit photosynthesis in aquatic plants and are toxic to fish and other organisms.

Treatment Strategies Physicochemical Methods, Coagulation-flocculation, adsorption, and advanced oxidation processes (AOPs) are effective in removing dyes and heavy metals [31]. Biological Treatments: Use of microbes such as *Bacillus subtilis* and fungi like *Aspergillus niger* to degrade dyes [58]. Membrane Technologies, Nanofiltration and reverse osmosis are employed for wastewater recycling [1]. Regulatory Framework and Sustainability Governments and international organizations have established guidelines for effluent discharge. Sustainable practices, such as zero-liquid discharge (ZLD) and cleaner production methods, are gaining traction [70].

Toxicity and Environmental Impact

The biological treatment of textile bleaching and printing effluents is a sustainable alternative to conventional methods, which often involve harsh chemicals. However, like any technology, it has its challenges, including potential toxicity and environmental impacts. Below is an overview of these issues with citations and references. Residual Toxicity in Treated Effluents Biological treatments, such as those involving microbial consortia or enzymes, may not completely degrade all toxic compounds. Residual byproducts, such as aromatic amines or partially degraded dyes, can persist in treated effluents and pose ecological and toxicological risks to aquatic organisms [33, 35]. Impact, These byproducts can affect aquatic flora and fauna, disrupting ecosystems and bioaccumulating in the food chain. SludgeGeneration, Biological treatments often produce sludge, which may contain concentrated amounts of toxic substances initially present in the effluents. Improper disposal of this sludge can lead to secondary environmental contamination [58]. Impact, Land and water pollution due to leaching of toxic substances from sludge. Energy and Resource Demand, Although biological treatments are often considered eco-friendly, some processes require aeration, temperature control, or additional nutrients, which can increase energy and resource consumption. This diminishes the overall environmental benefits if not managed sustainably [16]. Microbial Community Dynamics, The introduction of non-native microbial strains into biological treatment systems could disrupt local microbial ecosystems if not contained. Some engineered or introduced microbes may outcompete native species or transfer antibiotic resistance genes [27]. Impact, Long-term disruption of native microbial communities in natural water bodies. Incomplete Removal of Persistent Pollutants Biological processes may struggle with the degradation of recalcitrant compounds, such as certain azo dyes, heavy metals, and chlorinated organics [24]. Impact, These pollutants remain active in the environment, posing health risks to humans and wildlife.

3 Biological Treatment Processes

Biological treatment processes are methods used to treat wastewater or other polluted environments by utilizing microorganisms to break down contaminants (Table 2). These processes can be classified into several types, such as aerobic, anaerobic, and anoxic treatments, depending on the environmental conditions and the type of pollutants involved. Pre-Treatment of Effluents, Combining physical or chemical pre-treatments with biological methods can enhance degradation efficiency and reduce toxicity [43]. Enhanced Bioreactors, Using advanced bioreactor designs, like membrane bioreactors or immobilized enzyme systems, can improve pollutant breakdown and reduce sludge production [14]. Monitoring and Regulation, Regular monitoring of treated effluents for residual toxicity and implementing stringent discharge standards can minimize environmental risks.

Aerobic Biological Treatment

Aerobic treatment occurs in the presence of oxygen, where microorganisms use oxygen to break down organic pollutants into simpler compounds like carbon dioxide and water. This process is typically used for treating domestic and industrial wastewater that contains high levels of biodegradable organic matter. The most common aerobic processes are activated sludge systems, trickling filters, and oxidation ponds. In an activated sludge system, air is pumped into a tank containing microorganisms (mainly bacteria) which degrade organic pollutants. The treated water is then separated from the microorganisms, which are either returned to the tank or removed as excess sludge.

Anaerobic Biological Treatment

Anaerobic treatment takes place in environments without oxygen. In this process, microorganisms use alternative electron acceptors, like nitrates, sulfates, or carbon dioxide, to break down organic materials. Anaerobic systems are particularly useful in treating wastewater with high levels of organic matter, such as food processing or agricultural waste, and they produce biogas (methane), which can be harnessed as an energy source. The most common anaerobic processes include anaerobic digesters, where microorganisms degrade organic material in the absence of oxygen. These systems are efficient for reducing the volume of organic waste and generating renewable energy [48].

Table 2 Microorganisms commonly involved in biological treatment processes

Microorganism	Type	Function/Role	Application	References
Bacillus species	Bacteria	Degradation of organic pollutants, nitrogen cycle	Wastewater treatment, bioremediation of hydrocarbons	Kümmerer [46]
Pseudomonas species	Bacteria	Degradation of hydrocarbons, nitrification, denitrification	Bioremediation of oil spills, removal of nitrogen compounds from wastewater	Akpan et al. [3]
Nitrosomonas species	Bacteria	Nitrification (ammonia oxidation)	Wastewater treatment, nitrogen removal in effluent	Amann et al. [9]
Nitrobacter species	Bacteria	Nitrite oxidation	Wastewater treatment, nitrogen removal in effluent	Kuenen [42]
Trichoderma species	Fungi	Organic pollutant degradation, bioremediation of pesticides	Soil bioremediation, removal of toxic compounds from wastewater	Martínez et al. [47]
Azospirillum species	Bacteria	Plant growth promotion and biological nitrogen fixing	Agricultural bioremediation, bioremediation of organic pollutants	Okon et al. [50]
Saccharomyces cerevisiae	Yeast	Degradation of organic pollutants, heavy metal uptake	Wastewater treatment, bioremediation of heavy metals	Gadd [25]
Bacillus subtilis	Bacteria	Degradation of oils, hydrocarbons, organic pollutants	Bioremediation of petroleum-contaminated soils and wastewater	Siddique et al. [63]
Rhizobium species	Bacteria	Nitrogen fixation, degradation of organic compounds	Agricultural bioremediation, nitrogen removal in wastewater	Kosslak et al. [41]

Anoxic Biological Treatment

Anoxic treatment occurs in environments with low or no oxygen but still contains some dissolved oxygen from other sources, such as nitrate. In these systems, microorganisms use nitrate or nitrite as electron acceptors instead of oxygen to degrade organic matter. This process is especially important in the nitrogen removal part of wastewater treatment. An example of an anoxic process is the denitrification step in nitrogen removal, where bacteria reduce nitrates into nitrogen gas, which is then released into the atmosphere [34]. Anoxic treatment is often combined with aerobic processes in sequencing batch reactors (SBR) or integrated fixed-film activated sludge (IFAS) systems for more effective nutrient removal.

4 Challenges and Limitations in Biological Treatment

Biological treatments in the textile industry have gained attention as sustainable alternatives to conventional chemical methods. However, these approaches face several challenges and limitations that hinder their widespread adoption: Efficiency of Biological Agents, The effectiveness of biological treatments, such as enzymes and microorganisms, can be limited by the complexity of textile wastewater. High concentrations of pollutants, including dyes, surfactants, and heavy metals, can inhibit microbial activity or reduce enzyme efficiency [37]. Operational Conditions, Biological processes often require specific environmental conditions, such as optimal pH, temperature, and nutrient availability. Maintaining these conditions consistently in industrial settings can be costly and technically challenging [7]. Incomplete Degradation, Certain dyes and textile chemicals are resistant to biodegradation. As a result, biological treatments may not achieve complete removal of pollutants, necessitating additional treatment steps [58]. Biological methods typically require longer treatment durations compared to chemical treatments. This limitation can be a bottleneck in high-volume industrial operations that demand rapid processing [33]. While biological treatments show promise at the laboratory scale, scaling these processes to industrial levels often reveals inefficiencies. Factors such as maintaining microbial cultures and ensuring uniform pollutant exposure are more difficult at larger scales [51]. The initial costs associated with setting up biological treatment facilities, including bioreactors and monitoring systems, can be high. Additionally, the operational costs of maintaining optimal conditions may outweigh the benefits in some cases [4, 5]. Regulatory and Acceptance Barriers, Industries are often hesitant to adopt new technologies due to regulatory uncertainties and the perceived risks of deviating from established methods. Convincing stakeholders of the long-term benefits of biological treatments remains a challenge [19].

5 Applications of Biological Treatment

Biological treatment methods are essential for addressing various environmental pollutants in wastewater, air, and soil. These methods use microorganisms, such as bacteria, fungi, and algae, to degrade or remove harmful substances. Biological treatments are widely implemented due to their sustainability, cost-effectiveness, and environmental benefits.

Wastewater Treatment

Activated Sludge Process (ASP)—The activated sludge process is a commonly used biological treatment for municipal and industrial wastewater. In this process, microorganisms in the activated sludge degrade organic pollutants by consuming them as food. The treatment is effective in removing biochemical oxygen demand (BOD), suspended solids, and nutrients like nitrogen and phosphorus. The implementation of this process in cities like New York has successfully reduced the levels of pollutants in wastewater, meeting environmental regulations [48].

Bioremediation of Contaminated Sites

Oil Spill Remediation—Bioremediation is an effective biological treatment technique for cleaning up oil spills in marine and terrestrial environments. One of the most notable examples is the Exxon Valdez oil spill, where bioremediation techniques, including nutrient addition to stimulate microbial growth, helped accelerate the degradation of hydrocarbons in the impacted areas [10]. This method is cost-effective and environmentally friendly, as it relies on natural microbial processes to break down pollutants [10].

Textile Effluent Treatment

Dye Degradation Using Fungi—Textile industries often discharge effluents containing hazardous dyes and chemicals (Fig. 1). White rot fungi, such as *Phanerochaete chrysosporium*, have been found to degrade azo dyes commonly used in the textile industry. The application of this biological treatment has been successfully implemented in India, where fungal biodegradation processes have been used to treat textile effluents, reducing the environmental impact of dye discharge [58].

Air Pollution Control

Biofiltration for VOC Removal—Biofiltration is a biological treatment method used to remove volatile organic compounds (VOCs) and other air pollutants from industrial emissions. Biofilters use microorganisms to degrade pollutants as they pass through a biologically active medium. In Sweden, a chemical manufacturing plant successfully implemented biofiltration systems to remove VOCs, achieving an efficiency of over 90% in VOC removal [22].

Biodegradation of Heavy Metals in Industrial Effluents

Chromium Removal Using Bacteria—Industrial effluents, particularly from tannery and electroplating industries, often contain heavy metals such as chromium, which are toxic to the environment. Bacterial strains such as *Pseudomonas* spp. have been used for the bioremediation of chromium-contaminated effluents. Studies have demonstrated the ability of these bacteria to reduce chromium(VI) to less toxic chromium(III), effectively lowering the toxicity of the effluent [26].

Waste-to-Energy via Anaerobic Digestion

Anaerobic Digestion for Biogas Production—Anaerobic digestion is used to treat organic waste by converting it into biogas, a renewable energy source. This biological treatment method is commonly applied in wastewater treatment plants, agricultural waste management, and even in landfills. In Germany, anaerobic digesters have been implemented at wastewater treatment plants to produce biogas, which is used for electricity and heat generation, contributing to energy sustainability [67].

6 Conclusion

The biological treatment of textile bleaching and printing effluents has gained significant attention due to its environmentally sustainable approach and cost-effectiveness. Traditional chemical methods often result in the generation of harmful secondary pollutants, whereas biological treatments rely on the natural capacity of microorganisms to degrade toxic substances found in textile wastewater, such as azo dyes and surfactants. Studies have demonstrated that bacterial and fungal species can effectively degrade complex organic compounds, reducing the toxic load in effluents [53]. Furthermore, biological treatment processes, including activated sludge systems, biofilms, and membrane bioreactors, have shown promise in treating textile wastewater, reducing chemical oxygen demand (COD), and improving the effluent quality to meet regulatory standards [69]. These methods are particularly beneficial in terms of low energy consumption and minimal chemical usage, making them more sustainable compared to conventional methods [57]. However, the efficiency of biological treatment depends on several factors such as the type of effluent, microbial species, and operational conditions, necessitating further research to optimize these processes for large-scale applications [69]. Despite these challenges, biological treatment offers a viable and eco-friendly alternative for the textile industry, reducing its environmental footprint significantly.

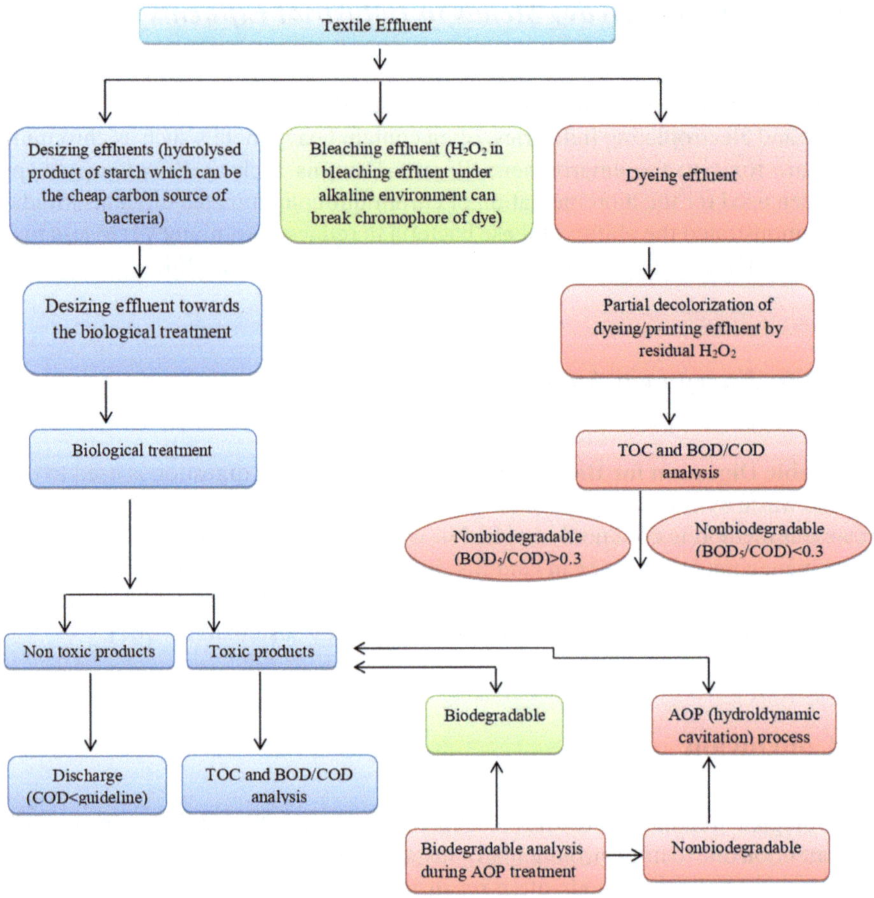

Fig. 1 Schematic representation of fate of textile effluents

References

1. Agarwal, M., Singh, K., & Gupta, R. (2011). Membrane technology in textile wastewater treatment. *Desalination and Water Treatment, 29*(1–3), 52–60. https://doi.org/10.5004/dwt.2011.2000
2. Ahluwalia, S. S., & Goyal, D. (2007). Microbial decolorization of textile dye-containing effluents: A review. *Waste Management, 27*(2), 207–217. https://doi.org/10.1016/j.wasman.2006.02.010
3. Akpan, I. R., et al. (2018). Pseudomonas as a bioremediation agent: Role in bioremediation. *Environmental Pollution, 243*, 198–209.
4. Aksu Z (2005) Application of biosorption for the removal of organic pollutants: A review. Biological Resources and Technology 97(10):101–124. https://doi.org/10.1016/j.biortech.2005.05.001
5. Aksu, Z. (2005). Application of biosorption for the removal of organic pollutants: A review. *Process Biochemistry, 40*(3–4), 997–1026. https://doi.org/10.1016/j.procbio.2004.04.008

6. Aksu, Z., & Tezer, S. (2005). A comparative study of biosorption characteristics of different biosorbents for the removal of chromium (VI) from aqueous solutions. *Journal of Hazardous Materials, 104*(1–3), 69–82. https://doi.org/10.1016/j.jhazmat.2003.12.023
7. Ali, N. (2010). Industrial applications of microbial enzymes. *Biotechnology Advances, 28*(6), 913–919. https://doi.org/10.1016/j.biotechadv.2010.08.003
8. Ali, N., Hameed, A., & Ahmed, S. (2021). Biodegradation of textile dyes by bacterial isolates: An eco-friendly approach. *Journal of Environmental Management, 277*, 111275. https://doi.org/10.1016/j.jenvman.2020.111275
9. Amann, R. I., et al. (2001). Nitrification processes in wastewater treatment. *Environmental Microbiology, 3*(11), 775–781.
10. Atlas, R. M., & Hazen, T. C. (2011). Oil biodegradation and bioremediation: A tale of the two worst spills in U.S. history. *Environmental Science & Technology, 45*(16), 6709–6715. https://doi.org/10.1021/es201104n
11. Bhat, R. M., & Bhat, P. (2012). Textile processing and the environmental impact. *Journal of Textile and Apparel Technology and Management, 7*(1), 1–10.
12. Bhatia, P., & Jain, M. (2021). *Thermal discharge effects on aquatic ecosystems.* Environmental Impact Reports.
13. Bitton, G. (2005). *Wastewater microbiology.* Wiley-Interscience.
14. Carliell, C. M., Barclay, S. J., Naidoo, N., Buckley, C. A., Mulholland, D. A., & Senior, E. (1995). Microbial decolorization of a reactive azo dye under anaerobic conditions. *Water SA, 21*(1), 61–69.
15. Garcia, V. S. G., Tominaga, F. K., Rosa, J. M., & Borrely, S. I. (2024). Emerging pollutants in textile wastewater: an ecotoxicological assessment focusing on surfactants. *Environmental Science and Pollution Research 31*(19), 27817–27828.
16. Chandra, R., Bharagava, R. N., Yadav, S., & Mohan, D. (2018). Toxicity and biodegradation of synthetic dyes: An overview. *Environmental Chemistry Letters, 16*(3), 1149–1166.
17. Chapman, D. (1996). *Water quality assessments: A guide to the use of biota, sediments, and water in environmental monitoring.* WHO.
18. Chen, G., He, Y., & Wang, X. (2020). Advances in wastewater treatment technologies for environmental sustainability. *Journal of Environmental Management, 263*, 110403. https://doi.org/10.1016/j.jenvman.2020.110403
19. Chowdhury, S., Mishra, R., & Saha, P. (2018). Textile dye wastewater treatment by biological methods. *Environmental Chemistry Letters, 16*(2), 551–563. https://doi.org/10.1007/s10311-018-0736-8
20. Correll, D. L. (1998). The role of phosphorus in the eutrophication of receiving waters: A review. *Journal of Environmental Quality, 27*(2), 261–266.
21. Tiwari, H., & Singh, R. S. (2022). Biotechnological approaches for microbial treatment of textile wastewater and resource recovery: Opportunities, challenges, and future perspectives. *Microbial Technologies for Wastewater Recycling and Management* 269–279.
22. Devinny, J. S., Deshusses, M. A., & Webster, T. S. (1999). *Biofiltration for air pollution control.* CRC Press.
23. Ellen MacArthur Foundation. (2021). *A new textiles economy: Redesigning fashion's future.*
24. Forgacs, E., Cserháti, T., & Oros, G. (2004). Removal of synthetic dyes from wastewaters: A review. *Environment International, 30*(7), 953–971.
25. Gadd, G. M. (1990). Heavy metal tolerance in fungi: A review. *Environmental Pollution, 66*(1), 21–38.
26. Gaur, A., & Kaur, S. (2009). Bioremediation of toxic metals from industrial effluents. *Journal of Environmental Management, 90*(9), 2769–2776. https://doi.org/10.1016/j.jenvman.2009.03.009
27. Gavrilescu, M., Demnerova, K., Aamand, J., Agathos, S., & Fava, F. (2015). Emerging pollutants in the environment: Present and future challenges in biomonitoring, ecological risks, and bioremediation. *New Biotechnology, 32*(1), 147–156.
28. Ghaedi, M., & Khataee, A. R. (2015). Recent advances in textile wastewater treatment. *Journal of Environmental Chemical Engineering, 3*(4), 2323–2334. https://doi.org/10.1016/j.jece.2015.09.014

29. Grand View Research. (2023). *Textile market size, share & trends analysis.* Report from https://www.grandviewresearch.com
30. Gupta, R., et al. (2020). Effect of salinity from textile effluents on aquatic life. *International Journal of Environmental Studies.*
31. Dutta, P., Rabbi, M., Sufian, M., & Mahjebin, S. (2022). Effects of textile dyeing effluent on the environment and its treatment: A review. *Engineering and Applied Science Letters (EASL) 5*, 1–17.
32. Gupta, V. K., & Suhas. (2009). Application of low-cost adsorbents for dye removal—A review. *Journal of Environmental Management, 90*(8), 2313–2342. https://doi.org/10.1016/j.jenvman.2008.11.017
33. Hai, F. I., Yamamoto, K., & Fukushi, K. (2007). Hybrid treatment systems for dye wastewater. *Critical Reviews in Environmental Science and Technology, 37*(4), 315–377. https://doi.org/10.1080/10643380601174723
34. Henze, M., van Loosdrecht, M. C. M., Ekama, G. A., & Brdjanovic, D. (2002). *Biological wastewater treatment: Principles, modelling, and design.* IWA Publishing.
35. Holkar, C. R., Jadhav, A. J., Pinjari, D. V., Mahamuni, N. M., & Pandit, A. B. (2016). A critical review on textile wastewater treatments: Possible approaches. *Journal of Environmental Management, 182*, 351–366.
36. ILO. (2022). *The role of textiles in global employment.*
37. Jadhav, J. P., Parshetti, G. K., Kalme, S. D., & Govindwar, S. P. (2016). Biodegradation of textile dye. *Journal of Hazardous Materials, 137*(3), 1347–1354. https://doi.org/10.1016/j.jhazmat.2006.05.022
38. Kant, R. (2012). Textile dyeing industry an environmental hazard. *Natural Science, 4*(1), 22–26. https://doi.org/10.4236/ns.2012.41004
39. Khandare, R. V., & Govindwar, S. P. (2015). Phytoremediation of textile dyes and effluents: Current scenario and future prospects. *Biotechnology Advances, 33*(8), 1697–1714. https://doi.org/10.1016/j.biotechadv.2015.09.003
40. Khandegar, V., & Saroha, A. K. (2013). Electrocoagulation for the treatment of textile industry effluent—A review. *Journal of Environmental Management, 128*, 949–963.
41. Kosslak, R. M., et al. (1987). Rhizobium: Symbiotic nitrogen fixation and bioremediation. *Journal of Environmental Quality, 16*(3), 193–199.
42. Kuenen, J. G. (2008). Nitrifying bacteria. In *Microbial diversity and bioprospecting* (pp. 203–214). ASM Press.
43. Kumar, A., Kumari, K., & Singh, K. (2021). Textile wastewater treatment by biological methods: A sustainable approach. *Water and Environment Journal, 35*(3), 538–552.
44. Kumar, A., Malik, A., & Verma, N. (2023). Algal-based textile wastewater treatment: Mechanisms and sustainability. *Environmental Research, 216*, 114567. https://doi.org/10.1016/j.envres.2023.114567
45. Kumar, N., & Sharma, T. (2022). Acidic discharges in textile industry. *Environmental Progress Journal.*
46. Sharma, R., Kumar, V., & Tyagi, S. K. (2025). Chemical characterization of textile industry effluents in the northern Ganga basin: A comparative study of Yamuna and non-Yamuna basin discharges. *International Journal of River Basin Management* 1–13.
47. Martínez, M. J., et al. (2002). Bioremediation of environmental pollutants by fungi. *Applied and Environmental Microbiology, 68*(8), 3653–3659.
48. Metcalf, L., & Eddy, H. (2014). *Wastewater engineering: Treatment and resource recovery* (5th ed.). McGraw-Hill Education.
49. Nigam, P., Suri, S., & Singh, D. (2000). Color removal from textile effluent using biosorption. *Bioresource Technology, 72*(1), 111–115. https://doi.org/10.1016/S0960-8524(99)00163-1
50. Okon, Y., et al. (2007). Azospirillum: A model for plant-bacteria interactions. *Biology and Fertility of Soils, 43*(1), 1–7.
51. Pandey, A., Singh, P., & Iyengar, L. (2007). Bacterial decolorization and degradation of azo dyes. *International Biodeterioration & Biodegradation, 59*(2), 73–84. https://doi.org/10.1016/j.ibiod.2006.08.006

52. Patel, H., & Mehta, A. (2019). Organic pollutants in textile effluents. *Journal of Cleaner Production.*
53. Kumar, A., Kapoor, A., Rathoure, A. K., Devnani, G. L., & Pal, D. B. (2025). Organic compounds removal using magnetic biochar from textile industries based wastewater-A comprehensive review. *Sustainable Processes Connect 1*(1), 1–10.
54. Rajasulochana, P., & Preethy, V. (2016). Comparison on efficiency of various techniques in treatment of waste and sewage water—A comprehensive review. *Resource-Efficient Technologies, 2*(4), 175–184. https://doi.org/10.1016/j.reffit.2016.09.004
55. Ramasamy, S., & Sivasankar, P. (2017). Environmental impact of textile industry and its treatment techniques. *Environmental Science and Pollution Research, 24*(36), 27239–27251. https://doi.org/10.1007/s11356-017-0551-6
56. Ramesh, V., et al. (2018). Impact of chlorine in textile effluents. *Journal of Industrial Pollution.*
57. Ramesh Kumar, M., Malathy, R., Paramasivam, S., Chung, I. M., Kim, S. H., & Mayakrishnan, P. (2022). pH neutralization of textile industry wastewater for effective recycling. *Pigment & Resin Technology 51*(1), 80–90.
58. Robinson, T., McMullan, G., Marchant, R., & Nigam, P. (2001). Remediation of dyes in textile effluent: A critical review on current treatment technologies with a proposed alternative. *Bioresource Technology, 77*(3), 247–255. https://doi.org/10.1016/S0960-8524(00)00080-8
59. Sawyer, C. N., McCarty, P. L., & Parkin, G. F. (2003). *Chemistry for environmental engineering and science.* McGraw-Hill Education.
60. Schwarzenbach, R. P., Egli, T., Hofstetter, T. B., von Gunten, U., & Wehrli, B. (2010). Global water pollution and human health. *Annual Review of Environment and Resources, 35*, 109–136.
61. Shah, A., & Bhatt, S. (2021). *Heavy metal contamination in dye wastewater.* Pollution Control Review.
62. Sharma, B., Dangi, A. K., & Shukla, P. (2020). Contemporary enzyme-based technologies for textile wastewater treatment. *Frontiers in Bioengineering and Biotechnology, 8*, 365. https://doi.org/10.3389/fbioe.2020.00365
63. Siddique, T., et al. (2006). Bacillus subtilis: A bacterium for bioremediation of environmental pollutants. *Environmental Microbiology, 8*(6), 1086–1092.
64. Singh, R., & Arora, S. (2019). Removal of synthetic textile dyes from wastewaters: A critical review on present treatment technologies. *Critical Reviews in Environmental Science and Technology, 49*(4), 347–405. https://doi.org/10.1080/10643389.2018.1540769
65. Singh, R., & Choudhary, K. (2021). *Dye wastewater analysis and mitigation.* Water Research Updates.
66. Smith, P., & Williams, S. (2018). Sustainable soil management: Practices and benefits. *Agriculture and Environment, 12*(4), 341–355.
67. Speece, R. E. (2008). *Anaerobic biotechnology for industrial wastewater treatment.* Springer.
68. Tchobanoglous, G., Burton, F. L., & Stensel, H. D. (2003). *Wastewater engineering: Treatment and reuse.* McGraw-Hill Education.
69. Tijing, L. D., Kim, H. J., & Shon, H. K. (2017). Textile wastewater treatment using membrane bioreactor and forward osmosis. *Environmental Science & Technology, 51*(22), 13672–13681.
70. UNEP. (2020). *Sustainability in textile industries: Guidelines for effluent management.* United Nations Environment Programme.
71. UNEP. (2019). Sustainability and circularity in the textile value chain.
72. World Bank. (2020). Textiles and the environment.
73. World Health Organization. (2021). Air pollution and health. Retrieved from https://www.who.int
74. WWF. (2023). *The environmental impact of cotton production.*
75. Yaseen, D. A., & Scholz, M. (2019). Textile dye wastewater characteristics and constituents of synthetic effluents: A critical review. *International journal of Environmental Science and Technology, 16*(2), 1193–1226.
76. Zhou, W., Gao, X., Zhao, Z., & Li, M. (2022). Phytoremediation of textile wastewater by aquatic plants. *Journal of Hazardous Materials, 424*, 127256. https://doi.org/10.1016/j.jhazmat.2021.127256

Phytogenic Nanoparticles for Waste Water Treatment

Bhavna D. Deshpande, Pratibha S. Agrawal, and Mahesh K. N. Yenkie

1 Introduction

First century of the new millennium showed an exponential increase in global economic growth on one hand whereas, at the same time, huge volumes of wastewater was released due to speedy urbanization and industrialization, causing various negative effects on human health as well as the environment especially aquatic communities have suffered greatly which is worrisome. Unquestionably, the food, tannery, printing, paper, paint, and textile industries are the main contributors to environmental pollution. The industry uses a lot of colours, dye, flavouring agents, organic and inorganic compounds for cleaning, washing, and other manufacturing processes, producing a large volume of wastewater. It is estimated that throughout the manufacture and processing stages of textile dyeing, approximately 15% of the 7×10^5 metric tonnes of dyes produced yearly are wasted as effluents [1–3]. Bio-accumulation of discharged waste (which includes dyes, stabilizers, emulsifiers, inorganic and organic compounds) in water bodies leads to high risk to aquatic flora and fauna, which is due to change in water turbidity, chemical oxygen demand, and light penetration leading to disrupts photosynthesis and endangers aquatic life. The metal ions, halogens, refractory organic compounds and aromatic rings in dyes used in textile or colouring industries have toxic effects and may lead to mutation, genotoxicity, endocrine disrupting, carcinogenesis in living beings.

The waste water from industries also contain dyes, suspended solids, acids, bases, salts and micronutrients [4, 5]. Hence, need to remove the pollutants before it is discharged in the large water bodies is imperative. For wastewater remediation, a number of conventional treatment techniques are available, including chemical coagulation/flocculation, cloud point extraction, ozonation [6], oxidation [7], chemical

B. D. Deshpande (✉) · P. S. Agrawal · M. K. N. Yenkie
Department of Applied Chemistry, Laxminarayan Innovation Technological University, Nagpur, India
e-mail: bhavna.dnd2025@gmail.com

R. A. Bhat et al. (eds.), *Biocentric Approaches in Textile Waste Management*, Textile Science and Clothing Technology, https://doi.org/10.1007/978-3-032-04974-2_17

precipitation, Fenton process [8]; photocatalytic processes [9, 10], sono-chemical process [11, 12], photo–Fenton processes [13], nano-filtration, ion exchange, reverse osmosis, filtration, ultra-filtration [14, 15], ultra sound with adsorption–desorption [16], chemical and biological treatment, adsorption, activated charcoal, and ion exchange resins [17]. Adsorption techniques has been used since ages for the decolourisation of sugar syrup by bone char and removal of nerve gas from battle field by using activated carbon. Adsorption process employ porous materials like clay, alumina, silica gel, activated carbon, synthetic resins and zeolite which was used as adsorbents for the industrial waste water treatment and purification [18, 19]. Though these processes are effective, yet high operation cost, excess amount of chemicals, problems regarding disposals and accumulation of sludge are areas to be overviewed. It has often been noted that contaminants resistant to biological treatments are highly chemically stable therefore, severe difficulty in fully mineralizing. The need for more advanced technologies to fully mineralize wastewaters was imperative. Hence, oxidation processes were used to break down the bio-refractory materials found in wastewater. The key factors for the selection of best oxidation process for mineralization of given organic compounds are operating conditions, process constraints, and pollution load [20]. High degradation efficiency for direct oxidation needs certain operating conditions which raises the process's operating costs [14].

2 Road-Map of Nanotechnology

Nanotechnology is not a new word, nanomaterials were used centuries ago. In Mesopotamia, nanomaterial were used as colouring pigments in glass technology, optical properties of copper and silver nanomaterial were used as outer lining in decoration by Glaze. Some remarkable artefacts and trophies by Romans, displayed in British Museum shows outstanding use of gold. Mixture of glass and quartz (Egyptian blue) was often used for decoration purpose. In medieval period stained glasses were used in churches, red colour was given, using gold colloids mixed into glass, and yellow was due to silver nanoparticles. In Japan, in mid nineteenth century copper nanoparticles were used to brighten Satsuma glass. In sixth–seventh century A.D, China used porcelain. In 5000 BC, Cyprus used clay in cosmetology, to bleach wool and clothes. In Egypt natural red ochre pigment extracted from clay along with hydrated Fe(III) was used to colour cheeks and lips. Clay was used as healing aid, dietary supplement and for the therapeutic applications in Central and South America, Africa and Australia.

Nanotechnology has captured market word wide in variety of commercial products. Researchers in the fields of biology, engineering, chemistry, and medicine are working at the molecular and cellular levels to produce advanced materials for the welfare of living being, thanks to nanotechnology [21]. Due to self-cleansing, biocompatible, nontoxic, skin-compatible, antimicrobial, and dermatological behaviors metal nanoparticles has wide applications.

In recent decade, noble metal derived nanoparticles specifically Ag, Pt, Au, and Pd are widely used in the areas of biological diagnostic probes, optoelectronics, catalysis, and display devices [22]. In 2003, Silver Nanoparticles, due to its antibacterial and anti-inflammatory properties were used in air conditionals, fridge, air purifier, vacuum cleaner, and washing machine. Silver nanoparticles plays important role in the electronics, bio-sensing, apparel, food processing, paints, sunscreens, cosmetics, and medical equipment [23, 24]. Silver and zinc were used in packaging industries and wound dressing due to its sterile property. Au and Ag metal NPs are used specific drug carriage, genetic disease and disorders diagnosis and therapy. Zinc and titanium, in cosmetics, UV blocking agents and processing industries. Cu and Pd NPs have been applied in polymers, optical limiting devices, batteries, and plastics plasmonic wave guides, biosensor [25, 26]. Nanosensors developed by research have been used for detection of crop pests and hormonal regulations, soil nutrient levels and stress factors of soil. Nanosensors are further developed for sensing auxin and oxygen distribution [27]. Later, scratch resistant and glossy nano-particles metallic paints were used by Mercedes Benz.

Nanoparticles are particles having the size ranging from 1 to 100 nm. Various chemicals under this range in different shapes and properties are suspended in environment incidental or manmade or engineered [28]. Incidental sources can be derived from numerous ways such as industrial activities, power plants, coal, oil, natural gas, incineration of solid waste, traffic emission, fossil fuels, military shooting, bomb explosions, and pulverization of rocks to soil [29]. Such nanomaterials, in metallic or inorganic form may be insoluble and non-biodegradable which can harm the entire environment hence, effects the biosphere [30].

Researches have studied iron, zinc, aluminium, silicon, titanium, metal and metal oxide [31] nanoparticles for waste water treatment, electronics, biocidal and antibacterial applications. Transition elements like iron, cobalt, nickel and their compounds are magnetic nanoparticles, which are used in analytic chemistry [32], antigen diagnosis, pathogen detection [33], hyperthermia [34] and tissue repair [35] along with water treatment processes. FeO NPs have been applied for numerous drug delivery therapy and medical practices. They are applied for hyperthermia, tissue repair, cell labelling, targeting and immunoassays. Further FeO NPs are used in detoxification of biological fluids, magnetic resonance imaging and magnetically responsive drugs [14–16]. Lately, polymeric NPs, lipid-based NPs, ceramic NPs, semiconductor NPs, metal NPs, and carbon-based NPs have shown diverse applications in the field of medicine, biomedical drugs, food, agriculture, paints, electronics, telecommunication, material, environmental, and physiochemical areas, energy, and sustainability [36].

Potential applications of nanomaterials have been made possible by nanotechnology, which offers to manipulate morphology, compositions, along with physical and chemical properties of materials in a controlled way for substantial progress. Significant modifications in the properties of metals can be achieved by the mode of synthesis and modulation of metals. Properties such as strength, lightweight and durability in nanoparticles has brought significant change in production of bulk material [36]. Moreover, the metallic NPs have high surface area, high ordered structure, mechanical strengths, thermal strength, magnetic properties, and optical properties which enables it toward many applications [37]. Engineered nanoparticles have found various applications in bio-medicals, diagnostic, clinical treatment, electronics and waste water treatment. These nanoparticles are prepared using chemicals, most prominently heavy metals like Ni, Cd, Mn, Zn, Ti, Au, Sb, Ag and their MO_x. Hence, in spite of encouraging and enthusiastic applications world-wide these heavy metals incidentally get released in environment and are the cause of concern [29].

Nanoparticles can enter living being, get accumulated at molecular level (tissues and organs) causing major troubles. The study of effects of NPs on human lives and environment is called nano-toxicology. Nanoparticles have believed to enter the body through inhalation, ingestion and permeation. The various organs such as gastro-intestinal system, circulatory system, heart, brains, lungs, skin, bone marrow, liver kidney, lymphatic system have potential risk on exposure to nanoparticles. Nanoparticles are used for diagnostics and therapeutics in cancer detection however, its accumulation can cause cancer [38]. However, in spite of numerous applications, synthesis of metal nanoparticles by physiochemical approach has environmental issues due to heavy metals (Fig. 1).

Thus, biological means of synthesis using microorganisms and plants has become a new trends due to its non-toxicity, low temperature and pressure, easily scalability and reproducibility, and well-defined morphology [37]. Considering toxicological aspect of nanoparticles, phytogenic synthesis of nanoparticles have grabbed attention lately.

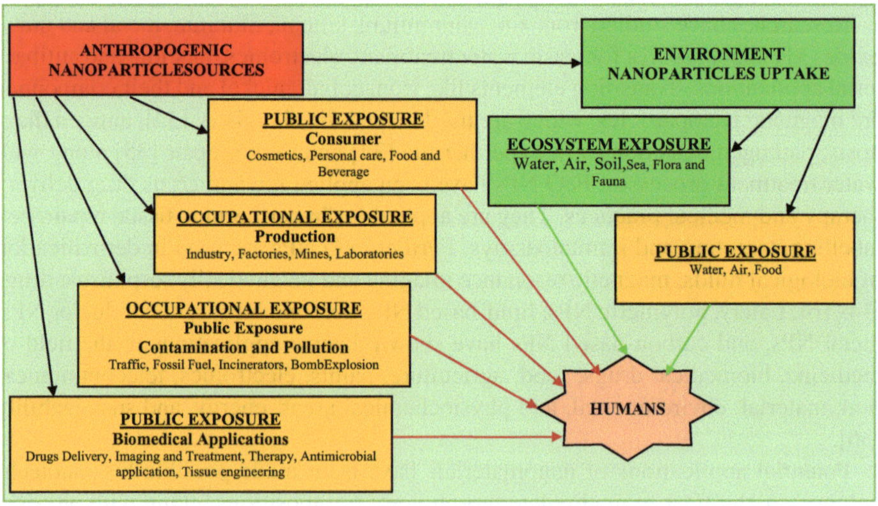

Fig. 1 Anthropogenic sources of NPs and its effects on human body

3 Phytogenic Synthesis of Iron Nanoparticles

Researchers have looked into relatively new synthesis techniques in an effort to counteract the negative effects of liquid phase chemical approach and use of unsafe surfactants. These techniques include the use of water soluble biopolymers like starch, Vitamin C, tannic acid, Vitamin B9 and gallic acid, glucose, amino acids use of microorganisms and substituting toxic chemicals with plant parts (Fig. 2) [39, 40]. Flavonoids, phenols, and terpenoids metabolites, rich in plant extracts have twin role in nanoparticles synthesis [41].

Phytogenic Nanoparticles can be synthesised by using microorganisms and plant extracts. Over the past few years metal nanoparticles have been synthesised using bacteria, fungi, yeast, biomass, plants extract and bio-derived components. For efficient and high yield of nanoparticles, certain parameters are required, which includes pH, incubation time, temperature, mixing ratio, salt concentration, redox conditions and irradiation (Fig. 3). The yield also depends on species and the extract from which it is derived [42, 43]. Flow Diagram for phytogenic synthesis is shown in Fig. 4.

Shape and size of nanoparticle is affected by pH. Studies show that at low pH, agglomeration of metal nanoparticle occurs due to very low rate of nucleation which results in large size of nanoparticles, whereas, at high pH, small sized nanoparticles are observed. Spherical and decahedral shapes were studied at higher pH. It was

Fig. 2 Phytogenic synthesis of NPs using different plant parts

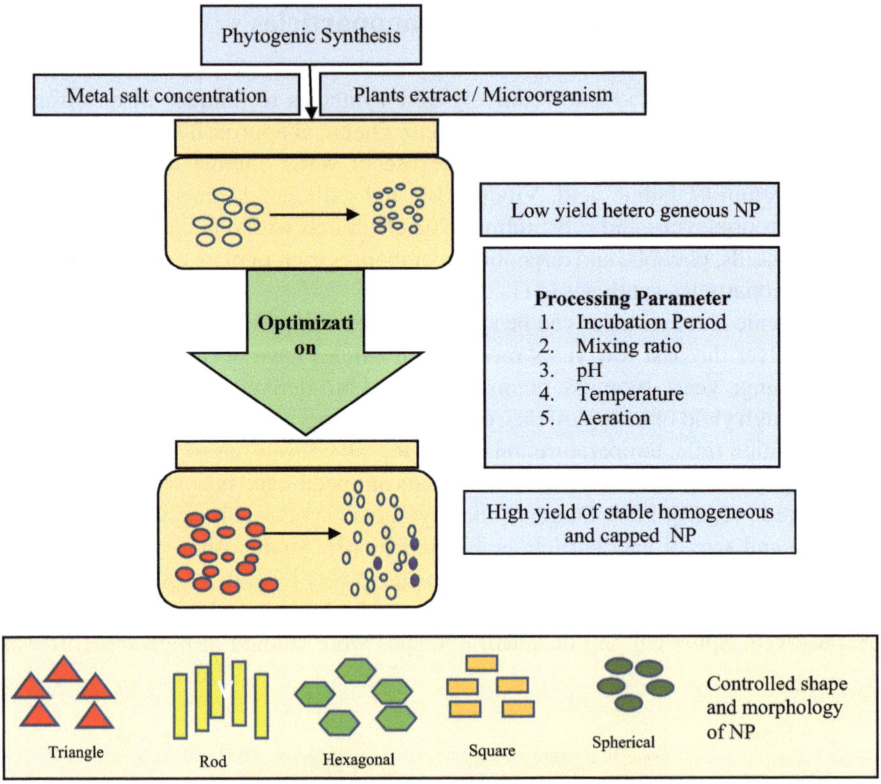

Fig. 3 Parameters for producing biological NP

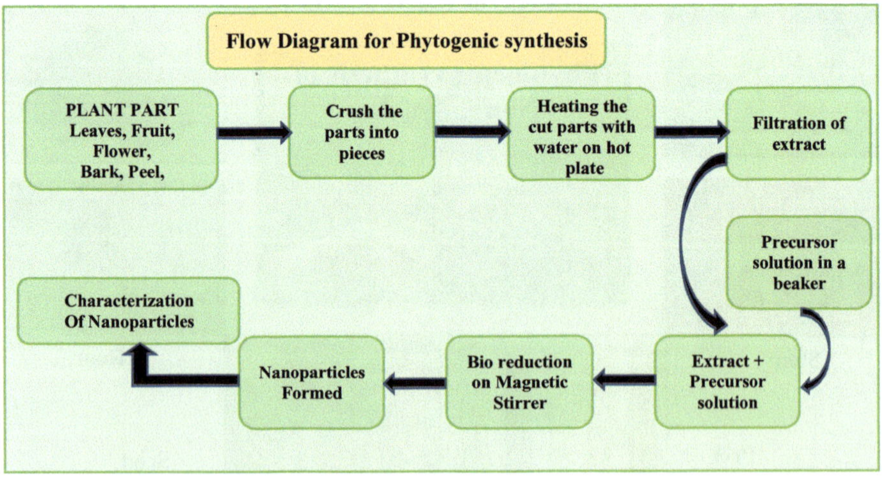

Fig. 4 Flow diagram for phytogenic synthesis

observed, that Ag NPs synthesised by phytogenic mode using *Curcuma longa*, has functional groups, resulting in huge binding sites for Ag ions to form large number of silver nanoparticles at higher pH than at low pH [39, 44].

Temperature affects the nucleation rate in phytogenic process. Rate of production, size and shape show variation with the change in temperature. The metal ion reduction and colour variation caused due to surface plasmon resonance are affected by incubation temperature. In a study carried on gold nanoparticles the temperature change between 60 and 100 °C showed colour variation as yellow-brown, purplish-pink and pinkish-brown [45] respectively. It is observed that accumulation, nucleation, coalescence, secondary reduction, secondary nucleation processes are temperature dependent which decides the morphology of NPs formed [46].

The synthesis mechanism of NPs by means of bacteria rest on upon enzymes, for example *B. licheniform* enzyme was responsible for reducing nitrate for silver nanoparticle synthesis. Mycosynthesis remains a better mode for biological NPs synthesis than bacteria. Studies show that Fungi has more tolerance and high yield production of NPs. Possible mechanisms proposed to explain the mycosynthesis of metal nanoparticles are nitrate reductase action, electron shuttle quinones or both. For instance, from Penicillium species and *Fusarium oxysporum*, enzymes, nitrate reductase, and /-NADPH-dependent reductases, significant role in nanoparticle synthesis was observed [35], which was similar to the mechanism in bacteria.

However, it has been studied that synthesis for nanoparticles via microorganism mode are slow, have low productivity, and downstream processing is required for nanoparticle recovery. Moreover, some difficulties related with microorganism mode of nanoparticles synthesis include the intricate steps, such as microbial sampling, isolation, culturing, and maintenance. Recently, phyto-nanotechnology, a simple, stable, rapid, eco-friendly, and cost-effective method has opened up a new pathway for the production of NPs [47].

Phytogenic synthesis of CeO_2, CuO, (TiO_2), (SiO_2), (Al_2O_3), (Fe_2O_3) have proved its potential in commercial products. TiO_2, Fe_2O_3, ZnO, have been used for manufacturing cosmetics and sunscreen, SiO_2 for dental filling, CeO_2 as fuel additives and TiO_2 as catalysts. Studies show leaf extract of *P. thonningii* was used to synthesize Ag NPs that helped to remove heavy metals such as Magnesium, Copper, Lead, Iron from waste water [48]. In another study, silver nanoparticles from *Ocimumtenuiflorum* (Black Tulsi) leaf extract was used to treat textile dye. A study reported that Iron nanoparticles synthesised using peel extract of tangerine were used for Cadmium removal from waste water [49]. Amongst these water soluble iron oxide nanoparticles (NPs) have gained interest in biomedical water treatment applications (Fig. 5).

Iron is an important and essential component of human body. The majority of living things need iron for a number of metabolic functions, such as drug metabolism, oxygen transport, DNA synthesis, ATP synthesis, and electron transport [50]. Nevertheless, there are two issues using iron in biological systems: the production of harmful oxides and the low solubility of free metal ions. These days, biomaterial are frequently used as stabilizing agents in experimental procedures due to its degradable and recyclable nature to protect the environment.

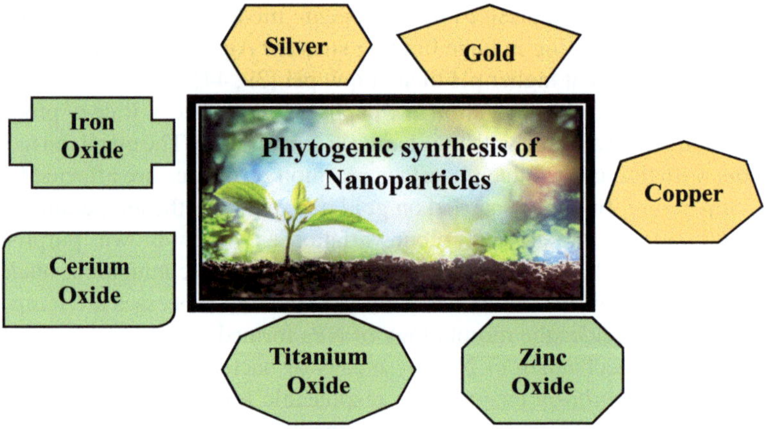

Fig. 5 Metallic NP mediated by phytogenesis

Phytogenic synthesis of iron oxide nanoparticles has gained popularity because it is economical and environmentally benign. Many of the detriments of known physical and chemical methods, such as their high energy and space requirements, high cost, and associated hazards, have been partially mitigated using phytogenic or biosynthesis method of nanoparticles. This study examined the applications of phytogenic iron oxide NPs in water and wastewater treatment, as well as their phytosynthesis. A lot of effort has been put into investigating an inexpensive, safe, and efficient dye removal technique using plant-mediated iron oxide nanoparticles [18].

Researchers have used a variety of techniques over the years to synthesize single phase α-Fe_2O_3 nanoparticles, including electrochemical, solution phase, vapor phase, and thermal oxidation methods to create nanostructures with various morphologies [51, 52]. Liquid phase synthesis methods are generally cost effect and offers high yield when variables like pH, ionic concentration, and temperature are adjusted. The large surface area to volume ratio of liquid phase produced nanoparticles makes them prone to aggregating in solution to decrease their surface energy.

Bare nanoparticle suspensions provide flexibility in adjusting their surface properties because they are extremely sensitive to external factors like pH and ionic concentration, this flexibility results due to lack of stability, which can be fixed by adding the right surfactants to the nanoparticles as stabilizing agents. The effectiveness of nanoparticles will be hampered by the presence of surfactants so formed. Nevertheless, the use of hazardous solvents and surfactants may have a detrimental effect on the produced NPs' biocompatibility [53, 54].

Iron oxides are being studied in several scientific fields, namely industrial chemistry, biology, geology, medicine, soil science, and mineralogy. Iron is in the trivalent state (Fe^{3+}) in the four oxides, α-Fe_2O_3 (hematite), γ-Fe_2O_3 (maghemite), β-Fe_2O_3, and ε-Fe_2O_3. In Fe_3O_4 (magnetite) it exist the tetravalent state (Fe^{4+}), where as in FeO (wüstite) it exist in the bivalent state (Fe^{2+}).Hematite, a very stable mineral, also referred as red ochre, specularite, kidney ore, and martite [55, 56]. Hematite nanoparticles have made significant contributions to research in interdisciplinary fields outside of biomedicine because they are stable and relatively less toxic. Small energy band gap, high solar activity, high stability, abundance, affordability, and environmental sustainability make it useful in a various applications, namely, targeted drug delivery, gas sensing, photo-electrochemical water splitting, catalysis, biomedical applications, magnetic storage devices, and lithium-ion battery production [57].

Hematite's low conduction properties, which lead to poor electrical and electrochemical performance of different hematite anodes, are its only notable drawback. By changing the size, shape, and morphology of the particles, one can improve their capacity for charge transfer at the nanoscale and adjust the optical band gap as well as magnetic and electronic characteristics. Therefore, depending on the synthesis method, nanosized hematite particles can acquire improved physical properties [58]. For the synthesis of hematite, a simple, effective, and environmentally friendly protocol has been created. Aqueous leaf extract of *Anacardiumoccidentale's* was used to synthesised iron nanoparticles [59]. Various sized nanoparticles produced by maximizing the ratio of precursor to extract have been studied for their size-dependent catalytic, nano-fluid, and antibacterial properties.

In another such study mesoporous α-Fe_2O_3 nanoparticles were synthesised using green tea (*Camellia sinensis*) leaf extract. This one-step process can be used for large-scale synthesis. 20 g of tea leaves (dry and green) were heated at 80 °C for 20 min in 1.5 L water on hot plate. The extract was filtered and stored at 10 °C. 5 g of $Fe(NO_3)_3 \cdot 9\ H_2O$ was mixed with 30 ml extract, later volume was adjusted to 50 ml. The solution was mixed and transferred into an autoclave, which was kept in oven for 13 h at 100–180 °C. The autoclave was cooled and the product was washed thoroughly with distilled water and kept for drying at 80 °C [60]. The mesoporous nanoparticles prepared were analysed using N_2 adsorption, Raman, XRD, XPS, SEM, and UV-visible spectroscopy. The average particle size of the well-crystallized, ultra-pure nanoparticles was 60 nm. Photo-catalysis activity under visible light irradiation was conducted using fluorescence spectroscopy. 60 mg of α-Fe_2O_3 extract was mixed with 50 ml of 0.5 mol/m^3 terephthalic acid, to this solution 2 mol/m^3 of NaOH solution is added. The terephthalic acid, easily reacted with OH to produce 2-hydroxyterephthalic acid, a highly fluorescent product. It is known that the quantity of OH formed is proportionate to the fluorescence intensity of this 2-hydroxyterephtalic acid [60, 61].

A variety of plants have been used by numerous researchers to carry out phytogenic iron nanoparticles. Iron nano-synthesis has been achieved using extracts from the following sources: Extracts of Neem leaves [58], Tangerine peel [60], *Acacia niloticap*ods [62], *Albizia lebbeck* leaves [63], *L. camara* fruit [64], *Eriobotrya japonica* leaves [56], orange peel [15], *Mangifera indica* [65], *Murrayakoenigii* [65], *Oolong tea* [66], *Zanthoxylumarmatum* leaves [67], green tea leaves [44, 68], Eucalyptus leaves [45], and *Lantana camara* leaves [69] have been used for Iron nano-synthesis (Tables 1, 2, 3 and 4).

Iron oxide Nanoparticles are prepared by stirring potato extract (25 ml), which works as starch template with $FeSO_4 \cdot 7\ H_2O$ (2 g). The highly viscous solution is heated at 60 °C for 4 h at constant temperature, than cooled at the rate of 10 °C/min. Red brown particles of iron oxides were obtained which was characterised using XRD and SEM. Crystalline nature of Iron oxide was studied using XRD. Debye Scherre formula was used to calculate particle size. The diffraction peaks matched with JCPDS card 84-0306, which proves hexagonal phase of hematite. The crystal size was observed between the range of 19 and 33 nm. The SEM micographs confirms hexagonal shape and diameter observed was 40–50 nm. Many studies have reported α-Fe_2O_3). It has been demonstrated that using these natural polymers as a template for the synthesis of FeO NPs is a less harmful method for the aquatic environment, particularly when freshwater fish like *Oreochromis mossambicus* are involved.

Such cost effective mode of iron oxide (FeO) synthesis was also studied using composite of *Psidiumguavaja-Moringaoleifera* (PMC). The leaf extract of the PMC were found to have six bacterial strain, as an antibacterial agent. PMC nanoparticles was used as antibacterial drug at low concentration. Photocatalytic degradation of methylene blue dye under visible light irradiation showed decrease of absorbance. Fresh PMC leaves were washed thoroughly under running water and distilled water, later air dried at normal temperature for 72 h. 10 g blended PMC was liquified with water for 1 h at 80 °C. The PMC extract was utilized for phytogenic synthesis and kept at 4 °C. 20 ml PMC extract was gently poured to liquid Iron(III) chloride in the ratio of 8:2 with vigorous stirring. Dark brown solid was pulverized, washed with distilled water, redried and sent for characterised. PMC Nanoparticles were study through UV–Visible Spectroscopy, Fourier Transform Infrared, X-ray Diffraction Spectroscopy (XRD), Transmission Electron, Microscopy (TEM) Spectroscopy (FT-IR), X-ray Diffraction Spectroscopy (XRD), and Vibrating Sample Magnetometer (VSM) [47].

Functionised magnetic nanoparticles (MNPs) of iron oxide were synthesised through a novel environmentally friendly method via phytogenesis that have the capacity to absorb various organic pollutants. Iron oxide MNPs was synthesised by crude plant extracts, to complete the reduction process, heat treatments was applied to ethanol at supercritical conditions which resulted in superparamagnetic iron oxide nanoparticles. Average crystalline size of 277, 29, and 45 nm in XRD results demonstrated the formation of spherical and cubic goethite (before treatment-YOPL), magnetite (after treatment-YOPR), and magnetite (after treatment and functionalization-YOPRP) [70]. The effectiveness of the nanoparticles against Methylene Blue (10–140 mg/L) and 4-Nitro Aniline (0.5–1 mM) was investigated. MNPs demonstrated a high pollutant removal efficiency towards 2-nitroaniline and methylene blue, despite their weak magnetic properties. On the other hand, bare nanoparticles reduce 4-NA more effectively. The efficiency of particles was well in pollutant solutions with higher pH, and the temperature between 10 and 30 °C, and shows efficient recyclability in actual waste water.

Nanoparticles were synthesised by washing, drying, crushing Neem leaves (*Azardica indica*), and the extract was then filtered twice with Whatman filter paper. To 0.1M of $Fe(Cl)_3 \cdot 6\,H_2O$ and 0.2M of $Fe(Cl)_2 \cdot 4\,H_2O$ (1:1) metal precursors solution, plant extract was added dropwise, black suspension of nanoparticles was obtained.

The suspension was divided into two. One part was centrifuged and washed immediately at 60 °C (YOPL—α-Geothite), second part was heated at 250 °C at the pressure of 1000 psi for 2 h (YOPR). The cooled particles from the reactor were collected in ethanol, then washed with distilled water and mixture of distilled water and ethanol. The nanoparticles formed were than characterised. As the magnetic property of YOPL—α-Geothite was very low, these particles were put to high pressure to enhance the magnetic property, however YOPR Fe_3O_4—magnetite lost their surfaces modifier at supercritical conditions, hence, functionalization was necessary. Measured amount of NPs were added to plant extract, it was stirred overnight, distilled water and ethanol mixture were used to wash the particles, and then NPs were dried at 60 °C (YOPRP). The nanoparticles formed were than characterised. The crystal structure of Magnetite for YOPR and YOPRP and Geothite for YOPL is revealed by XRD peaks. The Scherer formula, indicates a larger size for goethite FeO, it may be due to the concentration of solvent molecules used in the extraction process. Many substances coat the surface of nanoparticles [71]. With the majority of the spherical FeO NPs, a few cubic ones, are displayed in the SEM results, with an average size of 160 ± 60 nm. Result show that the size of these particles reduced to 100 ± 20 nm when exposed to high pressure and temperature, having larger surface area and higher magnetism. EDX was used to obtain elemental composition information. Prominent peaks of Fe and O, indicated the formation of iron oxide nanoparticles. FTIR studies shows intense hydroxyl vibrations at 1629 (bending) and 3392 cm^{-1} (stretching). After heat treatment, the deformations of 1398 and 1070 cm^{-1} nearly vanish from the magnetite spectra of the YOPR and YOPRP samples. The Fe–O bands at 687 and 501 cm^{-1}, (goethite structural) is replaced by FeIII-O vibrational modes at 584 cm^{-1} and 402 cm^{-1} (magnetite structure), A down shift of 584–574 cm^{-1} was observed with nanoparticle functionalization.

Magnetic properties are achieved by the heat treatment at supercritical conditions, addition of extracts to the metal precursors does not result in magnetic properties. Both the heat-treated nanoparticles (YOPR) and the functionalized after heat treatment (YOPRP) had saturation magnetization (Ms) and coercivity (Hc) of 44 and 36 emu/g and 123 and 133 Oe, respectively, which can be used for the intended application. The plant-based NPs (YOPL) has Ms = 13 emu/g and Hc = 44 Oe, which makes it impractical for use in hyperthermia, water treatment, or other applications [70]. UV studies displays maximum absorbance between 300 and 450 nm. YOPL, YOPR and YOPRP have similar absorbance maxima, but the peaks are of YOPL is narrow. The band gap for YOPL, YOPR and YOPRP was 3.17, 2.57, and 2.78 respectively.

FeO NPs were also been synthesised by the reduction of ferric chloride salt using *E. coli*. These phytogenic NPs were spherical, 27.7 nm in size, and are characterized using a various technique, including AFM and FTIR analysis. The phytogenic iron oxide nanoparticles by *E. coli* was optimized using pH (5) and temperature (37 °C) and average size observed was between 20 and 40 nm [72]. 10 ml of supernatant bacteria was combined with 5 ml of 10^{-3} M (1% V/V) $FeCl_2$ solution for the biosynthesis of Fe_2O_3 nanoparticles, another reaction mixture without salt was used as control reaction. Iron oxide nanoparticles are extracted after incubation of 24-h at

37 °C, following centrifugation at high rpm for 5 min. The particles formed were sent for characterization [73]. UV-Vis analysis show maximum peak at 473 nm (spherical iron oxide NPs). The NP synthesised was detected by colour change which was confirmed by the peak, which is in co-occurrence with absorbance. Fe_2O_3 nanoparticles was confirmed using atomic force microscopy (AFM), the particle size observed 27.7 nm approximately at 5 pH. The study show that average size of nanoparticles is affected with change in temperature and pH. FTIR spectrum shows stretching vibration at 1419.51 and 1579.59 cm^{-1} of O=C=O and O–H respectively, these are due to the carbon dioxide and water absorbed from surrounding. Low bands suggests the formation of Fe_2O_3 [41]. These photogenic nanoparticles have high surface area, improved enzyme activity, high and catalytic activity. Wang et al., synthesised iron nanoparticles from eucalyptus leaf extract which was used to treat swine wastewater. The spheroidal shape was studied using X-ray diffraction and FTIR spectrometer. The results indicated 84% COD removal [74].

Some researchers used green tea extract were used to synthesis Fe NPs, as catalysts to degrade mono-chlorobenzene using Fenton-like oxidation process. It was observed that Fe NPs produced from green extract showed best 81% degradation as it contains caffeine more than tea and black tea extract. The optimal results were observed at pH 3, 0.045 mol/L H_2O_2, and 0.6 g/L green tea iron nanoparticles [75]. Mystrioti et al., synthesised iron nano suspensions using 5 plant extracts and juices to reduce Cr(VI). The extracts and juices of red wine (RW), *Punica granatum* juice (pomegranate, PG), *Camellia sinensis* (green tea, GT), *Syzygium aromaticum* (clove, CL), and *Menthaspicata* (spearmint, SM) were used to synthesize iron suspension, amongst all higher concentration of iron suspension was observed in RW and PG which showed 500 mg removal capacity of Cr(VI) [76]. Ahmmad et al. reported synthesis of highly crystalline mesoporous hematite NPs through one-step hydrothermal process using extract of green tea leaves [60]. Similarly, using *Psoraleacorylifolia* seeds, Nagajyothi et al. demonstrated that α-Fe_2O_3 NPs could be produced using a phytogenic mode [77]. The methylene blue dye was effectively degraded by using catalytic activity and these nanoparticles were applicable in the treatment of industrial waste water. It has also been demonstrated that FeO NPs effectively eliminate As(V) from water [41].

α-Fe_2O_3 nanoparticles (Hematite) has been synthesised using *Rosmarinus officinalis* (rosemary) leave extract mediated by an entirely green process. To comply with green standards, no solvents or surfactants were used in the synthesis. The characteristics of the phytogenic α-Fe_2O_3 NPs was studied using FESEM which demonstrated that the NPs' are highly agglomerated, self-assembled nature, and the average size ranging at 6.531 ± 0.148 nm. Meanwhile, X-ray diffraction (XRD) demonstrated the rhombohedral structure and the crystallinity of the α-Fe_2O_3 NPs, with an average size falling between 7.412 and 10.742 nm, readings were in agreement with the FESEM readings [78].

Kaboli et al. used green rosemary leaves, boiled in deionised water at 60 °C for 2 h. The extract solution of pH 2.72 underwent filtration using the Whatman paper twice to get rid of any leftover solids [79]. To the filtered rosemary extract, iron(III) chloride hexahydrate was added and mixed evenly for 1 h on magnetic stirrer at 60 °C. The dried dark green powder after annealing turns to black colour. The nanoparticles were than subjected to characterization using EDS, FESEM, and FTIR. Phyto-synthesised α-Fe_2O_3 NPs were applicable to dielectric, nano-fluids, and solar energy.

Studies by Prasad et al., garlic vine leaf extract was used to reduce precursor iron(II) sulphate heptahydrate, resulting, nanocrystalline Fe(III) oxide. However, not much applications were studied in the field of biomedical and nanofluid [80]. Hematite nanoparticles were found to be synthesised using leaf extract of *Anacardiumoccidentale (A. occidentale*). The extractcontains anacardic acid, an important phytochemical, that contains a natural phenolic lipid which has widespread applications in industry and medicine. Anacardic acid has antiviral, antifungal, antibacterial, antioxidant, and even anti-carcinogenic properties in its biological potential.

Due to the abundance of alkaloids, tannins, saponins, phenol, steroids, and glycosides in *A. occidentale,* surface modification of nanoparticles formed, makes them appropriate for research and antimicrobial use. With excess of sodium borohydride ($NaBH_4$), chemo-catalytic ability of the synthesised hematite NPs was studied against dyes. These nanoparticles increase the thermal conductivity of base fluids which can be used in coolants for heat transfer systems [62].

In another study, 10 g *A. Occidentale*leaves are washed and boiled in 100 ml distilled water for 2 min. After cooling, filtering, the extract was used in nanoparticles synthesis. To 100 ml distilled water, 0.01 M $FeCl_3$, as precursor was added, followed by 2 min boiling. To the pale-yellow boiling solution 5 ml of green extract was slowly added with constant stirring. Deep brown solution turned to brown precipitate gradually. The precipitate was washed with deionised water several times and then with acetone, followed by drying. The powered precipitate was annealed for 3 h at 600 °C, deep red formed indicates hematite nanoparticles [63].

Researchers synthesised Hematite nanoparticles via phytogenic means using *Rhuspunjabensis* extract as a capping and reducing agent and $FeCl_3$ as precursor. Various characterizations techniques were used to elucidate structural, morphological, and compositional analyses were conducted [78]. The hematite (Fe_2O_3) NPs synthesized had a rhombohedral structure and measured approximately 41.5 ± 5 nm in size. Phase-pure Fe_2O_3 formation was indicated by TEM, FTIR, and TG/DTA. The synthetic NP has the ability to scavenge free radicals, have antioxidative potential, and possess reducing power which is shown by their biological activities. The phenolic and flavonoid-like characteristics resulting from the attachment of various functional groups to the NPs' surface are responsible for these activities. Whencompared to the corresponding plant extract, the NPs showed notable antileishmanial, antibacterial, and protein kinase inhibition activities. Hence, green Fe_2O_3 NP synthesis described, has potential therapeutic and diagnostic agent for cancer treatment [64].

Study by Valentin V. Makarov reported the synthesis of amorphous iron oxide NPs from the leaves of *Hordeum vulgare* (*upto* 30 nm) and *Rumexacetosa* (10–40 nm) plants. Stable Iron oxide and iron oxohydroxide NPs were produced from leaf extract. Both the extract contain nearly same protein and antioxidants however the pH values differed, 5.8 and 3.7 respectively [65].

T. Shahwan studied green tea leaves (GT-Fe NPs) to synthesiseiron oxide and ironoxohydroxide NPs. The nanoparticles were characterised and used as a Fenton-like catalyst for the decolorization of dyes [methylene blue (MB) and methyl orange] in aqueous solutions. The NP synthesized were irregular clusters ranging between 40 and 60 nm in size [66].

U. Ting Wang studied a one-stepbiosynthetic routeusing eucalyptus leaf extracts at room-temperature. Spheroidal iron NPs were used to treat swine waste water, resulting in the removal of 84.5% of COD and 71.7% of total nitrogen [68].

L. Huang, Xsynthesized NPs which were environmentally non-toxic using tea extracts of oolong and black teas. These NPs were used to degrade malachite green (MG). The NPs synthesised were characteristics using SEM, UV–vis, BET, EDS, and XRD [67].

In a similar study M. S. Al-RuqeishiGreen synthesisediron oxide nanorodsusing Omani mango tree leaves extract. The obtained Iron(III) oxide NPs were observed to have average diameter of 3.0 ± 0.2 nm and average length of 15 ± 2 nm. These nanorods synthesised has different diffraction planes indicating alpha phase, hematite (α-Fe_2O_3) and gamma phase, maghemite (γ-Fe_2O_3). These NPs when added to heavy oils reduces its viscosity [81].

In a study S. Machadosynthesised zero-valent iron NPs using oak, pomegranate and green tea leave extract. Characterization and analysis proved nZVIs, ranging between 10 and 20 nm [82].

Vidhisha Jassal used *Sapindus mukorossi* (raw reetha) as an earthy surfactant and water as a medium to synthesize rod like Iron Oxide nanoparticles (α-FeOOH, β-FeOOH and α-Fe_2O_3). These NPs were successfully utilized in as catalysts for removing 2, 3 and 4-Aminopyridines from aqueous waste [83].

Dan Cao investigatedhow the reactivity of iron oxide NPs can be enhanced usinga stabilizer, cetyltrimethylammonium bromide (CTAB). It was observed that removal of phosphate increased substantially from 71.0 to 97.3% using CTAB due to innerspherecomplexing mechanism and electrostatic attraction [68].

Imran Ali studies Green synthesis of iron nano particles used for the adsorption of ametryn from run off. Herbicide ametryn is broadly used to controlleaf weeds and grasses in sugarcane, bananas, maize and pineapple fields. The sorption method using NPs may be used for the removal of ametryn [84].

Several Phytogenic mode of iron nanoparticle syntheses and its morphology are listed in Table 1.

In a study by Buazar et al., potato extract was used to synthesize Fe_3O_4 NPs, on size 40 ± 2.2 nm used to remove organic pollutant methylene blue in 30 min [93]. In a similar study, Fe_3O_4 NPs were synthesised using *Zanthoxylumarmatum* DC leave extract to remove methylene blue. SEM and TEM characterization confirms the spherical shape and 17 nm size of nanoparticles. The process follows pseudo second order kinetics [94].

From the findings of Silveira et al. *Moringaoleifera* leaf extract synthesised FeO NPs, showed potential to remove fluoride ions [95].

Leaf extract of *Peltophorumpterocarpum* was used to synthesised rod shaped ferric oxide NPs, XRD spectrum confirms the presence of γ and α–Fe_2O_3 phases.

Table 1 Phytogenic synthesis of iron NPs and its morphology [69]

Phyto-sources	Plant part used	Resulted NPs	Size (nm)	Shape	References
Sorghum spp.	Power	FeOOH	50	Irregular cluster	[85]
Soya bean	Sprouts	Magnetite	8	Spherical	[86]
Watermelon	Rinds	Magnetite	20	Spherical	[87]
Mimosa pudica	Root	Magnetite	60–80	Spherical	[88]
Caricaya papaya	Leaves	Magnetite	Irregular	Irregular	[89]
Green tea Oolong tea Black tea	Leaves	ZVI, IONPs, FeOOH, and Fe_2O_3	40–50	Spherical	[67]
Andean blackberry	Leaves	Magnetite	40–70	Spherical	[81]
Amaranthus spinosus	Leaves	IONPs	58–530	Spherical	[90]
Eucalyptus	Leaves	Fe–p NPs	40–60	Cube	[82]
Green tea	Leaves	Fe–p NPs	70	Spherical	[83]
Orange Lime Lemon Mandarin	Pulp Peel Albedo	ZVI	3–300	Spherical Cylindrical Irregular	[91]
Oolong tea	Leaves	ZVI, Fe_3O_4 FeOOH, Fe_2O_3	40–50	Spherical	[84]
Camellia sinensis *Syzygium aromaticum* *Mentha spicata* *Punica granatum juice* Red wine	Leaves Fruit	FeNPs	60	Spherical	[32]
Mangifera indica (MI)	Leaves	Fe NPs	100–150	Spherical	
Tangerine	Peel	Fe NPs	50	Spherical	[49]
Oak, mulberry and cherry	Fruit	nZVI	10–30	Spherical	[92]

The average size 16.99 nm was observed. FTIR shows bands of Fe–O and compounds used in synthesis. Fenton-like catalytic degradation shows 90% removal of methylene blue in 220 min. The process followed second order kinetics [96].

Lung et al., investigated magnetite (Fe_3O_4) nanoparticles as adsorbents to remove divalent metal ions of Pb, Cd, and trivalent As. A comparison was drawn between NPs synthesizing by convention Fe_3O_4 NPs in aqueous solution and by peel extract of *Citrus limon* (lemon). The removal efficiencies of divalent Pb, Cd, and trivalent

As using phyogenic mode was high, 98.8 ± 5.6, 46.0 ± 1.3, and 48.2 ± 2.6% respectively as compared to conventional mode [97].

Ahmed et al., conducted an experiment to remove Alizarin yellow R dye from industrial wastewater by a solar driven photocatalyst process using iron NPs synthesised by fruit extract of *A. chinensis*. The NPs, as shown by TEM were cubic and rod shaped with average size 9.078–107 nm. The results showed 93.7% of dye removal in 42 h, the degradation rate followed pseudo first order reaction [98].

In another finding by Vasantharaj et al., studied aqueous leave extract of *Ruelliatuberosa* (*RT*) was used to synthesise FeONPs to degrade crystal violet dye under solar irradiation. The efficiency observed was 80%. TEM reveals hexagonal nano-rods with the average size 52.78 nm [99].

B. Kumar et al., studied maghemite (γ-Fe_2O_3) nanoparticles synthesised by aqueous fruit extract of *Ficuscarica.* TEM, SAED, UV–Vis, TGA, FTIR, XRD, and DLS techniques confirms size, nature, functionalization, purity, morphology, and optical property of the nanoparticle synthesised [100]. The Nanoparticles formed were spherical with the average size 4–6 nm. γ-Fe_2O_3NPs showed both free radical scavenging and catalytic activity. First was used against DPPH and later in presence of $NaBH_4$ showed reduction of 4-Nitrophenol [100].

4 Characterization of Nanoparticles

The measurement techniques applied to determine the important physical and chemical properties are essential to know the credibility of nanoparticles synthesised. Innovative products using nanoparticles made by cutting-edge industries in every field as they offer improved, distinctive qualities with significant societal and commercial value. High-performing nanomaterials needs more vigilance and need even-tougher methods to know whether they are compatible and sustainable to the environment. Hence, using physical and chemical techniques the risk assessment of novel nanoparticles is imperative [101].

The phytogenic mode of iron nanoparticles synthesised are characterised by any of these techniques UV, FTIR, SEM-EEX, TEM, DLS, TOF-SIMS, Zeta Pot, XPS, AES or STM (Fig. 7). These techniques confirm morphology, purity, nature, functional group, composition and optical properties.

Scanning Electron Microscopy (SEM)

For characterizing of size and shape of nanoparticles, the high-resolution SEM is a great tool, which quickly and easily does image acquisition and sample preparation. Even though the SEM image only shows a particular set of three-dimensional (3D) objects in two dimensions (2D). To achieve accurate particle morphology, it is crucial to select different imaging parameters while taking SEM pictures of nanoparticles and to select evaluation or analysis techniques, particularly thresholding algorithms. The operators of the SEM are responsible for choosing the right instrument parameters and algorithms and to assess the effects, and choose the right measurement conditions [102].

Transmission Electron Microscopy (TEM)

A variety of nanomaterials (NM) can be analyzed for size, shape, crystallographic structure, and chemical makeup using transmission electron microscopy (TEM), a flexible method [103]. TEM gives a spatial resolution covering entire size range from 1 to 100 nm. TEM produces 2D nanomaterial image. It allows to determine physical properties of nano materials quantitatively, based on 2D images. It helps to assess aggregation or agglomeration of constituents. TEM can incorporate EDS and EELS spectroscopic methods for elemental analysis and to characterise chemical bonding of nano materials. Despite of numerous applications TEM suffers some drawback. The TEM analysis depends upon transfer of sample to the TEM grid and specimen preparation. Acceleration Voltage of 100–300 kV is required for high resolution. Hard matter like zeolite, porous materials, polymers, and hybrid materials can damage beam required for analysis. TEM analysis are limited to those artifacts which results from high vacuum. TEM estimates the particles perpendicular and nor parallel to the electron beam [104].

Ultraviolet–Visible (UV–Vis) Spectrophotometry

An analytical method for measuring concentration of a give analyte in a transparent fluid is UV-Vis spectrophotometry. Analyte absorb light in the wavelength ranging between 190 and 1000 nm, 380 nm separates the visible and UV light boundaries, whereas, 780 nm is boundary separating the visible light and the near infrared (NIR) wavelength (Fig. 6).

UV-Vis spectrophotometry's main advantages are its low cost, wide availability, simple to use, and quantitative nature. Due to the transparent nature of most of the solution undergoing chemical reaction in visible range, UV–Vis spectrophotometry serve beneficial to monitor chemical reactions, where the reactants and products have different absorbance features. Beer–Lambert law is used for analytical purposes [105].

$$A = \log_{10}\frac{I_o}{I} = \varepsilon c l$$

where, I = Intensity of the transmitted light, I_0 = Reference Intensity, ε = Molar Extension coefficient, A = Absorbance, C = Concentration, l = Path length.

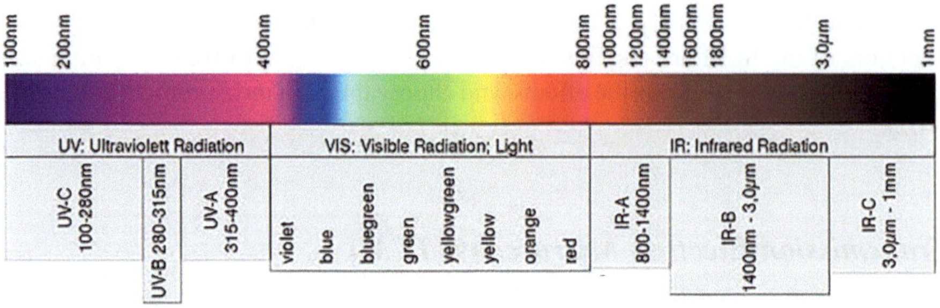

Fig. 6 Wavelength range of spectrophotometer

X-ray Photoelectron Spectroscopy

It is a surface sensitive quantitative method where by using photoelectric effect, XPS generate electrons, when the sample is illuminated with X-rays. It identifies the elements, composition, chemical state, density of electronic state and electronic structure. As long as the material can withstand the high vacuum environment, XPS can be applied to almost any surface for chemical surface measurement. XPS imaging and small spot analyses can be subjected to typically $\sim$ 10 μm resolution unlike AES, which can resolve a few tens of nanometres apart. The sample experiences low flux of electron in XPS, which helps to analyse sensitive materials that deteriorate quickly on exposure to electrons like organic materials and helps in insulating materials. XPS can be used in line profiling and depth profiling. It can be applied for the study of material in its normal state and can be used when the material undergoes cleavage, exposed to heat, reaction with gasses or solutions, scraping, UV light or ion implantation [106].

Energy-Dispersive X-ray Spectroscopy (EDS)

Elemental analysis or chemical characterization of a sample can be done using EDS. It works on the principle that each element has its own atomic spectra which gives different peaks when exposed to X-ray excitation. EDS can capture smallest structure due to its high detection sensitivity. It is attached to SEM due to same working principle. All the elements known can be detected by EDS [107].

Brunauer, Emmett, and Teller (BET)

Properties of matter can be changed significantly with the change in particle and pore size, which will affect its performance, hence, specific surface area (SSA) of any material is a crucial parameter to be studied. Hence, SSA analysis of porous materials is frequently utilized in industry and typically carried out by applying the BET method. SSA is basic physical and chemical property listed in the safety data sheets which is required for safety assessment of nanomaterials [108]. SSA of a given material is total surface area by mass (m^2/g).

$$SSA = \frac{SA}{M}$$

Zeta-Potential Measurements

Zeta potential is an indirect measurable parameter. Stability of colloidal particles or electrophoretic mobility are determined by Zeta-potential. It works on the basis of dynamic light scattering (DLS). Zeta potential can be calculated from electrophoretic mobility μ applying equation [109].

$$\mu = \frac{\varepsilon_0 \varepsilon_m}{\eta}$$

where, ε_0 = dielectric constants in vacuum, ε_m = dielectric constants in solution, η = dynamic viscosity.

Fourier Transform Infrared Spectroscopy

FTIR analysis scans the samples with infrared light to determine which materials are organic, inorganic, and polymeric. Changes in the distinctive absorption band pattern denotes the modification in the composition of the material. FTIR is helpful for characterizing and identifying unknown materials as well as for spotting pollutants.

The sample absorbs some of the 10,000–100 cm^{-1} infrared radiation and converts into vibrational or rotational energy while with the remaining portion passing through. The signals detected at the detector are ranging from 4000 to 400 cm^{-1} spectrum. FTIR is a useful tool for chemical identification as every molecule has a unique fingerprint [110].

X-rays Diffraction (XRD)

A non-contact and non-destructive method used to study crystalline phases, structural properties and polymeric forms of material is called X-ray diffraction (XRD). X rays are electromagnetic radiations of short wavelength, they are produced when charged particles with adequate energy is slowed. Every crystalline substance gives definite pattern and independent patterns are observed with the mixture of substances, whereas, same pattern is given by same substance. XRD determines orientation of single crystal, measures the morphology and internal stress of crystal and unknown structure.

XRD works on Bragg's law

$$n\lambda = 2d \sin \theta$$

The average distance between rows of atoms, orientation, crystal structure of an unknown material, dimensions, form, and internal stress of small crystalline can be ascertained with the help of X-ray diffraction (XRD).

XRD helps to determine the average spacing between layers or rows of atoms, to determine the orientation of a single crystal, to find the crystal structure of an unknown material, and to measure the size, shape and internal stress of small crystalline regions [111].

Energy Dispersive X-ray Analysis

This method determines the elemental proportion and provides an overall mapping of the sample. When an electron beam with high energy strikes the conducting surface of sample, X-rays are released from the sample and its intensity depends on the conducting sample. X-rays generated in sample due to EDX is 2 μm in depth, hence, the intensity of X ray is low and the image take long time to acquire [112]. Any heavy metal ion present on the surface of NPs can be detected by EDX. Limitation of EXD is detection of elements with low atomic numbers.

Atomic Force Microscopy

AFM is a powerful and versatile microscopy method used in the study of materials at the nanoscale. By taking an image in three dimensions, engineers and scientists can use it to obtain a range of surface measurements. AFM can create images with atomic and angstrom scale resolution with little sample preparation. It is helpful for displaying surface texture of wide range of materials. It has high three-dimensional spatial resolution and is a non-destructive method [113].

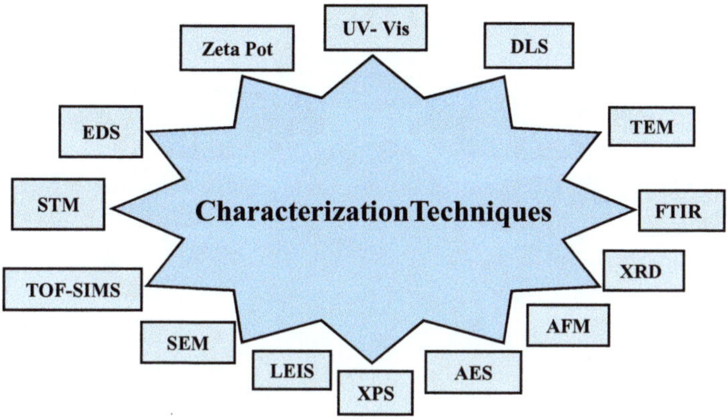

Fig. 7 Characterization of NPs [114]

Raman Spectroscopy

Raman spectroscopy is used to investigate a system's vibrational, rotational, and other modes using monochromatic laser light. Energy is shifted as result of the interaction between the laser light and the photons. Usually, a laser beam is used to illuminate the sample. The electromagnetic radiation from the laser hit spot is collected by a lens and then directed through a collimator. In stokes Raman scattering molecule at excited state drops to vibrational state with less energy of scattered photon then incident. However, if the molecule existed in vibrational state already and drops to the ground state after scattering with high energy it is anti-strokes Raman scattering. Raman shift depends on vibrational level and polarization alters the intensity of scattering [110] (Fig. 7).

5 Potential Benefits of Photogenic Nanoparticles in Waste Water Treatment

The benefits of Phyto-nanotechnology include biocompatibility, scalability, and potential medicinal applications. The growing demand for nanoparticles can be satisfied by synthesising plant-derived nanoparticles, which are easily accessible and harmless hence, an eco-friendly approach. Extracts from a variety of plant parts and water as reducing medium are utilized to synthesis MNPs [43, 47–61]. The use of medicinal herb, *Panax Ginseng* has been suggested to successfully synthesis Au and Ag NPs, using the leaf and root extract [44–46]. The precise process responsible for phyto-based nano synthesis needs further research. According to certain theories biomolecules and secondary metabolites aid metal salt reduction and serve as

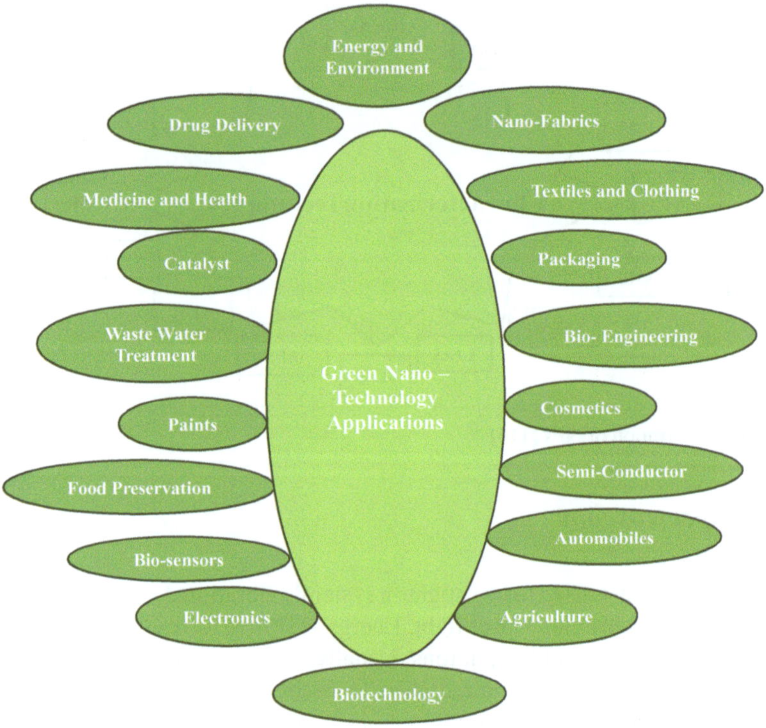

Fig. 8 Application of green nano-technology

capping and stabilizing agents for artificially generated NPs [70]. Number of applications have been studied using phytogenic synthesis, which is also termed as green nanotechnology as shown in Fig. 8.

Phytogenic mode of nanoparticle synthesis environmentally friendly, involves simple process, cost effective, requires less energy, and has eco-friendly approach. These are more stable and has proved more efficient. Biological sources can be microbes or plant based. However, plant based or phytogenic mode considered to be more advantageous, economical and simple [115]. Plant extracts are used as reducing and coating agents. The diverse functional groups that comprise plant secondary metabolites are recognized for their biological activity and their reducing abilities [116–118]. Beyond their capacity for reduction, those phytochemicals can function as coating agents for multidimensional absorption in addition to stabilizing the nanoparticles. The plants are rich in carbohydrates, amino acids, nitrogen base and reducing sugars. Plant's extracts include a variety of metabolites [119, 120]. Hence, parts of plants, leaf, stem, flora, peel, fruit, bark and seed can be used as a source of variety of metabolites. In the phytogenic synthesis of NPs metal ions are formed due to the reduction of metal salts. Consequently, the NPs has been applied in different medicinal applications and magnetic nanoparticles for water treatment [21, 120, 121]. Magnetic FeO NPs were allowed to grow directly on a solid medium to preserve active ingredients, magneticity, which increase the reusability of NPs. Therefore, to enable alternative recovery, particles were grown on a solid PVA surface in addition to being manipulated magnetically. Because of its exceptional acidity and basicity tolerance, as well as its thermal stability up to 354.8 °C, polyvinyl alcohol (PVA) is selected as a solid support [122]. One step Green synthesis is typically recommended for economy, ecology, and health reasons. Furthermore, the surface of the nanoparticles containing multifunctional group enhances the absorption of various organic pollutants and inorganic ions present in the contaminated water as indicated in the different study in Table 2.

Wastewater treatment is one of the most crucial solutions to the global water crisis and other related environmental problems.

Solution to the world's water crisis and other associated environmental issues is wastewater treatment. Nanotechnology has shown promising effects on reducing waste, industrial contaminants and improving efficacy of energy production. An important application of nanotechnology has been for waste water treatment using zero-valent metal (Zn, Ag, Fe), ferrites (Zn, Ni, Mn, Co, Mg) [128], metal oxide TiO_2 [129], ZnO [130], SiO_2 [131], Fe_2O_3 [132], Fe_2O_4 nanoparticles, metal organic framework (MOF), carbon nanotubes, and nanocomposites [53]. Nanosorbents, photocatalysis, nanofiltration, are certain promising techniques used for water treatment in nanotechnology [39]. In the process of photocatalysis, radiation causes a change in the rate of reaction (UV, Visible, IR). Photocatalysis can be used in two different ways, one as solar or as artificial UV light. Semiconductors used as nanomaterials are suitable for photocatalytic activity. This is due to the availability electrons and holes produced by irradiation. Photocatalysts help in removing refractory pollutants from effluent [133]. The method used depends on the further utility of treated water along with treatment efficiency, eco-friendly approach, cost effectiveness, reusability if any. It is imperative that disposal of untreated water shall largely affect the aquatic flora and fauna causing many mutations in living organisms.

Inorganic and organic contaminants were efficiently removed by plant mediated iron nanoparticles. Due to the catalytic properties INPs are used extensively in many studies. They have been used numerously in removing dyes (MB, MO, CR, MG [67]) Inorganics (nitrates [74], phosphates [134]), pesticides [135], transition metals (Cr, Cu, Co) [76], and Crude oil [136]. Zerovalent iron nanoparticles when coated with phytochemicals enhances physiochemical characteristics of newly formed NP [65].

In a study, the efficiency of pollutant removal using Fe_3O_4/α-FeOOH@*Azadirachta indica* was compared with other nanoparticles [70]. It was observed that synthesised nanoparticles showed better efficiency to remove methylene blue than 4-Nitro Aniline (Table 4).

Table 2 Phytogenic synthesis of iron NPs and its efficiency in the adsorption of different adsorbates [39]

Phytogenic-source	Size (nm)	Morphology	Adsorbate	Adsorption capacity (mg/g)/ Percentage efficiency (%)	References
Neem leaf	21.0–21.03	Spherical	DDT	90.00–90.20	[18]
Tangerine peel	20–50	Spherical	Cd	90.00	[49]
Eriobotrya japonica	88–89	Spherical	Cr(VI)	312.5/–	[18]
Acacia nilotica pods	210–230	Irregular	Methyl orange	19/–	[123]
Albizia lebbeck leaf	–	–	Crystal violet	–/96.59	[124]
Albizia lebbeck leaf	–	–	Congo red	–/96.59	[124]
L. camara fruit	1.1–1.9	Spherical	Ni (II)	227.20/–	[44]
Eriobotrya japonica leaf	171.2	Irregular Spherical	Basic red 46	–	[125]
Orange peel	500	Spherical	Cd	–/71.43	[3]
Mangifera indica	100–150	Spherical	Total phosphates	–/91.89	[126]
Murraya koenigii (MK)	100–150	Spherical	Phosphates	–/97.68	
MK	100–150	Spherical	Ammonia Nitrogen	–/87.08	
MK	100–150	Spherical	COD	–/88.24	
Azadirachta indica (AI)	96–110	Spherical	Phosphates	–/98.08	
Azadirachta indica (AI)	96–110	Spherical	Ammonia nitrogen	–/84.32	
Azadirachta indica (AI)	96–110	Spherical	COD	–/82.35	
Magnolia champaca (MC)	99–129	Spherical	Phosphates	–/98.84	
MC	99–129	Spherical	Ammonia nitrogen	–/74.78	
MC	99–129	Spherical	COD	–/88.23	
MC	99–129	Spherical	BOD	–/88.23	
Oolong tea extract	40–50	Spherical	Malachite green	–/75.5	[84]
Zanthoxylum armatum leaf	17	Spherical	Methylene blue	0.4712/–	[94]

(continued)

Table 2 (continued)

Phytogenic-source	Size (nm)	Morphology	Adsorbate	Adsorption capacity (mg/g)/ Percentage efficiency (%)	References
Tea leaf	5–15	Spherical	Degradation of bromothymol blue	Effective	[127]
Green tea leaf	40–60	Irregular clusters	Methylene blue	–/99	[66]
Green tea leaf	40–60	Irregular clusters	Methyl orange	–/99	
Eucalyptus leaf extract	20–80	Spherical	Total N	–/71.7	[74]
Eucalyptus leaf extract	20–80	Spherical	COD	–/84.5	

6 Challenges and Future Prospects

Bio-waste (peels, seeds, leaves, bark, flower) is unavoidable, about 80% of biomass post-harvest is waste, which is typically burned and disposed off, resulting in the emissions of greenhouse gases and other pollutants that have a negative impact on public health and environment. But, if properly handled and disposed it can save a large amount of energy, assisting in the long-term use of resources and lowering the waste deposits. One such step towards the sustainability is synthesis of phytogenic NPs using the bio-mass [147], which can be regenerated, recyclable, reusable, and economical. Main drawback of handling bio-waste is that it composition varies worldwide, which will lead to an uneven large-scale synthesis of nanoparticles, but if, segregated, it can be an energy saver [36].

Phytogenic NPs so synthesised are non-hazards and eco-friendly. The phyto-extract contains metabolites, used as reducing or capping agent in the synthesis of NPs [119]. Phytogenic NPs can create zero-valent wastewater discharge. They can be effectively used in the eliminating pollutants from effluents. However, it's important to take into account a few difficulties and potential outcomes which should be overviewed [148].

(a) Prior to application, it is important to evaluate the effects of persistent exposure to phytogenic synthesized nanoparticles on the environment and on humans
(b) Prior to the application to the real pollutant, optimization of all the parameters must be done to avoid any alteration in the magnetic behaviour of the phyto-synthesized NPs.
(c) For effective and efficient pollutant removal from water and wastewaters, the stability of phytogenic NPs should be maintained.

Table 3 Plant mediated iron nanoparticles and their potential to remove organic contaminants [69]

Phyto-sources	Type of NP	Contaminants	Removal efficiency (%)	References
Shirazi thyme Pistachio green	Zero-valent iron	Phosphorus	97.72–35.79	[134]
Green tea eucalyptus	FeO-iron oxide core–shell	Nitrate	59.00–59.70	[74]
			41.00–41.45	
Amaranthus spinosus	IONPs	Methylene blue (MB)	69–69.4	[90]
		Methyl orange	75–75.25	
Andean blackberry	Magnetite	Congo red	–	[81]
		Methylene blue		
		Methyl orange		
Sapindus mukorossi	α-FeOOH	2-Aminopyridine (AP)	75.00–75.21	[137]
		3-AP	92.00–94.27	
		4-AP	80.00–81.78	
	β-FeOOH	2-Aminopyridine (AP)	72.45–73.92	
		3-AP	90.50–92.36	
		4-AP	80.00–80.12	
	Fe_2O_3	2-Aminopyridine (AP)	70.45–71.03	
		3-AP	85.00–88.31	
		4-AP	75.50–78.99	
Camellia sinensis Red wine	FeNPs	Cr(VI)	93–95	[76]
			89–90	
Omani mango	IONPs	Heavy crude oil	48–49	[136]
Green tea (*Camellia sinensis*) pomegranate (Punica granatum)	Fe NPs	Color removal	90–95	[138]
		Dissolved organic carbon removal from textile wastewater	75–80	
Azadirachta indica (AI) Magnolia champaca (MC), Mangifera indica (MI) Murraya koenigii (MK)	Fe NPs	AI-Fe NPs—phosphate	98.0–98.1	[119]
Tangerine	Iron oxide NPs	Cd removal	85–90	[49]

Table 4 Pollutant removal comparison of NPs [1]

Material	Concentration		Optimum conditions		% Removal/ reduction	Contact time (min)	References
	Adsorbent	Adsorbate (MB/4-NA)	pH	Temp (°C)			
γ-Fe_2O_3@*Cynometra rami flora* fruit	30 mg	20 mg/L (MB)[a]			94	120	[139]
γ-Fe_2O_3@cellulose	0.4 mg/ml	20 mg/L (MB)	5.2–5.7		~ 40	420	[140]
Fe_3O_4@graphene oxide	15 mg/L	50 mg/L (MB)		30	90	180	[141]
Fe_3O_4/Fe/Fe_3C@PCNF1	40 mg + H_2O_2 catalyst (0.32 M)	100 mg/L		85		30	[142]
Fe_3O_4-GO@Polyacrylic acid (PAA)	8 mg/ml	20 mg/L	10–11		100	150	[36]
IL/Fe_3O_4/Ze	0.2 mg/ml + $NaBH_4$ (0.5 M)	0.05 M (4-NA)	–		~ 28	–	[143]
		0.005 M (MB)	–	–	~ 47	–	
NiO	5 mg	0.0145 mM (4-NA) + $NaBH_4$ (13.4 mM)		40.15	97	100	[144]
Ag@polymer microgels	25.2 μg/mL	0.06 mM (4-NA) + $NaBH_4$ (12.64 mM)	9.0–9.42	21.1	~ 90	22	[145]
AgNPs-rGO AgNPs	1 mM	2 mM (4-NA) + $NaBH_4$ (0.2 mM)			~ 87	20	[146]
					~ 88	55	
Fe_3O_4/ α-FeOOH@Azadirachta indica extract	5 mg/ml	140 mg/L (MB)	6.5–12	10–30	99.96	4 (Chip)/20 (powder)	[70]
		0.5 mM (4-NA)			83	160 (powder)	

[a] Photocatalytic degradation

(d) Research should be aimed to synthesize new phytogenic NPs, which can remove multiple pollutants from wastewaters.
(e) Wide range of phytogenic NPs can be synthesised using different functional groups by manipulating plant metabolites.
(f) Commercial viability of the phytogenic NPs synthesised should be analysed prior its use, no such data available to date.
(g) For biomedical applications, a biocompatibility test should be mandatory.
(h) Further research to produce low cost NPs of desired size and shape with less expenditure is must.
(i) The yield of the phytogenic nanoparticles corresponding to conventional way of synthesis needs elucidation.
(j) Phytogenic synthesis lack knowledge of mechanism, stability, biocompatibility and chemical components.
(k) More research is needed to comprehend how active groups are attached to the surface of nanomaterials to achieve greater efficacy.

7 Conclusion

This review covered the recent surge in treating water and wastewater with plant-mediated iron nanoparticles. Phytogenic mode opens a pathway to an eco-friendly approach replacing the conventional physical and chemical methods synthesis, which are toxic and uneconomical. The several benefits offered by phytogenic route of nanoparticle synthesis includes, the ability to produce stable nanoparticles with controlled sizes and shapes, lessens toxic contaminants, rapid synthesis using variety of microorganisms and medicinal plants, and lessens the probability of forming complex chemicals during synthesis.

Plant-mediated iron nanoparticles have shown significant role in removing organic pollutants even at the higher concentrations. These nanoparticles have shown removal of heavy metals and dyes using photocatalytic and adsorption techniques. The potential of iron nanoparticles in numerous fields has subjected their need for large scale production. Thus, phytogenic mode of synthesis of iron nanoparticles can be promising substitute for waste water and water treatment on commercial scale.

References

1. Pathania, D., Sharma, S., & Singh, P. (2013). Removal of methylene blue by adsorption onto activated carbon developed from Ficus carica bast. *Arabian Journal of Chemistry*. https://doi.org/10.1016/j.arabjc.2013.04.021
2. Auta, M., & Hameed, B. H. (2011). Preparation of waste tea activated carbon using potassium acetate as an activating agent for adsorption of acid blue 25 dye. *Chemical Engineering Journal, 171*(2), 502–509. https://doi.org/10.1016/j.cej.2011.04.017

3. Gupta, V. K., et al. (2012). Photo-catalytic degradation of toxic dye amaranth on TiO_2/UV in aqueous suspensions. *Materials Science and Engineering: C, 32*(1), 12–17. https://doi.org/10.1016/j.msec.2011.08.018
4. Buzzini, A. P., & Pires, E. C. (2007). Evaluation of a upflow anaerobic sludge blanket reactor with partial recirculation of effluent used to treat wastewaters from pulp and paper plants. *98*, 1838–1848. https://doi.org/10.1016/j.biortech.2006.06.030
5. Ajaz, M., Rehman, A., Khan, Z., Nisar, M. A., & Hussain, S. (2019). Degradation of azo dyes by Alcaligenes aquatilis 3c and its potential use in the wastewater treatment. *AMB Express*. https://doi.org/10.1186/s13568-019-0788-3
6. Bower, K. C., Duirk, S., & Miller, C. M. (1998). Ozone-hydrogen peroxide advanced oxidation process (No. 330, pp. 167–176).
7. Munter, R. (2001). Advanced oxidation processes—Current status and prospect. *Proceedings of the Estonian Academy of Sciences Chemistry, 50*(2), 59–80. https://doi.org/10.1002/9780470561331.ch18
8. Deshpande, B. D., Agrawal, P. S., & Yenkie, M. K. N. (2020). Nanoparticles aided AOP for degradation of p-nitro benzoic acid. *Materials Today Proceedings*, 2–6. https://doi.org/10.1016/j.matpr.2020.02.924
9. Deshpande, B. D., Agrawal, P. S., & Yenkie, M. K. N. (2019). *Advanced oxidative degradation of benzoic acid and 4-nitro benzoic acid—A comparative study* (Vol. 210003).
10. Divyapriya, G., Singh, S., Martínez-huitle, C. A., Scaria, J., Karim, A. V., & Nidheesh, P. V. (2021). Chemosphere treatment of real wastewater by photoelectrochemical methods: An overview. *Chemosphere, 276*, 130188. https://doi.org/10.1016/j.chemosphere.2021.130188
11. Kodavatiganti, S., Bhat, A. P., & Gogate, P. R. (2021). Intensified degradation of acid violet 7 dye using ultrasound combined with hydrogen peroxide, fenton, and persulfate. *Separation and Purification Technology, 279*, 119673. https://doi.org/10.1016/j.seppur.2021.119673
12. Moradi, M., Elahinia, A., Mousavi, A., Vasseghian, Y., Dragoi, E., & Omidi, F. (2020). A review on pollutants removal by Sono-photo-fenton processes. *Journal of Environmental Chemical Engineering, 8*(5), 104330. https://doi.org/10.1016/j.jece.2020.104330
13. Deshpande, B. D., Agrawal, P. S., & Yenkie, M. K. N. (2019). AOP as a degradative tool for oxidation of 4-hydroxybenzoic acid. *AIP Conference Proceedings 2104, 020034*, 3–9.
14. Ameta, R., Chohadia, A. K., Jain, A., & Punjabi, P. B. (2018). Fenton and photo-fenton processes. https://doi.org/10.1016/B978-0-12-810499-6.00003-6
15. Talaeipour, M., Nouri, J., Hassani, A. H., & Mahvi, A. H. (2017). An investigation of desalination by nanofiltration, reverse osmosis and integrated (hybrid NF/RO) membranes employed in brackish water treatment. 1–9. https://doi.org/10.1186/s40201-017-0279-x
16. Wang, L., Boussetta, N., Lebovka, N., & Vorobiev, E. (2019). Ultrasonics—Sonochemistry ultrasound assisted purification of polyphenols of apple skins by adsorption/desorption procedure. *Ultrasonics Sonochemistry, 55*, 18–24. https://doi.org/10.1016/j.ultsonch.2019.03.002
17. Yin, H., Guo, H., Qiu, P., Yi, L., & Li, J. (2017). Case analysis on textile wastewater subjected to combined physicochemical—Biological treatment and ozonation. *66*, 140–148. https://doi.org/10.5004/dwt.2016.1619
18. Abegunde, S. M., Idowu, K. S., & Sulaimon, A. O. (2020). Plant-mediated iron nanoparticles and their applications as adsorbents for water treatment—A review. *Journal of Chemical Reviews, 2*(2), 103–113.
19. Rida, K., Bouraoui, S., & Hadnine, S. (2019). Adsorption of methylene blue from aqueous solution by kaolin and zeolite. *Applied Clay Science, 83–84*, 99–105. https://doi.org/10.1016/j.clay.2013.08.015
20. Lofrano, G., & Meriç, S. (2019). A review on occurrence, measurement, toxicity and tannin removal processes from wastewaters.*Environmental Engineering & Management Journal (EEMJ), 18*(1), 109–123.
21. Akbarzadeh, A., Samiei, M., & Davaran, S. (2012). Magnetic nanoparticles: Preparation, physical properties, and applications in biomedicine. *Nanoscale Research Letters Feb 21;7*(1):144. http://doi.org/10.1186/1556-276X-7-1441.

22. Leela, A., & Vivekanandan, M. (2008). Tapping the unexploited plant resources for the synthesis of silver nanoparticles. *African Journal of Biotechnology, 7*(17), 3162–3165. http://www.academicjournals.org/AJB.
23. Rai, M., Ingle, A. P., Birla, S., Yadav, A., Alves, C., & Santos, D. (2015). Strategic role of selected noble metal nanoparticles in medicine. *Critical Reviews in Microbiology*, 1–24. https://doi.org/10.3109/1040841X.2015.1018131
24. Ahamed, M., Alsalhi, M. S., & Siddiqui, M. K. J. (2010). Silver nanoparticle applications and human health. *Clinica Chimica Acta, 411*(23–24), 1841–1848. https://doi.org/10.1016/j.cca.2010.08.016
25. Momeni, S., & Nabipour, I. (2015). A simple green synthesis of palladium nanoparticles with *Sargassum alga* and their electrocatalytic activities towards hydrogen peroxide. https://doi.org/10.1007/s12010-015-1690-3
26. Nasrollahzadeh, M., & Sajadi, S. M. (2015). Green synthesis of copper nanoparticles using *Ginkgo biloba* L. leaf extract and their catalytic activity for the Huisgen [3 + 2] cycloaddition of azides. *Journal of Colloid and Interface Science*. https://doi.org/10.1016/j.jcis.2015.07.004
27. Koren, K., & Brodersen, K. E. (2015). Optical sensor nanoparticles in artificial sediments—A new tool to visualize O_2 dynamics around the rhizome and roots of seagrasses. https://doi.org/10.1021/es505734b
28. Mohanpuria, P., Rana, N. K., & Yadav, S. K. (2008). Biosynthesis of nanoparticles: Technological concepts and future applications. *Journal of Nanoparticle Research, 10*(3), 507–517. https://doi.org/10.1007/s11051-007-9275-x
29. Nel, A., Xia, T., Madler, L., & Li, N. (2006). Toxic potential of materials at the nanolevel. *Science, 311*(5761), 622–627 (Online). Available: https://www.jstor.org/stable/3843500
30. Zoroddu, M. A., Medici, S., Ledda, A., Nurchi, V. M., Lachowicz, J. I., & Peana, M. (2014). Toxicity of nanoparticles. *Current Medicinal Chemistry*, 3837–3853.
31. Behera, S. S., Jha, S., Arakha, M., & Panigrahi, T. K. (2013). Synthesis of silver nanoparticles from microbial source—A green synthesis approach, and evaluation of its antimicrobial activity against *Escherichia coli. International Journal of Engineering Research and Applications, 3*(2), 58–62.
32. Gu, H., Xu, K., & Xu, B. (2006). Biofunctional magnetic nanoparticles for protein separation and pathogen detection. 941–949. https://doi.org/10.1039/b514130c
33. Lin, P., et al. (2007). Surface modification of magnetic nanoparticle via Cu (I)-catalyzed alkyne-azide [2 + 3] cycloaddition. (I), 3508–3509.
34. Mahmoudi, M., Hofmann, H., Rothen-rutishauser, B., & Petri-fink, A. (2011). Assessing the in vitro and in vivo toxicity of superparamagnetic iron oxide nanoparticles. *Chemical Reviews, 112*(4), 2323–2338.
35. Gupta, A. K., & Curtis, A. S. G. (2004). Lactoferrin and ceruloplasmin derivatized superparamagnetic iron oxide nanoparticles for targeting cell surface receptors. *Biomaterials, 25*, 3029–3040. https://doi.org/10.1016/j.biomaterials.2003.09.095
36. Singh, P., Kim, Y. J., Zhang, D., & Yang, D. C. (2016). Biological synthesis of nanoparticles from plants and microorganisms. *Trends in Biotechnology, 34*(7), 588–599. https://doi.org/10.1016/j.tibtech.2016.02.006
37. Khan, I., Saeed, K., & Khan, I. (2017). Nanoparticles: Properties, applications and toxicities. *Arabian Journal of Chemistry*. https://doi.org/10.1016/j.arabjc.2017.05.011
38. Parveen, S., Misra, R., & Sahoo, S. K. (2017). Nanoparticles: A boon to drug delivery, therapeutics, diagnostics and imaging. *Nanomedicine in Cancer*, 47–98.
39. Goutam, S. P., Saxena, G., Roy, D., Yadav, A. K., & Bharagava, R. N. (2020). Green synthesis of nanoparticles and their applications in water and wastewater treatment. In *Bioremediation of industrial waste for environmental safety* (pp. 349–379). Springer Singapore. https://doi.org/10.1007/978-981-13-1891-7_16
40. Saif, S., Tahir, A., & Chen, Y. (2016). Green synthesis of iron nanoparticles and their environmental applications and implications. 1–26. https://doi.org/10.3390/nano6110209
41. Martínez-Cabanas, M., López-García, M., Barriada, J. L., Herrero, R., & Sastre de Vicente, M. E. (2016). Green synthesis of iron oxide nanoparticles. Development of magnetic hybrid

materials for efficient As(V) removal. *Chemical Engineering Journal, 301*, 83–91. https://doi.org/10.1016/j.cej.2016.04.149
42. Pereira, L., Mehboob, F., Stams, A. J. M., Mota, M. M., Rijnaarts, H. H. M., & Alves, M. M. (2013). Metallic nanoparticles: Microbial synthesis and unique properties for biotechnological applications, bioavailability and biotransformation. *8551*, 1–15. https://doi.org/10.3109/07388551.2013.819484
43. Kora, A.J., Beedu, S.R., & Jayaraman, A. (2012). Size-controlled green synthesis of silver nanoparticles mediated by gum ghatti (*Anogeissus latifolia*) and its biological activity. 1–10.
44. Nithya, K., Sathish, A., Senthil Kumar, P., & Ramachandran, T. (2018). Fast kinetics and high adsorption capacity of green extract capped superparamagnetic iron oxide nanoparticles for the adsorption of Ni(II) ions. *Journal of Industrial and Engineering Chemistry, 59*, 230–241. https://doi.org/10.1016/j.jiec.2017.10.028
45. Bankar, A., Joshi, B., Ravi, A., & Zinjarde, S. (2010). Banana peel extract mediated synthesis of gold nanoparticles. *Colloids and Surfaces B: Biointerfaces, 80*, 45–50. https://doi.org/10.1016/j.colsurfb.2010.05.029
46. Kumar, R., & Nayanmoni, D. (2011). Green synthesis of gold nanoparticles using *Nyctanthes arbortristis* flower extract. 615–619. https://doi.org/10.1007/s00449-010-0510-y
47. Madubuonu, N., et al. (2019). Biosynthesis of iron oxide nanoparticles via a composite of Psidium guavaja–Moringa oleifera and their antibacterial and photocatalytic study. *Journal of Photochemistry and Photobiology B: Biology, 199*. https://doi.org/10.1016/j.jphotobiol.2019.111601
48. Shittu, K. O., & Ihebunna, O. (2017). Purification of simulated waste water using green synthesized silver nanoparticles of *Piliostigma thonningii* aqueous leave extract. *Advances in Natural Sciences: Nanoscience and Nanotechnology, 8*(4), 045003.
49. Ehrampoush, M. H., Miria, M., Salmani, M. H., & Mahvi, A. H. (2015). Cadmium removal from aqueous solution by green synthesis iron oxide nanoparticles with tangerine peel extract. *Journal of Environmental Health Science and Engineering, 13*(1), 84. https://doi.org/10.1186/s40201-015-0237-4
50. Karthikeyeni, S., Sakthivel, V., & Ganesh, A. Biosynthesis of iron oxide nanoparticles and its haematological effects on fresh water fish *Oreochromis mossambicus*. [Online]. Available: https://www.researchgate.net/publication/258935456
51. Yang, Y., Ma, H., Zhuang, J., & Wang, X. (2011). Morphology-controlled synthesis of hematite nanocrystals and their facet effects on gas-sensing properties. *Inorganic Chemistry, 50*(20), 10143–10151.
52. Lin, Y., Zhou, S., Sheehan, W., & Wang, D. (2011). Nanonet-based hematite heteronanostructures for efficient solar water splitting. *Journal of the American Chemical Society, 133*(8), 2398–2401.
53. Deshpande, B. D., Agrawal, P. S., Yenkie, M. K. N., Dhoble, S. J., & Belekar, R. M. (2020). Nano-structures & nano-objects prospective of nanotechnology in degradation of waste water: A new challenges. *Nano-Structures & Nano-Objects, 22*, 100442. https://doi.org/10.1016/j.nanoso.2020.100442
54. Ganapathe, L. S., Mohamed, M. A., & Yunus, R. M. (2020). Magnetite (Fe_3O_4) nanoparticles in biomedical application: From synthesis to surface functionalisation. *Magnetochemistry, 6*(4), 68.
55. Zia, M., Phull, A. R., & Ali, J. S. (2016). Challenges of iron oxide nanoparticles. 49–67.
56. Noukelag, S. K., Arendse, C. J., & Maaza, M. (2020). Biosynthesis of hematite phase a-Fe_2O_3 nanoparticles using an aqueous extract of Rosmarinus officinalis leaves. *Materials Today: Proceedings*, 3679–3683. https://doi.org/10.1016/j.matpr.2020.10.977
57. Xu, P., et al. (2012). Use of iron oxide nanomaterials in wastewater treatment: A review. *Science of the Total Environment, 424*, 1–10. https://doi.org/10.1016/j.scitotenv.2012.02.023
58. Wheeler, D. A., Wang, G., Ling, Y., Li, Y., & Zhang, J. Z. (2012). Nanostructured hematite: Synthesis, characterization, charge carrier dynamics, and photoelectrochemical properties. *Energy & Environmental Science*, 6682–6702. https://doi.org/10.1039/c2ee00001f

59. Ekwumemgbo, P. A., Shallangwa, G. A., & Okon, I. E. (2023). Green synthesis and characterization of iron oxide nanoparticles using Prosopis Africana leaf extract. *Communication in Physical Sciences, 9*(2), 125–136.
60. Ahmmad, B., Leonard, K., Islam, S., & Kurawaki, J. (2013). Green synthesis of mesoporous hematite (α-Fe_2O_3) nanoparticles and their photocatalytic activity. *Advanced Powder Technology, 24*(1), 160–167. https://doi.org/10.1016/j.apt.2012.04.005
61. Hashimoto, K. (2000). Detection of active oxidative species in TiO_2 photocatalysis using the fluorescence technique. *2*, 207–210.
62. Encarnac, S., et al. (2016). Total phenolic content, antioxidant activity and pre-clinical safety evaluation of an *Anacardium occidentale* stem bark Portuguese hypoglycemic traditional herbal preparation. *82*, 171–178. https://doi.org/10.1016/j.indcrop.2015.11.001
63. Rufus, A., Sreeju, N., Vilas, V., & Philip, D. (2017). Biosynthesis of hematite (α-Fe_2O_3) nanostructures: Size effects on applications in thermal conductivity, catalysis, and antibacterial activity. *Journal of Molecular Liquids, 242*, 537–549. https://doi.org/10.1016/j.molliq.2017.07.057
64. Venkatraman, S. (2014). Has nanomedicine lived up to its promise? https://doi.org/10.1088/0957-4484/25/37/372501
65. Makarov, V. V., et al. (2014). Biosynthesis of stable iron oxide nanoparticles in aqueous extracts of *Hordeum vulgare* and *Rumex acetosa* plants. *Langmuir, 30*(20), 5982–5988. https://doi.org/10.1021/la5011924
66. Shahwan, T., et al. (2011). Green synthesis of iron nanoparticles and their application as a Fenton-like catalyst for the degradation of aqueous cationic and anionic dyes. *Chemical Engineering Journal, 172*(1), 258–266. https://doi.org/10.1016/j.cej.2011.05.103
67. Huang, L., Weng, X., Chen, Z., Megharaj, M., & Naidu, R. (2014). Green synthesis of iron nanoparticles by various tea extracts: Comparative study of the reactivity. *Spectrochimica Acta Part A: Molecular and Biomolecular Spectroscopy, 130*, 295–301. https://doi.org/10.1016/j.saa.2014.04.037.
68. Cao, D., Jin, X., Gan, L., Wang, T., & Chen, Z. (2016). Removal of phosphate using iron oxide nanoparticles synthesized by eucalyptus leaf extract in the presence of CTAB surfactant. *Chemosphere, 159*, 23–31. https://doi.org/10.1016/j.chemosphere.2016.05.080
69. Ebrahiminezhad, A., Zare-Hoseinabadi, A., Sarmah, A. K., Taghizadeh, S., Ghasemi, Y., & Berenjian, A. (2018). Plant-mediated synthesis and applications of iron nanoparticles. *Molecular Biotechnology, 60*(2), 154–168. https://doi.org/10.1007/s12033-017-0053-4.
70. Getahun, Y., Gardea-Torresdey, J., Manciu, F. S., Li, X., & El-Gendy, A. A. (2022). Green synthesized superparamagnetic iron oxide nanoparticles for water treatment with alternative recyclability. *Journal of Molecular Liquids, 356*, 118983.
71. Borowik, A., et al. (2019). The impact of surface functionalization on the biophysical properties of silver nanoparticles. *Nanomaterials, 9*(7), 973.
72. Hassan, D. F., & Mahmood, M. B. (2019). Biosynthesis of iron oxide nanoparticles using *Escherichia coli*. *Iraqi Journal of Science, 60*(3), 453–459. https://doi.org/10.24996/ijs.2019.60.3.5
73. Brenner, D. J., Staley, J. T., & Krieg, N. R. (1992). *Classification of procaryotic organisms and the concept of bacterial speciation* (pp. 27–32).
74. Wang, T., Lin, J., Chen, Z., Megharaj, M., & Naidu, R. (2014). Green synthesized iron nanoparticles by green tea and eucalyptus leaves extracts used for removal of nitrate in aqueous solution. *Journal of Cleaner Production, 83*, 413–419. https://doi.org/10.1016/j.jclepro.2014.07.006
75. Kuang, Y., Wang, Q., Chen, Z., Megharaj, M., & Naidu, R. (2013). Heterogeneous Fenton-like oxidation of monochlorobenzene using green synthesis of iron nanoparticles. *Journal of Colloid and Interface Science, 410*, 67–73.
76. Mystrioti, C., Xanthopoulou, T. D., Tsakiridis, P., Papassiopi, N., & Xenidis, A. (2016). Comparative evaluation of five plant extracts and juices for nanoiron synthesis and application for hexavalent chromium reduction. *Science of the Total Environment, 539*, 105–113. https://doi.org/10.1016/j.scitotenv.2015.08.091

77. Pandurangan, P. C. N. M. (2016). Green synthesis of iron oxide nanoparticles and their catalytic and in vitro anticancer activities. *Journal of Cluster Science*. https://doi.org/10.1007/s10876-016-1082-z
78. Naz, S., Islam, M., Tabassum, S., Fernandes, N. F., Carcache de Blanco, E. J., & Zia, M. (2019). Green synthesis of hematite (α-Fe_2O_3) nanoparticles using Rhus punjabensis extract and their biomedical prospect in pathogenic diseases and cancer. *Journal of Molecular Structure, 1185*, 1–7. https://doi.org/10.1016/j.molstruc.2019.02.088
79. Kaboli, H., Azizi, M., Reza, M., & Hossien, S. (2018). Biocatalysis and agricultural biotechnology green synthesis of iron nanoparticles by Rosemary extract and cytotoxicity effect evaluation on cancer cell lines. *Biocatalysis and Agricultural Biotechnology, 16*, 54–62. https://doi.org/10.1016/j.bcab.2018.07.017
80. Prasad, A. S. (2016). Iron oxide nanoparticles synthesized by controlled bio-precipitation using leaf extract of Garlic Vine (*Mansoa alliacea*). *Materials Science in Semiconductor Processing, 53*, 79–83. https://doi.org/10.1016/j.mssp.2016.06.009
81. Kumar, B., Smita, K., Cumbal, L., Debut, A., Galeas, S., & Guerrero, V. H. (2016). Phytosynthesis and photocatalytic activity of magnetite (Fe_3O_4) nanoparticles using the Andean blackberry leaf. *Materials Chemistry and Physics, 179*, 310–315. https://doi.org/10.1016/j.matchemphys.2016.05.045
82. Wang, Z. (2013). Iron complex nanoparticles synthesized by eucalyptus leaves. *ACS Sustainable Chemistry & Engineering, 1*(12), 1551–1554. https://doi.org/10.1021/sc400174a
83. Markova, Z., et al. (2014). Iron(II,III)–polyphenol complex nanoparticles derived from green tea with remarkable ecotoxicological impact. *ACS Sustainable Chemistry & Engineering, 2*(7), 1674–1680. https://doi.org/10.1021/sc5001435
84. Huang, L., Weng, X., Chen, Z., Megharaj, M., & Naidu, R. (2014). Synthesis of iron-based nanoparticles using oolong tea extract for the degradation of malachite green. *Spectrochimica Acta, Part A: Molecular and Biomolecular Spectroscopy, 117*, 801–804. https://doi.org/10.1016/j.saa.2013.09.054
85. Njagi, E. C., et al. (2011). Biosynthesis of iron and silver nanoparticles at room temperature using aqueous sorghum bran extracts. *Langmuir, 27*(1), 264–271. https://doi.org/10.1021/la103190n
86. Cai, Y., Shen, Y., Xie, A., Li, S., & Wang, X. (2010). Green synthesis of soya bean sprouts-mediated superparamagnetic Fe_3O_4 nanoparticles. *Journal of Magnetism and Magnetic Materials, 322*(19), 2938–2943. https://doi.org/10.1016/j.jmmm.2010.05.009
87. Prasad, C., Gangadhara, S., & Venkateswarlu, P. (2016). Bio-inspired green synthesis of Fe_3O_4 magnetic nanoparticles using watermelon rinds and their catalytic activity. *Applied Nanoscience, 6*(6), 797–802. https://doi.org/10.1007/s13204-015-0485-8
88. Niraimathee, V. A. (2016). Green synthesis of iron oxide nanoparticles from *Mimosa pudica* root extract. *International Journal of Environment and Sustainable Development, 15*(3).
89. Latha, M. G. N. (2014). Biosynthesis and characterisation of Fe_3O_4 nanoparticles using Caricaya papaya leaves extract. *International Journal of Science and Research, 3*(11),1551-1556.
90. Muthukumar, H., & Matheswaran, M. (2015). *Amaranthus spinosus* leaf extract mediated FeO nanoparticles: Physicochemical traits, photocatalytic and antioxidant activity. *ACS Sustainable Chemistry & Engineering, 3*(12), 3149–3156. https://doi.org/10.1021/acssuschemeng.5b00722
91. Machado, S., Grosso, J. P., Nouws, H. P. A., Albergaria, J. T., & Delerue-Matos, C. (2014). Utilization of food industry wastes for the production of zero-valent iron nanoparticles. *Science of the Total Environment, 496*, 233–240. https://doi.org/10.1016/j.scitotenv.2014.07.058
92. Poguberović, S. S., et al. (2016). Removal of As(III) and Cr(VI) from aqueous solutions using 'green' zero-valent iron nanoparticles produced by oak, mulberry and cherry leaf extracts. *Ecological Engineering, 90*, 42–49. https://doi.org/10.1016/j.ecoleng.2016.01.083
93. Buazar, F., Baghlani-Nejazd, M. H., Badri, M., Kashisaz, M., Khaledi-Nasab, A., & Kroushawi, F. (2016). Facile one-pot phytosynthesis of magnetic nanoparticles using potato

extract and their catalytic activity. *Starch/Staerke, 68*(7–8), 796–804. https://doi.org/10.1002/star.201500347
94. Ramesh, A. V., Rama Devi, D., Mohan Botsa, S., & Basavaiah, K. (2018). Facile green synthesis of Fe_3O_4 nanoparticles using aqueous leaf extract of *Zanthoxylum armatum* DC. for efficient adsorption of methylene blue. *Journal of Asian Ceramic Societies, 6*(2), 145–155. https://doi.org/10.1080/21870764.2018.1459335
95. Saim, A. K., Adu, P. C. O., Amankwah, R. K., Oppong, M. N., Darteh, F. K., & Mamudu, A. W. (2021). Review of catalytic activities of biosynthesized metallic nanoparticles in wastewater treatment. *Environmental Technology Reviews, 10*(1), 111–130. https://doi.org/10.1080/21622515.2021.1893831
96. Anchan, S., Pai, S., Sridevi, H., Varadavenkatesan, T., Vinayagam, R., & Selvaraj, R. (2019). Biogenic synthesis of ferric oxide nanoparticles using the leaf extract of *Peltophorum pterocarpum* and their catalytic dye degradation potential. *Biocatalysis and Agricultural Biotechnology, 20*, 101251. https://doi.org/10.1016/j.bcab.2019.101251
97. Lung, I., Stan, M., Opris, O., Soran, M. L., Senila, M., & Stefan, M. (2018). Removal of Lead(II), Cadmium(II), and Arsenic(III) from aqueous solution using magnetite nanoparticles prepared by green synthesis with Box–Behnken design. *Analytical Letters, 51*(16), 2517–2529. https://doi.org/10.1080/00032719.2018.1446974
98. Ahmed, A., et al. (2020). Efficient photocatalytic degradation of toxic Alizarin yellow R dye from industrial wastewater using biosynthesized Fe nanoparticle and study of factors affecting the degradation rate. *Journal of Photochemistry and Photobiology B: Biology, 202*, 111682. https://doi.org/10.1016/j.jphotobiol.2019.111682
99. Vasantharaj, S., Sathiyavimal, S., Senthilkumar, P., LewisOscar, F., & Pugazhendhi, A. (2019). Biosynthesis of iron oxide nanoparticles using leaf extract of *Ruellia tuberosa*: Antimicrobial properties and their applications in photocatalytic degradation. *Journal of Photochemistry and Photobiology B: Biology, 192*, 74–82. https://doi.org/10.1016/j.jphotobiol.2018.12.025.
100. Kumar, B., Smita, K., Galeas, S., Guerrero, V. H., Debut, A., & Cumbal, L. (2020). One-pot biosynthesis of maghemite (γ-Fe_2O_3) nanoparticles in aqueous extract of Ficus carica fruit and their application for antioxidant and 4-nitrophenol reduction. *Waste and Biomass Valorization, 0123456789*. https://doi.org/10.1007/s12649-020-01279-9
101. Shard, G., Hodoroaba, V., & Unger, W. E. S. (2020). *Introduction*. Elsevier Inc. https://doi.org/10.1016/B978-0-12-814182-3.00001-8
102. Vladár, A. E., & Hodoroaba, V. (2020). *Characterization of nanoparticles by scanning electron microscopy*. Elsevier Inc. https://doi.org/10.1016/B978-0-12-814182-3.00002-X
103. Mast, J., Verleysen, E., Hodoroaba, V., & Kaegi, R. (2020). *Characterization of nanomaterials by transmission electron microscopy: Measurement procedures*. Elsevier Inc. https://doi.org/10.1016/B978-0-12-814182-3.00004-3
104. Xiaoxing Ke, C. B., & Van Tendeloo, G. (2015). Possibilities and limitations of advanced transmission electron microscopy for carbon-based nanomaterials. *Beilstein Journal of Nanotechnology, 6*, 1541–1557. https://doi.org/10.3762/bjnano.6.158
105. Shard, A. G., Schofield, R. C., & Minelli, C. (2020). *Ultraviolet–Visible spectrophotometry* (Vol. 1). Elsevier Inc. https://doi.org/10.1016/B978-0-12-814182-3.00012-2
106. Shard, A. G. (2020). *X-ray photoelectron spectroscopy*. Elsevier Inc. https://doi.org/10.1016/B978-0-12-814182-3.00019-5
107. Hodoroaba, V. (2020). *Energy-dispersive X-ray spectroscopy (EDS)*. Elsevier Inc. https://doi.org/10.1016/B978-0-12-814182-3.00021-3
108. Gibson, N., Kuchenbecker, P., Rasmussen, K., & Hodoroaba, V. (2020). *Volume-specific surface area by gas adsorption analysis with the BET method*. Elsevier Inc. https://doi.org/10.1016/B978-0-12-814182-3.00017-1
109. Dukhin, A. S., & Xu, R. (2020). *Zeta-potential measurements*. Elsevier Inc. https://doi.org/10.1016/B978-0-12-814182-3.00014-6
110. Titus, D., Samuel, E. J. J., & Roopan, S. M. (2019). *Chapter 12—Nanoparticle characterization techniques*. Elsevier Inc. https://doi.org/10.1016/B978-0-08-102579-6.00012-5

111. Sathyanarayana, D. (2015). *Vibrational spectroscopy: Theory and applications*. New Age International.
112. Joshi, M., Bhattacharyya, A., & Ali, S. W. (2008). Characterization techniques for nanotechnology applications in textiles. *Indian Journal of Fibre & Textile Research (IJFTR), 33*, 304–317.
113. Edwards, D. T., Leblanc, M., & Perkins, T. T. (2021). Modulation of a protein-folding landscape revealed by AFM-based force spectroscopy notwithstanding instrumental limitations. https://doi.org/10.1073/pnas.2015728118
114. Rajan, R., Chandran, K., Harper, S. L., Yun, S. I., & Kalaichelvan, P. T. (2015). Plant extract synthesized silver nanoparticles: An ongoing source of novel biocompatible materials. *Industrial Crops and Products, 70*, 356–373. https://doi.org/10.1016/j.indcrop.2015.03.015.
115. Sathya, K., Saravanathamizhan, R., & Baskar, G. (2018). Green synthesised iron and iron-based nanoparticle in environmental and biomedical application—A review. *IET Nanobiotechnology, 13*(45), 2–15.
116. Arakha, M., Pal, S., Samantarrai, D., & Panigrahi, T. K. Antimicrobial activity of iron oxide nanoparticle upon modulation of nanoparticle-bacteria interface. *Scientific Reports*, 1–12. https://doi.org/10.1038/srep14813
117. Nanoparticles, M., et al. Secondary metabolites in the green synthesis of metallic nanoparticles (Figure 1), 1–25. https://doi.org/10.3390/ma11060940
118. Mohan, K., Kumar, B., Siva, K., & Sreedhara, P. (2013). Biobased green method to synthesise palladium and iron nanoparticles using *Terminalia chebula* aqueous extract. *Spectrochimica Acta Part A: Molecular and Biomolecular Spectroscopy, 102*, 128–133. https://doi.org/10.1016/j.saa.2012.10.015
119. Devatha, C. P., et al. (2017). Biosynthesis of iron oxide (Fe_2O_3) nanoparticles via aqueous extracts of Sageretia thea (Osbeck.) and their pharmacognostic properties. *Green Chemistry Letters and Reviews, 10*(4), 186–201. https://doi.org/10.33945/SAMI/JCR.2020.2.3
120. Samrot, A. V., Senthilkumar, P., Rashmitha, S., Veera, P., & Sahithya, C. S. (2018). Azadirachta indica influenced biosynthesis of super-paramagnetic iron-oxide nanoparticles and their applications in tannery water treatment and X-ray imaging. *Journal of Nanostructure in Chemistry, 8*(3). https://doi.org/10.1007/s40097-018-0279-0
121. Kostyukova, D., & Chung, Y. H. (2016). Synthesis of iron oxide nanoparticles using isobutanol. *Journal of Nanomaterials, 2016*.
122. Feldman, D. (2020). Poly (vinyl alcohol) recent contributions to engineering and medicine. 1–11.
123. Da'na, E., Taha, A., & Afkar, E. (2018). Green synthesis of iron nanoparticles by acacia nilotica pods extract and its catalytic, adsorption, and antibacterial activities. *Applied Sciences, 8*(10), 1922. https://doi.org/10.3390/app8101922
124. Raju, C. H., Bharadwaj, M. S., Prem, K., & Satyanandam, K. (2016). Green synthesis of iron nanoparticles using Albizia lebbeck leaves for synthetic dyes decolorization. *International Journal of Engineering Science Technology And Research (IJESTR), 5*(12).
125. Önal, E. S., Yatkin, T., Ergüt, M., & Özer, A. (2017). Green synthesis of iron nanoparticles by aqueous extract of Eriobotrya japonica leaves as a heterogeneous fenton-like catalyst: Degradation of Basic Red 46. *International Journal of Chemical Engineering and Applications, 8*(5), 327–333. https://doi.org/10.18178/ijcea.2017.8.5.678
126. Devatha, C. P., Thalla, A. K., & Katte, S. Y. (2016). Green synthesis of iron nanoparticles using different leaf extracts for treatment of domestic waste water. *Journal of Cleaner Production, 139*, 1425–1435. https://doi.org/10.1016/j.jclepro.2016.09.019
127. Hoag, G. E., Collins, J. B., Holcomb, J. L., Hoag, J. R., Nadagouda, M. N., & Varma, R. S. (2009). Degradation of bromothymol blue by 'greener' nano-scale zero-valent iron synthesized using tea polyphenols. *Journal of Materials Chemistry, 19*(45), 8671. https://doi.org/10.1039/b909148c
128. Paswan, S. K., Kumar, P., Singh, R. K., Shukla, S. K., & Kumar, L. (2021). Spinel ferrite magnetic nanoparticles: An alternative for wastewater treatment. In *Pollutants and water management: Resources, strategies and scarcity* (pp. 273–305).

129. Lee, S., & Park, S. (2013). TiO_2 photocatalyst for water treatment applications. *Journal of Industrial and Engineering Chemistry, 19*, 1761–1769. https://doi.org/10.1016/j.jiec.2013.07.012
130. Mun, K., Wei, C., Sing, K., & Ching, J. (2016). Recent developments of zinc oxide based photocatalyst in water treatment technology: A review. *Water Research, 88*, 428–448. https://doi.org/10.1016/j.watres.2015.09.045
131. Paper, C., Munasir, M., Surabaya, U. N., & Hidayat, N. (2020). Amorphous-SiO_2 nanoparticles for water treatment materials. https://doi.org/10.1063/5.0015673
132. Al-Saad, K. A., Amr, M. A., Hadi, D. T., Arar, R. S., Al-Sulaiti, M. M., Abdulmalik, T. A., Alsahamary, N. M., & Kwak, J. C. (2012). Iron oxide nanoparticles: Applicability for heavy metal removal from contaminated water. *Arab Journal of Nuclear Sciences and Applications, 45*(2), 335–346.
133. Deshpande, B. D. Water pollution sources and purification: Challenges and scope. *Bentham Books*. https://doi.org/10.2174/97898150506841220101
134. Soliemanzadeh, A., Fekri, M., Bakhtiary, S., & Mehrizi, M. H. (2016). Biosynthesis of iron nanoparticles and their application in removing phosphorus from aqueous solutions. *Chemistry and Ecology, 32*(3), 286–300. https://doi.org/10.1080/02757540.2016.1139091
135. Pushkar, B., & Sevak, P. (2019). Adsorption study of green synthesized Fe-oxide nanoparticle for DDT removal. *International Journal of Pharmaceutical Sciences Review and Research, 55*(2), 84–90.
136. Al-Ruqeishi, M. S., Mohiuddin, T., & Al-Saadi, L. K. (2019). Green synthesis of iron oxide nanorods from deciduous Omani mango tree leaves for heavy oil viscosity treatment. *Arabian Journal of Chemistry, 12*(8), 4084–4090. https://doi.org/10.1016/j.arabjc.2016.04.003
137. Jassal, V., Shanker, U., & Gahlot, S. (2016). Green synthesis of some iron oxide nanoparticles and their interaction with 2-amino, 3-amino and 4-aminopyridines. *Materials Today: Proceedings, 3*(6), 1874–1882. https://doi.org/10.1016/j.matpr.2016.04.087
138. Ozkan, Z. Y., Cakirgoz, M., Kaymak, E. S., & Erdim, E. (2018). Rapid decolorization of textile wastewater by green synthesized iron nanoparticles. *Water Science and Technology, 77*(2), 511–517. https://doi.org/10.2166/wst.2017.559
139. Bishnoi, S., Kumar, A., & Selvaraj, R. (2018). Facile synthesis of magnetic iron oxide nanoparticles using inedible Cynometra ramiflora fruit extract waste and their photocatalytic degradation of methylene blue dye. *Materials Research Bulletin, 97*, 121–127. https://doi.org/10.1016/j.materresbull.2017.08.040
140. Xiong, R., Wang, Y., Zhang, X., & Lu, C. (2014). Facile synthesis of magnetic nanocomposites of cellulose@ultrasmall iron oxide nanoparticles for water treatment. *RSC Advances, 4*(43), 22632–22641. https://doi.org/10.1039/C4RA01397B
141. Othman, N. H., Alias, N. H., Shahruddin, M. Z., Abu Bakar, N. F., Nik Him, N. R., & Lau, W. J. (2018). Adsorption kinetics of methylene blue dyes onto magnetic graphene oxide. *Journal of Environmental Chemical Engineering, 6*(2), 2803–2811. https://doi.org/10.1016/j.jece.2018.04.024
142. Shin, S. R. (2016). Graphene-based materials for tissue engineering. *Advanced Drug Delivery Reviews, 105*(Part B), 255–274.
143. Arumugam, V., et al. (2021). Ionic liquid covered iron-oxide magnetic nanoparticles decorated zeolite nanocomposite for excellent catalytic reduction and degradation of environmental toxic organic pollutants and dyes. *Journal of Molecular Liquids, 342*, 117492. https://doi.org/10.1016/j.molliq.2021.117492
144. Jeon, K. W. B. (2020). Catalytic reduction of ortho- and meta-nitroaniline by nickel oxide nanoparticles. *Elastomers and Composites, 55*(3), 191–198.
145. Farooqi, Z. H., Naseem, K., Begum, R., & Ijaz, A. (2015). Catalytic reduction of 2-nitroaniline in aqueous medium using silver nanoparticles functionalized polymer microgels. *Journal of Inorganic and Organometallic Polymers and Materials, 25*(6), 1554–1568. https://doi.org/10.1007/s10904-015-0275-5
146. Li, Y., et al. (2001). Effect of long-term exposure to fluoride in drinking water on risks of bone fractures. *Journal of Bone and Mineral Research, 16*(5), 932–939.

147. Aswathi, V. P., Meera, S., Maria, C. G. A., & Nidhin, M. (2023). Green synthesis of nanoparticles from biodegradable waste extracts and their applications: A critical review. *Nanotechnology for Environmental Engineering, 8*(2), 377–397. https://doi.org/10.1007/s41204-022-00276-8.
148. Das, R. K., et al. (2017). Biological synthesis of metallic nanoparticles: Plants, animals and microbial aspects. *Nanotechnology for Environmental Engineering, 2*(1). https://doi.org/10.1007/s41204-017-0029-4

Metabolic Pathways of Microbial Detoxification of Toxic Dyes

Jitendra Kumar Sharma, Mahendra Kumar Gupta, Mir Sajad Rabani, Anjali Pathak, Shivani Tripathi, Charu Gupta, and Meenakshi Shrivastav

1 Introduction

The textile dyeing industry plays a significant role in the world economy by providing rich bright colors that enhance the aesthetic value of fabrics. However, this sector is also one of the largest polluters in the environment, especially through the release of untreated effluents from dyeing processes. Studies show that 20% of industrial water pollution occurs because of textile dyeing, making it the second-largest source of water pollution after agriculture [4]. These harmful effluents contain substances that are related to the products of dyeing processes such as toxic metals and non-biodegradable dyes. Such pollutants threaten not only aquatic ecosystems but also human health. These may cause significant environmental impacts such as reduced biodiversity, soil degradation, and interference with vital ecological processes. The frequent exposure of textile dye effluents poses a risk to human health and has been linked directly or indirectly to several health issues, including immune system suppression, autoimmune diseases, blood circulation issues, central nervous system damage, neurobehavioral disorders, allergic reactions, various cancerous diseases, vomiting, excessive diarrhea, tissue damage, eye skin disorders, eye infections, and lung-related issues are names to few [89]. Given the currently acute environmental challenges, there is an apparent need for innovative and non-pollutant solutions to mitigate the problems of dye pollution [57, 104]. The current method of microbial degradation has emerged as a possible solution to this problem. This bioremediation approach uses microorganisms to break down complex dye chemicals into less harmful products. Thereby this approach has provide an environmentally

J. K. Sharma · M. K. Gupta (✉) · M. S. Rabani · A. Pathak · S. Tripathi
Microbiology Research Lab, School of Studies in Botany, Jiwaji University, Gwalior, Madhya Pradesh, India
e-mail: mkgsac@yahoo.com

C. Gupta · M. Shrivastav
School of Studies in Microbiology, Jiwaji University, Gwalior, Madhya Pradesh, India

R. A. Bhat et al. (eds.), *Biocentric Approaches in Textile Waste Management*, Textile Science and Clothing Technology, https://doi.org/10.1007/978-3-032-04974-2_18

friendly alternative to the established physical and chemical remediation technologies [108]. Microbial processes are significant in the detoxification of harmful dyes while providing an economically viable approach to the treatment of wastewater in the textile industry through these processes [51]. Many bacterial and fungal speciesmay use dyes as carbon sources, which highlights the potential of using microbial metabolism to remove dyes effectively. This chapter will aim to discuss in detail the underlying mechanisms of microbial degradation of textile dyes and their practical applications for wastewater treatment. Hence, we will observe the types of microorganisms involved, their enzymatic pathways for dye degradation and various factors influencing this process. Additionally, this study discusses the current advancements in biotechnological methods to enhance microbial degradation processes and their implications for sustainable practices in the textile industry. This work is so designed to illustrate the relevance of integrating the use of microbiological approaches into the context of environmental management strategies developed to control the toxic ecological effects of textile dyeing.

2 Types of Textile Dyes

The composition of a dye is generally divided into the chromophore part and the auxochrome part. Chromophores can be known as the dye molecule's light absorption part, and auxochromes can be known as the color intensifier part of the dye molecules. The auxochrome improves the ability of dye molecules to dissolve in water and/or to bond with the fiber more tightly, through chemical reduction to the chromophore [5, 42]. Dyes are classified into triphenylmethane, azo, nitro, anthraquinone, indigoid, nitroso and phthalein dyes based on the chemical constitution of their molecules. Further classification of the dyes can be made based on the chromophore as pointed out in Table 1. Depending on the charge of the particle they are further classified into nonionic (dispersed dyes), anionic (direct, acidic, and reactive), and cationic (basic) dyes [49, 102, 103, 107].

The primary classifications include Azo dyes, anthraquinone dyes and triphenylmethane dyes. These are some of the most used synthetic dyes in the textile industry. These exhibit variations in characteristics, applications and possible impacts on the environment.

Azo Dyes

Azo dyes are a class of synthetic colorants where one or more azo functional groups (–N=N–) are present in the molecular structure [93]. The annual global production of azo dyes is estimated to be around 1 million tons. These compounds are significant because they find extensive use, constituting approximately 60–70% of the synthetic dyes employed in the textile industry, as well for their bright colors and excellent

Table 1 Classification and examples of dyes based on chromophore present in their structure

S.No.	Class	Chromophore	Example
1	Azo dyes	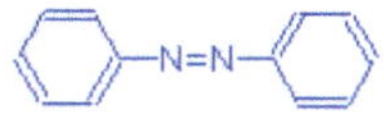	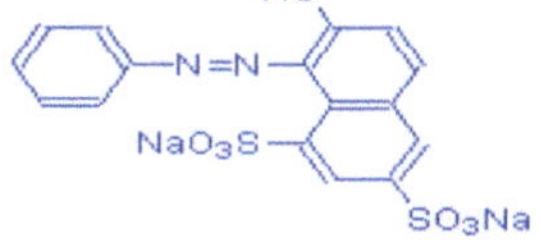 Orange G
2	Acridine	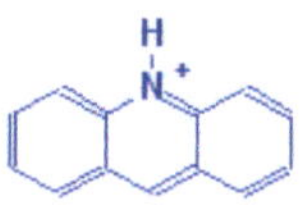	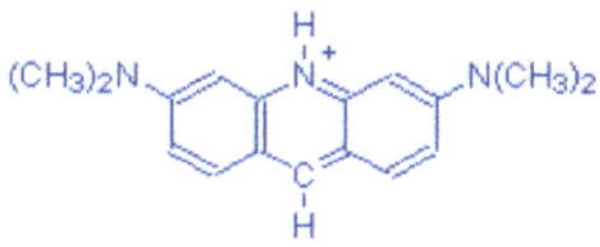 Acridine Orange
3	Anthraquinone	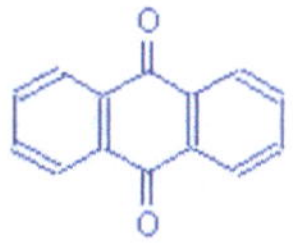	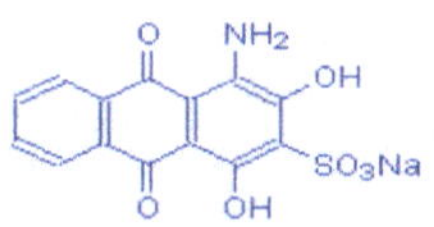 Nuclear Fast Red
4	Nitro dyes	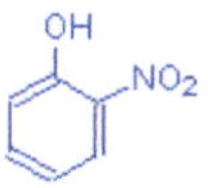	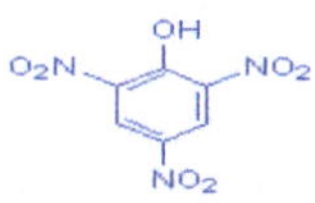 Picric Acid
5	Aryl methane dyes	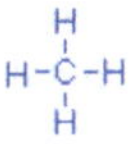	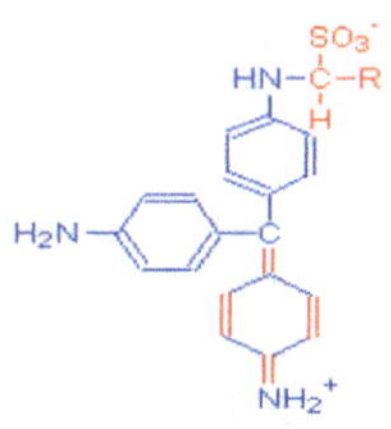 Schiff's Reagent

(continued)

Table 1 (continued)

S.No.	Class	Chromophore	Example
6	Thiazole dye	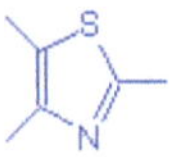	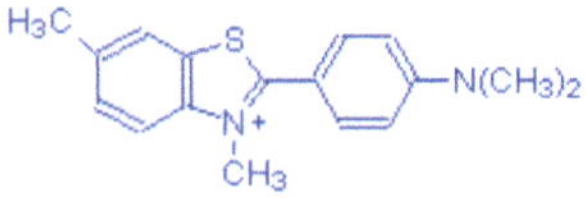 Thioflavine T
7	Triphenylmethane dyes	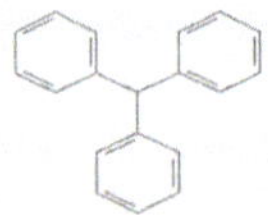	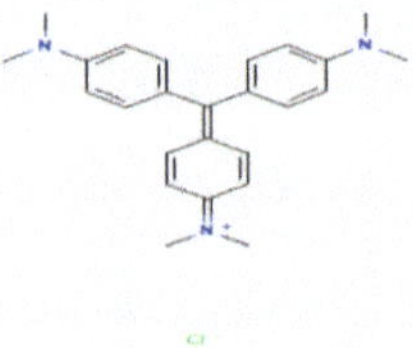 C.I. Basic Violet 3

dyeing properties. Examples include Reactive Red, Methyl Orange, and Allura Red, all of which are widely used in various industrial applications for textiles, food coloring and cosmetics [48].

Azo dyes are widely used due to their chemical stability, and their resistance to a variety of dyeing techniques makes them an indispensable part of modern textile production [22]. These dyes vary and can also be applied to use with other types of fabrics such as cotton, wool, and artificial fibers [40]. Their high coloring strength and the compatibility of dyeing techniques with a wide range of methods such as exhaust dyeing that allows the coloring agent to penetrate the fabric gradually, and continuous dyeing techniques that ensure uniform color distribution enhances their functional use [77]. In addition, relatively easy synthesis of azo dyes renders them mostly accessible and economically sound in industrial applications [85, 93]. Despite these benefits, the environmental impacts of azo dyes, especially their carcinogenic intermediates as a result of biotransformation, have been largely a health risk [22]. Many azo dyes form potentially harmful aromatic amines during reduction, which pose potential health and environmental risks and therefore increase interest in their biodegradability and removal from wastewater [10].

Anthraquinone Dyes

Anthraquinone dyes are an important class of synthetic dyes, that have gained attention for their high durability as well as excellent colorfastness property. Synthesized from the anthraquinone skeleton, such dyes have two carbonyl groups on either side of a benzene ring [25]. This arrangement not only gives a variety of intense colors, from yellow to violet but also gives them enhanced resistance towards environmental degradation effects, such as light and wash. These dyes are stable, with notable stability in their structure that is associated with their complex configuration of molecules. Stability is necessary for these dyes in textiles, especially where the materials are used more frequently outdoors and washed even more frequently [25]. The longevity of color intensity suggests that the dyes will be useful in textile uses that require considerable resistance both to fading and wash-out [64, 88]. Dyeing normally involves the application of these dyes on cellulose-based fibers such as cotton with methods to increase their affinity for the fabric. Anthraquinone skeleton enables numerous possible modifications via the introduction of substituents such as amino, hydroxyl, halogen, and sulfonic acid groups. These changes could significantly affect the color characteristics and solubility of the dye, thereby expanding its potential uses [95]. Anthraquinone dyes are commercially used in conjunction with traditional printing techniques for printing purposes. These dyes are versatile beyond cotton. They can be printed on other fibers like wool and silk if formulated appropriately [95]. The use of anthraquinone dyes is not limited to the textile industry, some other industries that apply these dyes include cosmetics and foodstuffs, mainly because the bright colors will enhance cosmetic attractiveness and always meet the safety requirements in place [25].

Triphenylmethane Dyes

Triphenylmethane dyes are one of the important classes of synthetic organic compounds, in which triphenylmethane forms the core structure. These two most representative dyes of this class, "Malachite Green" and "Crystal Violet" have been widely used in a variety of applications because of their very deep coloring and good aqueous solubility. In triphenylmethane structure, a central carbon atom is covalently bonded to three phenyl groups. The structure of the molecule leads to the very intense coloration of the dyes. Both Basic Green 4 (Malachite Green) and Basic Violet 3 (Crystal Violet) are widely used in wool, silk, and some synthetic fiber dyeing due to their ability to yield deep, saturated colors, hence extremely popular in the textile industry [24, 75].

It shows bright colors because of excellent solubility; however, it exhibits poor lightfastness in comparison to other classes of dyes. In that respect, the use of these dyes is susceptible to fading after passing long periods of exposure to light and its radiation, limiting their utility for items wherein retention of color becomes crucial, such as outdoor fabrics [108]. Research has shown that the lightfastness of these dyes is far inferior to other classes of dyes and thus less suitable for such applications that need lasting color [53]. The use of Malachite Green and Crystal Violet is still surrounded by controversy. These dyes were found to have antimicrobial properties, however, at the same time, these dyes are also toxic and mutagenic and hence create concerns regarding their use in various applications. As an example, it has been found that these dyes might be toxic to mammalian cells and thus pose ecological risks once released into water bodies [24].

Other Dyes

Besides azo, anthraquinone, and triphenylmethane dyes, there are other classes of dyes in use in the textile industry to produce different colors and properties in fabrics. These include acid dyes, basic dyes, direct dyes, dispersed dyes, sulfur dyes, and natural dyes.

Acid dyes are water-soluble synthetic dyes used principally for dyeing protein fibers like wool and silk and synthetic fibers such as nylon. They only give the best results when the conditions applied are acidic. They work through the interaction between acidic groups and amino groups of protein fibers. Acid dyes are bright and possess great color fastness properties. As a result of this property, they have considerable industrial utility. The pre-treatment with mordants may enhance fixation in fabric before the application; they also can be directly used under certain conditions [67].

Basic dyes also called cationic dyes, are primarily used to dye acrylic fibers but can be applied to other substrates like cotton and silk with the aid of mordants. These dyes are water-soluble and display brilliant colors but generally, they are not very lightfast or washfast. Basic dyes are often used in after-treatment processes to add color depth to fabrics previously treated with acid dyes [67, 101].

Direct dyes are another class of dyes in wide use in textile dyeing, mainly for cellulose fibers like cotton and viscose. They can attach directly to the fiber without the use of a mordant; however, they generally show lower colorfastness than other classes of dyes. Direct dyes are easy to apply and can produce bright colors but may fade quickly when exposed to washing or light [37, 67].

Disperse dyes are the largest class of dyes used mainly for dyeing synthetic fibers like polyester and acetate. They are water-insoluble and, therefore, must be reduced to a fine paste prior to application. In the process of dyeing, disperse dyes disperse in the matrix of the fiber, allowing bright hues even on such hydrophobic substrates. They generally exhibit good light fastness but often require specific condition, high temperatures to be applied effectively [67, 101, 106].

Sulfur dyesare used mainly on cotton fabrics because they have a very good wash fastness and resistance to fading. Sulfur dyes are insoluble but become soluble by chemical treatment with caustic soda or sodium sulfide during dyeing. Fabrics are oxidized after dyeing to develop the desired shades. Sulfur dyes are especially useful for producing darker shades [20, 67, 101].

Lastly, there has been a revival in natural dyes because consumers' demand for sustainable and eco-friendly products has grown. Derived from plant sources, roots, leaves, flowers or insects like cochineal, natural dyes give a special aesthetic with earthy tones that feel connected to traditional craftsmanship. However, they are mostly plagued by issues of color consistency and fastness when compared to synthetic counterparts. Despite these limitations, natural dyes maintain biodiversity and support local economies through traditional dyeing practices [37, 109].

3 Effects of the Dyes on the Environment and Human Health

The long-term effects of synthetic textile dyes on the environment, particularly their toxicity and the treatment of wastewater containing these compounds, require immediate global attention. There is an enormous dependence of the textile industry on synthetic dyes that may pose some negative implications for both ecosystems and human health.

Azo dyes and their degradation products (aromatic amines) are well known carcinogenic, mutagenic and teratogenic to humans and ecosystem [6, 7, 41]. Azo dyes can cause genetic changes and skin inflammation, hypersensitivity, as well as contact dermatitis and potential carcinogenic effects [41, 46]. Azo dyes contribute environmentally by increasing bio and chemical oxygen demand in water bodies bringing about alterations in aquatic ecosystems and flora and fauna by reducing photosynthesis and the growth of plants [7, 41, 46]. Some of these dyes are highly resistant to aerobic treatment which permits their passage through most wastewater plants with little change [22, 63].

Anthraquinone dyes, because of their complex and stable structures, these are generally more toxic than azo dyes and have raised serious environmental and health concerns in the textile industry [34]. This could retain in water bodies and decrease light penetration, inhibiting photosynthesis and adversely impacting aquatic ecosystems [59]. Thus, the ecotoxicity of anthraquinone dyes is variable: whereas some natural derivatives such as dermorubin show low toxicity, others like dermocybin may be significantly toxic to aquatic organisms such as *Daphnia similis* and *Danio rerio* embryos [36]. A promising means of reducing the ecological impact of these dyes has been the use of advanced oxidation processes, of sodium percarbonate systems activated by cobalt oxide, that are capable of effectively degrading Reactive Blue 19 [34]. Nevertheless, products from the photodegradation of these dyes can sometimes be more toxic than the parent compounds, as has occurred with anthracene, a related polycyclic aromatic hydrocarbon whose photodegradation products exacerbate dermo toxicity and induce inflammatory responses in human skin cells [18]. Furthermore, plant derived hydroxyanthraquinone dyes have been reported to be carcinogenic, whereas some of them have potential anti-cancer properties [19]. Thus, according to the above, these dyes, therefore, present a dual nature that require much needed comprehensive ecotoxicological studies that can help to better understand the relationship of synthetic and natural anthraquinone dye with the health implication and developing of safer alternative use in industry [36].

Industrial effluent with triphenylmethane dyes that pollutes water sources and harms aquatic life [55]. Through light absorption, the dyes affect the rate of photosynthesis of aquatic plants and affect the entire metabolism of aquatic organisms; the dyes also accumulate in the tissues of water creatures, and then humans get them through fish and other products [55]. Additionally, triphenylmethane dyes can chelate metal ions, increasing metal toxicity in aquatic organisms, and are known for their genotoxic and carcinogenic properties [74]. Ingestion of these dyes either through direct consumption or by eating fish that has been contaminated can cause problems such as dermatitis, allergies and probably even cancer [63, 74].

4 Advanced Treatment Strategies and Need for Sustainable Approaches

Treating textile dye effluent is one of the most serious environmental challenges faced due to hazardous effects of synthetic dyes on the ecosystems and human health. However, conventional physical and chemical techniques such as coagulation and advanced oxidation processes have been used, but are often plagued with high operating costs and problems with sludge disposal [105]. Recent progress is directed towards the sustainable, integrated physical/chemical/biological interactions involving greater efficiency and minimal environmental degradation. Microbial enzyme degradation and fungal decolorization are among the promising techniques of bioremediation in the decomposition of bioorganic constituents from otherwise non-degradable content present in wastewater. However, photocatalysis with catalysts like SnS (Tin Sulfide) have shown to have spectacular rate of dye degradation under natural or incident light making it an energy saving alternative to conventional processing. Another area with noteworthy interest is creation of nano adsorbent and composite adsorbent with exceptional characteristics and efficiency in selective removal of dyes from wastewater. The most effective approach of dye removal currently by these adsorbents in combination with enzyme based biological techniques. In order to address the multi-dimensional aspects of dye degradation, an integrated approach involving these advanced techniques can be used and would extend the path towards the environmental sustainability as well as clean the water resources. Such novel techniques are of extreme importance in developing cost effective and scalable solutions in reducing adverse impacts of textile dye effluent on the environment.

5 Microbial Groups Involved in Dye Degradation

In bioremediation processes, microorganisms such as bacteria, fungi, and algae degrade complex dyes into non-toxic compounds. These three groups of microorganisms have their distinctive metabolism and enzymatic capacity to efficiently break down various types of dyes.

Bacteria

Bacteria play a vital role in the degradation of toxic textile dyes, which have become a serious environmental problem due to their resistant nature. Textile industries generate dye-containing wastewater in large amounts, and such contaminants remain far from being satisfactorily removed by conventional treatment methodologies [30]. Bacterial degradation may thus be considered an efficient bioremediation approach, using the metabolic functions of bacteria naturally for breaking complex dye structures into much less harmful molecules.

The main mechanism, by which bacteria contribute to dye's destruction, involves the synthesis of particular enzymes that facilitate the destruction of dye molecules. Key enzymes in this process involve azoreductase, laccase, peroxidase, and phenol oxidase. All these enzymes are capable of cleaving azo bonds that are part of many synthetic dyes into intermediate metabolites subject to further degradation by bacterial metabolic pathways. For example, it was determined that bacterial strains such as *Enterobacter* sp. and *Bacillus subtilis* have high decolorizing activities of certain dyes, such as Crystal Violet and Malachite Green, with removal efficiencies of up to 100% under optimal conditions of pH 6.5 and temperature around 35 °C [86]. Specific bacterial isolates, such as *Serratia marcescens* and strain JW-22, have shown significant dye degradation capabilities, achieving up to 95% decolorization efficiency within hours [14, 82].

Some of the abiotic environmental factors influencing the effectiveness of bacterial degradation include pH, temperature, and dye concentration. For instance, the optimal pH and temperature for effective degradation of dye by *Enterobacter* sp. CV–S1 were 6.5–7.0 and 35 °C, respectively, at which the enzymes were optimally active [86]. The bacterial consortium found in recent studies could decolorize Acid Orange dye at pH 7.5 and 32 °C, which thus demonstrates that the most favorable conditions for dye degradation vary among bacterial types [12]. Apart from enzymatic action, bacteria are also able to use dyes as carbon sources for their growth, adding to their degradational efficiency. Such a dual role not only facilitates the breakdown of toxic compounds but also fosters the proliferation of bacteria in contaminated environments. For example, some isolates obtained from textile effluents have been reported to grow on dyes as their sole source of carbon, showing their potential in bioremediation applications [30]. It has also been observed that mixed bacterial consortia generally act better than their single counterparts. The interaction among bacterial species may lead to enhanced metabolic activities or result in an expanded scope regarding the types of dyes amenable to treatment. This has successfully been applied in several studies where a high rate of decolorization and mineralization of textile dyes was obtained with consortia compared to single-strain cultures [12]. A successful dye degradation is demonstrated by the reduction in chemical oxygen demand (COD) and biological oxygen demand (BOD) following bacterial treatment. As an example, an Acid Orange dye 3 degrading bacterial consortium was demonstrated to reduce BOD by 96% and COD by 79% [12]. Such a reduction not only represents effective dye removal but also represents improvements to the quality of the water that is better for discharge into natural water bodies. As shown in (Table 2) several bacterial strains have been reported for their ability to biodegrade harmful textile dyes.

Table 2 List of bacterial strains that effectively degrade various textile dyes

Bacterial strain	Dye degraded	Degradation efficiency	Key enzymes/ substrates	References
Bacillus albus DD1	RB5 (Azo dye)	98%	Not specific	[78]
Nivibacillus thermophiles SG-1	Azo dye (Orange I)	100%	Gene encoding riboflavin biosynthesis protein	[97]
Geobacillus stearothermophilus ATCC 10149	Remazol Brilliant Blue R	90%	Extracellular laccase	[39]
Bacillus licheniformis LS04	Reactive black 5	More than 80%	Laccase	[65]
Halopigeraswanensis	Malachite green, methyl orange	Up to 93%	Lignin peroxidase (LiP), manganese peroxidase (MnP), and laccase	[21]
Bacillus subtilis	Mixed azo dyes	87.35%	Intracellular azoreductase	[60]
Pseudomonas mendocina	Mixture of reactive red (RR), reactive brown (RB), and reactive black (RBL)	58.40%	Laccase	[94]
Halomonas sp.	Acid brilliant scarlet GR	–	Azoreductase	[100]
Pseudomonas sp. SUK1	Red BLI	99.28	–	[54]
Pseudomonas putida	Crystal violet	50	–	[91]

Fungi

The diversified capacities of fungi to biodegrade a wide range of synthetic textile dyes can be recognized mainly due to their highly sophisticated mechanisms of enzyme production [81]. Notably, the production of extracellular enzymes plays an important role in the degradation of the complex and structurally elaborate compounds within the dyes [81]. Certain dominant fungal species have been identified as having the capacity to break down complex dyes. A filamentous fungus *Aspergillus niger* possesses an extremely high range of decomposing activity, such as several dyes like Basic Blue 9 and Acid Blue 29. The laccases and peroxidases produced are used for the aid of oxidation-based dye breakage process [12] [71], *Trichoderma reesei* has been studied for their ligninolytic enzymes which break down effectively lignin and allied compounds, therefore playing a role in decolorization processes of dyeing [12]. White rot fungus (*Phanerochaetechrysosporium)* has been considered effective in the degradation of complex organic pollutants, including synthetic dyes through lignin peroxidase activity [71]. Certain single-celled fungi (Yeasts) have been also recognized within the biotechnological realm regarding their role, wherein they are part of biodegradation processes; these have their specific metabolic pathways and diversity that tend to favor organic and some inorganic wastes made of various pollutants and different by-products within industries about cleanup efforts towards environmental sustainability [81]. *Candida tropicalis* is also effective in decolorizing azo dyes. Studies have been carried out showing that the azo bonds are reduced to less harmful compounds through azoreductase activity [51] and *Saccharomyces cerevisiae* degrades many reactive dyes by producing multiple enzymes including laccase and azoreductase. The growth of this yeast on molasses containing reactive dyes showed bioaccumulation properties [51, 69]. *Debaryomyces polymorphous* species of yeast have been reported to decolorize a range of azo dyes successfully, thus helping in bioremediation of textile wastewater activity [51].

Microalgae

Microalgae have emerged as a crucial component in the degradation of textile dyes, thus offering a very environmentally friendly, feasible route for the increasing problem of dye-laden wastewater coming out of the textile industry and this phenomenon is also called phycoremediation. Higher demand for bright colors has resulted in the excessive use of synthetic dyes, many of which are toxic and non-biodegradable with conventional treatment methods. Generally, microalgae remove dye pollutants through two major mechanisms: biosorption and biodegradation demonstrated in Fig. 1. Biosorption is defined to be the passive sequestration of dye molecules on the cell wall of microalgae and could lead to a positive contribution in reducing the concentration of the dye in wastewater. It is affected by operational parameters like pH, temperature, and dye concentration. For example, one of the most used microalgae is *Chlorella vulgaris*, which, under optimum conditions, can remove azo dyes at a rate as high as 91.7%. The high surface area-to-volume ratio of microalgae increases their efficiency towards pollutant adsorption, thus making them one of the most viable agents in wastewater treatment [47]. On the other hand, biodegradation is the enzymatic cleavage of a complex dye molecule into simpler, less toxic molecules. Specific enzymes, like azoreductase, can be produced by microalgae that cleave azo bonds in synthetic dyes. This enzymatic activity converts a complex dye structure to simple, non-toxic compounds that could be fully mineralized into carbon dioxide and water. Some studies have demonstrated that the active metabolism of such dyes by certain microalgal species improves the general quality of water [51, 70]. Among the well-known species of microalgae, *Scenedesmus* sp. has been recognized due to its high resistance and efficiency in degrading a wide variety of dyes [70], while in the case of *Codium decorticatum*, the degradation rate for Crystal Violet and Congo Red was as high as 96.9% [70]. Similarly, *Chlamydomonas variabilis* excellent performance was shown by this species as a better Methylene Blue degrader with peaks of efficiency as high as 98% [70]. Nevertheless, the application of microalgae as a tool for the degradation of textile dyes has advantages other than the removal of pollutants, which include inexpensive treatment of wastewater. The growth of microalgae using wastewater as a nutrient feedstock greatly lowers the cost of wastewater treatment while yielding valuable biomass usable to produce various industrial products or biofuels. This idea is known as the idea of the circular economy as we are trying to produce less waste and recycle more of the resources available. Several fungal and algal strains have been reported for their ability to degrade various textile dyes. (Table 3).

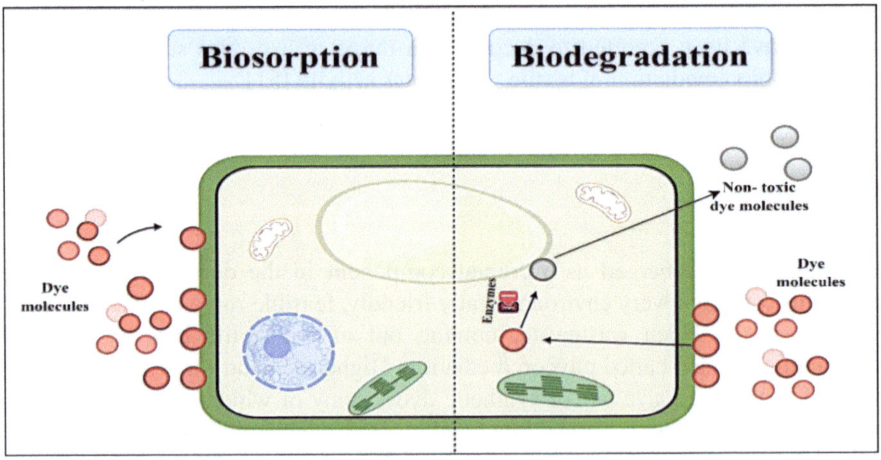

Fig. 1 Microbial mechanism of dye removal: biosorption (adsorption of dye molecules onto the cell surface) and biodegradation (enzymatic breakdown of dyes into non-toxic products)

Table 3 Biodegradation of textile dyes by different fungal and algal strains with efficiency and optimum conditions

Fungi/algae	Degraded dye	Efficiency (%)	Optimum conditions		References
			pH	Temperature (°C)	
Fungal strains					
Aspergillus tamari	Bromophenol blue	100	5	28	[87]
Aspergillus tamari	Coomasie brilliant blue	93	5	30	[87]
Coriolopsis sp.	Orange G	60	4.5	35	[23]
Fusarium oxysporum	Glycoconjugate azo dye	100	–	25	[80]
A. niger	Rose Bengal (18 g L^{-1})	100	5.8	28	[110]
A. flavus	Acid blue 29 (100 mg L^{-1})	70- 83.6	7.0	30–35	[8]
Trichoderma tomentosum	Acid red 3 R (85.5 mg L^{-1})	99.2	–	–	[45]
Algae (biosorption)					
Codium decorticatum	Crystal violet (CV)	96.9	10	25	[1]
Codium decorticatum	Congo red (CR)	89.8	4	25	[1]
Chlamydomonas variabilis	Methylene blue (MB)	98	7	25	[68]
Ulva lactuca	Congo red	91.92	8	25	[34]
Enteromorpha intestinalis	Malachite green	99.63	9.92	25	[43]
Spirulina platensis	Malachite green	94.12	7.57	–	[16]
Laminaria digitata	Methylene blue	95	5.6	25	[17]
Laminaria digitata	Reactive blue 19	60	1	25	[17]
Algae (biodegradation)					
Oscilltoria sp.	Malachite green (5 mg/L)	93	–	25	[38]
Oscilltoria sp.	Methylene blue (5 mg/L)	66	–	25	[38]
Haematococcus sp.	Congo red (10 mg/ L)	98	7	24	[66]
Scenedesmus obliquus	Methyl red (20 mg/ L)	55.45	7.4	25	[2]
Chlorella vulgaris	Reactive black 71 (200 mg/L)	80	5	40	[50]
Chlorella vulgaris	Direct blue (300 mg/L)	78	8	40	[50]

6 Microbial Mechanisms of Dye Degradation

Microbial degradation of textile dyes can occur through three main mechanisms: Biodegradation. Biosorption and Bioaccumulationare shown in Fig. 2.

Biodegradation

Textile dye biodegradation is an effective and sustainable technique for treating textile dye-laden wastewater, utilizing microorganisms to change harmful dyes to non-toxic substances. Their ability to degrade synthetic dyes is primarily attributed to two mechanisms: biodegradation and biosorption.

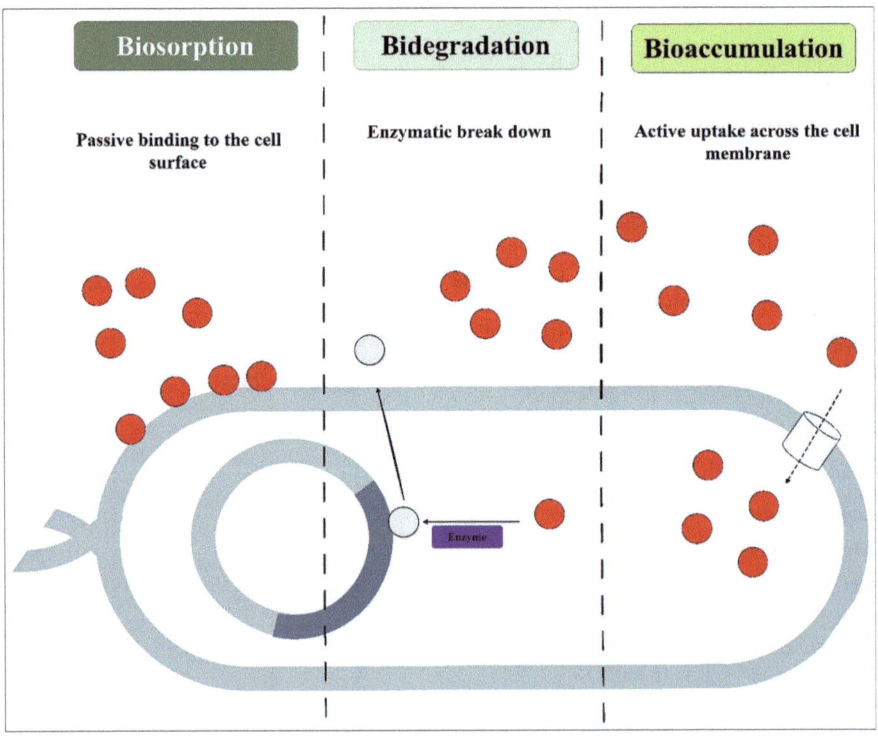

Fig. 2 Mechanisms of dye degradation emphasizing biosorption, biodegradation, and bioaccumulation

Enzymatic Action

The breakdown of the dye molecules involves some specific enzymes secreted by microorganisms. Several extracellular microbial enzymes- mainly laccases, peroxidases, azoreductases, proteases, and lipases are greatly in the limelight as they have been known to assume possible functions in supporting degradation as well as detoxification processes through a large group of synthetic dyes [51, 71]. As illustrated in Figure 3 dye degradation involves multiple microbial enzymes acting through reductive and oxidative pathways. These enzymes present a sustainable and effective substitute for traditional chemical and physical techniques employed in dye removal, which frequently require considerable energy, acquire high expenses, and produce secondary pollutants [93].

- **Laccases: Role in oxidative degradation**

The multicopper oxidases, such as laccases, catalyze the oxidation of a very wide range of substrates that include both phenolic and non-phenolic aromatic compounds. The ability of these enzymes to oxidize a range of dyes, and up to 99% dye removal efficiency, is extensively studied [96]. This enzyme is highly efficient in the degradation of synthetic dyes. The oxidation catalysis by laccase through one-electron transfer leads to the formation of reactive radicals that later break down non-enzymatically.

Laccases are enzymes with excellent efficiency, especially in the degradation of anthraquinone dyes, characterized by stiff structures and highly conjugated systems. They also have excellent efficiency against azo dyes, the largest class in the domain of synthetic dyes. These enzymes have been used with good results in the decolorization process of different classes of dyes, namely reactive, vat, and direct dyes, which makes them versatile and suitable for this domain. More than that, laccases have been demonstrated to have synergistic interaction with some mediators, such as ABTS or syringaldehyde, which increases dramatically their substrate specificity, allowing them to work on a more extended list of substrates, that includes non-phenolic and larger aromatic compounds [26]. The application of laccases has notable advantages mainly due to their versatile substrates, which can facilitate an effective interaction with any compounds. Furthermore, these enzymes can work under relatively rather mild operational conditions, meaning they are more accessible in practice. Laccases are also capable of functioning in both free and immobilized forms, which brings an added effectiveness to be used in dye degradation in industrial effluents generally.

- **Peroxidases**

Amongst these specific enzymes, there are the peroxidases. They contain heam, the iron-containing compound-hence is relevant in bio-reactions using hydrogen peroxide as their co-substrate. They are catalytically involved in catalyzing oxidative reactions in a process that is critical to breaking down large aromatic structures often associated with many synthetic dyes breaking []. There are several degradative peroxidases for this purpose [92]. They fall under three notable species. Amongst them is lignin peroxidase and it is abbreviated as LiP-the manganese peroxidase as MnP-and then versatile peroxidase often referred to as VP.

Lignin Peroxidase (LiP), mainly catalyzes the degradation of high-molecular-weight aromatic compounds. Under this class falls such dyes as oxidatively degrading non-phenolic structures. LiPexhibits exceptional potential for the oxidative breaking down of dyes, which are hard to break with commonly used methods, including methylene blue and crystal violet. Manganese Peroxidase (MnP), primarily targets phenolic substrates, oxidizing Mn^{2+} to Mn^{3+} that acts as a diffusible mediator to indirectly break the dyes []. MnP acts against azo, reactive, and heterocyclic dyes. Versatile Peroxidase (VP), provides the substrate range of both LiP and MnP for the degradation of phenolic and non-phenolic dye molecules, respectively [44]. Peroxidases have been widely used for the treatment of industrial effluent-containing dyes, although immobilized forms exhibit better stability along with reusability [].

- **Azoreductase**

Azoreductases are flavoproteins that specifically demonstrate affinity towards azo dyes. These dyes are chemical compounds that can be characterized by the presence of azo bonds, that is, the nitrogen-nitrogen double bond (–N=N–) linking together aromatic rings [93]. These enzymes start the reductive cleavage process by breaking down azo bonds in such a way that aromatic amines are produced. Following such biochemical transformation, these aromatic amines get decomposed and mineralized due to microbial consortia or other enzymatic systems present in the environment [92]. The importance of the activity of azoreductases for biochemical reduction reactions stems from its dependency on electron donors like NADH or NADPH. This activity by enzymes is very relevant in the process because many times, products formed due to this reaction are much less toxic than the compounds [111]. The ability of azoreductases has been demonstrated to degrade many azo dyes, including methyl orange, Congo red, and acid black, among others. The microorganisms are commonly applied in sequential anaerobic–aerobic treatment processes for the complete mineralization of azo dyes [93].

- **Proteases**

Proteases are, however, enzymes that break proteins into peptides. In terms of dye degradation, they can help break those dye molecules down, for example, protein-based or peptide-like dye molecules. The mechanism of action of proteases is of two types, The first is hydrolysis of dye structure in which proteases can cleave dye molecules to cut peptide bonds in the dye because the large dye compounds degrade into smaller more toxic units [93]. Therefore, such a process can increase the total degradation efficiency of the dye to other enzymes or the metabolism of the microorganisms. On the other hand, we found that some textile dyes can be efficiently degraded by alkaline proteases-specific strains. For instance, the use of bacterial consortia comprising *Bacillus cereus* and *Bacillus firmus* to produce alkaline proteases that decolorized Direct Red-81 and Direct Orange-34 dyes to about 68.88% and 70.78%, respectively under optimum conditions [83].

- **Lipases**

Lipases are specialized enzymes that play a very important role in the biochemical process of catalyzing the hydrolysis of fats and oils. They have been found to contribute indirectly to the degradation of dyes. This is done through the breakdown of lipids that are often present in various dye formulations or the effluents produced by textile industries [27]. The mechanism through which these enzymes function includes the breakdown of lipid-based dyes, whereby the ability of lipids to bind with the dyes themselves facilitates the hydrolysis process, thereby releasing the bound dyes, which can then be available for further microbial activity that can contribute to their degradation. Beyond the role of lipases in improving microbial metabolism through enhancement of the overall metabolic rate of microorganisms involved in the degradation of dyes, it also increases the availability of substrates for which these microorganisms can make use. While it may be correct that there is a lack of well-defined lipase-producing strains related to the degradation of dyes, it must not be forgotten that general lipase-producing bacteria, which include various strains of the Pseudomonas genus, among others, have been recorded and studied in numerous processes of biodegradation [61] (Fig. 3).

- **Factors influencing enzymatic activity**

Enzymatic hydrolysis of textile dyes is significantly influenced by several factors such as pH, temperature, and nutrient availability [96]. All these variables are important to the optimization of conditions to make microbial enzymes degrade synthetic dyes efficiently in industrial applications. The effects of each of these factors on enzymatic activity are therefore outlined in the following sections.

The pH level of the environment has a significant impact on the activity of enzymes involved in dye degradation. Every enzyme has an optimal pH range within which its activity is at maximum. Most laccases have been reported to be optimally active at acidic pH levels. The pleurotus sajor-caju-produced laccase resulted in the maximum decolorization of various dyes at a pH of 3.2. At such a pH level, more bleaching activity was achieved with different kinds of dyes while much lower levels of decolorization occurred at pH values of 4.4 and 5.0 [15, 52]. Extreme pH high or low will denature the enzyme and thus reduce the enzyme activity. For example, laccases are stable and active in a range of slightly acidic to neutral pH. However, an increase beyond this range might retard its activity.

Another significant factor that influences the role played by enzymes is temperature. Enzymes usually have optimal temperatures within which they act. All the enzymes have optimum activity with the rise in temperature to a certain extent; above that activity reduces along with their denaturation. For example, in the case of *P. ostreatus* laccase, optimal activity is obtained about at 30 °C; but the decolorization process gets significantly reduced at increased temperatures of above 45 °C. Again, it is because of the instability of the enzymes [15, 51]. The kinetic energy of the

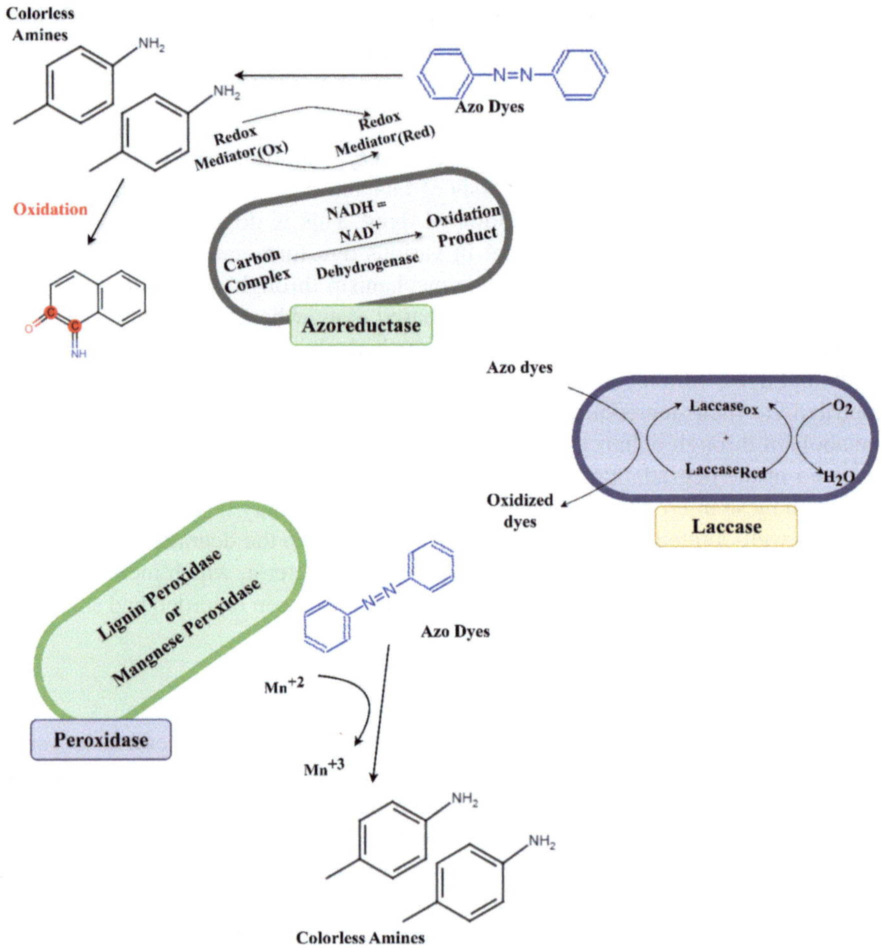

Fig. 3 Mechanism of azo dye degradation through different microbial enzymes

molecules increases with the increase in temperature. This results in more frequent hits between the enzymes and substrates. However, if the temperature crosses over the stability threshold of the enzymes (which is generally around 40–50 °C for most microbial enzymes), it will lead to irreversible denaturation.

The availability of nutrients in the environment may affect microbial growth and enzyme production. The availability of substrate (carbon and nitrogen sources) so that microbes can grow and larger amounts of enzymes that break down dyes are produced. For instance, nutrient-rich conditions generally elicit high concentrations of laccase from fungi like *Pleurotus ostreatus*, which is vital for the efficient breakdown of dyes [51, 52]. Nutrients can also serve as co-factors or inducers for enzyme synthesis. For example, some metal ions in the presence of specific ones, such as copper-can increase the production of laccases in fungi, thereby increasing their capability to degrade dyes [15, 51].

Metabolic Pathways Involved in Textile Dyes Degradation

Bacteria can use even the most complex dye's structure as the objects of their metabolism and transform it into simpler metabolites [96]. This ability is very common in the *Pseudomonas*, *Bacillus*, and *Proteus* species that degrade multiple dyes effectively under both aerobic as well as under anaerobic conditions [12, 101]. The microbial degradation of textile dyes involves complex metabolic pathways, varied by environmental conditions, but it is primarily anaerobic and aerobic.

- **Anaerobic degradation pathways**

Under anaerobic conditions, microorganisms cleave azo linkages in textile dyes by reductive mechanisms [73]. This is the first step to start the degradation process.

Azo Bond Reduction: Of all the pathways through which azo dyes are broken down by anaerobic microorganisms, it is most important that these are dominated by azo reductases. These catalyze the reduction of azo bonds (–N=N–) to aromatic amines. It proceeds more effectively in microaerophilic conditions where the partly absent oxygen environment favors the action of such reductive enzymes [56, 62].

This can then be followed with further metabolic processes and the yielding of aromatic amines that will be products coming from the azo dye reduction. Yet these intermediates could involve possible toxicity and therefore require further treatments for their possible detoxification. These produced aromatic amines also undergo several other enzymatic transformations by a group of microbial enzymes involving hydroxylation and ring cleavage [3, 73].

Microbial Consortia: Mixed microbial cultures or consortia have been proven to degrade azo dyes more efficiently compared with pure cultures. This is attributed to the diversity in enzymatic activity and metabolic pathways of such cultures, which increases overall efficiency. For instance, some bacteria specifically break azo bonds, whereas others can metabolize aromatic amines produced as a result [3, 12, 62].

- **Aerobic Degradation Pathways**

The azo bond breakage immediately and an aerobic atmosphere after degradation appears favorable because of aromatic amine formation.

Oxidative Enzymes: When exposed to aerobes, the action of microorganisms involves laccases, peroxidases, and veratryl alcohol oxidase among others for the enhancement of degradation reactions of aromatic amines into less dangerous chemicals. This oxidative process induces the degradation of toxic intermediates by hydroxylation and demethylation processes into less harmful intermediates [56, 73].

Full mineralization is, therefore, the ultimate goal of aerobic degradation. The entire dye constituents are completely oxidized to carbon dioxide and water. In many cases, this involves the TCA cycle, whereby simpler organic fragments derived from the breakdown of the dyes are oxidized to the fullest. In this cycle, the energy generated feeds microbial growth while ensuring that all the carbon derived from the dyes is converted into non-toxic products [13].

Sequential Treatment Processes: The overall efficiency of dye degradation increases if it is subjected to sequential anaerobic–aerobic treatment processes. The azo bond breaks during initial anaerobic treatment, and aromatic amines are produced; thus, these can be better detoxified in the subsequent aerobic treatment. Two steps minimize toxicity and maximize mineralization, making it an optimistic approach for treating textile wastewater [73].

Biosorption

Some bacteria can also physically bind the dye particles onto their surface without removing the dye. This process is characterized by the interaction of dye molecules with the microbial biomass through ion exchange with functional groups existing on the outer surface of microbial cells [71, 84]. Even though adsorption does not entail the breakdown of the dye, it is capable of diminishing color richness in wastewater [81].

Bioaccumulation

Bioaccumulation is an important mechanism in the bioremediation of textile dyes; it involves the uptake and transformation of dye molecules by microorganisms like bacteria, fungi, and algae [84, 90]. In contrast to passive adsorption, this process results in the degradation of dyes inside the organisms. Bacterial strains such as *Pseudomonas luteola* and *Bacillus cereus* can use dyes as carbon sources, whereas fungi produce enzymes that break down the complex structures of dyes [81]. Algae also contribute to dye removal through nutrient uptake. Environmental factors such as pH and nutrient availability influence efficacy, hence bioaccumulation is one of the promising strategies for sustainable remediation of textile dyes in wastewater management.

Conversion of Dyes to Nontoxic Metabolites

The overall aim of dye degradation methods is to change the harmful dye compounds into nontoxic metabolites. This transformation normally occurs in several steps:

(a) **Preliminary Deconstruction**: This usually starts with the enzymatic reactions that cleave complex dye structures into less harmful intermediates. This can be taken as a demonstration of how azoreductase's activity creates aromatic amines of azo dyes [9, 79].

(b) **Intermediate Transformation**: Those intermediates undergo transformation further through enzymatic reactions where oxidoreductases convert them into nontoxic products like carboxylic acids or alcohols. The productivity of such transformation is environment-dependent, aerobic, or anaerobic, and on the strains available in the microbial system [28, 99].

(c) **Detoxification**: Experiments of research have shown that full biodegradation results in much lower toxicity of its metabolites than water invertebrates such as *Daphnia magna*, thus effective detoxification mechanisms [9, 28].

7 Applications of Microbial Degradation

Microbial degradation of textile dyes emerged as a significant approach toward the efficient and safe treatment of problems associated with dye pollution by the microbial degradation of textile dyes. This method can be exploited in different fields like remediation of the environment, industry application, and development of novel microbial technology. The following is an overview of these applications.

Bioremediation Techniques

Some bacterial strains have been identified to break down toxic textile dyes effectively. For instance, the optimum conditions decolorization rate is highly achieved by close to 100% in some incubation periods by *Micrococcus luteus* and *Enterobacter* sp. This bacterium can easily break toxic dyes into non-toxic substances thus making it suitable to handle textile effluent before discharge into the environment [14, 86].

Fungi are also applied for the degradation of dyes due to their very strong enzymatic system. The laccases and peroxidases in fungi can oxidize the complex structures of dyes, thereby breaking them into much less simple nontoxic compounds. The above technique is particularly handy when dealing with recalcitrant dyes, which cannot be broken with the help of some other treatment techniques [58].

Algae and yeast have also been identified as a promising candidate for bioremediation in the recent past owing to their ability to absorb and metabolize dyes. These can be applied in constructed wetlands or biofilm systems to give enhanced removal of dyes from wastewater. They also enhance nutrient cycling [98]. Adding various kinds of microbes including bacteria, fungi, and algae increases the overall efficiency of dye-degrading. Consortium is the exploitation of a particular metabolic pathway by a group of microorganisms. Hence, leading to higher rates of decolorization and mineralization [35].

Industrial Applications

Wastewater Treatment of the colored wastewater produced by the textile industries is in which artificial dyes are found in quite high contents. It is, in fact, a cost-effective and ecologically sound method of microbial degradation treatment process for the effluents, which may then safely be reused in industrial practices [14, 86].

Environment-friendly dyeing techniques: The innovation in microbial technology has cemented a way for the eco-friendly dyeing process by microbial enzymes rather than dangerous chemicals. It decreases the footprint of textile production without any compromise with the quality color and durability [58]. It has been demonstrated that the micro-biological degradation method significantly minimizes sludge production compared to physicochemical methods, which therefore enhances the sustainability and feasibility of wastewater treatment [86, 98].

Advances in Microbial Technology

The new developments in genetic engineering have made it possible to generate microorganisms that show higher capacities in the degradation of dyes. Metabolic pathway modification has now made it possible for scientists to design strains with improved efficiency in the breakdown of particular classes of dyes [35, 98].

In recent biofilm technology has emerged as a significant technique in which contact between the microbe and pollutant increased and so also the efficiency of the process for degradation. Biofilms are created under optimally suited conditions for microbial growth and activity, which allow increased rates of dye removal from the wastewater [58, 84].

Apart from biofilm, the metagenomic approach would help in studying various microbial communities found in polluted areas. It would enable the finding of new strains with strong abilities to degrade dyes, which can then be used for cleaning up the environment [35, 58].

Phytotoxicity testing, the new techniques also entail carrying out a test on the phytotoxicity of the treated water in terms of plant tests. This means that bioremediation ensures both the removal of color and toxicant substances in the water before letting back the ecosystems [35].

In conclusion, it is one of the advanced approaches to solving pollution problems caused by the textile industry as it relates to microbial degradation of textile dyes. Such includes remediation strategies and identification of new industrial applications; there is also progress in microbial technology. Together, these deal with harmful effects on the environment and encourage sustainable practices in the industry. As shown in (Fig. 4) microbial degradation has diverse applications in the treatment of textile effluents.

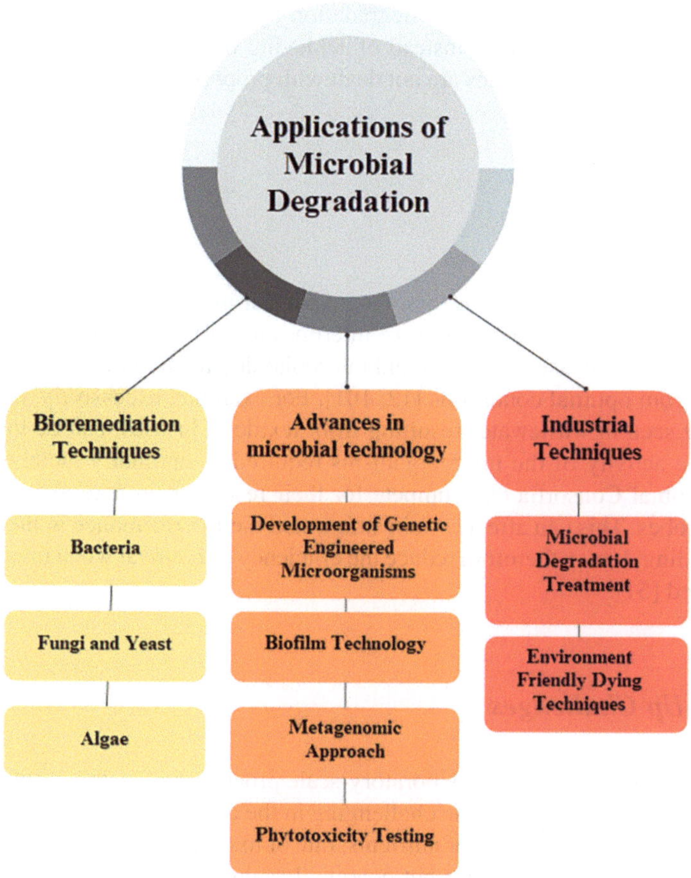

Fig. 4 Applications of microbial degradation

8 Challenges and Limitations

One of the promising approaches to solving the dye contamination environmental problems is microbial degradation of textile dyes. Nevertheless, its wide application & effectiveness is hampered by several obstacles and constraints. The major challenges related to the microbial degradation of textile dyes are as follows:

Resistance of Textiles to Dye

The complex and stable chemical structure of most synthetic dyes especially azo dyes is not biodegradable. Since they are synthetic, but cannot easily be degraded via natural biodegradation, they are hardy against environmental stresses [12, 76]. These dyes can be broken down through degradation by microorganisms that may form toxic intermediate metabolites instead of achieving full mineralization to harmless substances. If the intermediates are not dealt with properly, these are additional risks to the environment. [51].

Ecological Conditions

Under certain environmental factors, the activity of microbes is very sensitive; the parameters influencing the activities of microbes are pH, temperature, salinity, and oxygen. However, causes of fungal and microbial degradation depend greatly upon deviating from optimal conditions [12, 101]. For example, excessively salty levels, commonly seen in wastewater resulting from textile industries, can be too salty to support the activity of the microorganisms required to optimally metabolize dyes [12]. Microbial Consortia can compete for their resources in natural settings with other microbes. This can affect certain strains and their performance in the ability of dye-degrading. It may therefore reduce the efficiency of removal when mixed culture is employed [51].

Scaling-Up Challenges

The major studies indicate that laboratory-scale process works efficiently; however, extrapolation of the same is quite challenging in the context of field situations. Such factors include the availability of nutrients, rate of microbial survival, etc., which are likely present in real wastewater but potentially inhibit the efficiency found within controlled experiments [101]. The results from the research about microbial degradation are quite inhomogeneous and heavily depend upon the used strains, the given environment, and the chemical dyes. It creates difficulty in the predictability of results and the development of uniform processes in industries [51, 98].

Low Substrate Utilization

A large majority of the organisms employed in degrading dyes cannot metabolize these chemicals as the sole source of carbon or nitrogen. In addition, they require supplementary organic substrates, e.g. glucose or yeast extract to grow and to perform metabolic activities in the degradation process [51, 98]. This makes it complex in treatment schemes and raises operating costs. Dye-high concentrations are highly toxic to the microbial population, hence lesser decolorization rates as well as overall degradation efficiencies. Normally, the rate of decolorization declines directly in proportion to an increase in the concentration of dye because of the toxic effect or insufficient microbial biomass to properly treat the pollution [12, 51].

Technical Limitation

The number of required enzymes needed to break down the dyes can change depending on the stain, strain of microbes, and condition of growth. It is often the case that microorganisms do not have the equipment to produce large quantities of enzymes, such as azo reductases or laccases, to degrade the dyes. The efficiency of microbial degradation mostly depends on specific strains with unique metabolic properties. The procedure for recovering and preserving such strains is cumbersome and will not ensure identical performance for different dyes belonging to diverse categories or conditions [51, 76].

9 Future Perspectives

The potential for the microbial degradation of textile dyes is immense to address the environmental concerns posed by the dye contamination of the textile industry. Since people are becoming increasingly aware of the ecological impacts of synthetic dyes, demand for remediation strategies will continue to rise. The future critical areas of interest are going to be advanced bioremediation techniques, integration with new technologies, and understanding of microbial processes. One of the most promising areas for future work is bioremediation methods advanced by genetic engineering and synthetic biology. Scientists can modify microbial strains to improve their capacity in degrading dyes, thereby creating better biocatalysts that would break down a more varied range of dyes from highly resistant to nonresistant. This might be much greater levels of the enzymes produced by genetically modified bacteria or fungi, like laccases or azoreductases, leading to a large increase in the degradation rates of the most dangerous dyes. Additionally, this might increase more if a metagenomics study on natural microbial communities reveals new strains for novel mechanisms of degradation which industries would then exploit.

The integration of microbial degradation processes with the most advanced technologies, such as nanotechnology and bioelectrochemical systems, promises to create exciting opportunities. With the introduction of nanoparticles, the stability and efficiency toward more effective dye degradation of the enzyme is increased. However, Microbial Fuel Cells (MFCs) including dye-degrading microorganisms will find applications in wastewater purification and electricity generation as well. This could then result in much more sustainable wastewater management solutions that both generate energy and remove environmental contaminants. One major direction for future work is to improve environmental conditions to promote microbial activity. Explanation of the ability of microbes to degrade would be particularly useful if the responses of different parameters such as pH, temperature, and available nutrients were analyzed. If the development of bioreactor systems could be successfully realized, industrial applications of the removal efficiencies would be greatly improved, maintaining proper conditions for microbial and enzymatic growth.

Overcoming the constraints inherent in the existing bioremediation technologies is of prime importance for its popularization. The survival of toxic intermediate metabolites in the degradation process is still a problem. In the future, the objective of the research should be to achieve the complete mineralization of dyes into non-toxic products. If sequential treatment processes could be developed that include anaerobic and aerobic methods, or chemical oxidation combined with biological methods, overall effectiveness could be increased. As the discharge of wastewater with dyes becomes more and more stringent regulated, industries will ultimately opt to find environmentally friendly ways of removing the dyes. The application of microbial degradation technologies for the treatment of textile effluents will not only facilitate regulatory compliance but also aid corporate sustainability. In addition, biotechnological methods will gain importance for treating and reusing effluents in water-scarce areas.

In conclusion, the prospect for biodegradation of textile dyes by microbes appears encouraging with improvement in biotechnology techniques and increased interest for sustainable technologies. Further innovation and research in this context will pave the way towards the creation of effective technology that can help reduce consequences on the environment and thereby promote a circular economy within the textile industry. Coordination and teamwork between the researchers, industry players, and policymakers will determine the full realization of microbial technologies in the alleviation of textile dye pollution in the future.

10 Conclusion

Textile industry effluent is a serious global concern as it causes hazardous consequences on the ecosystem and human health. A variety of degradation approaches have been employed to degrade textile dyes. But they are cost-effective, toxic, and non-sustainable approaches. To combat these problems, microbial degradation of dye has emerged as sustainable, cost-effective and eco-friendly approach. However, the microbial degradation processes involve a variety of microbes such as bacteria fungi and algae. These microbes are exploited to break down complex dye molecules through enzymatic as well as adsorptive mechanisms. Studies show that certain strains of these microorganisms attain large decolorization rates within hours under ideal environmental conditions (pH, temperature and nutrient availability). Recent research emphasizes the ability of microalgae-based systems to remove not only dyes but also heavy metals and other pollutants from wastewater. The combination of algal systems with microbial consortia or conventional treatment processes might further improve overall efficiency in the treatment of wastewater. The metabolic processes that contribute to the biodegradation of textile wastewater are not yet fully understood.

Hence, with the identification of genes and metabolites participating in the degradation processes, it becomes important to target the involved mechanisms and metabolic pathways. Moreover, Advanced biotechnological methods, like biofilm technology, bio-nanotechnology and genetically engineered microbial strains or consortia, can boost degradation rates and increase the range of efficiently degradable dyes.

References

1. Abou Oualid, H., Abdellaoui, Y., Laabd, M., El Ouardi, M., Brahmi, Y., Iazza, M., & Abou Oualid, J. (2020). Eco-efficient green seaweed *Codium decorticatum* biosorbent for textile dyes: Characterization, mechanism, recyclability, and RSM optimization. *ACS Omega, 5*(35), 22192–22207.
2. Abou-El-Souod, G., Hamouda, R. A., & El-Sheekh, M. (2020). Influence of heavy metal as co-contamination on biodegradation of dyes by free and immobilized Scenedesmus obliquus. *Desalination and Water Treatment, 182*, 351–358.
3. Abubacker, M. N., & Ayesha, A. (2011). Decolourization of azo-dyes by bacterial consortium. *Biosciences Biotechnology Research Asia, 8*(2).
4. Akter, H. B. (2022). Textile dyeing effluents and their impact on the environment. *Textile Today*. Retrieved from Textile Today.
5. Alam, R., Mahmood, R. A., Islam, S., Ardiati, F. C., Solihat, N. N., Alam, M. B., et al. (2022). Understanding the biodegradation pathways of azo dyes by immobilized white-rot fungus *Trametes hirsuta* D7 using Uplc-Ft-Icr-Ms supported by in silico simulations and toxicity assessment. Available at SSRN 4154292.
6. Ali, A., Elsayid, Z., Zaman, C., Ramprosad, D., Md. Mostak, A., Md. Mahfujur, R., Khalisanni, K., Yasmin, A. W., Irfan, A. B., Sarfaraz, K., Md. Hussien, K., Husain, P., & Ajita, M. (2023). Removal of azo dyes from aqueous effluent using bio-based activated carbons: Toxicity aspects and environmental impact. *Separations*https://doi.org/10.3390/separations10090506
7. Alzain, H., & Mohammed Karim, H. (2023). A review of environmental impact of azo dyes. *International Journal of Research and Review, 10*(6), 64–689.https://doi.org/10.52403/ijrr.20230682
8. Ameen, F., Dawoud, T. M., Alshehrei, F., Alsamhary, K., & Almansob, A. (2021). Decolorization of acid blue 29, disperse red 1 and congo red by different indigenous fungal strains. *Chemosphere, 271*, Article 129532.
9. Aragaw, T. A., Bogale, F. M., & Gessesse, A. (2022). Adaptive response of thermophiles to redox stress and their role in the process of dye degradation from textile industry wastewater. *Frontiers in Physiology, 13*, Article 908370. https://doi.org/10.3389/fphys.2022.908370
10. Bafana, A., Devi, S. S., & Chakrabarti, T. (2011). Azo dyes: Past, present, and the future. *Environmental Reviews, 19*(NA), 350–370.
11. Bao-Kun, Lai., Sung, Te-Chin., Pan, Yuan-Pin. (2019). Acid dye composition, use thereof and method using the same for dyeing nylon textiles.
12. Bera, S. P., Shah, M. P., & Godhaniya, M. (2022). Microbial remediation of textile dye acid orange by a novel bacterial Consortium SPB92. *Frontiers in Environmental Science*. https://doi.org/10.3389/fenvs.2022.930616
13. Bera, S. P., & Tank, S. K. (2021). Microbial degradation of Procion red by *Pseudomonas stutzeri. Scientific Reports, 11*(1), 1–12. https://doi.org/10.1038/s41598-021-82494-9
14. Bernaitis, L., Arun, K., Bibin, K., Prejish, C., Jomol, P., & Anakha, V. (2024). The application and assessment of bacterial isolates for textile dye decolorization. *Indian Journal of Microbiology Research, 2024*, 22885. https://doi.org/10.18231/j.ijmr.2024.035

15. Bettin, F., Cousseau, F., Martins, K., Zaccaria, S., Girardi, V., da Silveira, M. M., & Dillon, A. J. P. (2019). Effects of pH, temperature, and agitation on the decolorization of dyes by laccase-containing enzyme preparation from *Pleurotus sajor-cajun. Brazilian Archives of Biology and Technology, 62*, Article e19180338. https://doi.org/10.1590/1678-4324-2019180338
16. Bonyadi, Z., Nasoudari, E., Ameri, M., Ghavami, V., & Shams, M. (2022). Sillanpää, M. Biosorption of Malachite green dye over spirulina platensis mass: Process modeling, factors optimization, kinetic, and isotherm studies. *Applied Water Science, 12*, 167.
17. Bouzikri, S., Ouasfi, N., & Khamliche, L. (2022). *Bifurcaria bifurcata* activated carbon for the adsorption enhancement of Acid Orange 7 and Basic Red 5 dyes: Kinetics, equilibrium and thermodynamics investigations. *Energy Nexus, 7*, Article 100138.
18. Brzezinski, M., Leisha, M., Martin, A., Kayla, S., Kaijun, L., Nin, G., Chi, H., Kaitlin, G., Zhanfei, L., & Wei, X. (2024). Photodegradation enhances the toxic effect of anthracene on skin. *Journal of Hazardous Materials, 471*, 134386–134386.https://doi.org/10.1016/j.jhazmat.2024.134386
19. Caro, Y., Thomas, P., Isabelle, G., Mireille, F., & Laurent, D. (2017). Hydroxyanthraquinone dyes from plants.
20. Chakraborty, J. N., (2011). *Dyeing with sulphur dye* (pp. 43–56). https://doi.org/10.1016/B978-93-80308-46-3.50005-4
21. Chauhan, A. K., & Choudhury, B. (2021). Synthetic dye degradation using lignolytic enzymes produced from *Halopigera swanensis* strain ABC_IITR by solid state fermentation. *Chemosphere, 273*, Article 129671.
22. Chauhan, V., Gautam, P., & Kanwar, S. S. (2022). Azo dyes. In *Elsevier eBooks* (pp. 433–448). https://doi.org/10.1016/b978-0-323-85657-7.00019-5
23. Cheng, W. N., Sim, H. K., Ahmad, S. A., Syed, M. A., Shukor, M. Y., & Yusof, M. T. (2016). Characterization of an azo-dye-degrading white rot fungus isolated from Malaysia. *Mycosphere, 7*(5), 560–569.
24. Cheriaa, J., & Bakhrouf, A. (2009). Triphenylmethanes, malachite green, and crystal violet dyes decolorization by *Sphingomonas paucimobilis. Annals of Microbiology, 59*(1), 57–61. https://doi.org/10.1007/BF03175599
25. Christie, R., & Abel, A. (2022). 1 Anthraquinonoid pigments. In *De Gruyter eBooks* (pp. 1–14). https://doi.org/10.1515/9783110588071-001
26. Couto, S. R., & Toca-Herrera, J. L. (2007). Laccase production at reactor scale by filamentous fungi. *Biotechnology Advances, 25*(6), 558–569. https://doi.org/10.1016/j.biotechadv.2007.07.002
27. da Luz, J. M. R., de Souza Lopes, L., da Silva, M. C. S., Vieira, N. A., Cardoso, W. S., & Kasuya, M. C. M. (2023). Lentinula edodes lingo cellulolases and lipases produced in Macaúba residue and use of the enzymatic extract in the degradation of textile dyes. 3 *Biotech, 13*(12), 406. https://doi.org/10.1007/s13205-023-03827-1
28. Das, S., Cherwoo, L., & Singh, R. (2022). Decoding dye degradation: Microbial remediation of textile industry effluents. *Biotechnology Notes, 4*, 64–76. https://doi.org/10.1016/j.biotno.2023.10.001
29. Das, S., Cherwoo, L., Singh, R. (2023). Decoding dye degradation: Microbial remediation of textile industry effluents. *Biotechnology notes*. https://doi.org/10.1016/j.biotno.2023.10.001
30. De Vasconcelos, G. M. D., Della-Flora, I. K., Kelbert, M., De Andrade, L. M., De Oliveira, D., De Arruda GuelliUlson De Souza, S. M., De Souza, A. a. U., & De Andrade, C. J. (2024). Screening of azo-dye-degrading bacteria from textile industry wastewater-activated sludge. *Eng—Advances in Engineering, 5*(1), 116–132. https://doi.org/10.3390/eng5010008
31. Dulo, B., Phan, K., Githaiga, J. et al. (2021). Natural Quinone Dyes: A Review on Structure, Extraction Techniques, Analysis and Application Potential. *Waste Biomass Valor 12*, 6339–6374. https://doi.org/10.1007/s12649-021-01443-9
32. El-Ahmady El-Naggar, N., Rabei, N.H., El-Malkey, S.E. (2020). Eco-Friendly Approach for Biosorption of Pb2+ and Carcinogenic Congo Red Dye from Binary Solution onto Sustainable Ulva Lactuca Biomass. *Sci. Rep. 10*, 16021.

33. Ezhilarasu, A. (2016). Textile industry Dye degrading by bacterial strain Bacillus sp. *International Journal of Advertisement Research Biological Science, 3*(3), 211-226.
34. Fan, H., Yanzhao, X., Cuizhen, S., Rupeng, L., Feiyong, C., Meng, L., Wei, Z., Yang, X., & Zhen, Z. (2024). Degradation of Anthraquinone dye wastewater by sodium percarbonate with CoO heterogeneous activation. *Journal of Chemical Technology & Biotechnology*. https://doi.org/10.1002/jctb.7684
35. Farhan, M. (2022). Microbial degradation of industrially important textile dyes. *Pakistan BioMedical Journal*. 09–13. https://doi.org/10.54393/pbmj.v5i10.809
36. Farias, N., Oliveira de, A., Fernandes de, A., Amanda dos, S., Gabriela, C., Fonseca, A., Harold, S., Freeman, R. R., & Gisela de, A. U. (2023). Is natural better? An ecotoxicity study of anthraquinone dyes. *Chemosphere*. https://doi.org/10.1016/j.chemosphere.2023.140174
37. Fried, R., Oprea, I., Fleck, K., & Rudroff, F. (2021). Biogenic colorants in the textile industry—A promising and sustainable alternative to synthetic dyes. *Green Chemistry, 24*(1), 13–35. https://doi.org/10.1039/d1gc02968a
38. Gelebo, G. G., Tessema, L. H., Kehshin, K. T., Gebremariam, H. H., Gebremikal, E. T., Motuma, M. T., et al. (2020). Phycoremediation of synthetic dyes in an aqueous solution using an indigenous *Oscillatoria* sp., from Ethiopia. *Ethiopian Journal of Science and Sustainable Development, 7*(2), 14–20.
39. Gianolini, J. E., Britos, C. N., Mulreedy, C. B., & Trelles, J. A. (2020). Hyperstabilization of a thermophile bacterial laccase and its application for industrial dyes degradation. *Biotech, 10*, 288. https://doi.org/10.1007/s13205-020-02277-3
40. Golka, K., Kopps, S., & Myslak, Z. W. (2004). Carcinogenicity of azo colorants: Influence of solubility and bioavailability. *Toxicology Letters, 151*(1), 203–210.
41. Goswami., D. C., Jayanti, M., Chanchal, M., & Biswanath, B. (2024). Bioremediation of azo dye: A review on strategies, toxicity assessment, mechanisms, bottlenecks and prospects. *Science of The Total Environment*, 176426.https://doi.org/10.1016/j.scitotenv.2024.176426
42. Gowri, A. K., Karunakaran, M. J., Muthunarayanan, V., Ravindran, B., Nguyen-Tri, P., Ngo, H. H., & Chandran, T. (2020). Evaluation of bioremediation competence of indigenous bacterial strains isolated from fabric dyeing effluent. *Bioresource Technology Reports, 11*, Article 100536.
43. Hamouda, R. A., El-Naggar, N. E. A., Doleib, N. M., & Saddiq, A. A. (2020). Bioprocessing strategies for cost-effective simultaneous removal of chromium and malachite green by marine alga *Enteromorpha intestinalis*. *Science and Reports, 2020*(10), 13479.
44. He, Y., Gao, J.-F., Feng, F.-Q., Liu, C., Peng, Y.-Z., & Wang, S.-Y. (2012). The comparative study on the rapid decolorization of azo, anthraquinone, and triphenylmethane dyes by zero-valent iron. *Chemical Engineering Journal—CHEM ENG J*. https://doi.org/10.1016/j.cej.2011.05.107
45. He, X. L., Song, C., Li, Y. Y., Wang, N., Xu, L., Han, X., & Wei, D. S. (2018). Efficient degradation of Azo dyes by a newly isolated fungus *Trichoderma tomentosum* under non-sterile conditions. *Ecotoxicology and Environmental Safety, 150*, 232–239.
46. Hemashenpagam., N., & Selvajeyanthi, S. (2023). *Textile Dyes and Their Effect on Human Beings*, 41–60. https://doi.org/10.1007/978-981-99-3901-5_3
47. Herrera, J. S., & Moreno, Z. R. (2022). Biodegradation of Erionyl Turquoise azoic textile dye from *Chlorella vulgaris* microalgae in simulated wastewater. *Journal of Textile Engineering & Fashion Technology, 8*(4), 121–124. https://doi.org/10.15406/jteft.2022.08.00309
48. Hunger, K. (2003). *Industrial dyes: Chemistry, properties*. Wiley-VCH.
49. Ihsanullah, I., Jamal, A., Ilyas, M., Zubair, M., Khan, G., & Atieh, M. A. (2020). Bioremediation of dyes: Current status and prospects. *Journal of Water Process Engineering, 38*, Article 101680.
50. Ishchi, T., & Sibi, G. (2020). Azo dye degradation by *Chlorella vulgaris:* Optimization and kinetics. *International Journal of Biological Chemistry, 14*(1), 1–7.
51. Jamee, R., & Siddique, R. (2019). Biodegradation of synthetic dyes of textile effluent by microorganisms: An environmentally and economically sustainable approach. *European Journal of Microbiology and Immunology, 9*(4), 114–118. https://doi.org/10.1556/1886.2019.00018

52. Johnnie, D. A., Issac, R., Prabha, M. L., & Gomez, L. A. (2023). Biodegradation assay of heavy metals and dyes decolorization in textile industrial effluent using laccase isolated from *Pleurotus ostreatus*. *Journal of Pure and Applied Microbiology, 17*(4), 2324–2343. https://doi.org/10.22207/JPAM.17.4.29
53. Jones, J. J., & Falkinham, J. O., 3rd. (2003). Decolorization of malachite green and crystal violet by waterborne pathogenic mycobacteria. *Antimicrobial Agents and Chemotherapy, 47*(7), 2323–2326. https://doi.org/10.1128/AAC.47.7.2323-2326.2003
54. Kalyani, D. C., Telke, A. A., Dhanve, R. S., & Jadhav, J. P. (2009). Ecofriendly biodegradation and detoxification of Reactive Red 2 textile dye by newly isolated *Pseudomonas* sp. SUK1. *Journal of Hazardous Materials, 163*(2–3), 735–742.
55. Kaur, G., & Surojit, B. (2020). Adverse effect of triphenylmethane dyes on environmental health and its detoxification for improved ecosystem. *Journal of Emerging Technologies and Innovative Research, 7*(11), 174–183.
56. Khandare, R., & Govindwar, S. (2015). *Microbial degradation mechanism of textile dye and its metabolic pathway for environmental safety.* https://doi.org/10.1201/b19243-15
57. Khattab, T. A., Abdelrahman, M. S., & Rehan, M. (2020). Textile dyeing industry: Environmental impacts and remediation. *Environmental Science and Pollution Research International, 27*(4), 3803–3818. https://doi.org/10.1007/s11356-019-07137-z
58. Khongthaw, B., Chauhan, P. K., Chishty, N., Kumar, D., Velmurugan, A., Singh, A., Bhtoya, R., Devi, N., Nene, A., Sadeghzade, S., & Ighalo, J. O. (2024). Aqueous phase textile dye degradation by microbes and nanoparticles: A review. *Advances in Materials Science and Engineering, 8845873*, 25 p. https://doi.org/10.1155/2024/8845873
59. Krawczyk, K., Stanisław, W., Edyta, K., Daniele, S., Tomasz, K., Klaudiusz, G., Vinod, V. T., & Padil Miroslav, Č. (2020). UV-catalyzed persulfate oxidation of an anthraquinone based dye. *Catalysts, 10*(4), 456. https://doi.org/10.3390/CATAL10040456
60. Krithika, A., Gayathri, K. V., Kumar, D. T., & Doss, C. G. P. (2021). Mixed azo dyes degradation by an intracellular azoreductase enzyme from alkaliphilic Bacillus subtilis: A molecular docking study. *Archives of Microbiology, 203*, 3033–3044.
61. Kumar, J., & Senthil Kumar, M. (2020). A study on improving dyeability of polyester fabric using lipase enzyme. *Autex Research Journal, 20*(3), 243–249. https://doi.org/10.2478/aut-2019-0030
62. Lade, H., Kadam, A., Paul, D., & Govindwar, S. (2015). Biodegradation and detoxification of textile azo dyes by a bacterial consortium under sequential microaerophilic/aerobic processes. *EXCLI Journal, 14*, 158–174. https://doi.org/10.17179/excli2014-642
63. Lara, L., Cabral, I., & Cunha, J. (2022). Ecological approaches to textile dyeing: A review. *Sustainability, 14*(14), 8353. https://doi.org/10.3390/su14148353
64. Li, H. H., Wang, Y. T., Wang, Y., Wang, H. X., Sun, K. K., & Lu, Z. M. (2019). Bacterial degradation of anthraquinone dyes. *Journal of Zhejiang University, Science. B, 20*(6), 528–540. https://doi.org/10.1631/jzus.B1900165
65. Lu, L., Zhao, M., Wang, T., Zhao, L., Du, M., Li, T., & Li, D. (2011). Characterization and dye decolorization ability of an alkaline resistant and organic solvents tolerant laccase from *Bacillus licheniformis* LS04. *Bioresource Technology, 115*, 35–40. https://doi.org/10.1016/j.biortech.2011.07.111
66. Mahalakshmi, S., Lakshmi, D., & Menaga, U. (2015). Biodegradation of different concentration of dye (Congo red dye) by using green and blue green algae. *Internatinal Journal of Environment Research, 9*(2), 735-744.
67. Mahapatra, N. (2018). Textile dyes. *Journal of Textile and Clothing Science, 1*(1), 01–02.
68. Moghazy, R. M. (2019). Activated biomass of the green microalga Chlamydomonas variabilis as an efficient biosorbent to remove methylene blue dye from aqueous solutions. *Water Sa, 45*(1), 20–28.
69. Moyo, S., Makhanya, B. P., & Zwane, P. E. (2022). Use of bacterial isolates in the treatment of textile dye wastewater: A review. *Heliyon, 8*(6), Article e09632. https://doi.org/10.1016/j.heliyon.2022.e09632

70. Nair, V. K., Selvaraju, K., Samuchiwal, S., Naaz, F., Malik, A., & Ghosh, P. (2023). Phycoremediation of synthetic dyes laden textile wastewater and recovery of bio-based pigments from residual biomass: An approach towards sustainable wastewater management. *Processes, 11*(6), 1793. https://doi.org/10.3390/pr11061793
71. Ngo, A. C. R., & Tischler, D. (2022). Microbial degradation of azo dyes: Approaches and prospects for a hazard-free conversion by microorganisms. *International Journal of Environmental Research and Public Health, 19*(8), 4740. https://doi.org/10.3390/ijerph190 84740
72. Ngo, A. C. R., & Tischler, D. (2022). Microbial Degradation of Azo Dyes: Approaches and Prospects for a Hazard-Free Conversion by Microorganisms. *International journal of environmental research and public health, 19*(8), 4740. https://doi.org/10.3390/ijerph190 84740.
73. O'Neill, C., Lopez, A., Esteves, S., Hawkes, F. R., Hawkes, D. L., & Wilcox, S. (2000). Azo-dye degradation in an anaerobic-aerobic treatment system operating on simulated textile effluent. *Applied Microbiology and Biotechnology, 53*(2), 249–254. https://doi.org/10.1007/ s002530050016
74. Oplatowska, M., Ryan, F. D., Rita, M. D., Glenn, K., & Christopher, T. E. (2011). The potential for human exposure, direct and indirect, to the suspected carcinogenic triphenylmethane dye Brilliant Green from green paper towels. *Food and Chemical Toxicology, 49*(8), 1870–1876.https://doi.org/10.1016/J.FCT.2011.05.005
75. Ouettar, L., Guechi, E. K., Hamdaoui, O., Fertikh, N., Saoudi, F., & Alghyamah, A. (2023). Biosorption of triphenyl methane dyes (malachite green and crystal violet) from aqueous media by Alfa (*Stipa tenacissima* L.) Leaf Powder. *Molecules (Basel, Switzerland), 28*(8), 3313. https://doi.org/10.3390/molecules28083313
76. Patil, N. P., Bholay, A. D., Kapadnis, B. P., & Gaikwad, V. B. (2016). Biodegradation of model azo dye methyl red and other textile dyes by isolate *Bacillus circulans* NPP1. *Journal of Pure and Applied Microbiology, 10*(4), 2793–2800. https://doi.org/10.22207/JPAM.10.4.38
77. Perkins, W. S. (1996). *Textile coloration and finishing*. Carolina Academic Press.
78. Pham, V. H. T., Kim, J., Chang, S., & Bang, D. (2023). Investigating bio-inspired degradation of toxic dyes using potential multi-enzyme producing extremophiles. *Microorganisms, 11*(5), 1273. https://doi.org/10.3390/microorganisms11051273
79. Pinheiro, L. R. S., Gradíssimo, D. G., Xavier, L. P., & Santos, A. V. (2022). Degradation of azo dyes: Bacterial potential for bioremediation. *Sustainability, 14*(3), 1510. https://doi.org/ 10.3390/su14031510
80. Porri, A., Baroncelli, R., Guglielminetti, L., Sarrocco, S., Guazzelli, L., Forti, M., et al. (2011). Fusarium oxysporum degradation and detoxification of a new textile-glycoconjugate azo dye (GAD). *Fungal Biology, 115*(1), 30–37.
81. Pundir, A., Thakur, M. S., Prakash, S., Kumari, N., Sharma, N., Parameswari, E., et al. (2024). Fungi as versatile biocatalytic tool for treatment of textile wastewater effluents. *Environmental Sciences Europe, 36*(1), 1–15.
82. Rajendran, S., Muthukumaragurupandian, G., Ramesh, K., & Rathinam, V. (2024). Preliminary screening, isolation, and identification of methylene blue dye degrading microorganisms from textile effluent contaminated site. *Environment and Ecology Research, 12*, 163–171. https://doi.org/10.13189/eer.2024.120207
83. Rathod, M. G., & Pathak, A. P. (2018). *Efficient decolorization of textile dyes by alkaline protease-producing bacterial consortia.*
84. Revathi, N., Raja, J. D., Rajesh, J., & Sankarganesh, M. (2023). *Recent techniques in dye degradation: A biological approach* (pp. 167–186). Bentham Science Publishers eBooks. https://doi.org/10.2174/9789815179545123010011
85. Robinson, T., McMullan, G., Marchant, R., & Nigam, P. (2001). Remediation of dyes in textile effluent: A critical review on current treatment technologies with a proposed alternative. *Bioresource Technology, 77*(3), 247–255.
86. Roy, D. C., Biswas, S. K., Saha, A. K., Sikdar, B., Rahman, M., Roy, A. K., Prodhan, Z. H., & Tang, S. S. (2018). Biodegradation of crystal violet dye by bacteria isolated from textile industry effluents. *PeerJ, 6*, Article e5015. https://doi.org/10.7717/peerj.5015

87. Saraswathy, N., Shanmugapriya, S., Shakthipriyadarshini, S., Sadasivam, S., & Shanmugaprakash, M. (2010). Decolorization of textile dyes by *Aspergillus tamarii*, mixed fungal culture and *Penicillium purpurogenum.*
88. Sen, S. K., Raut, S., & Raut, S. (2023). Mycoremediation of anthraquinone dyes from textile industries: A mini-review. *Biotechnologia, 104*(1), 85–91. https://doi.org/10.5114/bta.2023.125088
89. Sharif, A., Nasreen, Z., Bashir, R., & Kalsoom, S. (2020). 8. Microbial degradation of textile industry effluents: A review. *Pure and Applied Biology (PAB), 9*(4), 2361–2382.
90. Singh, S., et al. (2024). Role of microbes in dye degradation. In P. Chaudhary& A. Chaudhary (Eds.), *Microbial inoculants. Interdisciplinary biotechnological advances.* Springer, Singapore. https://doi.org/10.1007/978-981-97-0633-4_15
91. Sneha, U., Poornima, R., & Sridhar, S. (2013). *Decolorization of synthetic textile dyes using Pseudomonas putida.*
92. Souza, A. G. V., Maria, T. C., Saran, L. M., & Alves, L. M. C. (2022). Enzymatic bioremediation of dyes from textile industry effluents. In *The toxicity of environmental pollutants.* Intechopen.
93. Sreedharan, V., & Bhaskara Rao, K. V. (2019). Biodegradation of textile azo dyes. In K. Gothandam, S. Ranjan, N. Dasgupta & E. Lichtfouse (Eds.), *Nanoscience and biotechnology for environmental applications. Environmental chemistry for a sustainable world* (Vol. 22). Springer. https://doi.org/10.1007/978-3-319-97922-9_5
94. Sridharan, R., Krishnaswamy, V. G., Archana, K. M., Rajagopal, R., Thirumal Kumar, D., Doss, G. P., & C. (2021). Integrated approach on azo dyes degradation using laccase enzyme and Cul nanoparticle. *SN Applied Sciences, 3*, 1–12.
95. Stothers, J., & Abrahart, E. N. (2024). Dye. *Encyclopedia Britannica.* https://www.britannica.com/technology/dye
96. Tadele, A. A., Fekadu, M. B., & Eshet, L. T. (2024). *Ligninolytic oxidative enzymes and their role in textile dye biodegradation: A comprehensive review.* https://doi.org/10.2166/wpt.2024.229
97. Tang, J., Wang, Y., Yang, G., Luo, H., Zhuang, L., Yu, Z., & Zhou, S. (2018). Complete genome sequence of the dissimilatory azo reducing thermophilic bacterium *Novibacillus Thermophiles* SG-1. *Journal of Biotechnology, 284*, 6–10. https://doi.org/10.1016/j.jbiotec.2018.07.032
98. Thakur, V., Singh, R., & Chopra, C. (2021). Microbial degradation of textile dyes: A review. *Journal of Emerging Technologies and Innovative Research, 8*(8).
99. Thangaraj Sheela, S. K. S. (2020). Dye degradation potential and its degradative enzymes synthesis of *Bacillus cereus* SKB12 isolated from a textile industrial effluent. *Journal of Applied Biology & Biotechnology, 8*, 42–46, 3. https://jabonline.a2hosted.com/jab_php/abstract.php?article_id=jab469
100. Tian, F., Guo, G., Zhang, C., Yang, F., Hu, Z., Liu, C., & Wang, S. W. (2019). Isolation, cloning, and characterization of an azoreductase and the effect of salinity on its expression in a halophilic bacterium. *International Journal of Biological Macromolecules, 123*, 1062–1069.
101. Tripathi, M., Singh, S., Pathak, S., Kasaudhan, J., Mishra, A., Bala, S., Garg, D., Singh, R., Singh, P., Singh, P. K., Shukla, A. K., & Pathak, N. (2023). Recent strategies for the remediation of textile dyes from wastewater: A systematic review. *Toxics, 11*(11), 940. https://doi.org/10.3390/toxics11110940
102. Varjani, S., Rakholiya, P., Ng, H. Y., You, S., & Teixeira, J. A. (2020). Microbial degradation of dyes: An overview. *Bioresource Technology, 314*, Article 123728.
103. Vikrant, K., Giri, B. S., Raza, N., Roy, K., Kim, K. H., Rai, B. N., & Singh, R. S. (2018). Recent advancements in bioremediation of dye: Current status and challenges. *Bioresource Technology, 253*, 355–367.
104. Vividha, K. (2024). *Dyeing for change: Addressing the health and environmental risks of the textile industry. Fibre2Fashion.* Retrieved from Fibre2Fashion.
105. Wadhawan, G., & Gupta, A. (2024). Biological treatment of methyl orange dye and textile wastewater using halo-alkaliphilic bacteria under highly alkaline conditions. *Extremophiles: life under extreme conditions, 29*(1), 6. https://doi.org/10.1007/s00792-024-01369-9

106. Wei, W., Zhang, Y., Zhang, X., Gu, W., & Xie, Z. (2020). *Disperse dye composition, disperse dye as well as preparation method and application thereof*.
107. Yagub, M. T., Sen, T. K., Afroze, S., & Ang, H. M. (2014). Dye and its removal from aqueous solution by adsorption: A review. *Advances in Colloid and Interface Science, 209*, 172–184.
108. Yahuza, S., Sabo, I. A., & Abubakar, A. (2023). Biosorption of triphenylmethane (TPM) dyes by microbial biomass: A review. *Journal of Environmental Microbiology and Toxicology, 11*(2), 20–28. https://doi.org/10.54987/jemat.v11i2.888
109. Zannat, A., Uddin, M. N., Mahmud, S. T., Mia, R., & Ahmed, T. (2024). Natural dyes and pigments in functional finishing. In *Elsevier eBooks* (pp. 271–287). https://doi.org/10.1016/b978-0-443-15213-9.00012-0
110. Zhou, M., Zhang, Y., Chen, Y., Zhang, F., & Yang, D. (2022). Optimization of the decolorization conditions of Rose Bengal by using *Aspergillus niger* TF05 and a decolorization mechanism. *Microbiology, 168*(1), Article 001128.
111. Zimmermann, T., Kulla, H. G., & Leisinger, T. (1982). Properties of purified orange II Azoreductase, the enzyme initiating azo dye degradation by *Pseudomonas* KF46. *European Journal of Biochemistry, 129*(1), 197–203. https://doi.org/10.1111/j.1432-1033.1982.tb07040.x

MIX
Papier aus verantwortungsvollen Quellen
Paper from responsible sources
FSC® C105338

If you have any concerns about our products,
you can contact us on
ProductSafety@springernature.com

In case Publisher is established outside the EU,
the EU authorized representative is:
Springer Nature Customer Service Center GmbH
Europaplatz 3, 69115 Heidelberg, Germany

Printed by Libri Plureos GmbH
in Hamburg, Germany